CISM COURSES AND LECTURES

The series presents lecture notes, monographs, edited works and proceedings in the field of Mechanics, Engineering, Computer Science and Applied Mathematics.
Purpose of the series is to make known in the international scientific and technical community results obtained in some of the activities organized by CISM, the International Centre for Mechanical Sciences.

INTERNATIONAL CENTRE FOR MECHANICAL SCIENCES

COURSES AND LECTURES - No. 387

MODERN METHODS OF ANALYTICAL MECHANICS AND THEIR APPLICATIONS

EDITED BY

VALENTIN V. RUMYANTSEV AND ALEXANDER V. KARAPETYAN
RUSSIAN ACADEMY OF SCIENCES

Springer-Verlag Wien GmbH

Le spese di stampa di questo volume sono in parte coperte da
contributi del Consiglio Nazionale delle Ricerche.

This volume contains 82 illustrations

Originally published by Springer-Verlag Wien New York in 1998
SPIN 10691316

In order to make this volume available as economically and as rapidly as possible the authors' typescripts have been reproduced in their original forms. This method unfortunately has its typographical limitations but it is hoped that they in no way distract the reader.

ISBN 978-3-211-83138-0 ISBN 978-3-7091-2520-5 (eBook)
DOI 10.1007/978-3-7091-2520-5

PREFACE

The volume contains lectures given at CISM Advanced School course "Modern methods of analytical mechanics and their applications" (Udine, June 16-20, 1997).

The course aims at giving a comprehensive and up-to-date view of modern methods of analytical mechanics (general equations of analytical dynamics and problems of the integrability; invariant objects; the stability and bifurcations) and their applications (rigid body dynamics; celestial mechanics; multibody systems etc.).

The first part (V.Y. Rumyantsev) is devoted to the theoretical aspects of analytical mechanics: the theory of Poincaré-Chetayev equations and their generalizations for mechanical systems with holonomic and nonholonomic constraints including the cases of dependent variables and quazicoordinates, the theory of integration of equations of motion; generalization of the methods of Routh, Hamilton-Jacobi, Poisson, Whittaker, Suslov etc. General propositions are illustrated by mechanical examples, including a rigid body with cavities, containing fluid.

The second part (C. Simo) is devoted to the effective computations in celestial mechanics and astrodynamics: Hamiltonian formalism; invariant objects (periodic orbits, invariant tori, stable and unstable manifolds, center manifolds); symbolic and numerical computations of invariant objects; applications to some n-body problems and to spacecraft missions design.

The third part (L. Salvadori, F. Visentin) is devoted to some problems on stability, observability and bifurcation in mechanics and their applications: the direct Liapunov's method; one parameter families of Liapunov's functions (theory and applications); stability and Andronov-Hopf bifurcation; total stability and observability.

The fourth part (A.Y. Karapetyan) is devoted to problems of construction, stability and bifurcation of invariant sets (in particular, steady motions) of mechanical systems: invariant sets of mechanical systems with constants of motion; invariant sets of mechanical systems with symmetry (in particular, steady motions and relative equilibria of mechanical systems with cyclic coordinates). General propositions are illustrated with examples from rigid body dynamics (a top on a plane with friction, a body in the central gravitational field etc.).

The fifth part (M. Pascal) is devoted to analytical methods in dynamical simulation of flexible multibody systems: dynamical simulation and vibrational analysis of flexible multibody systems; dynamical simulation of a flexible manipulator arm; dynamical model of a large space structure by analytical methods; dynamical control laws for flexible robots.

The sixth part (P. Hagedorn, W. Seemann) is devoted to technical applications of the theory of vibrations to mechanical engineering systems: free vibrations of a string with small bending stiffness and damping; cylinder vibrating in a duct filled with incompressible fluid of low viscosity; variational principles in the derivation of equations of motion. The use of Lagrange multipliers in constrained systems.

The course is at an advanced level. It is designed for postgraduate students, research engineers and academics that are familiar with the basic concepts of analytical dynamics and stability theory. Although the course deals with mechanical problems, most of the concepts and methods involved are equally applied to general dynamical systems.

V.V. Rumyantsev
A.V. Karapetyan

CONTENTS

PART I

THE GENERAL EQUATIONS OF ANALYTICAL DYNAMICS

V.V. Rumyantsev
Russian Academy of Sciences, Moscow, Russia

ABSTRACT

Poincaré's remarkable idea [1], [2] to represent motion's equations of holonomic systems in terms of a certain transitive Lie group of infinitesimal transformations was extended by Chetayev [3]-[6] to the cases of rheonomic constraints and dependent variables, when transformation group is intransitive one. Chetayev transformed Poincaré's equations to canonical form and elaborated the theory of their integration. These fine results were developed in a number of papers [7]-[23]. Our lectures represent an introduction in the theory of generalized Poincaré's and Chetayev's equations based on closed systems of transformations. These equations include both the motion's equations in independent and dependent, holonomic and non-holonomic coordinates for holonomic and non-holonomic systems and in this sense are the general equations of analytical dynamics.

Chapter 1

Poincaré's Equations and Chetayev's Equations.

1.1 Poincaré's Equations.

In the paper [1] H.Poincaré has deduced new general equations of dynamics which represent motion's equations of holonomic systems with the help of some group of infinitesimal transformations.

Consider some continuous transitive group of infinitesimal transformations. Let be $X_i(f)$ some substitution of this group

$$X_i(f) = X_i^1 \frac{\partial f}{\partial x_1} + \cdots + X_i^n \frac{\partial f}{\partial x_n}$$

where $x_1, \ldots, x_n$ are independent variables which define the position of the system, for that T and U are the kinetic energy and the force function; $f(x_1, \ldots, x_n) \in C^2$.

The transformations form the group, therefore commutator

$$X_i X_k - X_k X_i = \sum_{s=1}^{n} c_{ik}^s X_s, \quad i, k = 1, \ldots, n, \tag{1.1}$$

where c_{ik}^s are structural constants.

It is possible to express the velocities by means of parameters η_μ of real displacements

$$\dot{x}_\mu = \frac{dx_\mu}{dt} = \eta_1 X_1^\mu + \cdots + \eta_n X_n^\mu, \quad \mu = 1, \ldots, n$$

The passage of the system from the position $(x_1, \ldots, x_n)$ in some infinitely close position $(x_1 + \dot{x}_1 dt, \ldots, x_n + \dot{x}_n dt)$ is fulfilled by means of group's transformations $\sum_i \eta_i dt X_i(f)$.

The kinetic energy as function of x and $\dot{x}$ it is possible to express in the form of the function $T(x, \eta)$. If we give to x and η virtual displacements δx and $\delta\eta$ then as result T and U obtain the additions

$$\delta T = \sum_i \left(\frac{\partial T}{\partial x_i} \delta x_i + \frac{\partial T}{\partial \eta_i} \delta \eta_i \right), \quad \delta U = \sum_i \frac{\partial U}{\partial x_i} \delta x_i$$

It is possible to pose

$$\delta x_\mu = \omega_1 X_1^\mu + \cdots + \omega_n X_n^\mu$$

The passage of the system from the position x in some infinitely close position $x + \delta x$ is fulfilled by means of group's transformations $\sum_i \omega_i X_i(f)$, where ω_i - parameters of a virtual displacement.

Poincaré believes

$$\sum_i \left(\frac{\partial T}{\partial x_i} + \frac{\partial U}{\partial x_i} \right) \delta x_i = \sum_i \Omega_i \omega_i, \tag{1.2}$$

where $\Omega_i = X_i(T + U)$.

For the Hamilton action $J = \int_{t_0}^{t_1} (T + U)\, dt$ we have

$$\delta J = \int_{t_0}^{t_1} \sum_i \left(\frac{\partial T}{\partial \eta_i} \delta \eta_i + \Omega_i \omega_i \right) dt$$

From the relation $d\delta f = \delta df$ it is easy to find

$$\delta \eta_i = \frac{d\omega_i}{dt} + \sum_{k,s} c_{ks}^i \eta_k \omega_s \tag{1.3}$$

Poincaré derived his equations

$$\frac{d}{dt} \frac{\partial T}{\partial \eta_s} = \sum_{k,i} c_{ks}^i \frac{\partial T}{\partial \eta_i} \eta_k + \Omega_s, \quad s = 1, \ldots, n \tag{1.4}$$

from Hamilton's principle of least action $\delta J = 0$.

Poincaré's equations (1.4) include as special cases:

1) the Lagrange's equations

$$\frac{d}{dt} \frac{\partial T}{\partial \dot{x}_i} - \frac{\partial T}{\partial x_i} = \frac{\partial U}{\partial x_i}, \quad i = 1, \ldots, n, \tag{1.5}$$

when the group reduces to the commutables transformations $X_i = \dfrac{\partial}{\partial x_i}$ which increase one from the variables x_i on a infinetely small constant and besides $\eta_i = \dot{x}_i$ and all $c_{ks}^i = 0$.

2) the Euler's equations for a rotation of a rigid body around a fixed point

$$A\frac{dp}{dt}+(C-B)\,qr=\Omega_1,\quad B\frac{dq}{dt}+(A-C)\,rp=\Omega_2,$$

$$C\frac{dr}{dt}+(B-A)\,pq=\Omega_3, \tag{1.6}$$

where the role of η_i play the projections p, q, r of instantaneous angular velocity ω of rigid body on his principal axes of inertia, and besides A, B, C - the moments of inertia, Ω_i - the moments of external forces,

$$T=\frac{1}{2}\left(Ap^2+Bq^2+Cr^2\right)$$

is the kinetic energy, X_1, X_2, X_3 - the group $SO(3)$, for which structural constants

$$c_{12}^3=c_{23}^1=c_{31}^2=1,\quad c_{21}^3=c_{32}^1=c_{13}^2=-1$$

and all another $c_{ks}^i=0$. Such group it is easy to construct with help of the kinematical Euler's equations [6]

$$p=\dot\psi\sin\theta\sin\varphi+\dot\theta\cos\varphi,\quad q=\dot\psi\sin\theta\cos\varphi-\dot\theta\sin\varphi,\quad r=\dot\varphi+\dot\psi\cos\theta$$

from which we find the expressions

$$\dot\theta=p\cos\varphi-q\sin\varphi,\quad \dot\psi=\frac{1}{\sin\theta}\left(p\sin\varphi+q\cos\varphi\right),$$

$$\dot\varphi=r-\mathrm{ctg}\theta\left(p\sin\varphi+q\cos\varphi\right) \tag{1.7}$$

Here θ, ψ, φ are the Euler angles. Hence

$$\frac{d}{dt}f\left(\theta,\psi,\varphi\right)=pX_1f+qX_2f+rX_3f$$

where

$$X_1=-\mathrm{ctg}\theta\sin\varphi\frac{\partial}{\partial\varphi}+\frac{\sin\varphi}{\sin\theta}\frac{\partial}{\partial\psi}+\cos\varphi\frac{\partial}{\partial\theta},$$

$$X_2=-\mathrm{ctg}\theta\cos\varphi\frac{\partial}{\partial\varphi}+\frac{\cos\varphi}{\sin\theta}\frac{\partial}{\partial\psi}-\sin\varphi\frac{\partial}{\partial\theta},\quad X_3=\frac{\partial}{\partial\varphi} \tag{1.8}$$

is the transitive Lie group $SO(3)$.

Poincaré has noted that the case, when $U=0$ and T depends only from η_i, is especially interesting. In this case the parameters η_i it is possible to consider as coordinates in the algebra Lie g of the group Lie G and the Poincaré's equations (1.4) are the closed system of differential equations on g.

1.2 Motion of a Rigid Body Containing a Fluid.

Poincaré used his equations himself in the paper [2] for the problem on a rotation of a rigid body with ellipsoidal cavity fulfilled by an ideal fluid. By his confession [1] he deduced the equations (1.4) for a investigation such motion.

Let the cavity be in the shape of the ellipsoid

$$\frac{x_1^2}{a_1^2}+\frac{x_2^2}{a_2^2}+\frac{x_3^2}{a_3^2}=1$$

where the a_i denote the lengths of the ellipsoid's semiaxes. Poincaré assumed that the velocities of the fluid particles are linear functions of their coordinates x_i. He called this motion simple and pointed out, on the basis of Helmholtz's theorems, that if the motion of the fluid in an ellipsoidal cavity is initially simple, it will remains so all of the time. The motion of the fluid then can be interpreted from the kinematical point of view. Each fluid particle can be made to correspond to some imagined particle with coordinates x_i' and velocities relative to the body u_i', where

$$x_i'=\frac{x_i}{a_i},\quad u_i'=\frac{u_i}{a_i},\quad i=1,2,3.$$

An ensemble of these imagined particles will fill a sphere S

$$\sum_{i=1}^{3}x_i'^2=1$$

In the case of the simple motion of the fluid, the imagined particles will move as points of a rigid body, i.e., the motion of the fluid will be represented by rotation of the sphere S as a single rigid body. Denoting the projections of the instantaneous angular velocity ω_1 of the sphere S on the axes of the ellipsoid by p_1, q_1, r_1, we find that the projections of the absolute velocity of the fluid on the same axes are defined by the expressions

$$v_1=\frac{a_1}{a_3}q_1x_3-\frac{a_1}{a_2}r_1x_2+qx_3-rx_2,\ v_2=\frac{a_2}{a_1}r_1x_1-\frac{a_2}{a_3}p_1x_3+rx_1-px_3,$$

$$v_3=\frac{a_3}{a_2}p_1x_2-\frac{a_3}{a_1}q_1x_1+px_2-qx_1 \tag{1.9}$$

The kinetic energy of the fluid's relative motion is

$$T_r=\frac{1}{2}\left(A_1p_1^2+B_1q_1^2+C_1r_1^2\right)$$

where

$$A_1=\frac{M}{5}\left(a_2^2+a_3^2\right),\quad B_1=\frac{M}{5}\left(a_3^2+a_1^2\right),\quad C_1=\frac{M}{5}\left(a_1^2+a_2^2\right),\quad M=\frac{4}{3}\pi\rho a_1a_2a_3$$

are the moments of inertia and the mass of the fluid, $\rho = \text{const}$ denotes of the fluid's density. The main moment of momentum in the fluid's relative motion has the components $A_1'p_1, B_1'q_1, C_1'r_1$; $A_1' = \frac{2}{5}Ma_2a_3, \quad B_1' = \frac{2}{5}Ma_3a_1, \quad C_1' = \frac{2}{5}Ma_1a_2$.

The kinetic energy of the whole system is

$$T = \frac{1}{2}\left(Ap^2 + Bq^2 + Cr^2 + 2\left(A_1'pp_1 + B_1'qq_1 + C_1'rr_1\right) + A_1p_1^2 + B_1q_1^2 + C_1r_1^2\right) \tag{1.10}$$

where A, B, C - the moments of inertia of the system. The projections of the system's kinetic moment are

$$\frac{\partial T}{\partial p} = Ap + A_1'p_1, \quad \frac{\partial T}{\partial q} = Bq + B_1'q_1, \quad \frac{\partial T}{\partial r} = Cr + C_1'r_1.$$

The moments of inertia of the rigid body are $A_0 = A - A_1, B_0 = B - B_1, C_0 = C - C_1$.

Our system has six degrees of freedom and six infinitesimal virtual displacements corresponding to

1) the rotations of the whole system as the single body around the ellipsoidal's axes,

2) the rotations of the sphere around the same axes.

Let X_i, X_{i+3}, $(i = 1, 2, 3)$ are these six transformations. The composition's rules of rotations give the group's structural equations

$$\begin{aligned} X_1X_2 - X_2X_1 &= X_3, & X_4X_5 - X_5X_4 &= -X_6, \\ X_2X_3 - X_3X_2 &= X_1, & X_5X_6 - X_6X_5 &= -X_4, \\ X_3X_1 - X_1X_3 &= X_2, & X_6X_4 - X_4X_6 &= -X_5 \end{aligned}$$

Any from transformations X_i are commutative with any transformations X_{i+3}, so that all structural constants are equal to $+1$, or -1, or 0.

In the coordinates of fluid's particles x_i we have

$$\begin{aligned} X_4f &= \sum\left(x_2\frac{a_3}{a_2}\frac{\partial f}{\partial x_3} - x_3\frac{a_2}{a_3}\frac{\partial f}{\partial x_2}\right), \\ X_5f &= \sum\left(x_3\frac{a_1}{a_3}\frac{\partial f}{\partial x_1} - x_1\frac{a_3}{a_1}\frac{\partial f}{\partial x_3}\right), \\ X_6f &= \sum\left(x_1\frac{a_2}{a_1}\frac{\partial f}{\partial x_2} - x_2\frac{a_1}{a_2}\frac{\partial f}{\partial x_1}\right) \end{aligned} \tag{1.11}$$

Poincaré's equations (1.4) for our system have the form

$$\begin{aligned} \frac{d}{dt}\frac{\partial T}{\partial p} &= r\frac{\partial T}{\partial q} - q\frac{\partial T}{\partial r} + L_1, \quad \frac{d}{dt}\frac{\partial T}{\partial q} = p\frac{\partial T}{\partial r} - r\frac{\partial T}{\partial p} + L_2, \\ \frac{d}{dt}\frac{\partial T}{\partial r} &= q\frac{\partial T}{\partial p} - p\frac{\partial T}{\partial q} + L_3, \end{aligned} \tag{1.12}$$

$$\frac{d}{dt}\frac{\partial T}{\partial p_1} = q_1\frac{\partial T}{\partial r_1} - r_1\frac{\partial T}{\partial q_1}, \quad \frac{d}{dt}\frac{\partial T}{\partial q_1} = r_1\frac{\partial T}{\partial p_1} - p_1\frac{\partial T}{\partial r_1},$$

$$\frac{d}{dt}\frac{\partial T}{\partial r_1} = p_1\frac{\partial T}{\partial q_1} - q_1\frac{\partial T}{\partial p_1}, \tag{1.13}$$

L_i - the projections of principal moment of external forces applied to the system.

The equations (1.12) express the theorem on moment of momentum of the system, the equations (1.13) are equivalent to the Helmholtz's equations on fluid's vortexs

$$\frac{d}{dt}\mathrm{rot}\mathbf{v} + \omega \times \mathrm{rot}\mathbf{v} = (\mathrm{rot}\mathbf{v} \cdot \nabla)\,\mathbf{v} \tag{1.14}$$

Indeed, the projections of rotv on axes x_1, x_2, x_3 are equal

$$\xi = \frac{5}{Ma_2a_3}\frac{\partial T}{\partial p_1}, \quad \eta = \frac{5}{Ma_3a_1}\frac{\partial T}{\partial q_1}, \quad \zeta = \frac{5}{Ma_1a_2}\frac{\partial T}{\partial r_1}$$

and the first from the equations (1.13)

$$\frac{d\xi}{dt} = q_1\frac{a_1}{a_3}\zeta - r_1\frac{a_1}{a_2}\eta$$

coincides with the projection on axis x_1 of the equation (1.14). By use the expression (1.10) the equations (1.12) and (1.13) receive the forms

$$A\dot{p} + A_1'\dot{p}_1 + q\,(Cr + C_1'r_1) - r\,(Bq + B_1'q_1) = L_1, \ldots$$
$$A_1'\dot{p} + A_1\dot{p}_1 + r_1\,(B_1'q + B_1q_1) - q_1\,(C_1'r + C_1r_1) = 0, \ldots$$

It is interesting that if the rigid body is hold immovable ($p = q = r = 0$) then the equations (1.13) become

$$A_1\dot{p}_1 = (C_1 - B_1)\,q_1r_1, \quad B_1\dot{q}_1 = (A_1 - C_1)\,r_1p_1, \quad C_1\dot{r}_1 = (B_1 - A_1)\,p_1q_1,$$

i.e., innere motion of the fluid in motionless body follows to the Poinsot's laws.

In the particular case of the spherical cavity ($a_1 = a_2 = a_3 = a$) the motion of the fluid in the cavity has no effect on the motion of the rigid body. In fact, in this case $A_1 = A_1' = B_1 = B_1' = C_1 = C_1' = \frac{2M}{5}a^2$ and by substraction equations (1.13) from equations (1.12) we get the equations

$$A_0\dot{p} + (C_0 - B_0)\,qr = L_1, \quad B_0\dot{q} + (A_0 - C_0)\,rp = L_2,$$

$$C_0\dot{r} + (B_0 - A_0)\,pq = L_3.$$

If $L_i = 0$ ($i = 1, 2, 3$) the equations (1.12), (1.13) have the first integrals

$$T = \mathrm{const}, \qquad \sum\left(\frac{\partial T}{\partial p}\right)^2 = \mathrm{const}, \qquad \sum\left(\frac{\partial T}{\partial p_1}\right)^2 = \mathrm{const}.$$

For the case of the symmetry with respect to the axis x_3 when

$$a_1 = a_2, \quad A = B, \quad A_1 = B_1, \quad C_1 = C_1', \quad A_1' = B_1', \quad L_3 = 0$$

there is also the first integral $r =$ const and from the equations (1.12) and(1.13) follows the equation

$$C_1 \dot{r}_1 + A_1' (pq_1 - p_1 q) = 0 \tag{1.15}$$

If we suppose $L_i = 0$ $(i = 1, 2, 3)$ then the motion's equations are integrire in quadratures. The first integrales give

$$p^2 + q^2, \qquad p_1^2 + q_1^2, \qquad pp_1 + qq_1, \qquad (qp_1 - pq_1)^2$$

in forms of polinomens second and forth degrees of r_1, and the equation (1.15) gives r_1 in a form of elliptical function of time t.

In the problem, just considered, the simple motion of the fluid is described by the finite number of the variables. It is impossible in general case of fluid's motion, but the Poincare's equations can be applicable also to mechanical systems with infinite degrees of freedom. As the example of such system we consider a motion of a rigid body with one fixed point O, having of arbitrary cavity partially or completely filled by an ideal homogeneous incompressible fluid. The rigid body and the fluid will be regarded as the single mechanical system [9].

We introduce two coordinate frames: 1) fixed $Ox_1'x_2'x_3'$ and 2) moving $Ox_1x_2x_3$, rigidly connected with the rigid body and axes directed along its principal axes of inertia for the point O. Motion's equations we shall refere to the moving coordinate frame.

A position of the body in the space $Ox_1'x_2'x_3'$ can be determined, for example, by Euler's angles θ, ψ, φ, and positions of the fluid's particles by their coordinates x_1, x_2, x_3 in the frame $Ox_1x_2x_3$.

Let τ denotes the volume of a region occupied by fluid at the given time instant, S' - a boundary of the region τ, σ - the surface of the cavity. If the fluid fills the cavity completely S' is identical with σ. However if the fluid fills the cavity partially, then the surface S' consists from a free surface S, a equation of which $f(t, x_1, x_2, x_3) = 0$ is unknown beforehand, and a part σ_1 of the surface σ, with which the fluid comes into contact at the given time instant. In this case the remaining part of the cavity is filled with a air, the pressure of which is $p_0 =$ const, while its mass is disregarded.

The absolute velocity of any point of the system

$$\mathbf{v} = \boldsymbol{\omega} \times \mathbf{r} + \mathbf{u}$$

where $\boldsymbol{\omega}(p, q, r)$ - angular velocity of the body, $\mathbf{r}(x_1, x_2, x_3)$ - radius-vector of the point, $\mathbf{u}(u_1, u_2, u_3) = d\mathbf{r}/dt$ is the relative velocity (local time derivative in the frame $Ox_1x_2x_3$). For any point of the body $\mathbf{u} = 0$. For the parameters of real displacements we take $\eta_1 = p$, $\eta_2 = q$, $\eta_3 = r$, $\tilde{\eta}_j = u_j$ $(j = 1, 2, 3)$.

A passage of our mechanical system from its instantaneous position to its close simultaneous position can be carried out by imparting to the entire system an infinitesimal rotation motion $\delta\boldsymbol{\theta}\,(\delta\theta_1, \delta\theta_2, \delta\theta_3)$ and by imparting to the fluid's particles a relative displacements $\delta_1\mathbf{r}\,(\delta x_1, \delta x_2, \delta x_3)$, which are independent of $\delta\boldsymbol{\theta}$, but are connected by incompressibility condition

$$\operatorname{div}\delta_1\mathbf{r} = \frac{\partial\delta x_1}{\partial x_1} + \frac{\partial\delta x_2}{\partial x_2} + \frac{\partial\delta x_3}{\partial x_3} = 0$$

inside of the region τ and by conditions

$$\delta_1\mathbf{r}\cdot\mathbf{n} = 0, \text{ on } \sigma_1, \quad p = p_0 \text{ on } S,$$

where $\mathbf{n}$ is the unit vector of the external normal to a surface.

There are the relations [25]

$$\delta\boldsymbol{\omega} = \frac{d\delta\,\boldsymbol{\theta}}{dt} + \boldsymbol{\omega}\times\delta\boldsymbol{\theta}, \quad \delta\mathbf{r} = \delta\boldsymbol{\theta}\times\mathbf{r} + \delta_1\mathbf{r}$$

For the parameters of virtual displacements we take

$$\omega_i = \delta\theta_i, \quad \tilde{\omega}_j = \delta x_j \quad (i, j = 1, 2, 3)$$

The group of operators defining of virtual displacementes consists from two groups: the group $SO(3)$ of operators (1.8) X_1, X_2, X_3, and of infinite Abelian group $\tilde{X}_j = \partial/\partial x_j$ $(j = 1, 2, 3)$, commutable with group $SO(3)$.

The kinetic energy of the system

$$T = T_1 + \rho\int_\tau \mathcal{T}\,d\tau = \frac{1}{2}\boldsymbol{\omega}\cdot\boldsymbol{\Theta}\cdot\boldsymbol{\omega} + \boldsymbol{\omega}\cdot\rho\int_\tau \mathbf{r}\times\mathbf{u}\,d\tau + \frac{1}{2}\int_\tau \mathbf{u}^2\,d\tau$$

Here M is the mass, $\boldsymbol{\Theta}$ is the inertia tensor of the system for the point O, T_1 - the kinetic energy of body, $\mathcal{T} = \frac{1}{2}v^2$ is the kinetic energy's density of the fluid. The symbol $\int_\tau(\ldots)\,d\tau$ denotes the triple integral taken over the region occupied by the fluid.

According to the Hamilton-Ostrogradskiy principle for absolute motion of the mechanical system

$$\int_{t_0}^{t_1}\left(\delta T + \sum_\nu \mathbf{F}_\nu\cdot\delta\mathbf{r}_\nu + \int_\tau p\,\operatorname{div}\delta_1\mathbf{r}\,d\tau\right)dt = 0$$

where $p\,(t, x_1, x_2, x_3)$ is Lagrange multiplier which is identical here with a hydrodynamic pressure, regarding as a reaction of the incompressibility condition, $\sum_\nu \mathbf{F}_\nu\cdot\delta\mathbf{r}_\nu$ is the sum of the elementary works applied to the system of active forces $\mathbf{F}_\nu$ on virtual displacements.

Among the active forces acting on the fluid we shall differentiate between mass forces $\mathbf{F}(F_1, F_2, F_3)$, acting on each element $\rho\, d\tau$ of the fluid mass, and surface forces $-p_0\mathbf{n}$, acting on each element dS of the fluid's free surface.

Referring to the above we obtain the following expressions

$$\delta T = \sum_{i=1}^{3}\left(\omega_i X_i T + \frac{\partial T}{\partial \eta_i}\delta\eta_i\right) + \rho\sum_{j=1}^{3}\int_\tau\left(\tilde{\omega}_j\tilde{X}_j T + \frac{\partial T}{\partial \tilde{\eta}_j}\delta\tilde{\eta}_j\right)d\tau =$$

$$= \mathbf{G}\cdot\left(\frac{d\delta\theta}{dt} + \omega\times\delta\theta\right) + \rho\int_\tau[\mathbf{v}\cdot\delta\mathbf{u} + (\mathbf{v}\times\omega)\cdot\delta_1\mathbf{r}]\,d\tau,$$

$$\sum_\nu \mathbf{F}_\nu\cdot\delta\mathbf{r}_\nu = \mathbf{L}\cdot\delta\theta + \rho\int_\tau \mathbf{F}\cdot\delta_1\mathbf{r}\,d\tau - p_0\int_S \mathbf{n}\cdot\delta_1\mathbf{r}\,dS,$$

$$\int_\tau p\,\mathrm{div}\delta_1\mathbf{r}\,d\tau = \int_S p\mathbf{n}\cdot\delta_1\mathbf{r}\,dS - \int_\tau \mathrm{grad}p\cdot\delta_1\mathbf{r}\,d\tau,$$

where

$$\mathbf{G} = \Theta\cdot\omega + \rho\int_\tau \mathbf{r}\times\mathbf{u}\,d\tau = \mathrm{grad}_\omega T$$

is the angular momentum of the system,

$$\mathbf{L} = \Sigma_1\mathbf{r}_\nu\times\mathbf{F}_\nu + \rho\int_\tau \mathbf{r}\times\mathbf{F}\,d\tau$$

denotes the principal moment of the active forces applied to the system, index 1 belongs to points of the rigid body.

Substitution these expressions in the Hamilton-Ostrogradskiy principle and integration by parts some terms led to the following Poincare's equations of motion of the system

$$\begin{aligned}\frac{d}{dt}\frac{\partial T}{\partial \eta_i} &= \sum_{\alpha,\beta} c_{\alpha i}^{\beta}\eta_\alpha\frac{\partial T}{\partial \eta_\beta} + X_i T + L_i \quad (i,\alpha,\beta = 1,2,3)\\ \frac{d}{dt}\frac{\partial T}{\partial \tilde{\eta}_j} &= \tilde{X}_j\left(T - \frac{p}{\rho}\right) + F_j \quad (j = 1,2,3)\end{aligned} \tag{1.16}$$

and to the boundary condition $p = p_0$ for the pressure on the free surface S. These equations and boundary condition must be supplemented by the incompressibility equation

$$\mathrm{div}\mathbf{u} = \frac{\partial\tilde{\eta}_1}{\partial x_1} + \frac{\partial\tilde{\eta}_2}{\partial x_2} + \frac{\partial\tilde{\eta}_3}{\partial x_3} = 0$$

and by the kinematic conditions

$$u_n = \mathbf{u}\cdot\mathbf{n} = 0 \text{ on } \sigma_1, \qquad \frac{\partial f}{\partial t} + \mathbf{u}\cdot\mathrm{grad}f = 0 \text{ on } S.$$

The equations (1.16) are equivalent to the vector equations

$$\frac{d\mathbf{G}}{dt} + \omega \times \mathbf{G} = \mathbf{L}, \quad \frac{d\mathbf{v}}{dt} + \omega \times \mathbf{v} = \mathbf{F} - \frac{1}{\rho}\mathrm{grad}p \tag{1.17}$$

It is evident that the first from equations (1.17) expresses the theorem of angular momentum of the system, the second - Euler's hydrodynamic equation, refered to the moving coordinate frame.

1.3 Chetayev's Equations.

In the paper [3] N.Chetayev transformed the Poincaré's equations to canonical form and to Jacobi's equations.

To transform Poincaré's equations to canonical form, Chetayev replaced the variables η_s and the function $T + U$ by new variables and function

$$y_s = \frac{\partial T}{\partial \eta_s}, \quad s = 1, \ldots, n \tag{1.18}$$

$$H(t, x_1, \ldots, x_n, y_1, \ldots, y_n) = \sum_s y_s \eta_s - T - U$$

supposed $\|\frac{\partial^2 T}{\partial \eta_i \partial \eta_j}\| \neq 0$. In the new variables the Poincaré's equations (1.4) have the canonical form

$$\frac{dy_s}{dt} = \sum_{k,i} c^i_{ks} \frac{\partial H}{\partial y_k} y_i - X_s H, \quad \frac{dx_s}{dt} = \sum_k \frac{\partial H}{\partial y_k} X^s_k, \tag{1.19}$$

Let the differential equation in partial derivatives of first order

$$\frac{\partial V}{\partial t} + H[t, x_i, X_i(V)] = 0 \tag{1.20}$$

defines V as a function of independent variables x_i, t. If it is known a complete integral $V(t, x_1, \ldots, x_n, a_1, \ldots, a_n)$ of the equation (1.20) with n arbitrary constants a_i, from which every do not is additive constant, then the final equations of motion will have the form

$$\frac{\partial V}{\partial a_i} = b_i, \quad y_i = X_i(V)$$

where b_i - arbitrary constants.

In the paper [4] Chetayev extended the Poincaré's results [1] and his own results [3] to the cases of non-stationary constraints and dependent variables, when the group of transformations is intransitive one [24].

Denote by $x_1, \ldots, x_n$ the variables defining a position of the system in moment t. If the system has k degrees of freedom then exists the intransitive group of infinitesimal transformations

$$X_0(f) = \frac{\partial f}{\partial t} + X_0^1 \frac{\partial f}{\partial x_1} + \cdots + X_0^n \frac{\partial f}{\partial x_n},$$
$$X_i(f) = X_i^1 \frac{\partial f}{\partial x_1} + \cdots + X_i^n \frac{\partial f}{\partial x_n}, \quad i = 1, \ldots, k \tag{1.21}$$

which transforms the system in the moment t from the position $(x_1, \ldots, x_n)$ in a real infinitely close position $(x_1 + \dot{x}_1 dt, \ldots, x_n + \dot{x}_n dt)$ by infinitesimal transformation of the group

$$X_0(f)dt + \sum_i \eta_i X_i(f)dt$$

and in a virtual infinitely close position $(x_1 + \delta x_1, \ldots, x_n + \delta x_n)$ by infinitesimal transformation of the subgroup

$$\sum_i \omega_i X_i(f)$$

Chetayev supposed that the transformation X_0 is commutable with all another transformations X_i. The equations of a motion have the Poincaré's form (1.4)

$$\frac{d}{dt}\frac{\partial T}{\partial \eta_i} = \sum_{s,m} c_{si}^m \frac{\partial T}{\partial \eta_m}\eta_s + X_i(T + U) \quad i = 1, \ldots, k$$

If introduce the new variables y_s and the function $H(t, x, y)$ (1.18) then the equations of motion reduce to the canonical form

$$\frac{dy_i}{dt} = \sum c_{si}^m \frac{\partial H}{\partial y_s} y_m - X_i(H), \quad \eta_i = \frac{\partial H}{\partial y_i} \tag{1.22}$$

or to another form

$$\frac{dy_i}{dt} = \sum c_{si}^m \frac{\partial H}{\partial y_s} y_m - X_i(H), \quad \frac{dx_j}{dt} = X_0^j + \sum_i \frac{\partial H}{\partial y_i} X_i^j \tag{1.23}$$

The integral

$$\int \sum_s y_s \eta_s \delta t - H \delta t$$

is the relative integral invariant of the system's differential equations of trajectories, consequent it is sufficient to find a complete integral

$$V(t, x_1, \ldots, x_n, a_1, \ldots, a_n)$$

of the differential equation in partial derivatives of first order

$$X_0(V) + H\left[t, x_1, \ldots, x_n, X_1(V), \ldots, X_k(V)\right] = 0 \tag{1.24}$$

the motion's equations will be

$$\frac{\partial V}{\partial a_i} = b_i, \quad y_s = X_s(V)$$

with constantes a_i, b_i.

The relation $f(t, x_i, y_s) = \text{const}$ is the first integral of the canonical equations (1.22), if it is fulfilled the equality

$$X_0(f) + (H, f) = 0$$

where the generalized Poisson's bracket (φ, ψ) is the expression

$$(\varphi, \psi) = \sum_s \left(\frac{\partial \varphi}{\partial y_s} X_s(\psi) - \frac{\partial \psi}{\partial y_s} X_s(\varphi) \right) + \sum_{i,s,k} c_{si}^k \frac{\partial \varphi}{\partial y_s} \frac{\partial \psi}{\partial y_i} y_k \tag{1.25}$$

From the identity

$$(f, (\varphi, \psi)) + (\varphi, (\psi, f)) + (\psi, (f, \varphi)) = 0$$

it is seen that if $\varphi(t, x_i, y_s) = a$ and $\psi(t, x_i, y_s) = b$ are two first integrals of the equations (1.22) then $(\varphi, \psi) = c$ is the third first integral.

The above results generalize:

1) the classical results of Lagrange, Hamilton, Jacobi and Poisson, when the group of transformations reduces to the group of commutable transformations;

2) the Poincaré equations and the theorems given in the paper [3] when $X_0^s = 0$, $(s = 1, \ldots, k)$.

Example 1. a) For a heavy rigid body with one fixed point (see 2) in 1.1) the force function

$$U(\theta, \varphi) = -Mg\left[x_1^0 \sin\theta \sin\varphi + x_2^0 \sin\theta \cos\varphi + x_3^0 \cos\theta\right]$$

where x_i^0 are the coordinates of the centre of mass. The Euler's equations have the form (1.6) with right-hand sides

$$\Omega_1 = Mg\left(x_3^0 \sin\theta \cos\varphi - x_2^0 \cos\theta\right), \quad \Omega_2 = Mg\left(x_1^0 \cos\theta - x_3^0 \sin\theta \sin\varphi\right),$$

$$\Omega_3 = Mg \sin\theta \left(x_2^0 \sin\varphi - x_1^0 \cos\varphi\right).$$

If introduce the new variables

$$y_1 = \frac{\partial T}{\partial p} = Ap, \qquad y_2 = \frac{\partial T}{\partial q} = Bq, \qquad y_3 = \frac{\partial T}{\partial r} = Cr$$

then

$$p = \frac{y_1}{A}, \qquad q = \frac{y_2}{B}, \qquad r = \frac{y_3}{C},$$

$$H(\theta, \varphi, y_1, y_2, y_3) = \frac{1}{2}\left(\frac{y_1^2}{A} + \frac{y_2^2}{B} + \frac{y_3^2}{C}\right) - U(\theta, \varphi)$$

and the canonical form (1.19) for equations (1.6) and (1.7) will be

$$\frac{dy_1}{dt} = \left(\frac{1}{C} - \frac{1}{B}\right) y_2 y_3 + X_1 U, \quad \frac{dy_2}{dt} = \left(\frac{1}{A} - \frac{1}{C}\right) y_3 y_1 + X_2 U,$$
$$\frac{dy_3}{dt} = \left(\frac{1}{B} - \frac{1}{A}\right) y_1 y_2 + X_3 U, \tag{1.26}$$

$$\frac{d\theta}{dt} = \frac{y_1}{A}\cos\varphi - \frac{y_2}{B}\sin\varphi, \quad \frac{d\psi}{dt} = \frac{1}{\sin\theta}\left(\frac{y_1}{A}\sin\varphi + \frac{y_2}{B}\cos\varphi\right),$$
$$\frac{d\varphi}{dt} = \frac{y_3}{C} - \operatorname{ctg}\theta\left(\frac{y_1}{A}\sin\varphi + \frac{y_2}{B}\cos\varphi\right) \tag{1.27}$$

where X_i are given by formulas (1.8).

b) If we take

$$\gamma_1 = \sin\theta\sin\varphi, \quad \gamma_2 = \sin\theta\cos\varphi, \quad \gamma_3 = \cos\theta \tag{1.28}$$

as the defining coordinates x_i, where γ_i satisfy the Poisson's equations

$$\frac{d\gamma_1}{dt} = r\gamma_2 - q\gamma_3, \quad \frac{d\gamma_2}{dt} = p\gamma_3 - r\gamma_1, \quad \frac{d\gamma_3}{dt} = q\gamma_1 - p\gamma_2 \tag{1.29}$$

then the group $SO(3)$ will be

$$X_1 = \gamma_3\frac{\partial}{\partial\gamma_2} - \gamma_2\frac{\partial}{\partial\gamma_3}, \quad X_2 = \gamma_1\frac{\partial}{\partial\gamma_3} - \gamma_3\frac{\partial}{\partial\gamma_1},$$
$$X_3 = \gamma_2\frac{\partial}{\partial\gamma_1} - \gamma_1\frac{\partial}{\partial\gamma_2}. \tag{1.30}$$

with commutator

$$[X_1, X_2]f = X_3 f, \quad [X_2, X_3]f = X_1 f, \quad [X_3, X_1]f = X_2 f$$

Consequently, the non-vanishing structure constants will be $c_{12}^3 = c_{23}^1 = c_{31}^2 = 1$, $c_{21}^3 = c_{32}^1 = c_{13}^2 = -1$. Poincaré's equations (1.4) reduce to Euler's equations (1.6) with right-hand sides

$$\Omega_1 = Mg\left(x_3^0\gamma_2 - x_2^0\gamma_3\right), \quad \Omega_2 = Mg\left(x_1^0\gamma_3 - x_3^0\gamma_1\right),$$

$$\Omega_3 = Mg\left(x_2^0\gamma_1 - x_1^0\gamma_2\right)$$

to which we must add the Poisson's equations (1.29).

If we replace p, q, r by the variables y_1, y_2, y_3 and consider the function

$$H\left(\gamma_1, \gamma_2, \gamma_3, y_1, y_2, y_3\right) = \frac{1}{2}\left(\frac{y_1^2}{A} + \frac{y_2^2}{B} + \frac{y_3^2}{C}\right) + Mg\left(x_1^0\gamma_1 + x_2^0\gamma_2 + x_3^0\gamma_3\right)$$

the Chetayev equations (1.19) take the form

$$\frac{dy_1}{dt} = \frac{B-C}{BC}y_2y_3 + Mg\left(x_3^0\gamma_2 - x_2^0\gamma_3\right), \quad \ldots,$$
$$\frac{d\gamma_1}{dt} = \gamma_2\frac{y_3}{C} - \gamma_3\frac{y_2}{B}, \quad \ldots \tag{1.31}$$

c) A position of a rigid body with one fixed point O in an inertial coordinates frame $O\bar{x}_1\bar{x}_2\bar{x}_3$ it is possible to determine also by means of nine cosines β_i^k of angles between $\bar{x}_i$ and axes x_k rigidly connected with the rigid body [6]. There are six relations

$$\sum_\alpha \beta_\alpha^k \beta_\alpha^s = \delta_{ks}, \quad \sum_\alpha \beta_k^\alpha \beta_s^\alpha = \delta_{ks}, \quad \alpha, k, s = 1, 2, 3 \tag{1.32}$$

Let

$$p_{ks} = \sum_\alpha \beta_\alpha^k \frac{d\beta_\alpha^s}{dt}$$

then

$$p_{ks} = -p_{sk}\ (s \neq k), \ p_{kk} = 0$$

because from the relation (1.29) we get

$$\sum_\alpha \beta_\alpha^k \frac{d\beta_\alpha^s}{dt} + \sum_\alpha \beta_\alpha^s \frac{d\beta_\alpha^k}{dt} = 0$$

Hence

$$\sum_k \beta_m^k p_{ks} = \sum_\alpha \left(\sum_k \beta_m^k \beta_\alpha^k\right)\frac{d\beta_\alpha^s}{dt} = \sum_\alpha \delta_{m\alpha}\frac{d\beta_\alpha^s}{dt} = \frac{d\beta_m^s}{dt}, \quad m, s = 1, 2, 3$$

These relations are the Poisson's equations

$$\frac{d\beta_m^1}{dt} = r\beta_m^2 - q\beta_m^3, \quad \frac{d\beta_m^2}{dt} = p\beta_m^3 - r\beta_m^1, \quad \frac{d\beta_m^3}{dt} = q\beta_m^1 - p\beta_m^2, \quad m = 1, 2, 3 \tag{1.33}$$

as

$$p = -p_{23} = p_{32}, \quad q = -p_{31} = p_{13}, \quad r = -p_{12} = p_{21} \quad .$$

The equations (1.33) for $m = 3$ give the equations (1.29): $\beta_3^i = \gamma_i$, $i = 1, 2, 3$.

Consider now a function of a position of the rigid body $f(t, \beta_k^s)$ in a moment t, for which

$$\frac{df}{dt} = \frac{\partial f}{\partial t} + \sum_{k,s} \frac{\partial f}{\partial \beta_k^s} \frac{d\beta_k^s}{dt} = \frac{\partial f}{\partial t} + \sum_{\alpha,k,s} \frac{\partial f}{\partial \beta_k^s} \beta_k^\alpha p_{\alpha s} =$$

$$\frac{\partial f}{\partial t} + p \sum_\alpha \left(\frac{\partial f}{\partial \beta_\alpha^2} \beta_\alpha^3 - \frac{\partial f}{\partial \beta_\alpha^3} \beta_\alpha^2 \right) + q \sum_\alpha \left(\frac{\partial f}{\partial \beta_\alpha^3} \beta_\alpha^1 - \frac{\partial f}{\partial \beta_\alpha^1} \beta_\alpha^3 \right) + r \sum_\alpha \left(\frac{\partial f}{\partial \beta_\alpha^1} \beta_\alpha^2 - \frac{\partial f}{\partial \beta_\alpha^2} \beta_\alpha^1 \right)$$

Consequently

$$X_0 = \frac{\partial}{\partial t}, \quad X_1 = \sum_\alpha \left(\beta_\alpha^3 \frac{\partial}{\partial \beta_\alpha^2} - \beta_\alpha^2 \frac{\partial}{\partial \beta_\alpha^3} \right),$$

$$X_2 = \sum_\alpha \left(\beta_\alpha^1 \frac{\partial}{\partial \beta_\alpha^3} - \beta_\alpha^3 \frac{\partial}{\partial \beta_\alpha^1} \right), \quad X_3 = \sum_\alpha \left(\beta_\alpha^2 \frac{\partial}{\partial \beta_\alpha^1} - \beta_\alpha^1 \frac{\partial}{\partial \beta_\alpha^2} \right)$$

and besides

$$[X_1, X_2] f = X_3 f, \ [X_2, X_3] f = X_1 f, \ [X_3, X_1] f = X_2 f$$

The operator $X_0 = \dfrac{\partial}{\partial t}$ is commutable with Lie's group X_1, X_2, X_3, i.e. $c_{0r}^s = 0 (r, s = 1, 2, 3)$. Poincares's equations (1.4) have the form Euler's equations (1.6), to which we must add the equations (1.33).

Example 2: Consider the canonical form for the equations (1.12,1.13).

We introduce the new variables

$$y_1 = \frac{\partial T}{\partial p} = Ap + A_1' p_1, \ y_2 = \frac{\partial T}{\partial q} = Bq + B_1' q_1, \ y_3 = \frac{\partial T}{\partial r} = Cr + C_1' r_1,$$

$$y_4 = \frac{\partial T}{\partial p_1} = A_1 p_1 + A_1' p, \ y_5 = \frac{\partial T}{\partial q_1} = B_1 q_1 + B_1' q, \ y_6 = \frac{\partial T}{\partial r_1} = C_1 r_1 + C_1' r \ .$$

From the first and the fourth of the equations we get

$$p = \frac{1}{\Delta_1}(A_1 y_1 - A_1' y_4), \ p_1 = \frac{1}{\Delta_1}(A y_4 - A_1' y_1), \ \Delta_1 = A A_1 - A_1'^2 \ .$$

From another pairs of equations we find

$$q = \frac{1}{\Delta_2}(B_1 y_2 - B_1' y_5),\ q_1 = \frac{1}{\Delta_2}(B y_5 - B_1' y_2),\ \Delta_2 = BB_1 - B_1'^2$$

$$r = \frac{1}{\Delta_3}(C_1 y_3 - C_1' y_6),\ r_1 = \frac{1}{\Delta_3}(C y_6 - C_1' y_3),\ \Delta_3 = CC_1 - C_1'^2 .$$

The kinetic energy in the new variables

$$2T^*(y_1, \cdots, y_6) = a_1 y_1^2 + b_1 y_2^2 + c_1 y_3^2 - 2(a_1' y_1 y_4 + b_1' y_2 y_5 + c_1' y_3 y_6) +$$
$$+ a y_4^2 + b y_5^2 + c y_6^2 ,$$

where

$$a_1 = \frac{A_1}{\Delta_1},\ b_1 = \frac{B_1}{\Delta_2},\ c_1 = \frac{C_1}{\Delta_3},\ a = \frac{A}{\Delta_1},\ b = \frac{B}{\Delta_2},\ c = \frac{C}{\Delta_3},$$

$$a_1' = \frac{A_1'}{\Delta_1},\ b_1' = \frac{B_1'}{\Delta_2},\ c_1' = \frac{C_1'}{\Delta_3} .$$

The canonical form of the equations (1.12,1.13) has the view

$$\frac{dy_1}{dt} = \frac{\partial T^*}{\partial y_3} y_2 - \frac{\partial T^*}{\partial y_2} y_3 + L_1,\quad \frac{dy_2}{dt} = \frac{\partial T^*}{\partial y_1} y_3 - \frac{\partial T^*}{\partial y_3} y_1 + L_2,$$

$$\frac{dy_3}{dt} = \frac{\partial T^*}{\partial y_2} y_1 - \frac{\partial T^*}{\partial y_1} y_2 + L_3,$$

$$\frac{dy_4}{dt} = \frac{\partial T^*}{\partial y_5} y_6 - \frac{\partial T^*}{\partial y_6} y_5,\quad \frac{dy_5}{dt} = \frac{\partial T^*}{\partial y_6} y_4 - \frac{\partial T^*}{\partial y_4} y_6, \tag{1.34}$$

$$\frac{dy_6}{dt} = \frac{\partial T^*}{\partial y_4} y_5 - \frac{\partial T^*}{\partial y_5} y_4 .$$

If $L_i = 0$ $(i = 1, 2, 3)$ the equations (1.34) have the first integrals

$$T^* = \text{const},\ \sum_i y_i^2 = \text{const},\ \sum_i y_{3+i}^2 = \text{const},\quad i = 1, 2, 3.$$

For the case of the symmetry with respect to the axis x_3 and by $L_3 = 0$ there is also the first integral $y_3 - y_6 = \text{const}$.

Chapter 2

The Generalized Poincaré's Equations.

2.1 Defining coordinates, parametrization of the constraints.

Consider a holonomic mechanical system with k degrees of freedom. Assume that for every instant of time t the position of the system is given by the defining coordinates [6] $x_i \quad (i = 1, \ldots, n \geq k)$

$$\mathbf{r}_\nu = \mathbf{r}_\nu (t, x_1, \ldots, x_n), \quad \nu = 1, \ldots, N,$$

where $\mathbf{r}_\nu$ is the radius vector of a point mass m_ν. When $n = k$ the variables x_i are independent Lagrange's coordinates, but when $n > k$ they are dependent, or redundant coordinates, subject to constraints, specified by the integrable system of differential equations

$$\eta_j \equiv a_{ji} (t, x_1, \ldots, x_n) \dot{x}_i + a_{j0} (t, x_1, \ldots, x_n) = 0, \quad j = k + 1, \ldots, n,$$

$\operatorname{rank}(a_{ji}) = n - k, \quad \dot{x}_i = \dfrac{dx_i}{dt}$. Everywhere summation is carried out over repeated subscripts.

The introduction of redundant coordinates is useful in some cases in order to simplify the expressions for the kinematic and dynamic quantities [25].

For symmetry and brevity we will conventionally put $t = x_0, \quad \dot{x}_0 = 1$, then the previous equations one can write down

$$\eta_j \equiv a_{ji} (x) \dot{x}_i = 0, \quad i = 0, 1, \ldots, n; \quad j = k + 1, \ldots, n \tag{2.1}$$

where $x = (x_0, x_1, \ldots, x_n)$.

The sufficient conditions for equations (2.1) to be integrable, as is well known [26], have the form

$$\frac{\partial a_{jr}}{\partial x_s} = \frac{\partial a_{js}}{\partial x_r}, \quad r, s = 0, 1, \ldots, n; \quad j = k+1, \ldots, n \tag{2.2}$$

We will complete (2.1) by the arbitrarily chosen linear forms

$$\eta_s \equiv a_{si}(x)\,\dot{x}_i, \quad s = 0, 1, \ldots, k; \quad i = 0, 1, \ldots, n; \quad \eta_0 = 1, \quad a_{0i} = \delta_{0i} \tag{2.3}$$

which are linearly independent both of one another and in relation to the forms (2.1), so that $\det(a_{ij}) \neq 0 \quad (i, j = 0, 1, \ldots, n)$, where δ_{ij} is the Kronecker delta.

Solving equations (2.1) and (2.3) for $\dot{x}_i$, we obtain the parametric representation of the constraints

$$\dot{x}_i = b_{is}(x)\,\eta_s, \quad s = 0, 1, \ldots, k; \quad i = 0, 1, \ldots, n; \quad b_{0s} = \delta_{0s}, \quad a_{si}b_{ir} = a_{ir}b_{si} = \delta_{sr}. \tag{2.4}$$

2.2 Systems of operators, Poincaré's parameters.

Parametrisation (2.4) enables us to construct a closed system of infinitesimal linear operators

$$X_s f = b_{is}\frac{\partial f}{\partial x_i}, \quad s = 0, 1, \ldots, k, \quad f(x) \in C^2 \tag{2.5}$$

defining the virtual and real displacements of the system

$$\delta f = \omega_r X_r f, \quad r = 1, \ldots, k; \quad df = \eta_s X_s f dt, \quad s = 0, 1, \ldots, k \tag{2.6}$$

respectively, where $\omega_r \equiv a_{ri}(x)\delta x_i \;\; (i = 1, \ldots, n; \; r = 1, \ldots, k)$ and η_s are the parameters of the virtual and real displacements introduced by Poincaré [1].

The system of operators (2.5) is a closed system in the sence that its commutator (the Poisson bracket) has the form

$$[X_r, X_s]\, f \equiv X_r X_s f - X_s X_r f = c_{rs}^m X_m f, \quad m, r, s = 0, 1, \ldots, k \tag{2.7}$$

where the structural coefficients

$$c_{rs}^m \equiv \left(\frac{\partial a_{mj}}{\partial x_i} - \frac{\partial a_{mi}}{\partial x_j}\right) b_{is} b_{jr} = a_{mj}\left(b_{ir}\frac{\partial b_{js}}{\partial x_i} - b_{is}\frac{\partial b_{jr}}{\partial x_i}\right), \quad i, j = 0, 1, \ldots, n \tag{2.8}$$

and besides $c_{rs}^m = -c_{sr}^m \quad (m, r, s = 0, 1, \ldots, k)$. The right side of the equality (2.7) is the linear combination of the operators (2.5), therefore the commutator (2.7) is the first-order differential operator.

Properties of a commutator [6], [27]; it is:

1) bilinear

$$\begin{aligned}[aX_\alpha + bX_\beta, X_\gamma] f &= (aX_\alpha + bX_\beta) X_\gamma f - X_\gamma (aX_\alpha + bX_\beta) f = \\ &= aX_\alpha X_\gamma f + bX_\beta X_\gamma f - aX_\gamma X_\alpha f - bX_\gamma X_\beta f = \\ &= a[X_\alpha, X_\gamma] f + b[X_\beta, X_\gamma] f\end{aligned}$$

where a and b are constants;

2) skew-symmetric

$$[X_\alpha, X_\beta] = -[X_\beta, X_\alpha]$$

3) satisfies the Jacobi identity

$$[X_\alpha, [X_\beta, X_\gamma]] + [X_\beta, [X_\gamma, X_\alpha]] + [X_\gamma, [X_\alpha, X_\beta]] = 0.$$

The system (2.5) is an infinitesimal group whose commutator satisfies propreties 1) - 3).

Since $X_1 f, \ldots, X_k f$ form an infinitesimal group of virtual displacements and do not contain the derivative $\frac{\partial f}{\partial t}$, and $X_0 f = \frac{\partial f}{\partial t} + b_{i0} \frac{\partial f}{\partial x_i}$, we infer that $[X_0, X_\alpha] f$ does not include $\frac{\partial f}{\partial t}$ and consequently $c^0_{0\alpha} = 0 \quad (\alpha = 1, \ldots, k)$, and besides $X_1, \ldots, X_k$ is a subgroup of the infinitesimal group $X_0, X_1, \ldots, X_k$ of real displacements.

By an appropriate choice of the auxiliary forms (2.3) we can simplify the closed system (2.5) and get a Lie infinitesimal group (Lie algebra in modern terminology), for which all $c^\gamma_{\alpha\beta}$ are constants $(\alpha, \beta, \gamma = 0, 1, \ldots, k)$.

Example [6]. Assume that in the system (2.1) $\det\|a_{rs}\| \neq 0, \quad r, s = k+1, \ldots, n$ and then

$$\dot{x}_j = b_{j\alpha} \dot{x}_\alpha + b_{j0}, \quad j = k+1, \ldots, n$$

If we accept that $\eta_0 = 1,\ \eta_\alpha = \dot{x}_\alpha \quad (\alpha = 1, \ldots, k)$, then

$$\frac{df}{dt} = \frac{\partial f}{\partial t} + \frac{\partial f}{\partial x_i} \dot{x}_i = \frac{\partial f}{\partial t} + b_{j0} \frac{\partial f}{\partial x_j} + \eta_\alpha \left(\frac{\partial f}{\partial x_\alpha} + b_{j\alpha} \frac{\partial f}{\partial x_j} \right)$$

and this means that

$$X_0 f = \frac{\partial f}{\partial t} + b_{j0} \frac{\partial f}{\partial x_j}, \quad X_\alpha f = \frac{\partial f}{\partial x_\alpha} + b_{j\alpha} \frac{\partial f}{\partial x_j}, \quad \alpha = 1, \ldots, k.$$

It follows that $[X_\alpha, X_\beta] f = 0$ and hence all $c^\gamma_{\alpha\beta} = 0 \quad (\alpha, \beta, \gamma = 0, 1, \ldots, k)$.

Assume that $\omega_\alpha = \delta x_\alpha \quad (\alpha = 1, \ldots, k)$ then

$$\delta f = \omega_\alpha X_\alpha f, \quad [X_\alpha, X_\beta] f = 0, \quad \alpha, \beta = 1, \ldots, k$$

The system of operators $X_\alpha f \quad (\alpha = 1, \ldots, k)$ forms an subgroup of the Lie group $X_0 f, X_1 f, \ldots, X_k f$.

The case in which all c_{sr}^m = const, was exactly what considered by Poincaré and by Chetayev. However in the general case of the closed system (2.5) the coefficients c_{sr}^m will, generally speaking, be variable, and this case is not excluded further from consideration.

Note that if the forms (2.3) are integrable, like forms (2.1), the conditions (2.2) are also satisfied for $j = 1, \ldots, k$, and functions of the form $\pi_s = \pi_s(x)$ then exist which can serve as new defining coordinates, and besides $\eta_s = \dot{\pi}_s \quad (s = 1, \ldots, k)$,

$$\frac{\partial \pi_s}{\partial x_i} = \frac{\partial \eta_s}{\partial \dot{x}_i} = a_{si}, \qquad \frac{\partial x_i}{\partial \pi_s} = \frac{\partial \dot{x}_i}{\partial \eta_s} = b_{is}.$$

In this case the system of operators (2.5) is an Abelian group of the functions

$$X_s f = b_{is}\frac{\partial f}{\partial x_i} = \frac{\partial f}{\partial \pi_s}, \qquad s = 0, 1, \ldots, k; \quad i = 0, 1, \ldots, n$$

for which all the coefficients $c_{rs}^m = 0 \quad (m, r, s = 0, 1, \ldots, k)$.

If the forms (2.3) are not integrable the quantities $\pi_s(x)$ as functions of time and the coordinates do not exist, but the symbols π_s are worth while introduced into consideration under the name of quasi-coordinates, using the conventional notation for the quasi-velocities $\eta_s = \dot{\pi}_s$ and the differentials of the quasi-coordinates $d\pi_s = \eta_s dt$ and also for "partial derivatives with respect to the quasi-coordinates" and for the inverse relations

$$\frac{\partial f}{\partial \pi_s} = b_{is}\frac{\partial f}{\partial x_i}, \quad \frac{\partial f}{\partial x_i} = a_{si}\frac{\partial f}{\partial \pi_s}, \quad i = 0, 1, \ldots, n; \quad s = 0, 1, \ldots, k \tag{2.9}$$

By (2.9) the operators (2.5) in this case can be represented in the form of

$$X_s f = \frac{\partial f}{\partial \pi_s}, \quad s = 0, 1, \ldots, k \tag{2.10}$$

with commutator (2.7), which take the form

$$[X_r, X_s] f \equiv \frac{\partial^2 f}{\partial \pi_r \partial \pi_s} - \frac{\partial^2 f}{\partial \pi_s \partial \pi_r} = c_{rs}^m \frac{\partial f}{\partial \pi_m}$$

where, in general, $c_{rs}^m \neq 0$.
If the constraints are holonomic and the operators d and δ are commutable, then [6]

$$\begin{aligned} 0 &= d\delta f - \delta df = d(\omega_\alpha X_\alpha f) - \delta(X_0 f + \eta_\alpha X_\alpha f)\, dt = \\ &= (d\omega_\alpha - \delta\eta_\alpha dt)\, X_\alpha f + \omega_\alpha (X_0 X_\alpha f + \eta_\beta X_\beta X_\alpha f)\, dt - \\ &- (\omega_\alpha X_\alpha X_0 f + \eta_\alpha \omega_\beta X_\beta X_\alpha f)\, dt = \end{aligned}$$

$$= (d\omega_\alpha - \delta\eta_\alpha dt) X_\alpha f + \omega_\alpha dt \,(X_0 X_\alpha f - X_\alpha X_0 f) +$$

$$+ \ \omega_\alpha \eta_\beta (X_\beta X_\alpha f - X_\alpha X_\beta f)\, dt =$$

$$= dt \left(\frac{d\omega_i}{dt} - \delta\eta_i + c^i_{\beta\alpha}\eta_\beta\omega_\alpha + c^i_{0\alpha}\omega_\alpha \right) X_i f$$

Since f is arbitrary, we have the relations

$$\frac{d\omega_i}{dt} - \delta\eta_i = c^i_{sr}\eta_r\omega_s + c^i_{0s}\omega_s; \quad i, r, s = 1, \ldots, k \tag{2.11}$$

initially established by Poincaré for the case when $c^i_{0s} = 0$.

Note that correlation (2.11) is similar to that for the bilinear covariant or the external derivative of the form $\omega_i = a_j \delta x_j$

$$d\omega_i\,(\delta x) - \delta\omega_i\,(dx) = d\omega_i\,(\delta x) - \delta\eta_i dt$$

taking into consideration (2.1), (2.4), (2.6) and (2.8).

The expressions for the coefficients c^i_{rs} are more easily obtained from relations (2.11) than by using the general formulae (2.8) [25].

2.3 The generalized Poincaré's Equations.

Poincaré [1] and Chetayev [3]-[5] used the Hamilton principle to derive the motion's equations, while Chetayev [6] also used the d'Alembert-Lagrange principle in its traditional form

$$(m_\nu \ddot{\mathbf{r}}_\nu - \mathbf{F}_\nu) \cdot \delta\mathbf{r}_\nu = 0, \quad \nu = 1, \ldots, N \tag{2.12}$$

We shall use the d'Alembert-Lagrange principle in the defining coordinates [6]

$$\left(\frac{d}{dt}\frac{\partial L}{\partial \dot{x}_i} - \frac{\partial L}{\partial x_i} - Q_i \right) \delta x_i = 0, \quad i = 1, \ldots, n \tag{2.13}$$

where $L\,(t, x, \dot{x}) = T\,(t, x, \dot{x}) + U\,(t, x)$ is the Lagrange function, $T\,(t, x, \dot{x})$ is the kinetic energy, $U\,(t, x)$ is the force function, $x = (x_1, \ldots, x_n)$, $\dot{x} = (\dot{x}_1, \ldots, \dot{x}_n)$, $Q_i = F^*_\nu \cdot \dfrac{\partial r_\nu}{\partial x_i}$ is the generalized non-potential force.

By (2.5) and (2.6) we have

$$\delta x_i = \omega_s X_s x_i = \omega_s b_{is}, \quad i = 1, \ldots, n; \quad s = 1, \ldots, k$$

as a consequence of which, by virtue of the arbitrariness of ω_s, the generalized Maggi's [28] equations of motion follow from (2.13), namely

$$\left(\frac{d}{dt}\frac{\partial L}{\partial \dot{x}_i} - \frac{\partial L}{\partial x_i} - Q_i\right) b_{is} = 0, \quad s = 1, \ldots, k \tag{2.14}$$

Using (2.4) we can express the Lagrange function in the form of the equality $L^*(t, x, \eta) = L(t, x, \dot{x})$, by differentiating which and using (2.3) we obtain the relations [29]

$$\frac{\partial L}{\partial \dot{x}_i} = \frac{\partial L^*}{\partial \eta_m} a_{mi}, \quad \frac{\partial L}{\partial x_i} = \frac{\partial L^*}{\partial x_i} + \frac{\partial L^*}{\partial \eta_m}\left(\frac{\partial a_{mj}}{\partial x_i} b_{jr}\eta_r + \frac{\partial a_{m0}}{\partial x_i}\right), \quad i, j = 1, \ldots, n;$$

$$\frac{d}{dt}\frac{\partial L}{\partial \dot{x}_i} = \frac{d}{dt}\left(\frac{\partial L^*}{\partial \eta_m}\right) a_{mi} + \frac{\partial L^*}{\partial \eta_m}\left(\frac{\partial a_{mi}}{\partial x_j} b_{jr}\eta_r + \frac{\partial a_{mi}}{\partial t}\right), \quad r = 0, 1, \ldots, k$$

substitution of which into (2.14) leads to generalized Poincaré's equations

$$\frac{d}{dt}\frac{\partial L^*}{\partial \eta_s} = c^m_{rs}\frac{\partial L^*}{\partial \eta_m}\eta_r + c^m_{0s}\frac{\partial L^*}{\partial \eta_m} + X_s L^* + Q^*_s, \quad m, r, s, = 1, \ldots, k \tag{2.15}$$

where $Q^*_s = Q_i b_{is}$. The structural coefficients

$$c^m_{rs} = \left(\frac{\partial a_{mj}}{\partial x_i} - \frac{\partial a_{mi}}{\partial x_j}\right) b_{is} b_{jr} \qquad i, j = 1, \ldots, n$$

$$c^m_{0s} = \left[\left(\frac{\partial a_{mj}}{\partial x_i} - \frac{\partial a_{mi}}{\partial x_j}\right) b_{j0} + \frac{\partial a_{m0}}{\partial x_i} - \frac{\partial a_{mi}}{\partial t}\right] b_{is} \tag{2.16}$$

elaborate expressions (2.8) for the explicit selection of $t = x_0$.

The equations (2.15) together with equations (2.4) form a compatible system of $k+n$ first-order defferential equations of motion, each with the same number of unknowns $\eta_1, \ldots, \eta_k, x_1, \ldots, x_n$. It is noteworthy that equations (2.15) in redundant coordinates contain no reaction forces of the constraint (2.1) and have the same outward appearance in both independent coordinates ($n = k$) and dependent coordinates ($n > k$).

The generalized Poincaré's equations (2.15) contain, as special cases, the equations (1.4), first given by Poincaré [1] for the cases $n = k$, $X_0 = \frac{\partial}{\partial t}$, $c^m_{0i} = 0$, $c^m_{rs} = \text{const}$, $Q^*_i = 0$ $(i, s = 1, \ldots, k)$; the Lagrange equations of the second kind

$$\frac{d}{dt}\frac{\partial L}{\partial \dot{x}_i} - \frac{\partial L}{\partial x_i} = Q_i, \quad i = 1, \ldots, k$$

when $n = k$, $\eta_s = \dot{x}_i$, $X_s = \frac{\partial}{\partial x_s}$ and all $c^m_{rs} = 0$; the generalized Boltzmann [30]-Hamel [31] equations in quasi-coordinates π_s

$$\frac{d}{dt}\frac{\partial L^*}{\partial \pi_s} = c_{rs}^m \pi_r \frac{\partial L^*}{\partial \pi_m} + c_{0s}^m \frac{\partial L^*}{\partial \pi_m} + \frac{\partial L^*}{\partial \pi_s} + Q_s^*, \quad m, r, s, = 1, \ldots, k \tag{2.17}$$

where $L^* = L^*(t, x_1, \ldots, x_n, \pi_1, \ldots, \pi_k)$, and besides it is supposed what the forms (2.3) are not integrable and the operators (2.5) have the form (2.10).

Equations (2.17) are the unique form of the Boltzmann equations in the case of dependent coordinates (after elimination indefinite coefficients), and the Boltzmann-Hamel equations in the case of independent coordinates. In particular, Euler's equations of motion of a rigid body around a fixed point follow from equations (2.17).

The equations (2.15) contain also as special case a generalized Lagrange equations in dependent coordinates [21], which we shall derive now.

Suppose that the constraints (2.1) are represented by differential equations

$$\dot{x}_j = b_{j\alpha}(t, x)\,\dot{x}_\alpha + b_j(t, x); \quad \alpha = 1, \ldots, k; \quad j = k+1, \ldots, n \tag{2.18}$$

which constitute the completely integrable system.

As parameters of the real and virtual displacements we take $\eta_\alpha = \dot{x}_\alpha$ and $\omega_\alpha = \delta x_\alpha$ respectively. Using (2.18) we find for the operators of the group (2.5)

$$X_0 f = \frac{\partial f}{\partial t} + b_j \frac{\partial f}{\partial x_j}, \qquad X_\alpha f = \frac{\partial f}{\partial x_\alpha} + b_{j\alpha} \frac{\partial f}{\partial x_j} \tag{2.19}$$

whose commutator vanishes because equations (2.18) are integrable. Consequently, all the structural constants of the group (2.19) vanish: $c_{\alpha\beta}^s = 0 \quad (\alpha, \beta, s = 0, 1, \ldots, k)$, i.e. the operators (2.19) form an Abelian group. The Poincaré's equations (2.15) in this case become

$$\frac{d}{dt}\frac{\partial L^*}{\partial \eta_\alpha} - X_\alpha L^* = Q_\alpha^*, \quad \alpha = 1, \ldots, k, \quad L^* = L^*(t, x_1, \ldots, x_n, \dot{x}_1, \ldots, \dot{x}_k)$$

or, in view of (2.19)

$$\frac{d}{dt}\frac{\partial L^*}{\partial \dot{x}_\alpha} - \frac{\partial L^*}{\partial x_\alpha} - b_{j\alpha}\frac{\partial L^*}{\partial x_j} = Q_\alpha^* \tag{2.20}$$

Equations (2.20), (2.18) form the system of differential equations of order $n+k$ in the same number of unknowns $x_1, \ldots, x_n, \dot{x}_1, \ldots, \dot{x}_k$.

Equations (2.20) are generalized Lagrange equations in the dependent (redundant) coordinates; these equations do not involve the reactions of the constraints (2.18). They are more convenient that the Lagrange equations with multipliers λ_j

$$\frac{d}{dt}\frac{\partial L}{\partial \dot{x}_i} - \frac{\partial L}{\partial x_i} = Q_i - b_{ji}\lambda_j, \quad \frac{d}{dt}\frac{\partial L}{\partial \dot{x}_j} - \frac{\partial L}{\partial x_j} = Q_j + \lambda_j, \tag{2.21}$$

$$L = L(t, x_1, \ldots, x_n, \dot{x}_1, \ldots, \dot{x}_n)$$

whose order is $2n > n + k$, considered together with equations (2.18), which are of order k. It is true that equations (2.21) and (2.18) enable one to determine not only $x_1, \ldots, x_n$ but also λ_j $(j = k+1, \ldots, n)$, and together with them the reactions of the constraints (2.18); but for large n they are not very tractable.

It can be shown, however, that the elimination of λ_j from equations (2.21) and the use of (2.18) lead to equations (2.20); we shall not go into details.

Example [6]. Similarity alteration of a body (D.N.Zeiliger [32]).

Assume that the center of gravity O of the body is the origin of the moving coordinate frame $Ox_1x_2x_3$, and that a_i are the coordinates of O in the fixed frame $\bar{O}\bar{x}_1\bar{x}_2\bar{x}_3$ and that β_i^k is the cosine of the angle between the fixed axis $\bar{x}_i$ and the moving axis x_k. The coordinates of the particles m of the body are

$$\bar{x}_i = a_i + \beta_i^k x_k \quad (i = 1, 2, 3)$$

We assume a_i, β_i^k, x_k to be the defining coordinates, and then

$$\frac{da_i}{dt} = u_i, \qquad \frac{d\beta_k^1}{dt} = r\beta_k^2 - q\beta_k^3, \qquad \frac{d\beta_k^2}{dt} = p\beta_k^3 - r\beta_k^1, \qquad \frac{d\beta_k^3}{dt} = q\beta_k^1 - p\beta_k^2.$$

If α is the velocity of the ray dilatation, then $\dfrac{dx_k}{dt} = \alpha x_k \quad (k = 1, 2, 3)$. Let $f = f\left(t, a_i, \beta_i^k, x_k\right)$. Then

$$\begin{aligned} \frac{df}{dt} &= \frac{\partial f}{\partial t} + \frac{\partial f}{\partial a_i} u_i + p\left(\frac{\partial f}{\partial \beta_i^2}\beta_i^3 - \frac{\partial f}{\partial \beta_i^3}\beta_i^2\right) + q\left(\frac{\partial f}{\partial \beta_i^3}\beta_i^1 - \frac{\partial f}{\partial \beta_i^1}\beta_i^3\right) + \\ &+ r\left(\frac{\partial f}{\partial \beta_i^1}\beta_i^2 - \frac{\partial f}{\partial \beta_i^2}\beta_i^1\right) + \alpha\left(\frac{\partial f}{\partial x_\nu}x_\nu + \frac{\partial f}{\partial y_\nu}y_\nu + \frac{\partial f}{\partial z_\nu}z_\nu\right), \quad \nu = 1, 2, \ldots. \end{aligned}$$

Take for real parameters the quantities u_i $(i = 1, 2, 3)$, p, q, r, α.

Consequently Lie's group of the real displacements of a body under similarity alteration consists of operators

$$X_0 = \frac{\partial}{\partial t}, \qquad A_i = \frac{\partial}{\partial a_i}, \qquad i = 1, 2, 3;$$

$$X = \frac{\partial}{\partial \beta_i^2}\beta_i^3 - \frac{\partial}{\partial \beta_i^3}\beta_i^2, \quad Y = \frac{\partial}{\partial \beta_i^3}\beta_i^1 - \frac{\partial}{\partial \beta_i^1}\beta_i^3,$$

$$Z = \frac{\partial}{\partial \beta_i^1}\beta_i^2 - \frac{\partial}{\partial \beta_i^2}\beta_i^1, \quad \Phi = x_\nu\frac{\partial}{\partial x_\nu} + y_\nu\frac{\partial}{\partial y_\nu} + z_\nu\frac{\partial}{\partial z_\nu}, \quad \nu = 1, 2 \ldots$$

The subgroup of virtual displacements A_i, X, Y, Z, Φ consists of three subgroups, namely, a subgroup of translations A_i, $[A_i, A_j] = 0$ $(i, j = 1, 2, 3)$, a subgroup of rotations X, Y, Z with the property

$$[X,Y]=Z,\quad [Y,Z]=X,\quad [Z,X]=Y,$$

a subgroup of ray dilatation consisting of only one Φ. These subgroups are commutable in the sense that

$$[A_i,X_j]=0,\quad [A_i,\Phi]=0,\quad [X_i,\Phi]=0\quad i,j=1,2,3$$

The kinetic energy, if the coordinate axes x_i are the principal axes of inertia, equal

$$T=\frac{M}{2}\left(\dot{a}_1^2+\dot{a}_2^2+\dot{a}_3^2\right)+\frac{1}{2}\left(Ap^2+Bq^2+Cr^2\right)+\frac{1}{2}J_0\alpha^2$$

where A, B, C, J_0 are the principal moments of inertia with reference to axes x_1, x_2, x_3 and point O, accordingly, M is the mass of the body.

Then Poincaré's equations of motion become

$$M\frac{d^2a_1}{dt^2}=A_1U,\quad M\frac{d^2a_2}{dt^2}=A_2U,\quad M\frac{d^2a_3}{dt^2}=A_3U,$$

$$\frac{d}{dt}Ap=(B-C)qr+XU,\quad \frac{d}{dt}Bq=(C-A)rp+YU, \tag{2.22}$$

$$\frac{d}{dt}Cr=(A-B)pq+ZU,\quad \frac{dJ_0\alpha}{dt}=\Phi\left(T+U\right)$$

where U is a force function.

But

$$\Phi A=\left(x_\nu\frac{\partial}{\partial x_\nu}+y_\nu\frac{\partial}{\partial y_\nu}+z_\nu\frac{\partial}{\partial z_\nu}\right)m_\nu\left(y_\nu^2+z_\nu^2\right)=2m_\nu\left(y_\nu^2+z_\nu^2\right)=2A$$

$$\Phi B=2B,\qquad \Phi C=2C,\qquad \Phi J_0=2J_0$$

Consequently

$$\Phi(T)=Ap^2+Bq^2+Cr^2+J_0\alpha^2$$

and the last from equations (2.22) become

$$\frac{d}{dt}J_0\alpha=Ap^2+Bq^2+Cr^2+J_0\alpha^2+\Phi(U),$$

$\Phi(U)$ is a virial. These equations were derived first by Professor D.N.Zeiliger [32].

2.4 The First Integrals of the Poincaré's Equations. Cyclic displacements.

The Poincaré's equations have some first integrals by certain conditions. If

$$X_0 = \frac{\partial}{\partial t}, \quad c^s_{0i} = 0, \quad \frac{\partial L}{\partial t} = 0, \quad Q_i \eta_i = 0 \quad i, s = 1, \ldots, k \tag{2.23}$$

then equations (2.15) have a generalized energy integral

$$\eta_i \frac{\partial L^*}{\partial \eta_i} - L^* = \text{const} \tag{2.24}$$

which gets by multiplying equations (2.15) by η_i and by addition the results together.

If $L^* = L^*_2 + L^*_1 + L^*_0$, where L^*_s are homogeneous forms of the variables η_i of degree s $(s = 0, 1, 2)$ then equality (2.24) has the form

$$L^*_2 - L^*_0 = h = \text{const} \tag{2.25}$$

This is the generalization of the kinetic energy integral. Its special cases are well-known, they are (1) Jacobi's integral and (2) the ordinary kinetic energy integral.

Note that conditions (2.23) are satisfied if all $a_{s0} = b_{i0} = 0$ in equations (2.3) and (2.4), while the coefficients a_{si} and b_{si} are explicitly independent of time, like the Lagrange function $L^*(x_1, \ldots, x_n, \eta_1, \ldots, \eta_k)$.

Considering the case in which there are not non-potential forces Q^*_i, Chetayev [5], [6] introduced the important concept of a cyclic displacement X_i $(i = r+1, \ldots, k)$, which satisfies the conditions

$$1^0. [X_\alpha, X_i] = 0; \quad c^s_{\alpha i} = 0 \quad \alpha, s = 0, 1, \ldots, k; \quad 2^0. X_i L^* = 0 \tag{2.26}$$

If cyclic displacements exist, equations (2.15) have the first integrals

$$\frac{\partial L^*}{\partial \eta_i} = b_i = \text{const}, \quad i = r+1, \ldots, k \tag{2.27}$$

By (2.27) the parameters $\eta_{r+1}, \ldots, \eta_k$ may be expressed as functions of $t, x, \eta_1, \ldots, \eta_r, b_{r+1}, \ldots, b_k$ and one can form a generalized Routh function

$$R(t, x_1, \ldots, x_n, \eta_1, \ldots, \eta_r, b_{r+1}, \ldots, b_k) = L^* - \eta_i \frac{\partial L^*}{\partial \eta_i} \tag{2.28}$$

Using the equalities

$$X_s R = X_s L^*, \quad X_\alpha R = 0,$$

$$\frac{\partial R}{\partial \eta_s} = \frac{\partial L^*}{\partial \eta_s}, \quad \frac{\partial R}{\partial b_\alpha} = -\eta_\alpha \quad s = 1, \ldots, r, \quad \alpha = r+1, \ldots, k$$

we can write equations (2.15) for non-cyclic displacements as generalized Routh equations

$$\frac{d}{dt}\frac{\partial R}{\partial \eta_i} = c^s_{\alpha i}\eta_\alpha \frac{\partial R}{\partial \eta_s} + c^j_{\alpha i}\eta_\alpha b_j + c^s_{0i}\frac{\partial R}{\partial \eta_s} + c^j_{0i} b_j + X_i R \qquad (2.29)$$

$$i, \alpha, s = 1, \ldots, r; \quad j = r+1, \ldots, k$$

after integration of which the quantities η_α are determined by the relations

$$\eta_\alpha = -\frac{\partial R}{\partial b_\alpha}, \quad \alpha = r+1, \ldots, k \qquad (2.30)$$

Elementary integrals like (2.27) were first given by Chaplygin [33] (see also [11]-[13]).

Remarks [5], [6]: *1) Integrals associated with cyclic coordinates or cyclic displacements are linear with respect to velocities or linear with respect to parameters η_α of real displacements. 2) It is possible to give also some generalizations of cyclic displacements*

2.5 The Generalized Jacobi-Whittaker's Equations.

Comparing conditions (2.23) and (2.26) we see that the operator $X_0 = \frac{\partial}{\partial t}$ satisfies conditions (2.26), i.e. it is possible to consider as an operator of cyclic displacement to which the energy integral corresponds.

Indeed, if we represent Poincaré's equations in parametric form [34], when the time $t = x_0$ and the coordinates $x_i \quad (i = 1, \ldots, n)$ are considered as variables x_α $(\alpha = 0, 1, \ldots, n)$ that are independent of one another, constrained by differential constraints (2.1) and specified by continuous differentiable functions of a certain parameter τ, $\; x_\alpha = x_\alpha(\tau)$, then energy integral of the Poincaré parametric equations with Lagrangian $L^*(x_i, \eta_s)\, x_0'$, $\; x_0' = \frac{dt}{d\tau}$, will correspond to the variable x_0.

Using the energy integral we can reduce the order of the system of equations by determining from the integral the variable $x_0' = t' = \varphi(x_i, \eta_s, h)$ and constructing the Routh function

$$R(x_i, \eta_s, h) = L^* x_0' - \left(L^* - \frac{\partial L^*}{\partial \eta_s}\eta_s\right) x_0' = \frac{\partial L^*}{\partial \eta_s}\eta_s x_0' \qquad (2.31)$$

on the right-hand side of which the variable x_0' is replaced by the function $\varphi(x_i, \eta_s, h)$.

The parametric Routh equations with $\tau = t$ take the form of Poincaré's equations (2.15) in which all $c^m_{0s} = 0$, $\quad Q^*_s = 0$.

If we take one of the quantities π_s, say π_1, as the parameter τ, we can obtain from the energy integral

$$\eta_1 = \dot{\pi}_1 = \frac{1}{t'} = \psi\left(x_i, \eta_r', h\right), \qquad r = 2, \ldots, k$$

where $\eta_r^{'} = \frac{d\pi_r}{d\pi_1} = \frac{\eta_r}{\eta_1}$, $t' = \frac{dt}{d\pi_1} = \frac{1}{\eta_1}$ and, substituting into (2.31), we obtain the new Routh function

$$R'\left(x_i, \eta_r^{'}, h\right) = \frac{\partial L^*}{\partial \eta_s}\frac{\eta_s}{\eta_1} \tag{2.32}$$

It easy to show [29], that the following equalities hold

$$\frac{\partial R'}{\partial \eta_r^{'}} = \frac{\partial L^*}{\partial \eta_r}, \quad \frac{\partial R'}{\partial x_i} = \frac{1}{\eta_1}\frac{\partial L^*}{\partial x_i}, \quad i = 1, \ldots, n; \quad r = 2, \ldots, k$$

substituting which into (2.17) for $s = 2, \ldots, k$ and all $c_{0s}^{m} = Q_s^* = 0$, we obtain the generalized Jacobi [35] - Whittaker [29] equations in quasi-coordinates

$$\frac{d}{dt}\frac{\partial R'}{\partial \eta_s^{'}} = c_{rs}^{m}\eta_r^{'}\frac{\partial R'}{\partial \eta_m^{'}} + \frac{\partial R'}{\partial \pi_s}, \quad s = 2, \ldots, k \tag{2.33}$$

If we put in (2.3) $\eta_1 = \dot{x}_1$ (then [31] $c_{rs}^{1} = 0$), equations (2.33) take the form

$$\frac{d}{dx_1}\frac{\partial R'}{\partial \eta_s^{'}} = c_{rs}^{m}\frac{\partial R'}{\partial \eta_m^{'}}\eta_r^{'} + \frac{\partial R'}{\partial \pi_s}, \quad m, r, s = 2, \ldots, k \tag{2.34}$$

As pointed out in Section 2.2, when equations (2.3) are integrable the variables π_s can serve as new defining coordinates. Then all $c_{rs}^{m} = 0$ and equations (2.34) take the form of the Jacobi-Whittaker equations.

Equations (2.34) need to be investigated in the general case together with the constraints equations (2.4), written in the following form

$$x_i^{'} = b_{is}\eta_s^{'}, \quad i = 2, \ldots, n; \quad s = 2, \ldots, k \tag{2.35}$$

Equations (2.34) and (2.35) can be regarded as the equations of motion of a new dynamical system with $k-1$ degrees of freedom, for which R' is the kinetic potential, η'_r are the parameters of the real displacements, while x_1 plays the part of time as an independent variable. The dependence of x_1 on the time t is established by quadrature [29].

Chapter 3

The Generalized Chetayev's Equations.

3.1 The Canonical Chetayev's Equations.

N.Chetayev [3]-[6] transformed Poincaré's equations to canonical form by introducing instead η_s and $L^*(t,x,\eta)$ new variables y_s and function $H^*(t,x,y)$ defined by the equations

$$y_s = \frac{\partial L^*}{\partial \eta_s}, \quad s = 1,\dots,k, \quad H^*(t,x,y) = y_s\eta_s - L^*(t,x,\eta) \tag{3.1}$$

from which the equations follow

$$X_s H^* = -X_s L^*, \quad \eta_s = \frac{\partial H^*}{\partial y_s}, \quad s = 1,\dots,k \tag{3.2}$$

The transformation (3.1) is the Legendre transformation, if we take into account the fact that $\|\frac{\partial^2 L^*}{\partial \eta_r \partial \eta_s}\| \neq 0$ $(r,s = 1,\dots,k)$. Since

$$y_s = \frac{\partial L^*}{\partial \eta_s} = \frac{\partial L}{\partial \dot{x}_i}\frac{\partial \dot{x}_i}{\partial \eta_s} = p_i b_{is}, \quad \eta_s = a_{sj}\dot{x}_j, \quad p_i = \frac{\partial L}{\partial \dot{x}_i}$$

it is obvious that the following equality holds

$$H^*(t,x,y) = p_i b_{is} a_{sj}\dot{x}_j - L(t,x,\dot{x}) = H(t,x,p)$$

(the formula $a_{sj}b_{is} = \delta_{ij}$ is taken into account).

Substituting (3.1), (3.2) into Poincaré's equations (2.15) we obtain the canonical

Chetayev's equations

$$\frac{dy_s}{dt} = c_{rs}^m \frac{\partial H^*}{\partial y_r} y_m + c_{0s}^m y_m - X_s H^* + Q_s^*,$$
$$\eta_s = \frac{\partial H^*}{\partial y_s}; \qquad m, r, s = 1, \ldots, k \tag{3.3}$$

These equations need to be investigated in the general case together with equations (2.4), by means of which to the second group of equations (3.3) can be given another form [6]

$$\frac{dx_i}{dt} = X_0 x_i + \frac{\partial H^*}{\partial y_s} X_s x_i, \quad i = 1, \ldots, n \tag{3.4}$$

Note that, like equations (2.15), the first group of equations of (3.3) can be derived directly from equations (2.14), rewritten in the form

$$\left(\frac{dp_i}{dt} + \frac{\partial H}{\partial x_i} - Q_i\right) b_{is} = 0, \qquad s = 1, \ldots, k$$

(equations of the form (3.2) are taken into account).

Chetayev's equations (3.3) contain as special cases:

1) the canonical Hamilton's equations

$$\frac{dq_i}{dt} = \frac{\partial H}{\partial p_i}, \quad \frac{dp_i}{dt} = -\frac{\partial H}{\partial q_i} + Q_i, \quad i = 1, \ldots, n$$

when the variables x_i are independent Lagrange's coordinates ($n = k$), while the group (2.5) is reduced to commutative transformations, where the Lagrange's generalized velocities $\dot{x}_i$ are taken as the parameters η_i of the real displacements, variables x_i, p_i will be canonical coordinates, while $H(t, x, p)$ is the classical Hamilton function;

2) the generalized Hamilton's equations in superfluous coordinates [21]

$$\frac{dy_s}{dt} = -\frac{\partial H^*}{\partial x_s} - b_{js} \frac{\partial H^*}{\partial x_j} + Q_s^*,$$

$$\frac{dx_s}{dt} = \frac{\partial H^*}{\partial y_s}, \quad s = 1, \ldots, k; \quad j = k+1, \ldots, n$$

3) the canonical Boltzmann-Hamel's equations in quasi-coordinates [21]

$$\frac{dy_s}{dt} = c_{rs}^m \frac{\partial H^*}{\partial y_r} y_m + c_{0s}^m y_m - \frac{\partial H^*}{\partial \pi_s} + Q_s^*, \quad \frac{d\pi_s}{dt} = \frac{\partial H^*}{\partial y_s}, \quad m, r, s = 1, \ldots, k;$$

By the conditions

$$X_0 = \frac{\partial}{\partial t}, \quad c^s_{0i} = 0, \quad i, s = 1, \ldots, k, \quad X_0 H^* = 0, \quad Q^*_i \frac{\partial H^*}{\partial y_i} = 0$$

which are equivalent to conditions (2.23), equations (3.3) have the energy integral

$$H^* (x_1, \ldots, x_n; \; y_1, \ldots, y_k) = \text{const}$$

equivalent to the energy integral (2.24) of equations (2.15).

When the cyclic displacements X_α $(\alpha = r+1, \ldots, k)$ exist under conditions (2.26), equations (3.3) with $Q^*_s = 0$ will allow of the integrals

$$y_\alpha = \beta_\alpha, \quad \alpha = r+1, \ldots, k$$

similar to integrals (2.27) of equations (2.15). For non-cyclic displacements X_i, the equations (3.3) take the form of the equations

$$\begin{gathered}\frac{dy_i}{dt} = c^s_{pi} \frac{\partial H^*}{\partial y_p} y_s + c^\alpha_{pi} \frac{\partial H^*}{\partial y_p} \beta_\alpha + c^s_{0i} y_s + c^\alpha_{0i} \beta_\alpha - X_i H^*, \\ \eta_i = \frac{\partial H^*}{\partial y_i}, \qquad i, p, s = 1, \ldots, r; \; \alpha = r+1, \ldots, k\end{gathered} \tag{3.5}$$

equivalent to the generalized Routh's equations (2.29), where

$$H^* = H^* (t, x_1, \ldots, x_k, y_1, \ldots, y_r, \beta_{r+1}, \ldots, \beta_k).$$

After integrating equations (3.5) the variables η_α are defined by the equations $\eta_\alpha = \dfrac{\partial H^*}{\partial \beta_\alpha}$ $(\alpha = r+1, \ldots, k)$.

Using the energy integral we can reduce Chetayev's equations by two orders. In fact, suppose we solve the integral $H^* (x, y) + h = 0$ for the variable y_1, so that

$$y_1 + K (x_1, \ldots, x_n, y_2, \ldots, y_k, h) = 0$$

Consider the Legendre transformation for equations (2.33) and (2.34)

$$y'_r = \frac{\partial R'}{\partial \eta'_r}, \quad K (x, y'_r, h) = y'_r \eta'_r - R' (x, \eta'_s, h), \quad r = 2, \ldots, k$$

from which we obtain the equalities

$$X_r K = -X_r R', \qquad \eta'_r = \frac{\partial K}{\partial y'_r}$$

Taking these equations into account, equations (2.33) can be written in the form of the generalized Whittaker equation [29]

$$\frac{dy'_s}{d\pi_1} = c^m_{rs}\frac{\partial K}{\partial y'_r}y'_m - \frac{\partial K}{\partial \pi_s}, \quad \frac{dt}{d\pi_1} = \frac{\partial K}{\partial h}, \quad \frac{dh}{d\pi_1} = 0, \quad m,r,s = 2,\dots,k \tag{3.6}$$

and also equations (2.34)

$$\frac{dy'_s}{dx_1} = c^m_{rs}\frac{\partial K}{\partial y'_r}y'_m - \frac{\partial K}{\partial \pi_s}, \quad \frac{dt}{dx_1} = \frac{\partial K}{\partial h}, \quad \frac{dh}{dx_1} = 0, \quad m,r,s = 2,\dots,k \tag{3.7}$$

The last pairs of equations (3.6) and (3.7) can be separated from the remaining equations since the first $2(k-1)$ equations do not contain t, while $h = \text{const}$. Hence, the first $2(k-1)$ equations of (3.6) or (3.7) can be regarded as the equations of motion of the reduced system with $k-1$ degrees of freedom [29].

Example. Consider the canonical equations for the Zeiliger's equations (2.22) (see example in 2.3). If we introduce

$$y_i = \frac{\partial L^*}{\partial u_i} = Mu_i \ (i = 1,2,3), \quad y_4 = \frac{\partial L^*}{\partial p} = Ap,$$

$$y_5 = \frac{\partial L^*}{\partial q} = Bq, \quad y_6 = \frac{\partial L^*}{\partial r} = Cr, \quad y_7 = \frac{\partial L^*}{\partial \alpha} = J_0\alpha$$

then we have

$$u_i = \frac{y_i}{M} \ (i = 1,2,3), \quad p = \frac{y_4}{A}, \quad q = \frac{y_5}{B}, \quad r = \frac{y_6}{C}, \quad \alpha = \frac{y_7}{J_0}$$

$$H^*(y_1,\dots,y_7) = \frac{1}{2M}\left(y_1^2 + y_2^2 + y_3^2\right) + \frac{1}{2}\left(\frac{y_4^2}{A} + \frac{y_5^2}{B} + \frac{y_6^2}{C}\right) + \frac{1}{2}\frac{y_7^2}{J_0}$$

Canonical equations for equations (2.22) have the form

$$\begin{aligned}
&\frac{dy_i}{dt} = A_i(U) \qquad\qquad (i = 1,2,3)\\
&\frac{dy_4}{dt} = \frac{B-C}{BC}y_5y_6 + X(U), \quad \frac{dy_5}{dt} = \frac{C-A}{AC}y_6y_4 + Y(U),\\
&\frac{dy_6}{dt} = \frac{A-B}{AB}y_4y_5 + Z(U), \quad \frac{dy_7}{dt} = \frac{y_4^2}{A} + \frac{y_5^2}{B} + \frac{y_6^2}{C} + \frac{y_7^2}{J_0} + \Phi(U),
\end{aligned} \tag{3.8}$$

3.2 Canonical Chetayev's equations are equations of Hamilton's type in non-canonical variables.

We shall assume in this paragraph that $Q_i^* = 0$ in equations (3.3), and $b_{i0} = 0$, $(i = 1, \dots, n)$ in equations (2.4). Then $X_0 = \frac{\partial}{\partial t}$, $c_{\alpha 0}^s = 0$ $(\alpha, s = 1, \dots, k)$.

Define the generalized Poisson bracket of smooth functions $f(t, x, y)$ and $\varphi(t, x, y)$ by equality

$$(f, \varphi) = \frac{\partial \varphi}{\partial y_\alpha} X_\alpha f - \frac{\partial f}{\partial y_\alpha} X_\alpha \varphi + c_{\alpha i}^s \frac{\partial f}{\partial y_i} \frac{\partial \varphi}{\partial y_\alpha} y_s \tag{3.9}$$

$$i, \alpha, s = 1, \dots, k$$

In the special case of canonical variables x_i, $y_i = \frac{\partial L}{\partial \dot{x}_i}$ $(i = 1, \dots, n = k)$ when the system (2.5) reduces to a permutation group, formula (3.9) reduces to the classical Poisson bracket

$$(f, \varphi) = \frac{\partial f}{\partial x_i} \frac{\partial \varphi}{\partial y_i} - \frac{\partial f}{\partial y_i} \frac{\partial \varphi}{\partial x_i}, \qquad i = 1, \dots, k \tag{3.10}$$

this being the reason for the choice of sign in (3.9); Chetayev [3]-[5] defined generalized Poisson bracket with the opposite sign on the right hand side of (3.9). Using (3.10), we can write the canonical Hamilton equations, as is well known, in the form

$$\begin{cases} \dot{x}_i = (x_i, H), \\ \dot{y}_i = (y_i, H) \end{cases} \iff \begin{cases} \dot{x}_i = \dfrac{\partial H}{\partial y_i}, \\ \dot{y}_i = -\dfrac{\partial H}{\partial x_i} \end{cases} \quad i = 1, \dots, k \tag{3.11}$$

where $H(t, x, y)$ is the classical Hamilton function.

It can be seen that the generalized Poisson brackets have the same properties as the classical Poisson bracket, namely [27], [36] they:

1) are skew symmetric $(f, \varphi) = -(\varphi, f)$,

2) are bilinear

$$(f, \lambda_1 \varphi_1 + \lambda_2 \varphi_2) = \lambda_1 (f, \varphi_1) + \lambda_2 (f, \varphi_2) \qquad (\lambda_i \in \mathbb{R})$$

3) satisfy the Jacobi identity

$$((f, \varphi), \psi) + ((\varphi, \psi), f) + ((\psi, f), \varphi) \equiv 0$$

4) obey the Leibniz rule

$$(f_1 f_2, \varphi) = f_1 (f_2, \varphi) + f_2 (f_1, \varphi).$$

Let $f(t,x,y) = \text{const}$ is the first integral of the canonical equations (3.3). Then, using definition (3.9), we have the identity

$$\begin{aligned}\frac{df}{dt} &= \frac{\partial f}{\partial t} + \frac{\partial H^*}{\partial y_\alpha} X_\alpha f + \frac{\partial f}{\partial y_\alpha}\left(c^s_{i\alpha}\frac{\partial H^*}{\partial y_i} y_s - X_\alpha H^*\right) = \\ &= X_0 f + (f, H^*) \equiv 0\end{aligned} \qquad (3.12)$$

The following generalization of Poisson's theorem [3] - [5] is true: if $\varphi(t,x,y) = a$ and $\psi(t,x,y) = b$ are the first two integrals of equations (3.3), then $(\varphi,\psi) = c$ will be the third first integral of these equations; a, b, c are arbitrary constants.

We shall now prove that the canonical Chetayev equations (3.3) and (3.4) may be expressed in the form

$$\frac{dy_i}{dt} = (y_i, H^*), \quad \frac{dx_j}{dt} = (x_j, H^*), \quad i = 1,\ldots,k; \quad j = 1,\ldots,n \qquad (3.13)$$

where $H^*(t,x,y)$ is the generalized Hamiltonian function (3.1).

Indeed, by definition (3.9)

$$\begin{aligned}(y_i, H^*) &= \frac{\partial H^*}{\partial y_\alpha} X_\alpha y_i - \frac{\partial y_i}{\partial y_\alpha} X_\alpha H^* + c^s_{\alpha r}\frac{\partial y_i}{\partial y_r}\frac{\partial H^*}{\partial y_\alpha} y_s = \\ &= -X_i H^* + c^s_{\alpha i}\frac{\partial H^*}{\partial y_\alpha} y_s \\ (x_j, H^*) &= \frac{\partial H^*}{\partial y_\alpha} X_\alpha x_j - \frac{\partial x_j}{\partial y_\alpha} X_\alpha H^* + c^s_{\alpha r}\frac{\partial x_j}{\partial y_r}\frac{\partial H^*}{\partial y_\alpha} y_s = \\ &= \frac{\partial H}{\partial y_\alpha} X_\alpha x_j\end{aligned} \qquad (3.14)$$

since $X_\alpha y_i = 0$, $\dfrac{\partial y_i}{\partial y_\alpha} = \delta_{i\alpha}$, $\dfrac{\partial x_j}{\partial y_\alpha} = 0$ by virtue of the fact that the variables x_j are independent of y_α $(i,\alpha,r = 1,\ldots,k; j = 1,\ldots,n)$ and vice versa, and that the variables y_α are also independent; $\delta_{i\alpha}$ is the Kronecker delta.

Comparing the right-hand sides of equations (3.3) and (3.4) with formulae (3.14), we confirm the correctness of (3.13). This implies that the canonical Chetayev equations are Hamiltonian equations in non-canonical variables. Equations of this kind, which are frequently more convenient than Hamiltonian equations in canonical variables, are studied in the modern theory of Hamiltonian systems [27], [37], [38], and consequently for such systems the results of [3] - [6] are applicable.

Example (see Example 1 in 1.3). a) For a heavy rigid body with one fixed point the canonical equations (1.26) and (1.27) may be written in the form

$$\dot{y}_i = (y_i, H^*)\ (i = 1, 2, 3), \quad \dot{\theta} = (\theta, H^*),$$

$$\dot{\psi} = (\psi, H^*), \qquad \dot{\varphi} = (\varphi, H^*)$$

where

$$H^*(\theta, \varphi, y_1, y_2, y_3) = \frac{1}{2}\left(\frac{y_1^2}{A} + \frac{y_2^2}{B} + \frac{y_3^2}{C}\right) + $$
$$+ \quad Mg\left(x_1^0 \sin\theta \sin\varphi + x_2^0 \sin\theta \cos\varphi + x_3^0 \cos\theta\right)$$

and transitive Lie group is given by formulaes (1.8).

b) If we take γ_i as the defining coordinates of body then equations (1.31) may be written

$$\dot{y}_i = (y_i, H^*), \quad \dot{\gamma}_i = (\gamma_i, H^*) \quad (i = 1, 2, 3)$$

where

$$H^* = \frac{1}{2}\left(\frac{y_1^2}{A} + \frac{y_2^2}{B} + \frac{y_3^2}{C}\right) + Mgx_i^0\gamma_i$$

and the group $\mathcal{SO}(3)$ is given by formulaes (1.30).

3.3 Different forms of Hamilton's principle of least action.

The Hamilton's principle is a significant principle of mechanics for the case of ideal holonomic constraints.

Assume the real positions P_0 and P_1 of a system to be known at the moments t_0 and t_1 in a certain motion; $\mathbf{r}_\nu$ $(\nu = 1, \ldots, N)$ are function of time which satisfy the constraints and for $t = t_0$ and $t = t_1$ have values corresponding to the positions P_0 and P_1. Let $\mathbf{r}_\nu + \delta\mathbf{r}_\nu$ be some functions of t which are sufficiently close to $\mathbf{r}_\nu$, satisfy the constraint equations and at $t = t_0$ and $t = t_1$ assume the same values as $\mathbf{r}_\nu$. Hence the expressions of $\delta\mathbf{r}_\nu$ have the sence of virtual displacements and cancel out at $t = t_0$ and $t = t_1$.

To derive Hamilton's principle we integrate the equality (2.12) or the D'Alembert-Lagrange principle with respect to time in the limits from t_0 to t_1. We have

$$\int_{t_0}^{t_1} m_\nu \ddot{\mathbf{r}}_\nu \cdot \delta\mathbf{r}_\nu dt = m_\nu \dot{\mathbf{r}}_\nu \cdot \delta\mathbf{r}_\nu \big|_{t_0}^{t_1} - \int_{t_0}^{t_1} m_\nu \dot{\mathbf{r}}_\nu \cdot \frac{d\delta\mathbf{r}_\nu}{dt} dt.$$

But

$$\delta \mathbf{r}_\nu|_{t_0}^{t_1} = \omega_s X_s \mathbf{r}_\nu|_{t_0}^{t_1} = 0, \quad \frac{d\delta\mathbf{r}_\nu}{dt} = \delta\dot{\mathbf{r}}_\nu,$$

$$m_\nu \dot{\mathbf{r}}_\nu \cdot \delta\dot{\mathbf{r}}_\nu = \delta\left(m_\nu \frac{\dot{\mathbf{r}}_\nu^2}{2}\right) = \delta T$$

where T is the kinetic energy of the system, and we get

$$\int_{t_0}^{t_1} (\delta T + \mathbf{F}_\nu \cdot \delta\mathbf{r}_\nu)\, dt = 0, \quad \omega_s = 0 \quad \text{at} \quad t = t_0, t_1 \tag{3.15}$$

The expression (3.15) is Hamilton-Ostrogradskiy's principle for the general case of the given forces $\mathbf{F}_\nu$.

When the forces $\mathbf{F}_\nu$ have a force function U, $\mathbf{F}_\nu = \text{grad}U$, then $\mathbf{F}_\nu \cdot \delta\mathbf{r}_\nu = \delta U$ and (3.15) assumes the form of variational Hamilton's principle

$$\delta \int_{t_0}^{t_1} (T + U)\, dt = \delta \int_{t_0}^{t_1} L^* dt = 0, \quad \omega_s = 0 \quad \text{at} \quad t = t_0, t_1 \tag{3.16}$$

A more thorough study shows that $\int_{t_0}^{t_1} L^* dt$ has a minimum if the time interval $t_1 - t_0$ is sufficiently small.

For systems subjected to an ideal holonomic constraints and to the potential forces it is can to write Hamilton's principle (3.16) in the second (Poincaré's) form [6]

$$\delta \int_{t_0}^{t_1} (y_s \eta_s - H^*)\, dt = 0 \tag{3.17}$$

when $\omega_s = 0$ for $t = t_0, t_1$ at the endpoints and when ω_s and δy_s $(s = 1, \ldots, k)$ are arbitrary and independent within the interval (t_0, t_1).

The Chetayev's canonical equations (3.3) by $Q_s^* = 0$ can be derived from the condition (3.17), taking into consideration equations (2.11). Indeed

$$\delta \int_{t_0}^{t_1} (y_s \eta_s - H^*)\, dt = \int_{t_0}^{t_1} \left(y_s \delta\eta_s + \eta_s \delta y_s - \frac{\partial H^*}{\partial y_s} \delta y_s - \omega_s X_s H^* \right) dt =$$

$$= y_s \omega_s|_{t_0}^{t_1} +$$

$$+ \int_{t_0}^{t_1} \left[\left(\eta_s - \frac{\partial H^*}{\partial y_s} \right) \delta y_s + \left(-\frac{dy_s}{dt} + c_{rs}^m \eta_r y_m + c_{0s}^m y_m - X_s H^* \right) \omega_s \right] dt = 0$$

The expression outside the integral is zero by virtue of the conditions at the endpoints. Because of the assumption that ω_s and δy_s are independent and arbutrary within the interval (t_0, t_1), we get the equations (3.3) for the case $Q_s^* = 0$.

For the special case, when the Lagrange's function $L^*(t,x,\eta)$ equals

$$L^*(t,x,\eta) = \Lambda(t,x,\eta) + \frac{d}{dt}f(t,x), \quad f(t,x) \in C^2 \tag{3.18}$$

the Hamilton's principle (3.16) is equivalent to the form

$$\delta \int_{t_0}^{t_1} \Lambda dt = 0, \quad \omega_s = 0 \quad \text{at} \quad t = t_0, t_1 \tag{3.19}$$

because of

$$\delta \int_{t_0}^{t_1} \frac{df}{dt} dt = \delta f|_{t_0}^{t_1} = 0$$

Hence, a addition (a asideon) of the addendum $\dot{f}$ to the Lagrange's function $L^*(t,x,\eta)$ do not exert influence on the Poincaré's equations (2.15). Such confirmation is analogous to given one by Pars [39] for Lagrange's equations. Analogous the second form of Hamilton's principle (3.17) is equivalent to the form

$$\delta \int_{t_0}^{t_1} (Y_s\eta_s - K)\,dt = 0, \quad \omega_s = 0 \quad \text{at} \quad t = t_0, t_1 \tag{3.20}$$

where

$$Y_s = \frac{\partial \Lambda}{\partial \eta_s} \qquad K(t,x,Y) = Y_s\eta_s - \Lambda = H^*(t,x,y) + X_0 f \qquad (s = 1,\ldots,k)$$

Therefore the Chetayev's equations (3.3) are equivalent to the equations

$$\begin{aligned} \frac{dY_s}{dt} &= c_{rs}^m \frac{\partial K}{\partial Y_r} Y_m + c_{0s}^m Y_m - X_s K + Q_s^*, \\ \eta_s &= \frac{\partial K}{\partial Y_s}; \qquad m, r, s = 1,\ldots,k \end{aligned} \tag{3.21}$$

As an example of function in the form (3.18) may be Lagrange's function for a holonomic system in its relative motion with respect to the moving coordinate frame $Ox_1x_2x_3$ [25], [39].

3.4 Hamilton's principal function or function of action. Jacobi's theorem.

A function of action is the function

$$V(t,x_1,\ldots,x_n;t_0,x_1^0,\ldots,x_n^0) = \int_{t_0}^{t} L^* dt = \int_{t_0}^{t} (y_s\eta_s - H^*)\,dt \tag{3.22}$$

where the integration takes place along a true trajectory of a mechanical system. From the definition (3.22) we have

$$\delta V = \omega_s X_s V + \omega_s^0 X_s^0 V = y_s \omega_s - y_s^0 \omega_s^0 +$$

$$+ \int_{t_0}^{t} \left[-\omega_s \left(\frac{dy_s}{dt} - c_{\alpha s}^{\beta} \eta_\alpha y_\beta + c_{0s}^{\beta} y_\beta + X_s H^* \right) + \delta y_s \left(\eta_s - \frac{\partial H^*}{\partial y_s} \right) \right] dt$$

where ω_s^0 denote the parameters of initial variation of trajectory, X_s^0 - infinitesimal operators X_s belong to initial moment of time t_0 and to initial position of the system $x_1^0, \dots, x_n^0$.

In the expression (3.22) the integration takes place along the true trajectory therefore the integral vanishes in the last relation according to canonical equations (3.3) (by $Q_s^* = 0$), and

$$\delta V = \omega_s X_s V + \omega_s^0 X_s^0 V = y_s \omega_s - y_s^0 \omega_s^0 \tag{3.23}$$

The comparison of coefficients by arbitraries the same ω_s, ω_s^0 gives the equations

$$X_s V = y_s, \qquad X_s^0 V = -y_s^0, \qquad s = 1, \dots, k \tag{3.24}$$

If the function of action V is known then the equations (3.24) solve the problem of mechanics. Second group of equations (3.24) defines of motions law in non evident shape.

Find the full derivative relative to time from the function V

$$\frac{dV}{dt} = X_0 V + \eta_s X_s V = L^*$$

or

$$X_0 V + \eta_s y_s - L^* = 0$$

From this relation according to equations (3.24) we deduce the first-order partial differential equation for the function of action

$$X_0 V + H^* (t, x_1, \dots, x_n, X_1 V, \dots, X_k V) = 0 \tag{3.25}$$

Hence the function of action V is one from complete integrals of the generalized Hamilton-Jacobi's equation (3.25).

And inversly, if it is known the complete integral

$$V (t, x_1, \dots, x_n, a_1, \dots, a_n) + a_{n+1}$$

of the equation (3.25), then the solution of the canonical equations (3.3) (with all $Q_s^* = 0$) is defined by the system

$$\frac{\partial V}{\partial a_j} = b_j, \quad y_s = X_s V \qquad j = 1, \dots, n; s = 1, \dots, k \tag{3.26}$$

where a_j, b_j are arbitraries constants.

Indeed for the complete integral the functional determinant $\|\frac{\partial^2 V}{\partial x_j \partial a_r}\| \neq 0$. We differentiate (3.26) with respect to t

$$\frac{\partial^2 V}{\partial t \partial a_j} + \frac{\partial^2 V}{\partial x_r \partial a_j}\frac{dx_r}{dt} = 0, \quad j = 1, \ldots, n$$

Then substitute the complete integral into the result and differentiate the obtained identity with respect to a_j:

$$\frac{\partial^2 V}{\partial a_j \partial t} + \frac{\partial^2 V}{\partial a_j \partial x_r}\left(X_0 x_r + \frac{\partial H^*}{\partial y_s} X_s x_r\right) = 0, \quad j = 1, \ldots, n$$

Therefore last two systems have a unique solution

$$\frac{dx_r}{dt} = X_0 x_r + \frac{\partial H^*}{\partial y_s} X_s x_r$$

Thus x_r, found from relations (3.26), satisfy the second system of the canonical equations (3.4). Furthermore

$$\begin{aligned}
\frac{dy_s}{dt} &= \frac{d}{dt} X_s V = X_0 (X_s V) + \eta_\alpha X_\alpha (X_s V) = \\
&= X_0 (X_s V) + \frac{\partial H^*}{\partial y_\alpha} X_\alpha (X_s V) = \\
&= c_{0s}^{\beta} X_\beta V + X_s (X_0 V) + \frac{\partial H^*}{\partial y_\alpha} \left(c_{\alpha s}^{\beta} X_\beta V + X_s (X_\alpha V)\right) + \\
&+ X_s H^* - X_s H^* = c_{\alpha s}^{\beta} \frac{\partial H^*}{\partial y_\alpha} y_\beta + c_{0s}^{\beta} y_\beta - X_s H^* + \\
&+ X_s (X_0 V + H^* (t, x_1, \ldots, x_n, X_1 V, \ldots, X_k V))
\end{aligned}$$

So we proved the generalized Jacobi's theorem.

At first sight, the first group of integrals (3.26) yields a general solution of system (3.3), (3.4), that depends on the $2n$ arbitrary constants a_j, b_j, while the order of the system is $k+n$. In reality, however, the solution will depend on only $k+n$ constants. In fact, since the constraints imposed on the system are represented by the completely integrable system of equations (2.1), and this system may be reduced to the form $d\Phi_j = 0$ $(j = k+1, \ldots, n)$. By (2.6) these equalities imply the relations

$$X_s \Phi_j = 0, \qquad s = 0, 1, \ldots, k; \; j = k+1, \ldots, n,$$

in view of which we can add terms $c_j\Phi_j$ with arbitrary c_j = const to the complete integral of equation (3.25). Consequently of the n essential constants a_j, not one of which is additive, $n-k$ will be the coefficients $c_{k+1},\ldots,c_n$, so the complete integral will have the following structure

$$V = W(t, x_1, \ldots, x_n; a_1, \ldots, a_k) + a_j\Phi_j + a_{n+1},$$

$$\frac{\partial(X_1W, \ldots, X_kW)}{\partial(a_1, \ldots, a_k)} \neq 0 \tag{3.27}$$

It follows from (3.27) that the integrals (3.26) may be written

$$\frac{\partial V}{\partial a_\alpha} = \frac{\partial W}{\partial a_\alpha} = b_\alpha, \quad y_\alpha = X_\alpha V = X_\alpha W,$$

$$\frac{\partial V}{\partial a_j} = \Phi_j = b_j, \qquad \alpha = 1, \ldots, k; j = k+1, \ldots, n \tag{3.28}$$

The third group of integrals (3.28) relates to the determination of the constants b_j of the holonomic constraints, and when these are added to the first group of (3.28) the solution is uniquely defined:

$$x_r = x_r(t, a_1, \ldots, a_k, b_1, \ldots, b_k, b_{k+1}, \ldots, b_n)$$

while the second group in (3.28) defines the variables y_s $(s = 1, \ldots, k)$.

It is interesting to compare the generalized Jacobi theorem with the Suslov's theorem [36], chap. XLIII. Suslov considered the Hamilton's equations with multipliers conjugate to equations (2.21) and instead of (3.25) he obtained the partial differential equation [36], formula (43.25)

$$\frac{\partial V}{\partial t} - \Lambda_j\frac{\partial f_j}{\partial t} + H\left(t, x_1, \ldots, x_n, \frac{\partial V}{\partial x_1}, \ldots, \frac{\partial V}{\partial x_n}\right) = 0 \tag{3.29}$$

where $f_j(t,x) = c_j$ are the integrated equations of the constraints (2.1), $\Lambda_j = -\int \lambda_j dt$ are impulse factors. He proved that if one knowns a complete integral of equation (3.29), then the equations

$$\frac{\partial V}{\partial a_s} = b_s, \quad \frac{\partial V}{\partial x_s} = p_s + \Lambda_j\frac{\partial f_j}{\partial x_s}, \quad s = 1, \ldots, n$$

are integrals of the Hamilton's equations conjugate to equations (2.21). Proposing to eliminate the impulse factors with the help of the differentiated equations of the constraints, Suslov then obtained an equation of the form (3.25), taking into account

(2.1) and integrals (3.26), just as the elimination of λ_j from equations (2.21) leads to equations (2.20).

The relation (3.23) for the function of action V permits to establish of relative integral invariant. Indeed, consider some closed way C_0 of virtual displacements ω^0 in moment t_0, which in moment t will be C. The integration of (3.23) by synonymous function of action gives

$$\int_C y_s\omega_s = \int_{C_0} y_s^0\omega_s^0$$

Thus the linear form $\Omega = y_s\omega_s$ defines main relative integral invariant of dynamics.

The relation (3.23) permits also to establish the canonical equations of motion. Indeed, let moment t is infinitely near to initial moment t_0, $t = t_0 + dt$, then $V = L^* dt$ and

$$\omega_s^0 = \omega_s - \frac{d\omega_s}{dt}dt, \quad y_s^0 = y_s - \frac{dy_s}{dt}dt$$

The relation (3.23) gives with precision to quantities of second order relative dt the equality

$$\delta L^* dt = \left(y_s\frac{d\omega_s}{dt} + \frac{dy_s}{dt}\omega_s\right)dt$$

Cancel on dt and introduce in the consideration of Hamiltonian function $H^* = y_s\eta_s - L^*$, we have the correlation

$$\left(\frac{\partial H^*}{\partial y_s} - \eta_s\right)\delta y_s + \left(\frac{dy_s}{dt} - c_{\alpha s}^{\beta}\eta_\alpha y_\beta - c_{0s}^{\beta}y_\beta + X_s H^*\right)\omega_s = 0$$

from which the canonical equations (3.3) are settled in consequence of the arbitrary of δy_s and ω_s.

Example. Let us consider Suslov's Example 134 in his notations [36]: two heavy particles of masses m_i with coordinates y_i, z_i $(i = 1, 2)$, are moving in the yz plane with axis Oz on vertical upwards, subject to the constraints

$$m_1\dot{y}_1 + m_2\dot{y}_2 = 0, \quad (y_1 - y_2)(\dot{y}_1 - \dot{y}_2) + (z_1 - z_2)(\dot{z}_1 - \dot{z}_2) = 0$$

$$(M = m_1 + m_2)$$

The position of the system is determined by the coordinates y_c, z_c of the centre of mass and by the quantities $\eta = y_1 - y_2$ and $\phi = \mathrm{arctg}\left[\frac{z_1 - z_2}{y_1 - y_2}\right]$, in terms of which the integrated equations of the constraints become

$$y_c = c_1 = \mathrm{const}, \qquad \eta\sec\varphi = c_2 = \mathrm{const}$$

The Lagrange's function is

$$L = \frac{M}{2}\dot{z}_c^2 + \frac{m_1 m_2}{2M}\sec^2\varphi\,(\dot{\eta}^2 + 2\eta\mathrm{tg}\varphi\cdot\dot{\eta}\dot{\varphi} + \eta^2\sec^2\varphi\cdot\dot{\varphi}^2) - Mgz_c$$

If we take the quantities $\dot{z}_c$ and $\dot{\varphi}$ as the parameters η_s $(s = 1, 2)$ of the real displacements, and y_c, z_c, η, φ as the defining coordinates x_i $(i = 1, \ldots, 4)$, then in equations of type (2.4) the nonzero coefficients b_{is} are $b_{21} = 1$, $b_{32} = -\eta\,\mathrm{tg}\varphi$, $b_{42} = 1$ and the operators (2.5) will be

$$X_0 = \frac{\partial}{\partial t}, \quad X_1 = \frac{\partial}{\partial z_c}, \quad X_2 = \frac{\partial}{\partial \varphi} - \eta \mathrm{tg}\varphi \frac{\partial}{\partial \eta}$$

Take into account the correlation $\dot{\eta} = -\eta\dot{\varphi}\mathrm{tg}\varphi$ we have

$$L^* = \frac{M}{2}\dot{z}_c^2 + \frac{m_1 m_2}{2M}\eta^2\dot{\varphi}^2\sec^2\varphi - Mgz_c$$

so as the equations (2.20) have the form

$$\ddot{z}_c = -g, \quad \ddot{\varphi} = 0$$

and we get

$$y_c = c_1, \qquad z_c = -\frac{gt^2}{2} + A_1 t + A_2,$$
$$\varphi = A_3 t + A_4, \quad \eta = c_2 \cos(A_3 t + A_4)$$

In the variables $y_1 = M\dot{z}_c$, $y_2 = \dfrac{m_1 m_2}{M}\eta^2\dot{\varphi}\sec^2\varphi$ the Hamilton's function is

$$H^* = \frac{y_1^2}{2M} + \frac{M}{2m_1 m_2}\frac{\cos^2\varphi}{\eta^2}y_2^2 + Mgz_c$$

and the equation (3.25) will be

$$\frac{\partial V}{\partial t} + \frac{1}{2}\left[\frac{1}{M}\left(\frac{\partial V}{\partial z_c}\right)^2 + \frac{M}{m_1 m_2}\frac{\cos^2}{\eta^2}\left(\frac{\partial V}{\partial \varphi} - \eta\,\mathrm{tg}\varphi\frac{\partial V}{\partial \eta}\right)^2\right] + Mgz_c = 0$$

The complete integral is [36]

$$\begin{aligned} V &= a_1 t + a_2 y_c + \frac{2M}{3}\sqrt{2g}\left(-z_c - \frac{2a_1 + \frac{M}{m_1 m_2}a_3^2}{2Mg}\right)^{3/2} + \\ &+ \eta\,(a_3\varphi + a_4)\sec\varphi \end{aligned}$$

Chapter 4

Dynamics of Non-Holonomic systems.

4.1 Poincaré's and Chetayev's equations for non-holonomic systems.

Poincaré's equations, like the Boltzmann-Hamel's equations in quasi-coordinates, are used to describe both holonomic and non-holonomic systems. The problem has already been investigated in [14]-[17], as well as in [21]-[23], where Chetayev's equations were also considered in this sense.

We must, however, emphasize, that the system of operators of virtual displacements is not closed for non-holonomic systems [16], [17], as a result of which one must use operators of the corresponding holonomic system obtained from non-holonomic system considered by mentally discarding non-integrable constraints.

We will consider the case of a non-holonomic system with l degrees of freedom in defining coordinates x_i ($i = 1, \ldots, n$), constrained by integrable equations

$$\eta_j \equiv a_{ji}(x)\dot{x}_i = 0, \quad \text{rank}\,(a_{ji}) = n - k, \qquad i = 0, 1, \ldots, n; \qquad j = k + 1, \ldots, n \tag{4.1}$$

and non-integrable equations

$$\eta_\alpha \equiv a_{\alpha i}(x)\dot{x}_i = 0, \quad \text{rank}\,(a_{\alpha i}) = k - l, \quad \alpha = l + 1, \ldots, k \tag{4.2}$$

We shall arbitrarily choose linear differential forms

$$\eta_s \equiv a_{si}(x)\,\dot{x}_i, \quad s = 0, 1, \ldots, l; \quad a_{0i} = \delta_{0i} \tag{4.3}$$

independent of one another, and also with forms (4.1) and (4.2),

$$\det\,(a_{ji}) \neq 0, \quad i, j = 0, 1, \ldots, n.$$

In particular, we can take the generalized velocities $\dot{x}_i$ as the quantities η_s ($s = 1, \ldots, l$).

Solution of forms (4.1)-(4.3) leads to the equations

$$\dot{x}_i = b_{is}(x)\eta_s, \quad i = 0, 1, \ldots, n; \quad s = 0, 1, \ldots, l; \quad b_{io} = \delta_{io} \tag{4.4}$$

For the corresponding holonomic system with k degrees of freedom, obtained by mentally discarding the non-integrable constraints (4.2) considered, i.e. essuming $\eta_\alpha \neq 0$ ($\alpha = l+1, \ldots, k$), instead of (4.4) we obtain the equations

$$\dot{x}_i = b_{ir}(x)\eta_r, \quad i = 0, 1, \ldots, n; \quad r = 0, 1, \ldots, k, \tag{4.5}$$

and we construct the closed system of operators (2.5).

Since the parameters of the virtual displacements $\omega_\alpha = 0$ when the constraints (4.2) are present from D'Alembert-Lagrange principle (2.13) we can derive the equations of motion of a non-holonomic system in the form (2.15)

$$\frac{d}{dt}\frac{\partial L^*}{\partial \eta_s} = c^m_{rs}\eta_r \frac{\partial L^*}{\partial \eta_m} + c^m_{0s}\frac{\partial L^*}{\partial \eta_m} + X_s L^* + Q^*_s \tag{4.6}$$

$$r, s = 1, \ldots, l; \qquad m = 1, \ldots, k$$

the number of which is less than the number of equations (2.15) by $k - l$. The structural coefficients c^m_{rs}, as previously, are given by (2.17), in which, however, the subscripts $r, s = 0, 1, \ldots, l$. We must add the equations (4.4) to equations (4.6), as a result of which we obtain a joint system of $l + n$ equations of motion with the same number of unknowns $x_1, \ldots, x_n, \eta_1, \ldots, \eta_l$.

Note that the function $L^*(t, x, \eta)$, which occurs in equations (4.6), constructed for the corresponding holonomic system, may, in general, depends on all the parameters η_r ($r = 1, \ldots, k$), as a consequence of which the constraint equations $\eta_\alpha = 0$ ($\alpha = l+1, \ldots, k$) need only be taken into account after setting up equations (4.6) [25], [31], [40].

Remark. Equations (4.6), when $Q^*_s = 0$, are equivalent to equations (3.14) [14] and (1.13) [15], but are superficially somewhat simpler due to the choise of the parameters η_α, which vanish by virtue of the equations of the non-integrable constraints (4.2).

For the cases when the generalized velocities $\dot{x}_s$ ($s = 1, \ldots, l$) are taken as the parameters η_s (4.3), i.e. when $a_{si} = \delta_{si}$ ($i = 1, \ldots, n$), all the structural coefficients $c^m_{rs} = 0$ for $m \leq l$ [31], and equations (4.6) take the form

$$\frac{d}{dt}\frac{\partial L^*}{\partial \dot{x}_s} = c^m_{rs}\frac{\partial L^*}{\partial \eta_m}\dot{x}_r + c^m_{0s}\frac{\partial L^*}{\partial \eta_m} + X_s L^* + Q^*_s, \quad r, s = 1, \ldots, l; \quad m = l+1, \ldots, k \tag{4.7}$$

where $L^* = L^*(t, x_1, \ldots, x_n, \dot{x}_1, \ldots, \dot{x}_l, \eta_{l+1}, \ldots, \eta_k)$.

If in the function $L^*(t, x, \eta)$ from equations (4.6) we replace the kinetic energy $T^*(t, x, \eta_1, \ldots, \eta_k)$ of the corresponding holonomic system by the kinetic energy

$\Theta(t, x, \eta_1, \ldots, \eta_l)$ of the non-holonomic system with constraints (4.2), equations (4.6) take the form of equations (5.5) [21]

$$\frac{d}{dt}\frac{\partial\Theta}{\partial\eta_s} = (c^m_{rs}\eta_r + c^m_{0s})\frac{\partial\Theta}{\partial\eta_m} + (c^p_{rs}\eta_r + c^p_{0s})\left(\frac{\partial T^*}{\partial\eta_p}\right) + X_s(\Theta + U) + Q^*_s \tag{4.8}$$

$$m, r, s = 1, \ldots, l; \quad p = l+1, \ldots, k$$

where $\left(\frac{\partial T^*}{\partial\eta_p}\right)$ denote the expressions $\frac{\partial T^*}{\partial\eta_p}$ with $\eta_s = 0$ $(p, s = l+1, \ldots, k)$.

Using the Legendre transformation (3.1) of equations (4.6), the motions equations of the non-holonomic system can be written in the form of Chetayev's canonical equations

$$\frac{dy_s}{dt} = c^m_{rs}\frac{\partial H^*}{\partial y_r}y_m + c^m_{0s}y_m - X_s H^* + Q^*_s, \eta_s = \frac{\partial H^*}{\partial y_s};$$
$$r, s = 1, \ldots, l; \qquad m = 1, \ldots, k \tag{4.9}$$

to which we must add the constraint equations (4.2) and relations (4.4), rewritten in the form

$$\frac{\partial H^*}{\partial y_\alpha} = 0, \qquad \alpha = l+1, \ldots, k$$
$$\frac{dx_i}{dt} = b_{ij}\frac{\partial H^*}{\partial y_r}; \quad i = 1, \ldots, n; \qquad r = 0, 1, \ldots, l \tag{4.10}$$

Equations (4.9) and (4.10) form a joint system of $n + k + l$ equations with the same number of unknowns $x_1, \ldots, x_n, y_1, \ldots, y_k, \eta_1, \ldots, \eta_l$. Equations (4.9) and (4.10) by $Q^*_s = 0$ and $n = k$ take the form of equations (7.17)-(7.19) of [40].

The canonical equations of motion of non-holonomic systems, equivalent to equations (4.8), have the form

$$\frac{dy_s}{dt} = \left(c^m_{rs}\frac{\partial H^*}{\partial y_r} + c^m_{0s}\right)y_m + \left(c^p_{rs}\frac{\partial H^*}{\partial y_r} + c^p_{0s}\right)\left(\frac{\partial T^*}{\partial\eta_p}\right) -$$
$$- X_s H^* + Q^*_s, \tag{4.11}$$

$$\eta_s = \frac{\partial H^*}{\partial y_s}; \qquad m, r, s = 1, \ldots, l; \quad p = l+1, \ldots, k$$

where $H^*(t, x, y) = y_s\eta_s - \Theta - U$.

Example. Working from equations (4.7), let us derive the equations of motion in Voronet's form [41] for a system with Lagrangian coordinates $x_1, \ldots, x_n$ and non-holonomic constraints

$$\dot{x}_s = \alpha_{si}(t, x)\dot{x}_i + \alpha_{s0}(t, x), \quad i = 1, \ldots, l; \quad s = l+1, \ldots, n \tag{4.12}$$

Set

$$\eta_i = \dot{x}_i, \quad x_0 = t, \quad \dot{x}_0 = 1, \quad \eta_s = \dot{x}_s - \alpha_{si}\dot{x}_i, \quad i = 0, 1, \ldots, l, \quad s = l+1, \ldots, n,$$

so that $\dot{x}_i = \eta_i$, $\dot{x}_s = \eta_s + \alpha_{si}\eta_i$ and, in view (4.2), (4.3), (2.19), we have the relations

$$a_{ij} = b_{ij} = \delta_{ij}, \quad a_{is} = b_{is} = 0, \qquad b_{si} = -a_{si} = \alpha_{si}, a_{sr} = b_{sr} = \delta_{sr}$$

$$i, j = 0, 1, \ldots, l; \qquad s, r = l+1, \ldots, n$$

$$c^r_{ji} = \frac{\partial \alpha_{ri}}{\partial x_j} - \frac{\partial \alpha_{rj}}{\partial x_i} + \frac{\partial \alpha_{ri}}{\partial x_k}\alpha_{kj} - \frac{\partial \alpha_{rj}}{\partial x_k}\alpha_{ki}, \quad i, j = 1, \ldots, l; \tag{4.13}$$

$$c^r_{0i} = \frac{\partial \alpha_{ri}}{\partial t} - \frac{\partial \alpha_{r0}}{\partial x_i} + \frac{\partial \alpha_{ri}}{\partial x_k}\alpha_{k0} - \frac{\partial \alpha_{r0}}{\partial x_k}\alpha_{ki}, \quad k, r = l+1, \ldots, n$$

Noting (4.12) and (4.13), we conclude that equations (4.7) take the form of Voronets's equations [41]

$$\frac{d}{dt}\frac{\partial \Theta}{\partial \dot{x}_i} = \frac{\partial(\Theta + U)}{\partial x_i} + \alpha_{ri}\frac{\partial(\Theta + U)}{\partial x_r} + \left(c^r_{ji}\dot{x}_j + c^r_{0i}\right)\left(\frac{\partial L^*}{\partial \dot{x}_r}\right) + Q^*_i \tag{4.14}$$

$$i, j = 1, \ldots, l; \quad r = l+1, \ldots, n$$

where $\Theta = \Theta(t, x_1, \ldots, x_n, \dot{x}_1, \ldots, \dot{x}_l)$ is the expression of the kinetic energy $T^*(t, x_1, \ldots, x_n, \dot{x}_1, \ldots, \dot{x}_n)$ after replacing $\dot{x}_s$ $(s = l+1, \ldots, n)$ by (4.12). Equations (4.14) and (4.12) are joint system of n equations.

The canonical form of the Voronets's equations (4.14) is

$$\frac{dy_i}{dt} = -\frac{\partial H^*}{\partial x_i} - \alpha_{ri}\frac{\partial H^*}{\partial x_r} + \left(c^r_{ji}\frac{\partial H^*}{\partial y_j} + c^r_{0i}\right)\left(\frac{\partial T^*}{\partial \dot{x}_r}\right) + Q^*_i, \tag{4.15}$$

$$\frac{dx_i}{dt} = \frac{\partial H^*}{\partial y_i}, \qquad i, j = 1, \ldots, l; \quad r = l+1, \ldots, n$$

where $y_i = \dfrac{\partial \Theta}{\partial \dot{x}_i}$, $H^*(t, x_1, \ldots, x_n, y_1, \ldots, y_l) = y_i\dot{x}_i - \Theta - U$.

In the special case in which the functions $T^*(t, x_1, \ldots, x_l, \dot{x}_1, \ldots, \dot{x}_n)$, $U(t, x_1, \ldots, x_l)$, $\alpha_{si}(t, x_1, \ldots, x_l)$, $Q^*_i(t, x_1, \ldots, x_l)$ do not depend explicitly on the coordinates x_s $(s = l+1, \ldots, n)$, equations (4.14) are identical with the closed system of l Chaplygin's equations [42]

$$\frac{d}{dt}\frac{\partial \Theta}{\partial \dot{x}_i} = \frac{\partial(\Theta + U)}{\partial x_i} + \left(c^r_{ji}\dot{x}_j + c^r_{0i}\right)\left(\frac{\partial L^*}{\partial \dot{x}_r}\right) + Q^*_i \quad i, j = 1, \ldots, l; \quad r = l+1, \ldots, n$$

where $c^r_{ji} = \dfrac{\partial \alpha_{ri}}{\partial x_j} - \dfrac{\partial \alpha_{rj}}{\partial x_i}$, $c^r_{0i} = \dfrac{\partial \alpha_{ri}}{\partial t} - \dfrac{\partial \alpha_{r0}}{\partial x_i}$.

Remark. The canonical form of Chaplygin's equations has the view of the equations(4.15) in which all terms $\dfrac{\partial H^*}{\partial x_r} = 0 \quad (r = l+1, \ldots, n)$ [40].

4.2 Determination of constraints reactions.

By the method of Chetayev [6] to determine the reactions of constraints the remaining $k-l$ equations (2.15) with the terms $b_{is}R_i$ added to their right-hand sides, enable us to find the reactions R_i of the constraints (4.2).

Indeed, if we mentally free the considered system from the constraints (4.2), replacing their effects by the reactions R_i $(i=1,\ldots,n)$, the result will be a holonomic system, for which equations

$$\frac{d}{dt}\frac{\partial L^*}{\partial \eta_s} = c_{rs}^m \frac{\partial L^*}{\partial \eta_m}\eta_r + c_{0s}^m \frac{\partial L^*}{\partial \eta_m} + X_s L^* + Q_s^* + b_{is}R_i \qquad (4.16)$$

$$m, r, s = 1,\ldots,k; \quad i = 1,\ldots,n$$

of type (2.15) are valid.

Since the constraints (4.2) are assumed to be ideal, the work done by their reactions in virtual displacements will vanish

$$R_i\delta x_i = R_i b_{ij}\omega_j = 0, \qquad i = 1,\ldots,n; \qquad j = 1,\ldots,l$$

Hence, since ω_j are arbitrary, it follows [17] that

$$R_i b_{ij} = 0, \quad (j = 1,\ldots,l) \qquad (4.17)$$

which implies that the first l equations of motion of the "free" system (4.16) by conditions (4.2) coincide with equations (4.6). Remaining $k-l$ equations (4.16)

$$\frac{d}{dt}\frac{\partial L^*}{\partial \eta_s} = c_{\alpha s}^r \eta_\alpha \frac{\partial L^*}{\partial \eta_r} + c_{0s}^r \frac{\partial L^*}{\partial \eta_r} + \frac{\partial L^*}{\partial \pi_s} + Q_s^* + b_{is}R_i \quad s = l+1,\ldots,k \qquad (4.18)$$

together with l equations (4.17) by conditions (4.2) enable us to determine the reactions R_i provided that $\det(b_{is}) \neq 0$.

Example: A plate with a edge on inclined plane under angle α to horizon.

Let a rigid plate leans on the plane by three legs, two of which can slide on plane without friction, the third finishes by a sharp semicircular edge, rigidly connected with the plate. It is supposed that a fulcrum of the edge is able to move free on plane along the tangent to the edge but not in the perpendicular direction; projection of the centre mass of the plate on plane coincides with the fulcrum A of the edge [40].

Denote by x, y the coordinates of the point A in the coordinate frame rigidly connected with the plane (axis x directed along slope down), φ is the angle between axis x and tangent to the edge. In these coordinates the Lagrangian function and non-holonomic constraint are

$$L = \frac{1}{2}\left(\dot{x}^2 + \dot{y}^2\right) + \frac{1}{2}k^2\dot{\varphi}^2 + gx\sin\alpha, \quad \eta_3 \equiv \dot{x}\sin\varphi - \dot{y}\cos\varphi = 0$$

According to equations (2.4), (2.5), (2.7) we have

$$\dot{x} = \eta_1 \cos\varphi - \eta_3 \sin\varphi, \quad \dot{y} = \eta_1 \sin\varphi + \eta_3 \cos\varphi, \quad \dot{\varphi} = \eta_2 \tag{4.19}$$

$$X_1 = \cos\varphi \frac{\partial}{\partial x} + \sin\varphi \frac{\partial}{\partial y}, \quad X_2 = \frac{\partial}{\partial \varphi}, \qquad X_3 = -\sin\varphi \frac{\partial}{\partial x} + \cos\varphi \frac{\partial}{\partial y}$$

$$[X_1, X_2] = -X_3, \qquad [X_1, X_3] = 0, \ [X_3, X_2] = X_1$$

The generalized Lagrangian function

$$L^* = \frac{1}{2}\left(\eta_1^2 + \eta_3^2\right) + \frac{1}{2}k^2\eta_2^2 + gx\sin\alpha$$

For our example the equations (4.7), (4.2), (4.19) receive the form

$$\dot{\eta}_1 = g \sin\alpha \cos\varphi, \quad \dot{\eta}_2 = 0, \quad \eta_3 = 0, \quad \dot{x} = \eta_1 \cos\varphi, \quad \dot{y} = \eta_1 \sin\varphi$$

and their general solution is [17]

$$\eta_1 = C + \frac{g \sin\alpha}{\omega} \sin\varphi, \quad \eta_2 = \omega = \text{const}, \quad \varphi = \omega t + \varphi_0, \quad C = \text{const}$$

$$x = x_0 + \frac{C}{\omega}(\sin\varphi - \sin\varphi_0) + \frac{g\sin\alpha}{4\omega^2}(\cos 2\varphi_0 - \cos 2\varphi)$$

$$y = y_0 + \frac{C}{\omega}(\cos\varphi_0 - \cos\varphi) + \frac{g\sin\alpha}{4\omega^2}(\sin 2\varphi_0 - \sin 2\varphi) + \frac{g\sin\alpha}{2\omega}t$$

From the equations (4.17), (4.18)

$$R_1 \cos\varphi + R_2 \sin\varphi = 0, \quad R_3 = 0, \quad -R_1 \sin\varphi + R_2 \cos\varphi = g \sin\alpha \sin\varphi + \eta_1\eta_2$$

we get

$$R_1 = -(g\sin\alpha\sin\varphi + \eta_1\eta_2)\sin\varphi, \quad R_2 = (g\sin\alpha\sin\varphi + \eta_1\eta_2)\cos\varphi$$

4.3 The Equivalence of Poincaré's Equations to Different Forms of the Equations Motions.

Previously [14]-[16] it was shown by direct calculations that Poincaré's equations of motion of non-holonomic systems are equivalent to Chaplygin's, Appell's, Hamel's, Volterra's, Ferrer's and certain other equations. The equivalence of the equations in quasi-coordinates to Appell's equations, and also to Chaplygin's equations was proved in [40] by deriving these groups of equations from the D'Alembert-Lagrange principle.

The Voronet's equations were derived from Poincar'e's equations (5.6) in [21] (see example in 4.1).

We shall show that Poincaré's equations are equivalent to certain other forms of motion's equations of non-holonomic systems.

In Section 2.3 Poincaré's equations were derived from Maggi's equations [28] (2.14). Similarly, equations (4.6) are equivalent to equations (2.14) when (4.2) are taken into account.

As Maggi showed [28] both Appell's equations and Volterra's equations follow from his equations.

Maggi considered a mechanical system with coordinates x_i $(i = 1, \dots, n)$ subject to m linear constraints, which can be both holonomic and non-holonomic, and explicitly dependent or independent of time. By solving the constraints equations for $\dot{x}_i$ he presented them in the form (4.4), referring to the quantities η_s (in his notation - e_s) as the characteristics of the motion of the system considered, where

$$b_{is} = \frac{\partial \dot{x}_i}{\partial \eta_s} = \frac{\partial \ddot{x}_i}{\partial \dot{\eta}_s}, \qquad s = 1, \dots, l = n - m.$$

Proceeding to the derivation of Volterra's equations, Maggi converted his equations of the form (2.14) (in which the kinetic energy T occurs instead of L, while Q_i denotes all the active forces applied to the system) to the form

$$\frac{d}{dt}\frac{\partial T}{\partial \eta_r} = \frac{db_{ir}}{dt}\frac{\partial T}{\partial \dot{x}_i} + b_{ir}\frac{\partial T}{\partial x_i} + P_r, \quad r = 1, \dots, l; \quad P_r = Q_i b_{ir} \tag{4.20}$$

Volterra [43] considered a system with N point masses, the velocities of which in a Cartesian system of coordinates are related to the characteristics of the motion by relations of the form (4.4)

$$\dot{x}_i = b_{is}\eta_s, \qquad i = 1, \dots, 3N, \qquad s = 1, \dots, l$$

where $b_{is} = b_{is}(x_1, \dots, x_{3N})$. Here Maggi's equations (4.20) take the form of Volterra's equations

$$\frac{d}{dt}\frac{\partial T}{\partial \eta_r} = c_{rs}^{(k)}\eta_k\eta_r + P_r, \qquad k, s = 1, \dots, l$$

$$\left(c_{rs}^{(k)} = m_i b_{ik}\frac{\partial b_{ir}}{\partial x_j}b_{js}; \quad m_i = m_{i+1} = m_{i+2}; \quad i, j = 1, \dots, 3N \right)$$

where $T(x_1, \dots, x_{3N}, \eta_1, \dots, \eta_l)$ is the kinetic energy.

Without giving Maggi's derivation of Appell's equations from equations (2.14) here we note that they are simpler to derive directly from equations (2.12). Differentiating equations (4.4) with respect to time we have $\ddot{x}_i = b_{is}\dot{\eta}_s + \dots$, where the dots denote terms not containing $\dot{\eta}_s$. Hence we find that $\dfrac{\partial \ddot{x}_i}{\partial \dot{\eta}_s} = b_{is}$, as a result of which, from (2.12), we obtain Appell's equations

$$\frac{\partial S}{\partial \dot{\eta}_s} = \Pi_s, \quad s = 1, \dots, l \tag{4.21}$$

where $S = \dfrac{1}{2} m_\nu \ddot{\mathbf{r}}_\nu^2$ is the energy of the accelerations and $\Pi_s = \mathbf{F}_\nu \cdot \mathbf{b}_{s\nu}$ is the generalized force referred to the quasi-coordinate π_s [40].

We shall show, finally, that Kane's equations [44] are equivalent to Poincaré's equations. Kane considered a dynamical system whose configuration in a inertial reference frame can be specified by Lagrangian coordinates $x_1, \dots, x_n$ and which is subject to constraints such that $n - l$ nonintegrable differential equations of the form (4.12).

As parameters of the real and virtual displacements we take $\eta_i = \dot{x}_i$ and $\omega_i = \delta x_i$ respectively. Using (4.12) we construct the operators

$$X_i = \frac{\partial}{\partial x_i} + b_{ji} \frac{\partial}{\partial x_j}, \quad i = 1, \dots, l; \quad j = l + 1, \dots, n \tag{4.22}$$

According to the relations (2.6)

$$\delta \mathbf{r}_\nu = \omega_i X_i \mathbf{r}_\nu, \quad \nu = 1, \dots, N$$

Substitution these expressions into D'Alambert-Lagrange principle (2.12) then gives the equations of motion

$$m_\nu \ddot{\mathbf{r}}_\nu \cdot X_i \mathbf{r}_\nu = \mathbf{F}_\nu \cdot X_i \mathbf{r}_\nu, \quad i = 1, \dots, l \tag{4.23}$$

Take into account (4.22) one can see that the equations (4.23) coincide with Kane's equations (19) [44]

$$K_{x_i} + K'_{x_i} = 0,$$

where

$$K_{x_i} = \left(\frac{\partial \mathbf{r}_\nu}{\partial x_i} + b_{ji} \frac{\partial \mathbf{r}_\nu}{\partial x_j} \right) \cdot \mathbf{F}_\nu, \quad K'_{x_i} = \left(\frac{\partial \mathbf{r}_\nu}{\partial x_i} + b_{ji} \frac{\partial \mathbf{r}_\nu}{\partial x_j} \right) \cdot \mathbf{F}'_\nu, \quad \nu = 1, \dots, N$$

according to Kane are generalized active force and generalized inertia force ($\mathbf{F}'_\nu = -m_\nu \ddot{\mathbf{r}}_\nu$) respectively correspond to x_i.

This research was carried out with financial support from the Russian Foundation for Basic research (96-01-00261) and from the Alexander von Humboldt Foundation.

Bibliography

[1] Poincaré, H.: Sur une forme nouvelle des équations de la mécanique, C.R.Acad.Sci.Paris, 132 (1901), 369-371.

[2] Poincaré, H.: Sur la precession des corps deformables, Bull. astronomique, 27 (1910), 321-356.

[3] Chetayev, N.: Sur les equations de Poincaré, C.R.Acad.Sci.Paris, 185 (1927), 1577-1578.

[4] Chetayev, N.: Sur les equations de Poincaré, Dokl.Acad.Nauk SSSR, 7 (1928), 103-104.

[5] Chetayev, N.: On Poincaré's equations. Prikl.Mat.Mech., 5, 2 (1941), 253-262.

[6] Chetayev, N.G.: Poincaré's equations, in: Theoretical Mechanics (Ed. V.Rumyantsev and K.Yakimova), Mir Publishers, Moscow, 1989, 316-355.

[7] Shurova, K.E.: A variation of Poincaré's equations, Prikl.Mat.Mech, 17, 1 (1953), 123-124; 18, 3 (1954), 384.

[8] Shurova, K.E.: The principal invariant of equations in variations, Vestnik MGU, Ser. Mat. Mekh. Astron. Fiz, 3 (1958), 47-49.

[9] Rumyantsev, V.V.: The equations of motion of a solid with a cavity filled with liquid, Prikl.Mat.Mekh, 19, 1 (1955), 3-12.

[10] Aminov, M.Sh.: The construction of groups of possible displacements, in: Proceedings of the Interuniversity Conference on the Applied Theory of Stability and Analytic Mechanics, KAI, Kazan', 1962, 21-30.

[11] Bogoyavlenskii, A.A.: Cyclic displacements for the generalized area integral, Prikl.Mat.Mekh., 25, 4 (1961), 774-777.

[12] Bogoyavlenskii, A.A.: Theorems of the interaction between the parts of a mechanical system, Prikl.Mat.Mekh., 30, 1 (1966), 203-208.

[13] Bogoyavlenskii, A.A.: The properties of possible displacements for theorems of the interaction between the parts of a mechanical system, Prikl.Mat.Mekh, 31, 2 (1967), 377-384.

[14] Fam Guen: The equations of motion of non-holonomic mechanical systems in Poincaré-Chetayev variables, Prikl.Mat.Mekh, 31, 2 (1967), 253-259.

[15] Fam Guen: The equations of motion of non-holonomic mechanical systems in Poincaré-Chetayev variables, Prikl.Mat.Mekh, 32, 5 (1968), 804-814.

[16] Fam Guen: One form of the equations of motion of mechanical systems, Prikl.Mat.Mekh, 33, 2 (1969), 397-402.

[17] Markhashov, L.M.: The Poincaré and Poincaré-Chetayev equations, Prikl.Mat.Mekh, 49, 1 (1985), 43-55.

[18] Markhashov, L.M.: A generalization of the canonical form of Poincaré's equations, Prikl.Mat.Mekh., 51, 1 (1987), 157-160.

[19] Markhashov, L.M.: On a remark of Poincaré, Prikl.Mat.Mekh., 51, 5 (1987), 724-734.

[20] Markhashov, L.M.: Particular solutions of the equations of motion and their stability, Izv.Acad.Nauk SSSR, MTT, 6 (1987), 26-32.

[21] Rumyantsev, V.V.: On the Poincaré - Chetayev equations, Prikl. Mat. Mekh.,58, 3 (1994), 3-16.

[22] Rumyantsev, V.V.: On the Poincaré - Chetayev equations,, Dokl. Ross. Acad. Nauk, 338, 1 (1994), 51-53.

[23] Rumyantsev, V.V.: General equations of analytical dynamics, Prikl.Mat.Mekh., 60, 6 (1996), 929-940.

[24] Eisenhart, L.P.: Continous Groups of Transformations, Princeton University, 1933.

[25] Lur'ye, A.I.: Analytical Mechanics, Fizmatgiz, Moscow, 1961.

[26] de la Vallé-Poissin, S.J.: Cours d'Analyse Infinitésimale, vol. 2, Gauthier-Villars, Paris, 1912.

[27] Arnold, V.I.: Mathematical Methods of Classical Mechanics, Nauka, Moscow, 1989.

[28] Maggi, G.A.: Da alcune nuove forma della equazioni della dinamica applicabile ai sistemi anolonomi, Atti della Reale Acad. Naz. dei Lincei.Rend.Cl.fis. e math. Ser. 5. Vol.10. Fasc. 2 (1901), 287-291.

[29] Whittaker, E.T.: A Treatise on the Analytical Dynamics of Particles and Rigid Bodies, Cambridge University Press, Cambridge, 1927.

[30] Boltzmann, L.: Uber die Form der Lagrange'schen gleichungen für nicht holonome, generalisierte koordinaten, Sitzungsb. der Wiener Acad. der wissenschaften. Math.-Naturwiss. Kl., Bd. 140, H.1-2 (1902), 1603-1614.

[31] Hamel, G.: Die Lagrange-Eulerschen gleichungen der mechanik, Z.Math, und Phys., Bd. 50 (1904), 1-57.

[32] Zeiliger, D.N.: Motion's theory of a similarity transformated body. Kazan', 1892.

[33] Chaplygin, S.A.: On a possible generalization of the areas theorem with application to the rolling of balls, in: Collected Works, Vol 1, Gostekhizdat, Moscow (1948), 26-56.

[34] Rumyantsev, V.V.: Parametric examination of dynamical non-holonomic systems and two problems of dynamics, in: Differential Equations: Qualitative Theory, Vol.2, Amsterdam North-Holland (1987), 883-919.

[35] Jacobi, C.G.J.: Vorlesungen über Dynamik. Verlag von G.Reimer, Berlin, 1884.

[36] Suslov, G.K.: Theoretical Mechanics, Gostekhizdat, Moscow, 1946.

[37] Marsden, J.E.:Lectures on Mechanics. London Math.Soc. Lecture Note Series, 174. Cambridge University Press, Cambridge, 1992.

[38] Arnol'd, V.I., Kozlov, V.V. and A.I.Neishtadt: Mathematical Aspects of Classical and Celestial Mechanics, in: Itogi Nauki i Tekhniki, Ser. Sovr. Problemy Matematiki. Fundamental'nye Napravleniya, 3, VINITI, Moscow, 1985.

[39] Pars, L.A.: A Treatise on Analytical Dynamics. Heinemann, London, 1964.

[40] Neimark, Yu.I. and N.A.Fufayev: Dynamics of Non-holonomic Systems. Nauka, Moscow, 1967.

[41] Voronets, P.V.: On the equations of motion for non-holonomic systems. Mat. Sbornik, 22, 4 (1902), 659-686.

[42] Chaplygin, S.A.: On the motion of a heavy body of rotation upon the horizontal plane, in: Collected Works, Vol.1, Gostekhizdat, Moscow, (1948), 57-75.

[43] Volterra, V.: Sopra una classe di equazioni dinamiche. Atti della R.Accad.delle Sci.di Torino, t.33 (1897), 451-475.

[44] Kane, T.R.: Dynamics of non-holonomic systems. Trans. ASME. Ser. E. J.Appl.Mech., 28, 4 (1961), 574-578.

PART II

EFFECTIVE COMPUTATIONS IN CELESTIAL MECHANICS AND ASTRODYNAMICS

C. Simó
University of Barcelona, Barcelona, Spain

Abstract

Problems in Celestial Mechanics and Astrodynamics are considered under the point of view of Hamiltonian Dynamical Systems. A main tool to analyze the dynamics consists in studying the *skeleton* of the system, that is, the invariant objects (fixed points, periodic orbits and invariant tori) as well as their related stable, unstable and centre manifolds. Methods to compute these objects are presented. They are analytical, with the symbolic implementacion, purely numerical or a combination of both. Several examples are presented.

1 Introduction

First we shall introduce some elementary ideas which are useful for the study of simple mathematical models of any physical problem. The reader familiar with these ideas can skip this section and proceed to the next one. For convenience, some definitions are given in both sections, in a slightly different language, making easier to skip this section.

1.1 Preliminaries on differential equations

Assume the problem to be studied is characterized by n variables, $v_1, v_2, ..., v_n$, or, shortly, v. To describe the evolution of v with respect to the time, we shall have a relation as

$$\dot{v} = f(t, v), \tag{1}$$

where $\dot{v}$ denotes the derivative of v with respect to the time t, or, in other words, the velocity of v as a function of v itself and time. The equation (1) is, in that case, the *nature law* that we assume describes the behaviour of the studied phenomenon. The equation (1) is an *ordinary differential equation.* If for a given time, $t^{(0)}$, the value, $v^{(0)}$, of the v variables is known, then the value of v, as a function of time, remains determined as far as (1) has a meaning. More concretely, we remember the

Theorem. *(Picard) Consider the following initial value problem: To determine a function* $\varphi(t; t^{(0)}, v^{(0)})$ *solution of (1) (that is, such that* $\frac{d}{dt}\varphi(t; t^{(0)}, v^{(0)}) = f(t, \varphi(t; t^{(0)}, v^{(0)}))$ *is satisfied for all* t *in a suitable interval) and such that the initial conditions hold:* $\varphi(t^{(0)}; t^{(0)}, v^{(0)}) = v^{(0)}$. *Then, if* f *is a continuous bounded function and it is locally Lipschitz with respect to* v*, such a function* φ *exists and is unique.*

An equation like (1) depends explicitely on time. When this is not the case ($\dot{v} = f(v)$) the equation is said to be autonomous. We can always reduce an equation to autonomous by adding an additional variable. Indeed, consider w with $n+1$ components $(w_0, w_1, \ldots, w_n)$ and take $w_0 = t, w_j = v_j$, if $j > 0$. Then

$$\begin{aligned} \dot{w}_0 &= 1, \\ \dot{w}_j &= f_j(w_0, w_1, ..., w_n), \ \ j = 1, ..., n, \end{aligned}$$

where f_j is the j–th component of f, that is, we have

$$\dot{w} = g(w), \tag{2}$$

which is autonomous. For autonomous equations the initial value of t, $t^{(0)}$, is irrelevant and can be taken as 0. Then the solution is simply written as $\varphi(t; x^{(0)})$ or $\varphi_t(x^{(0)})$ and is denoted as flow. If the solutions are defined, the following holds $\varphi_t(\varphi_s(x^{(0)})) = \varphi_{t+s}(x^{(0)})$ or, simply, $\varphi_t\varphi_s = \varphi_{t+s}$.

Let us consider how the solution of (2) depends on the initial conditions. If the solutions that for $t = 0$ start at $x^{(0)}$ and $y^{(0)}$ are defined for a given t, then

$$|\varphi(t, x^{(0)}) - \varphi(t, y^{(0)})| \leq e^{L|t-t_0|}|x^{(0)} - y^{(0)}|, \tag{3}$$

where L denotes the Lipschitz constant with respect to v, an inequality which follows from *Gronwall's Lemma.* In some cases the bound given by (3) is quite pessimistic. We would like to have something better and, furthermore, to know the behaviour of the flow, $\varphi_t(x^{(0)})$ with respect to $x^{(0)}$ when in (2) the function g is r times differentiable and $D^r g$ is continuous (that is, $g \in C^r$). One has the result:

Theorem. *If g is C^r then φ_t is a diffeomorphism of class C^r (that is, φ_t is a 1 to 1 map and both φ_t and φ_t^{-1} are C^r). Furthermore, the first differential, $D_x\varphi_t(x)$, satisfies the equation*

$$\frac{d}{dt}(D_x\varphi_t(x)) = Dg(\varphi_t(x))D_x\varphi_t(x) \tag{4}$$

with the initial condition $D_x\varphi_t(x)|_{t=0} = Id$, where Id denotes the identity matrix in $\mathbb{R}^n$.

Equation (4) is called first variational equation. Then

$$\varphi_t(y) - \varphi_t(x) = D_x\varphi_t(x)(y - x) + o(|y - x|),$$

which, quite often, gives bounds much better than (3) for $|y - x|$ small. Note that (4) is linear with time depending coefficients.

1.2 Continuous and discrete systems

If φ_t exists for all t real, then the property $\varphi_t\varphi_s = \varphi_{t+s}$ tells us that, if g is at least C^1, the set of maps $\{\varphi_t, t \in \mathbb{R}\}$ is a one parameter group of diffeomorphisms. This suggest the following definition:

A *dynamical system* is a couple (M, φ_t), where M is a manifold (for instance, a surface, an open set of $\mathbb{R}^n$, etc) and $\{\varphi_t\}$ is a one parameter group of diffeomorphisms acting on M and satisfying $\varphi_t\varphi_s = \varphi_{t \circ s}$, where $t \circ s$ denotes the group operation. If, furthermore, μ is a measure in M and φ_t preserves μ (that is, $\mu(\varphi_t A) = \mu(A)$ for all measurable sets $A \in M$), then the triple (M, μ, φ_t) is called a *measure preserving dynamical system.*

Quite often the uniparametric groups are $\mathbb{R}$ or $\mathbb{Z}$. In the first case we shall talk about a *continuous dynamical system* and a differential equation like (2) can be associated to it, by defining $g(x) = \frac{d}{dt}\varphi_t(x)|_{t=0}$. If the group is $\mathbb{Z}$ we talk about a *discrete system* and it is only necessary to know φ_1, because the group property gives $\varphi_n = \varphi_1^n$ for all $n \in \mathbb{Z}$. The map φ_1 is denoted the generator of the group.

Of course, it is also possible to define dynamical systems (either continuous or discrete) when φ_t is simply a group of homeomorphisms, without any differentiability condition. If (M,μ) is a couple where M is an abstract space and μ a measure on it, one can also define an (abstract) dynamical system (M,μ,φ_t), where now φ_t are 1 to 1 maps, except by zero measure sets, preserving μ: $\varphi_t^{-1}(A)$ and A have the same measure for all measurable A.

If the maps φ_t have no inverse then we talk about *dynamical semisystems* (either continuous or discrete). If the semigroup is $\mathbb{R}_+$ (that is, the non negative reals), we shall still have $\varphi_t\varphi_s = \varphi_{t+s}$ if $t,s \geq 0$, but, in general, $\varphi_{-t} = \varphi_t^{-1}$ do not exists for $t > 0$. In what follows we shall be interested only in the case of systems.

It is quite simple to go from a continuous to a discrete system. Let T a fixed time. Then we define, as generator, ϕ_1, of the discrete group, the map $\varphi_{t=T}$ of the continuous system. If we start with an equation as (1) and f is $T-$periodic with respect to t ($f(t+T,x) = f(t,x)$ for all $t \in \mathbb{R}$ and allt x, provided f is defined), then there is no need to increase first the dimension of the space from n to $n+1$, as we did before to transform an equation like (1) to the form (2). Indeed, going to dimension $n+1$, the flow φ_T produces values of t which are coincident modulus the period. Hence, there is no need to give the "time" component of the w variables. It is true that, in this case, φ_T depends on the initial value of the time, $t^{(0)}$, but the maps produced by different initial times, as $t^{(0)}$ and $t^{(1)}$, are conjugated by the relation $\varphi(t^{(0)}+T;t^{(0)},\cdot)\,\varphi(t^{(0)};t^{(1)},\cdot) = \varphi(t^{(0)}+T;t^{(1)}+T,\cdot)\,\varphi(t^{(1)}+T;t^{(1)},\cdot)$, because, thanks to the periodicity one has $\varphi(t^{(0)}+T;t^{(1)}+T,\cdot) = \varphi(t^{(0)};t^{(1)},\cdot)$.

Another procedure to define a discrete system from a continuous one is the so called *Poincaré map* (or *return map*) when the initial differential equation is autonomous. Let $\mathcal{S}$ be an hypersurface in M (that is, of dimension one less that the one of M) transversal to the flow. This means that for every point $p \in \mathcal{S}$ the vector field, $g(p)$, is not contained in the tangent plane to $\mathcal{S}$ at p. Assume it exists a time, $\tau(p) > 0$, such that $\varphi_{\tau(p)} \in \mathcal{S}$. Then, by the implicit function theorem it exists a map defined in a neighbourhood V of p in $\mathcal{S}$, $P : V \to P(V)$ by $P(q) = \varphi_{\tau(q)}(q), \forall q \in V$. Of course, it can happen that $P(q) \notin V$ and then the new image $P(P(q))$ will not be defined. This can happen because, in general, the Poincaré sections, $\mathcal{S}$, are not "global". On the other hand, if we consider $t < 0$ there is no guarantee that for some $r \in V$, the set $\{\varphi_t(r), t < 0\}$ intersects $\mathcal{S}$. If it happens that, for q in a neighbourhood, perhaps smaller than V, V_0, we have $P^j(q)$ defined for all $j \in \mathbb{Z}$, then we have obtained a discrete system, with generator P, from the continuous one $\{\varphi_t\}$. Note that, in this case, the dimension of the space to which the variables belong (or *phase space*) has decreased by one unit. Obviously, if the initial differential equation, $\dot{x} = g(x)$, has several independent *first integrals* $F_1, F_2, ..., F_k$ (that is, functions such that $(\nabla F_j(x), g(x)) = 0$ with linearly independent gradients) a further reduction of the dimension is possible. If $p \in \mathcal{S}$ and $F_j(p) = c_j$ then the map P can be restricted to $V_0 \cap F_1^{-1}(c_1) \cap \ldots \cap F_k^{-1}(c_k)$, decreasing the dimension by k addicional units. This is the case of the Hamiltonian systems, which have the Hamiltonian itself, H, as first integral.

Reciprocally, given a discrete system of generator ϕ it is possible to definer a continuous system having ϕ as Poincaré map or as $\varphi_{t=1}$ map. The procedure is denoted as *suspension*. But one has to increase the dimension of the phase space by one unit or to introduce a non autonomous differencial equation (but 1-periodic in t). Furthermore, the new manifold containing the flow can be very complicated. It is possible to obtain in explicit form, at least if the discrete system is close to the identity, simple 1-periodic differentiable vector fields (and even C^∞ if the map is regular enough), but it is not so easy to have them analytic. We shall not use this technic in what follows, despite it is quite useful conceptually and for many proofs.

1.3 Invariant objects. Hyperbolicity

We shall use indistinctly flows or diffeomorphisms. To simplify the presentation we shall assume that we work on an open set of $\mathbb{R}^n$. First we look for "simple" solutions of (2). The simplests ones are the fixed points. A point, w^*, is called *fixed point* of (2) if $g(w^*) = 0$. Hence $\varphi_t(w^*) = w^*\ \forall t \in \mathbb{R}$. The analysis of the behaviour close to a fixed point can be done by using (4), which, in the present case, has constant coefficients:

$$\frac{d}{dt}(D_x\varphi_t(x)) = Dg(w^*)D_x\varphi_t(x) \tag{5}$$

It is clear that, as this is a linear approximation, it will be necessary to take into account the terms of higher order for a full discussion. However, as we shall see, there are cases in which the information given by the linear part is enough to describe the behaviour close to w^*. The analysis of (5) is carried out by putting $Dg(w^*)$ in canonical form (Jordan form, for instance).

To present the cases in which the linear part gives enough information, a definition is required. We shall say that a matrix is *hyperbolic* (in the context of the differential equations) if it has all the eigenvalues with real part different from zero. In the case of dimension 2, if $Dg(w^*)$ is regular, only the case of *centers* is excluded. A *fixed point*, w^*, of (2) es called *hyperbolic* if $Dg(w^*)$ is an hyperbolic matrix (we assume $g \in C^1$, of course). A basic result is given by

Theorem. *(Hartman) Let w^* be an hyperbolic fixed point of a differential equation $\dot{w} = g(w)$. Let φ_t and ψ_t the flows associated to $\dot{w} = g(w)$ and $\dot{w} = Dg(w^*)w$, respectively. Then there exists a neighbourhood, U, of w^* and an homeomorphisme, h, defined on U, with $h(w^*) = w^*$ such that it conjugates φ_t and ψ_t (that is, $h\varphi_t = \psi_t h$ for all t while the points remain in the corresponding neighbourhood).*

In other words, letting aside the transformation h (which can be regarded as a change of variables) the flows φ_t and ψ_t are the same in a neighbourhood of w^*. Notice, however, that in general h is not a differentiable map. Only the number of eigenvalues with positive real part (or with negative real part) is enough to decide if two systems of the same dimension and with hyperbolic matrices, are conjugated.

In the cases of eigenvalues with zero linear part (including, in particular, the case of zero eigenvalues) the contribution of the nonlinear part is essential to describe the behaviour close to a fixed point.

For a linear system $\dot{y} = Ay$ in $\mathbb{R}^n$, if s is the number of eigenvalues with negative real part and $u = n - s$ the number of those with positive real part, there exist two linear subspaces, E^s and E^u, having dimensions s and u, respectively (one of them can be zero) which are invariant under the linear flow. E^s and E^u are denoted as *stable subspace* and *unstable subspace*, respectively. This is the situation for the linear system $\dot{w} = Dg(w^*)w$ if w^* is an hyperbolic fixed point. For the initial equation $\dot{w} = g(w)$ there are similar objects, ${W^s}_l, {W^u}_l$, of respective dimensions s and u, called *stable and unstable local manifolds*, respectively, with the properties

$$ {W^s}_l = \{x \in U | \lim_{t\to\infty} \varphi_t(x) = w^*\}, \quad {W^u}_l = \{x \in U | \lim_{t\to-\infty} \varphi_t(x) = w^*\}, \tag{6} $$

Of course, E^s and E^u are tangent to ${W^s}_l$ and ${W^u}_l$, respectively, at w^*. By transporting the objects of of (6) under the flow we have the global manifolds:

$$ W^s = \bigcup \varphi_t({W^s}_l), \quad W^u = \cup \varphi_t({W^u}_l), $$

where the unions are done $\forall t \in \mathbb{R}$, despite for the first one is it enough to take $t < 0$ (and $t > 0$ for the second one).

In a similar way we can study the *fixed points of diffeomorphisms*: $\phi(w^*) = w^*$. If we want to study *periodic points of strict period* $k > 1$ (a point w^* is called periodic under ϕ of strict period k if $\phi^k(w^*) = w^*$, and $\phi^j(w^*) \neq w^*$ for $j = 1, ..., k-1$) it is enough to use ϕ^k instead ϕ. As the passage from differential equations to diffeomorphisms can be seen as an *exponentiation* (remember that the solution ψ_t of the linear system $\dot{x} = Ax$ is $\psi_t = e^{At}$), we shall say that a fixed point is hyperbolic if *all the eigenvalues of* $D\phi(w^*)$ *have modulus different from 1.* In this case Hartman's theorem assures the conjugacy of ϕ and $D\phi(w^*)$ on a neighbourhood of w^*. In a similar form we have stable and unstable manifolds. Note, however, that as we deal with a diffeomorphism, the points on both manifolds move on them by "jumping" and not in a continuous way.

Hartman's theorem shows the existence of W^s and W^u, but it only assures that the manifolds are continuous (remember that h is, in general, just a homeomorphism). But if $g \in C^r$ or $\phi \in C^r$ (in the continuous and discrete cases, respectively) the *stable/unstable invariant manifold Theorem* asserts that W^s and W^u are also of class C^r.

It is important to remark that an hyperbolic fixed point is *persistent to perturbacions.* Indeed, reasoning for a diffeomorphism, let us consider a family of diffeomorphisms, $\phi(w, \lambda)$, depending on a parameter λ (it can be a vectorial parameter, with components $\lambda_1, ..., \lambda_p$). Assume that for $\lambda = \lambda^*$ the diffeomorphism $\phi(w, \lambda^*)$ has an hyperbolic fixed point, w^*. Hence $1 \notin SpecD_w\phi(w^*, \lambda^*)$, where $SpecA$ denotes the set of eigenvalues of the matrix A. The fixed point condition when we vary λ is

$$G(w, \lambda) = \phi(w, \lambda) - w = 0. \tag{7}$$

As $G(w^*, \lambda^*) = 0$ and $D_wG(w^*, \lambda^*) = D_w\phi(w^*, \lambda^*) - Id$ has no eigenvalue equal to zero, it is invertible. The *implicit function Theorem* tells us that, in a neighbourhood of λ^*, there exists a funtion $w = w(\lambda)$ such that $w^* = w(\lambda^*)$ and $G(w(\lambda), \lambda) = 0$. Furthermore $D_w\phi(w(\lambda), \lambda)$ is close to $D_w\phi(w(\lambda^*), \lambda^*)$ by the continuity of $D_w\phi$ and, hence, it is also hyperbolic if the neighbourhood is small enough.

Let us pass to other "simple" solutions. For an equation like (2) we can consider periodic solucions, satisfying $\varphi_t(w) = \varphi_{t+T}(w)$ for some $T > 0$. Notice that the knowledge of the solution for $t \in [0, T)$ determines it $\forall t \in \mathbb{R}$. The problem can be reduced to look for fixed points of a diffeomorphism. Indeed, let p be a point in the periodic orbit and $\mathcal{S}$ surface transversal to $g(p)$ at p. Then, the Poincaré map will be defined in a neighbourhood of p on $\mathcal{S}$ and $P(p) = p$. Note theat another possibility would be to ask for $\varphi_T(p) = p$, but it is immediate to see that the vector $\frac{d}{dt}\varphi_t(p) = g(\varphi_t(p))$ is a solution of (4). Hence $D_p\varphi_T(p)$ has 1 as eigenvalue, and we can not apply the implicit function Theorem. If the vector field is not autonomous, but T-periodic, we can look for T-periodic solutions by requiring $\varphi(T; 0, p) - p = 0$. Generically $D_p\varphi(T; 0, p)$ will not have eigenvalues equal to 1, despite other difficulties can appear (for instance, bad numerical conditioning if the field is a perturbation of an autonomous one and the part depending periodically on time is small).

Let us return to periodic orbits of an autonomous system. We say they are hyperbolic if, looking at them as fixed points of a Poincaré map, these fixed points are hyperbolic. Which Poincaré section we use is irrelevant, because using $\mathcal{S}_1$ or $\mathcal{S}_2$ one obtains conjugated differential matrices.

Consider now other *invariant objects.* A differential equation leaves invariant an object, A, if $\varphi_t(x) \in A, \forall x \in A, \forall t \in \mathbb{R}$. If $\phi^n(x) \in A, \forall x \in A, \forall n \in \mathbb{Z}$ we say that the diffeomorphism ϕ leaves A invariant. Fixed points, periodic orbits and their related stable and unstable manifolds, if they are hyperbolic, are examples of invariant objects. But other invariant objects can occur: k-dimensional tori (topologically equivalent to the product of k circles), spheres, Cantor sets, etc. An invariant object, A, es called *normally hyperbolic* under a diffeomorphism, ϕ, if for all point $x \in A$ there exists a decomposicion of $\mathbb{R}^n$ as $T_xA \oplus E_x^s \oplus E_x^u$, where T_xA is the tangent space to A at x, such that the differential of the diffeomorphism at x, $D\phi(x)$, send T_xA, E_x^s and E_x^u into $T_{\phi(x)}A$, $E_{\phi(x)}^s$ and $E_{\phi(x)}^u$, respectivelly and, furthermore, there are constants $C > 0$ and $\lambda > 0$ such that $\forall n \in \mathbb{N}$ the following holds:

a) $|D\phi^n(x)\xi| \leq Ce^{-\lambda n}|\xi|$ for all vectors $\xi \in {E_x}^s$,

b) $|D\phi^{-n}(x)\eta| \leq Ce^{-\lambda n}|\eta|$ for all vectors $\eta \in {E_x}^u$.

E^s_x gives the *stable directions* at x and E^u_x the *unstable directions* at x.

An analogous definition holds for flows of vector fields. If T_xA, E^s_x and E^u_x, which must depend on x in a continuous way, have dimensions d_A, d_s and d_u, respectivelly, then *the stable* and *unstable manifolds of* A, defined by

i) $W^s{}_A = \{x \in \mathbb{R}^n | \lim_{k\to\infty} d(\phi^k(x), A) = 0\}$,

ii) $W^u{}_A = \{x \in \mathbb{R}^n | \lim_{k\to\infty} d(\phi^{-k}(x), A) = 0\}$,

where $d(y, A)$ denotes the distance of the point y to the set A, have the following dimensions: $dim\, W^s_A = d_A + d_s$, $dim\, W^u_A = d_A + d_u$. It is clear that $d_A + d_s + d_u = n$. For instance, a normally hyperbolic periodic orbits of a flow in $\mathbb{R}^3$ has stable and unstable manifolds of dimension 2. Locally one can imagine these manifolds as cylinders.

1.4 Homoclinic phenomena. Unpredictability.

To simplify the presentation let us assume that we study planar diffeomorphisms. Consider, for instance, the so called *conservative Hénon map*, which can be written as

$$\phi_\varepsilon(x, y) = (\overline{x}, \overline{y}),$$

where

$$\begin{aligned} \overline{x} &= x + \varepsilon \overline{y}, \\ \overline{y} &= y + \varepsilon x(1 - x). \end{aligned}$$

This map depends on a parameter, ε, and it is *conservative* because the determinant $det\, D\phi_\varepsilon$ is equal to 1 for all ε, x and y, that is, it preserves the area in $\mathbb{R}^2$.

Consider now the Hamiltonian vector field $\dot{x} = y$, $\dot{y} = x(1-x)$, with energy function (or Hamiltonian) $H = y^2/2 - x^2/2 + x^3/3$. The related *phase portrait* (that is, the global picture of the orbits) is shown in figure 1.

Note that the stable and unstable manifolds of the hyperbolic fixed point ($p \equiv (0, 0)$) coincide and they are on the energy level $H = 0$. Of course, if we consider the time ε flow, φ_ε, of this vector field, the phase portrait is the same (but given a non fixed point, q, their image $\varphi_\varepsilon(q)$ is at a small distance of the point, if ε is also small). It is immediate (by using Taylor's formula) that

$$\varphi_\varepsilon(x, y) = (x + \varepsilon y + \frac{\varepsilon^2}{2} x(1 - x) + O(\varepsilon^3),\;\; y + \varepsilon x(1 - x) + \frac{\varepsilon^2}{2}(1 - 2x) + O(\varepsilon^3)).$$

In particular ϕ_ε and φ_ε coincide except by term $O(\varepsilon^3)$. Hence, we can consider the former Hamiltonian field as the *limit field* of the Hénon diffeomorphism when ε goes to zero. Then p is also a fixed point of ϕ_ε (in a general case the fixed points of ϕ_ε and φ_ε can differ in $O(\varepsilon)$). The invariant manifolds of p under ϕ_ε, $W^s{}_{p,\phi_\varepsilon}$ and $W^u{}_{p,\phi_\varepsilon}$, are

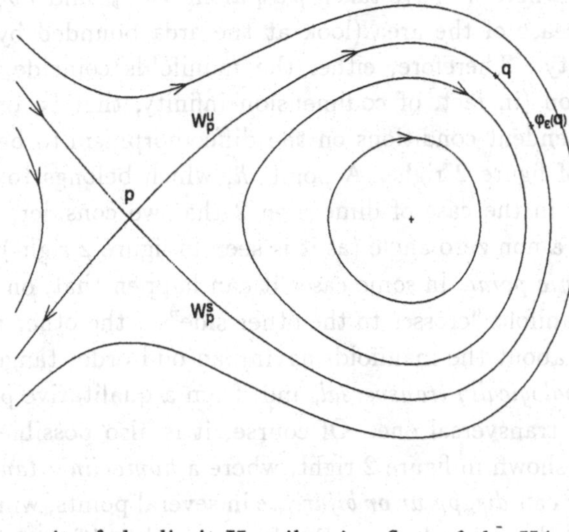

Figure 1: Phase portrait of the limit Hamiltonian flow of the Hénon map when ε goes to zero.

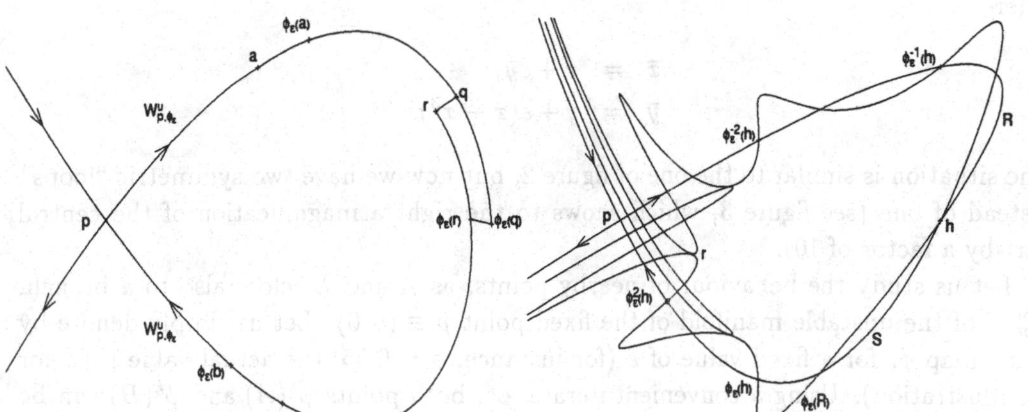

Figure 2: An illustration of the existence of homoclinic points in area preserving maps.

close to the manifolds, W^s_p and W^u_p of the flow. It is clear that W^s_{p,ϕ_ε} and W^u_{p,ϕ_ε} must intersect. Otherwise (see figure 2 left) let us consider a path as $p\,a\,q\,r\,b\,p$. The image is $p\,\phi_\varepsilon a\,\phi_\varepsilon q\ \phi_\varepsilon r\,\phi_\varepsilon b\,p$, where we have taken $p\,a\,q$ along W^u_{p,ϕ_ε} and $r\,b\,p$ along W^s_{p,ϕ_ε}. This leads to an increase of the area (look at the area bounded by $r\,q\,\phi_\varepsilon q\,\phi_\varepsilon r$) and, hence, to an absurdity. Therefore, either the manifolds coincide, $W^u_{p,\phi_\varepsilon} = W^s_{p,\phi_\varepsilon}$, a non generic situation (in fact, of codimension infinity, that is, one should impose infinitely many independent conditions on the diffeomorphism to be satisfied), or we are in the situation of figure 2 right. A point, h, which belongs to $W^u_{p,\phi_\varepsilon} \cap W^s_{p,\phi_\varepsilon}$ is called *homoclinic*. If, in the case of dimension 2 that we consider, W^u_{p,ϕ_ε} and W^s_{p,ϕ_ε} intersect in h making a non zero angle (as it is seen in figure 2 right), we say that h is a *transversal homoclinic point*. In some cases it can happen that, on a neighbourhood of h, one invariant manifold "crosses to the other side" of the other manifold, but the angle is zero. (Think about the manifolds having an odd order tangency at h). Then it is said that h is *topologically transversal*, and, from a qualitative point of view, this case is similar to the transversal one. Of course, it is also possible that there exist points, as the point r shown in figure 2 right, where a *homoclinic tangency* occurs. By changing ε the point r can *disappear or bifurcate* in several points, while h *persists* (and it is locally unique) under perturbation. Furthermore, the different "lobes", as $\mathcal{R}$, $\mathcal{S}$, $\phi_\varepsilon\mathcal{R}$, etc, have all the same area.

To see the dynamical consequences of this behaviour, we shall use a family of maps different from the Hénon one, the family being cubic instead of quadratic:

$$\phi_\varepsilon(x,y) = (\overline{x}, \overline{y}),$$

where

$$\begin{aligned} \overline{x} &= x + \varepsilon\overline{y}, \\ \overline{y} &= y + \varepsilon(x - x^3). \end{aligned}$$

The situation is similar to the one of figure 2, but now we have two symmetric "loops" instead of one (see figure 3, which shows to the right a magnification of the central part by a factor of 10).

Let us study the behaviour of nearby points, as A and B, close also to a branch, $W^{u,+}_p$, of the unstable manifold of the fixed point $p \equiv (0,0)$. Let us simply denote by ϕ the map ϕ_ε for a fixed value of ε (for instance, $\varepsilon = 0.55$, the actual value used for the illustration). Using a convenient iterate, ϕ^k, both points $\phi^k(A)$ and $\phi^k(B)$ can be seen on different sides of $W^{s,+}_p$ (see figure 3). Of course, this means that they already were on different sides since the very beginning, but this is much hard to see. The successives iterates will take them again close to p, being close to $W^{s,+}_p$ but on *different sides*. Hence, after a passage close to p one of them will follow closely the branch $W^{u,+}_p$, while the other will follow close to $W^{u,-}_p$. It is clear that both the images of A and B, will return once and again close to p. At every passage small deviations will allow to

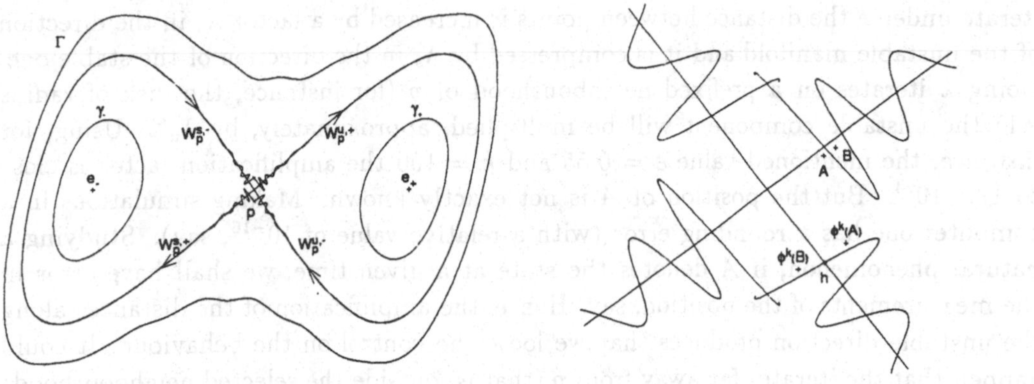

Figure 3: Phase portrait of the cubic area preserving diffeomorphism showing invariant curves and transversal homoclinic points. The right part is a magnification of a neighbourhood of the hyperbolic fixed point by a factor of 10.

"choose" which side of W^s_p they follow. We remark that, in the situation of figure 3 the following result can be proved:

Theorem. *For a given value of $\varepsilon > 0$, $\varepsilon < \varepsilon_0$, fixed, and using the map ϕ_ε, there exists $N \in \mathbb{N}$ such that, per every couple of sequences (doubly infinite, that is, infinite at both sides) $(..., n_{-3}, n_{-2}, n_{-1}, n_0, n_1, ...)$, $(..., s_{-3}, s_{-2}, s_{-1}, s_0, s_1, ...)$ where $n_j \geq N \, \forall j \in \mathbb{Z}$ and s_j is a sign (+ or -), there is at least an initial point, A, close to p, such that under iteration it goes to the right if s_0 is + and to the left if s_0 is −, and which returns at a local minimum distance to p after n_0 iterates; then it goes to the right or to the left according to the sign of s_1, returning again to a local minimum distance to p after n_1 additional iterates, etc. The same happens by using the inverse iteration with signs and numbers of iterates given by s_{-1} and n_{-1}; s_{-2} and n_{-2}, etc.*

Of course, this tells us that we can have all kinds of possible behaviour. This type of dynamics is called *quasirandom*, because any sequence $(..., n_{-3}, n_{-2}, n_{-1}, n_0, n_1, , , ,)$, $(..., s_{-3}, s_{-2}, s_{-1}, s_0, s_1, ...)$ *can be realized,* as if we could choose at random to go right or left every time the iterates pass close to p. In other words: the behaviour is *unpredictible* despite the system is deterministic. This can be a little bit surprising and requires a more detailed explanation.

First, and due to the determinism, it is clear that, given any initial point, A, the orbit is uniquely determined, and the related sequences of n and s is unique. Let us examine more closely the passages near p. There we can approximate ϕ by the linear part, the matrix $\begin{pmatrix} 1+\varepsilon^2 & \varepsilon \\ \varepsilon & 1 \end{pmatrix}$, which has an eigenvalue $\lambda_u = 1 + \varepsilon\sqrt{1+\varepsilon^2/4} + \varepsilon^2/2$ such that the corresponding eigenvector is tangent to W^u_p. In a similar way, there is an eigenvalue $\lambda_s = 1/\lambda_u$ with eigenvector tangent to W^s_p. Close to p, every time we iterate under ϕ the distance between points is increased by a factor λ_u in the direction of the unstable manifold and it is compressed by λ_s in the direction of the stable one. Doing k iterates on a prefixed neighbourhood of p (for instance, the disk of radius 0.1) the unstable component will be multiplied, approximately, by ${\lambda_u}^k$. Using, for instance, the mentioned value $\varepsilon = 0.55$ and $k = 100$ the amplification factor is close to $1.2 \cdot 10^{24}$. But the position of A is not exactly known. Making simulations in a computer one has a rounding error (with a relative value of 10^{-16}, say). Studying a natural phenomenon, if A denotes the state at a given time, we shall have error in the measurements of the position, say. Hence, the amplification of the distances along the unstable direction produces that we loose the control on the behaviour. It could happen that the iterates far away from p (that is, outside the selected neighbourhood) cancel exactly the expansions and compressions produced inside it. This really happens if instead of ϕ_ε we use φ_ε, the time ε flow described above. But this *integrability* is against the existence of a transversal homoclinic point as h in figure 3.

The studied map is area preserving and close, if ε is small, to the flow φ_ε. The fixed points e_+ and e_-, having coordinates $(1,0)$ and $(-1,0)$, respectively are of elliptic type. Locally φ_ε (and ϕ_ε)) behaves like a rotation. We recall that, for an area preserving map, ϕ, if an eigenvalue of $D\phi(x^*)$ is λ, where we assume x^* to be a fixed point of ϕ, then the other eigenvalue is $1/\lambda$. The fixed point is called *elliptic* if λ is a complex number of modulus 1, different from ± 1.

Under the flow φ_ε the limit period when going to $e_{\pm e}$, is $\pi\sqrt{2}$. When approaching the separatrix (the level $H = 0$) the period increases monotonically to infinity. The *rotation number* is defined, on these orbits, as the frequency: $2\pi/\mathit{period}$. As ϕ_ε is a relatively small perturbation ($O(\varepsilon^2)$) of φ_ε and φ_ε is a *twist map* (the derivative of the rotation number with respect to the energy is different from zero), Moser's twist Theorem (or, as more general, the Kolmogorov-Arnol'd-Moser, *KAM* Theorem) asserts that there exist closed curves invariant under ϕ_ε. Figure 3 shows three of these curves, as $\gamma_\pm$ around $e_\pm$, respectively, and Γ in the outer part. This implies that the successive iterates of points like A are confined. This is, of course, true if ε is small enough. think that in the present example it is required to have $|\varepsilon| < 0.56...$. By plotting iterates of a suitable point close to p, it seems that the iterates cover a domain of positive measure. The natural question is whether *the closure of the orbit has positive measure or not.* This is an open question up to now.

Some suitable references for this introduction are listed below.

- Arnold, V.I.:*Geometric theory of ordinary differential equations*, Springer 1982.
- Guckenheimer, J., Holmes, P.:*Nonlinear oscillations, dynamical systems and bifurcations of vector fields*, Applied Math. Sci. 42, Springer 1983.
- Palis, J., de Melo, W.:*Geometric theory of dynamical systems: an introduction*, Springer 1982.

- Palis, J., Takens, F.:*Hiperbolicity and sensitive chaotic dynamics at homoclinic bifurcations*, Cambridge Univ. Press 1993.
- Wiggins, S.:*Global bifurcations and chaos*, Applied Math. Sci. 73 Springer 1988.

In what follows we shall pass to higher dimension to be able to deal with models closer to the requirements of the applications. The situation, from a theoretical point of view, is much less clear than in the case of area preserving maps (or, equivalently, Hamiltonians with two degrees of freedom). From a practical point of view, the difficulties of the numerical computations and of the representations are increased, but still there are lots of methods which can be used. Some of them will be presented in the forthcoming sections.

2 The invariant objects: periodic orbits, invariant tori, invariant stable and unstable manifolds, centre manifolds

Celestial Mechanics is a part of Analytical Mechanics dealing with the study of the motion of natural and artificial bodies under the action of gravitation. Usually the number of interacting bodies is considered not too large. Other problems, like the dynamics of galaxies and clusters, have usually a different approach. The study of the motion of artificial bodies has some peculiarities (design of missions by means of suitable manoeuvres) and is sometimes denoted as Astrodynamics.

Celestial Mechanics and Astrodynamics can be seen under the point of view of Dynamical Systems. We shall consider, in what follows, that they admit a Hamiltonian formulation, despite in many cases one should allow for non conservative forces. Hence we shall assume that we are given a Hamiltonian $H = H(q,p)$, where q and p denote positions and momenta. In principle and for simplicity, we shall assume that H is an analytical function defined on a $2n$–dimensional space, that typically will be $\mathbb{R}^n$ with the obvious exceptions to avoid collisions of bodies. To present the main ideas we shall restrict our exposition to the interaction of punctual masses under gravitational attraction. The reader can easily adapt most of the methods to other classes of problems (including non conservative ones for several topics). Hence the equations of motion are

$$\frac{d}{dt}q_i = \dot{q}_i = \frac{\partial H}{\partial p_i}, \quad \frac{d}{dt}p_i = \dot{p}_i = -\frac{\partial H}{\partial q_i}, \quad i = 1,\ldots,n.$$

The number n is known as the *number of degrees of freedom.* The $2n$–dimensional space, $\mathcal{E}$, where the variables q and p live is known as the *phase space.* A key idea of the theory of Dynamical Systems is to look at any problem concerning the dynamics as a *geometrical problem* in the phase space. Hence, we look always for objects which have an intrinsic geometrical meaning, independent of the actual parametrizations used to describe them.

From a geometrical point of view a key point consists in trying to understand the structure of the phase space. We can find in it fixed points, periodic orbits and invariant tori. Some of these objects can be (at least partially) of *hyperbolic* character. The related stable and unstable manifolds form a kind of network of connections (the so called *homoclinic tangle*), which together with the previous invariant objects constitute the *skeleton* of the system. The dynamics follows closely the "paths" marked by these connections. The objects which are of purely *elliptic* type, under suitable conditions on the related frequencies, have the nearby phase space "almost foliated" by invariant tori. Here "almost" means that the we have a set of tori depending on parameters (say, amplitudes or "actions") and the parameters range on a Cantorian set of big relative measure. The "gaps" in the Cantor set are due to resonances, but if these resonances are of high order the gaps can be very small. In them there is a mild lack of predictability (or, if you want, some "chaos"). Notice that these gaps can be connected, leading to a *diffusion* of the action variables, but at a very slow rate. Other objects are partially elliptic and partially hyperbolic. Among them we can consider the ones having just an hyperbolic plane, that is, only one positive and one negative eigenvalues (with the same modulus because of the Hamiltonian character). These objects have *centre manifolds* of codimension 2, and the stable and unstable manifolds of these centre manifolds have codimension 1. They are ideal candidates to act as "barriers" or *confiners* to make difficult the escape of points. In fact these manifolds do not prevent escape: stable and unstable manifolds do not match but, generically, they intersect transversally. The manifolds are highly folded and this, together with small angles at the intersections, produces that the particles of the phase space have to wander for a long time before finding the way out.

The simplest solutions of the equations are the *fixed points*, where the vector field given by the right hand side of the equations above becomes zero. The next simpler solutions are the *periodic solutions.* If, for shortness, we denote by z the vector containing all the components of q and p, then a periodic solution if of the form $z(t) = w(\theta)$, where θ is an angle in $\mathbb{S}^1$ moving with time as $\theta = \omega t + \theta_0$, $\omega \in \mathbb{R}$ being the *frequency* and $\theta_0 \in \mathbb{S}^1$ being an *initial phase.* In a similar way we can consider *quasiperiodic solutions*, formally expressed as the periodic ones, but θ belonging now to $\mathbb{T}^k$, a k–dimensional torus, $\omega \in \mathbb{R}^k$ being a vector of frequencies and the initial phase, θ_0, being also in $\mathbb{T}^k$. The frequencies in the vector ω are denoted as the *basic frequencies* of the quasiperiodic solution. We shall assume that they satisfy a *non resonance condition*, that is,

$$(m, \omega) \neq 0, \ \ \forall m \in \mathbb{Z}^k \setminus \{0\}.$$

Here (,) denotes the scalar product. Otherwise the frequencies are resonant and some of them can be put as a function of the remaining ones, i.e., the number of basic frequencies is less than k.

Now consider Hamiltonians *close to integrable.* An n–degrees of freedom Hamiltonian is called *integrable* in the Liouville sense if it has n *functionally independent*, except, perhaps, on some manifolds of lower dimension, *first integrals*, F_i, $i = 1, \ldots, n$, one of them being, eventually, the Hamiltonian H itself, and such that they are *in involution.* That is, the *Poisson bracket*

$$[F_i, F_j] = \sum_{k=1}^{n} \frac{\partial F_i}{\partial q_k}\frac{\partial F_j}{\partial p_k} - \frac{\partial F_i}{\partial p_k}\frac{\partial F_j}{\partial q_k}$$

is zero for all choices of i and j. Then, consider the level surfaces of the first integrals,

$$\mathcal{M}_c = \{(q,p) \in \mathcal{E} | F_i(q,p) = c_i,\ i = 1, \ldots, n\},$$

the vector c having components c_i. If $\mathcal{M}_c$ is compact, then the celebrated Liouville–Arnol'd theorem assures that it is diffeomorphic to an n–dimensional torus. There are many other definitions of integrability (in Hamiltonian and other problems). The ones having an interesting dynamical meaning are those who guarantee a nice behavior of the solutions, so that there is no impredictability of the motion.

An integrable Hamiltonian, H_0, in the sense above can be put in *action-angle* variables (I, φ), and then $H_0 = H_0(I)$. Hence, the solutions are trivial:

$$I(t) = I(0),\ \ \varphi(t) = \varphi(0) + \omega t,$$

where the frequencies ω are given by $\omega_j = \omega_j(I) = \frac{\partial H_0}{\partial I_j}$. There are open sets of the phase space foliated by invariant tori. If we look now at a perturbation

$$H(I, \varphi) = H_0(I) + \varepsilon H_1(I, \varphi, \varepsilon)$$

of the Hamiltonian H_0, under some generic conditions, ensuring that the frequencies of the "unperturbed" Hamiltonian *do* change with the actions (i.e., $\frac{\partial \omega(I)}{\partial I}$ is an invertible matrix), the well known KAM theorem (Kolmogorov–Arnol'd–Moser) assures that the perturbed Hamiltonian still has most of the invariant tori, but they do not constitute a foliation at all. Between any two tori there are extremely narrow domains where some amount of "chaoticity" takes place. It can be completely irrelevant from the physical point of view and numerically undetectable, unless one is willing to do computations with a large number of digits. The "robust" tori, surviving the perturbation for ε small enough, are those whose frequencies satisfy some *Diophantine condition.* More concretely, positive values of c and τ must exist so that

$$|(m, \omega)| > c||m||^{-\tau},\ \forall m \in \mathbb{Z}^n \setminus \{0\}.$$

If $\tau > n - 1$ the set of frequencies ω for which this condition is not satisfied for any $c > 0$, has zero Lebesgue measure.

In some problems we have to allow for time depending Hamiltonians, $H = H(q, p, t)$. The cases that we shall consider here come from the consideration of the motion of a small particle under the action of more massive bodies. The particle does not affect the motion of the other bodies. So their motion can be taken as a known function of time. Usually the periodic or quasiperiodic approximation are good enough models. Then the dependence of H on t is periodic or quasiperiodic. The autonomous character of the Hamiltonian can be recovered by the usual method to add new variables. For instance, if $H(q,p,t)$ is a Hamiltonian depending on t in a quasiperiodic way through $\theta \in \mathbb{T}^d$ with vector of basic frequencies $\omega \in \mathbb{R}^d$, we can introduce a new Hamiltonian

$$\mathcal{H}(q,p,\theta,J) = H(q,p,\theta) + (\omega, J),$$

where $J \in \mathbb{R}^d$ are additional momenta, canonically conjugated to the angles θ. Of course, then we have $\dot{\theta}_j = \frac{\partial \mathcal{H}}{\partial J_j} = \omega_j, \;\; j = 1,\ldots,d$. The formulation is reduced to the autonomous case, but there is a main difference. The derivatives of the new Hamiltonian with respect to these new momenta are constant (i.e., they are the "external frequencies") and this gives some kind of degeneracy, or *lack of free actions*, when trying to apply some KAM-like results.

When a fixed point has been computed, the first thing to study is the linear stability. This is easily obtained from the eigenvalues of the linear part of the vector field at the point. If, as before, z denotes all the variables of the Hamiltonian, we have $\dot{z} = X_H(z)$, where X_H denotes the Hamiltonian vector field. Then, if the fixed point is denoted as z^*, one has to look for the eigenvalues of $DX_H(z^*)$. It is well known that, due to the infinitesimal symplectic character of $DX_H(z^*)$, if λ is an eigenvalue, so is $-\lambda$, and if λ is not real, also $\pm\bar{\lambda}$ are eigenvalues, where $\bar{\lambda}$ denotes the complex conjugate of λ. Furthermore, if 0 is an eigenvalue, then is has even multiplicity.

If we only look at the linear behavior around z^*, we can classify it according to the number of eigenvalues of $DX_H(z^*)$ having positive, negative or zero real part. The related eigenvectors generate linear subspaces known as the unstable, stable and central subspaces, respectively. Of course, the dimension of the stable and unstable subspaces are the same. The stable and unstable eigenvalues give the *hyperbolic part* of the local behavior, that is, the part associated to eigenvalues with non zero real part. The non zero eigenvalues with zero real part give rise to the *elliptic part* of the local behavior. The name *parabolic part* can be reserved for the subspaces associated to zero eigenvalues.

When we turn to the nonlinear behavior, a generalization of the previous description holds. There are *local stable and unstable manifolds*, which are invariant manifolds, under the flow of X_H, tangent to the corresponding linear subspaces. They are also characterized as the set of points, in a given neighborhood of z^*, such that under the flow for positive (respectively, negative) time go to z^* without leaving the neighborhood. By globalization under the flow (that is, by looking for the images of the local stable/unstable manifolds for all time for which the flow is defined) they give the *global stable and unstable manifolds*. Concerning regularity, as we assumed the Hamiltonian to be analytic, the stable and unstable manifolds are also analytic.

The central part is more subtle. As it follows from the formal computation, there is always a *formal* centre manifold. That is, a manifold which can be expressed, for instance, as the graph of a function given by a formal power series. But, despite the Hamiltonian is assumed to be analytic, the centre manifold is, in general, not more than C^∞. Hence, we can add to this function any C^∞-flat function and it will be also a formal centre manifold. We are faced here to a problem of *non uniqueness.* Of course, if the formal power series turns out to be convergent, then there exists an *unique C^ω centre manifold.* Another problem is the *globalization of the centre manifold.* In general it is a hard problem to decide about "the boundary of the centre manifold". I know examples where the centre manifold "ends" or "losses dimension" at some separatrix, but also of examples where it "increases dimension" by "absorbing" some directions that were previously hyperbolic. These changes are related to *global bifurcations* and they are typically hard to study but very relevant for the geometry of the phase space (and, hence, for the dynamics).

Taking now a centre manifold as an invariant object, we can look, if it do not has full dimension, to the stable/unstable manifolds of this centre manifold. If we are dealing with the centre manifold of a fixed point, for instance, the stable manifold of the centre manifold is usually denoted as *centre-stable manifold* of the point. In a similar way we can talk about *centre-unstable manifolds.* Assume that the hyperbolic part has dimension 2, that is, only one stable and one unstable directions exist. Then the centre-stable and centre-unstable manifolds play a key role in some aspects of the dynamics, as sketched above, because they have *codimension 1.*

Now we turn to the behaviour around periodic and quasiperiodic solutions. As usual, we can turn the orbit into a fixed point by introducing new (time dependent) variables. Let $z(t)$ a solution. Then the change $w = z - z(t)$ converts the solution into a fixed point. Of course, this change is symplectic and, hence, it keeps the Hamiltonian character of the equations of motion. The problem is, however, that even if the the initial Hamiltonian is autonomous, the new one, $H(w + z(t)$, will have the same dependence with respect to time as $z(t)$ has. Consider, first the case of a periodic solution.

The point $w = 0$ being fixed, the Hamiltonian around it has a power series expansion starting with terms of degree 2. We shall denote as H_2 these terms. Of course, all the homogeneous parts of the series have coefficients which depend periodically on t, with the same period of $z(t)$. If we linearize around $w = 0$, only the terms coming from H_2 are relevant. We know that this linear equation has periodic coefficients in t. It is well known that, by the Floquet theory, there is a linear change of variables, with the same period as the coefficients, which turns the linear equation to constant coefficients. This change can be taken to be symplectic, to preserve the Hamiltonian. Hence, the local character of the fixed point $w = 0$ is well defined. We can compute the eigenvalues and decide about the dimensions of the invariant manifolds. One should take into account that, if the initial Hamiltonian is autonomous, there will be always a zero eigenvalue associated to the phase shift along the periodic orbit. It must appear with even multiplicity. Another eigenvalue zero is associated to the conservation of the energy. Generically, only these two zeros appear as eigenvalues, but there are cases where the multiplicity can be higher (e.g., in the case of existence of additional first integrals). Also generically, when there are only two zero eigenvalues, the matrix of constant coefficients of the linearized equations reduced to the related subspace do not diagonalizes. It has a Jordan block structure. It follows easily from the implicit function theorem that there exist a *natural continuation* of the periodic orbit to a *family of periodic orbits.* Typically this family is, locally, parametrized by the energy. The out of diagonal element in the Jordan block is associated to the *variation of the period with respect to the energy level.*

It is useful to consider, to study the vicinity of a periodic orbit, a *transversal section* or *Poincaré section.* Then the dimension of the phase space is reduced by 1, and the flow is replaced by a *discrete dynamical system*, the so called *Poincaré map*, to be denoted as P. We can look at it when all the (local) levels of energy are considered, or when we fix the level at the one of the periodic orbit. In this last case the dimension will be reduced by 2, and the Poincaré map turns out to be a symplectic map on the corresponding space. Let us denote as $\bar{w}$ a local variable around 0 for this $2n-2$ Poincaré map and keep the same name, P, for the map. We can look at the local properties of P around $\bar{w}=0$. Now the stable, unstable and central directions are associated, respectively, to the eigenvalues of $DP(0)$ with modulus less than 1, greater than 1 or equal to 1. If the full $(2n-1)$ Poincaré map is considered, one should increase the central dimension by 1. To return to the behavior around the full periodic orbit, one should transport what happens in the Poincaré section under the flow. Again the central dimension is increased in an additional unit.

The quasiperiodic case is not so easy. The linearized equations can not be reduced to constant coefficients, in general. Anyway, if the linear system can be put as a constant coefficients one plus a small perturbation, depending on time quasiperiodically, then the reduction is possible for a relativelly large (but Cantorian) set of values of the perturbation parameter, provided some non degeneracy conditions hold. To avoid to require it at each sentence, we shall assume, from now on, that a *reducibility condition holds.* That is, that there exist a linear change of variables, depending quasiperiodically on time with the same set of basic frequencies as the $z(t)$ solution considered, such that it reduces the linearized equations to constant coefficients. There are ways to check, approximately, if this condition is satisfied. Under this assumption, the local character of the quasiperiodic solution is well defined. We can talk about hyperbolicity, stable, unstable and central directions, etc. Just note that if the dimension of the torus is k, then there are $2k$ eigenvalues equal to zero. One half of them is associated to the *phases shift* along the angles of the torus. The other half is associated to the formal preservation of the associated momenta. As mentioned before, if we are given an $n-1$-dimensional torus and it turns out to have just one stable and one unstable directions, the centre manifold associated to it has dimension $2n-2$ (think about a family of $n-1$–dimensional tori, depending on $n-1$ parameters on a Cantor set with very narrow gaps) and, hence, the centre–stable and centre–unstable manifolds have codimension 1.

Some of the usual techniques to deal with problems in Celestial Mechanics can be considered inside perturbation theory. An approximation to the problem can be integrable, so that the solutions can be expressed in a simple form (at least in theory!) and we can look for solutions of the full problem expanded in power series of some parameter. This requires analyticity of the equations of motion and their dependence with respect to parameters. The solution of the integrable system gives the zero order terms of this expansion. It can be also possible to try to perform changes of variable such that the problem is reduced to a simpler form, hopefully integrable. This is not possible, of course, but averaging methods or normal form techniques often succeed in carrying out such a reduction at the formal level. The lack of convergence of these procedures, due to small divisors problems in most of the cases, asks for bounds on the errors done when several steps of an iterative process are performed. Furthermore, one can derive information about the number of steps to be used so that the estimate of the total error is minimized. Typically this minimum is exponentially small in the small parameter involved in the expansions. More refined information requires the use of time as a complex variable.

However, most of the problems in Celestial Mechanics are beyond the present limits of the purely analytical techniques. Quite often one has to rely on numerical simulations to understand the behaviour of the solutions of some problem. On the other hand, purely numerical methods are not enough when the number of variables and parameters involved is too large. I consider that, at the present epoch, reliable and complete enough simulations can be done with at most 6 or 7 total variables. A combination of symbolic and numerical techniques, guided by careful analytical studies and the suitable geometric insight, can give very good and useful results.

In what follows we shall present some methods for the computation of different objects. They are presented through concrete examples, worked out in detail. The examples concern the motion in the 3D Restricted Three Body Problem (RTBP) around relative equilibrium (or *libration points*) of collinear and triangular type. The study is useful both to understand the dynamics of small bodies in the solar system and for the design of spacecraft missions.

At the end of these notes there appear some references for most of the topics presented and some other useful complements.

3 Global Description of the Orbits Near the L_2 Point of the Earth–Moon System in the RTBP model

Part of the methodology presented in this section and in the next one was done in collaboration with G. Gómez, À. Jorba and J. Masdemont under contracts of the European Space Agency. It was applied to the L_1 point of the RTBP in the Sun–Earth case and to the L_4 and L_5 points of the Earth–Moon system. Later on has been applied to some improved models of the physical problem and, finally, to the best available model: the one derived from the numerical JPL ephemeris.

Let us consider the spatial Restricted Three Body Problem, that is the motion of a massless particle (to be taken as a spacecraft) under the gravitational influence of two massive bodies (to be taken as Earth and Moon). Last two are assumed to be on circular orbits around their center of masses. Using a rotating system of reference with period the one of the circular orbits and center at the mass center, the two massive bodies can be considered at rest. Scaling masses and distances, the sum of the masses of Earth and Moon and their distance can be taken equal to one. Then mass of the Moon is then the parameter μ_{Moon} equal to 0.012150582.

In the rotating system of reference (synodical system) there are 5 equilibrium points (or libration points in the classical terminology). We are interested in the translunar libration point, which is aligned with Earth and Moon and placed beyond the Moon. L_1 denotes the equilibrium point between the Earth and the Moon, which is close to the Moon.

In the spatial RTBP the L_1 and L_2 libration points have a linear behaviour of the type saddle × center × center. Under generic assumptions there exist nearby two dimensional tori. Because of the small divisors problems, on those tori the frequencies should satisfy a diophantine condition. If we are interested in short time intervals the small divisors problems do not show up unless they are associated to low order resonances. Let ω_1, ω_2 be the two basic frequencies at L_i, i=1, 2. For small values of μ the frequencies are rather close. For the limit Hill's problem, $\mu = 0$, they are $(28^{1/2} - 1)^{1/2}$ and 2, respectively. For $\mu = \mu_{\text{Moon}}$, at the L_2 point, the values of ω_1 and ω_2 are near 1.86264588 and 1.78617616, and beyond the 1 to 1 resonance the next strong resonance appears only at order 47. A very strong resonance appears at order 8398.

When the amplitudes of the orbit change, so do the frequencies. It is known that the planar periodic Lyapunov orbits emanating from the libration point, bifurcate to halo orbits when the x amplitude (x is in the direction of the primaries) is roughly 0.1845, when the unit of distance is the one from the small mass to the libration point. In the present case this equals the value $\gamma = 0.16783273$ times the Earth–Moon distance. Hence this is the distance corresponding to the 1 to 1 resonance giving rise to the halo orbits. If we keep our study in a smaller region the effect of this resonance is not too strong. Higher order resonances have small amplitude and long period. Therefore we can try to obtain the invariant tori formally to some order (less than 47). In the concrete example presented below the order has been kept to 45.

Cutting the expressions to this order we obtain a foliation by invariant tori in some region around the equilibrium point. For some of the amplitudes the tori certainly do not exist, but the small size of the related stochastic zone makes this fact irrelevant for short time interval applications, in our case a spacecraft running during a few years.

3.1 The Equations of Motion

We consider the following system of reference: The origin is located at L_2. The positive x axis is directed from L_2 away from the smaller primary. The y axis is in the plane of sidereal motion of the primaries at $\pi/2$ from the x axis in counterclockwise sense. The z axis completes a positively oriented coordinate system. Let γ be the distance between m_2 and L_2, where m_2 denotes the position of the smaller primary. γ will be the new unit of distance. The unit of time is such that the period of revolution of the primaries in the sidereal system is 2π. Let μ be the mass of the smaller primary (in our case the Moon as given above) and $1-\mu$ the one of the bigger primary (the Earth in our application). The value of γ is obtained from the Euler quintic equation:

$$\gamma^5 + (3-\mu)\gamma^4 + (3-2\mu)\gamma^3 - \mu\gamma^2 - 2\mu\gamma - \mu = 0.$$

It can be solved by Newton method starting at $(\mu/3)^{1/3}$.

Let X, Y, Z be the usual synodical coordinates centered at the center of mass. The equations of motion are

$$\ddot{X} - 2\dot{Y} = \Omega_X, \qquad \ddot{Y} + 2\dot{X} = \Omega_Y, \qquad \ddot{Z} = \Omega_Z, \tag{8}$$

where

$$\Omega = \frac{1}{2}(X^2+Y^2) + (1-\mu)r_1^{-1} + \mu r_2^{-1},$$

being

$$r_1^2 = (X-\mu)^2 + Y^2 + Z^2, \qquad r_2^2 = (X-\mu+1)^2 + Y^2 + Z^2.$$

As we have described we make the change of coordinates

$$x = -\frac{1}{\gamma}(X-\mu+1+\gamma), \qquad y = -\frac{1}{\gamma}Y, \qquad z = \frac{1}{\gamma}Z. \tag{9}$$

Then

$$r_1^{-1} = (1-\gamma)^{-1}\left(1 + \frac{2\gamma}{1-\gamma}x + \frac{\gamma^2}{(1-\gamma)^2}\rho^2\right)^{-1/2} = (1-\gamma)^{-1}\sum_{n\geq 0}\left(\frac{\gamma\rho}{1-\gamma}\right)^n(-1)^n P_n\left(\frac{x}{\rho}\right),$$

$$r_2^{-1} = \gamma^{-1}(1-2x+\rho^2)^{-1/2} = \gamma^{-1}\sum_{n\geq 0}\rho^n P_n(\frac{x}{\rho}),$$

where $\rho^2 = x^2+y^2+z^2$ and P_n denote the Legendre polynomials.

Let

$$c_n = \gamma^{-3}\left(\mu + (-1)^n(1-\mu)(\frac{\gamma}{1-\gamma})^{n+1}\right). \tag{10}$$

auxiliary coefficients depending only on μ. From (10) it is easily seen that, in the Hill's case, when μ tends to zero, $c_2 = 4$ and $c_k = 3\times(-1)^k$ if $k>2$. Then the equations of motion are written as

$$\begin{aligned} \ddot{x} - 2\dot{y} - (1+2c_2)x &= \frac{\partial}{\partial x}\sum_{n\geq 3} c_n\rho^n P_n(\frac{x}{\rho}), \\ \ddot{y} + 2\dot{x} + (c_2-1)y &= \frac{\partial}{\partial y}\sum_{n\geq 3} c_n\rho^n P_n(\frac{x}{\rho}), \\ \ddot{z} + c_2 z &= \frac{\partial}{\partial z}\sum_{n\geq 3} c_n\rho^n P_n(\frac{x}{\rho}). \end{aligned} \tag{11}$$

We recall also, for further use, that (8) can be written in Hamiltonian form. Introducing

$$P_X = \dot{X} - Y, \qquad P_Y = \dot{Y} + X, \qquad P_Z = \dot{Z},$$

the Hamiltonian is

$$\begin{aligned} H &= \frac{1}{2}(P_X^2+P_Y^2+P_Z^2) + YP_X - XP_Y - (1-\mu)r_1^{-1} - \mu r_2^{-1} = \\ &= \frac{1}{2}(\dot{X}^2+\dot{Y}^2+\dot{Z}^2) - \Omega. \end{aligned} \tag{12}$$

The equations (11) can also be put in Hamiltonian form. Let $p_x = \dot{x}-y$, $p_y = \dot{y}+x$, $p_z = \dot{z}$. Then the corresponding Hamiltonian is

$$K = \frac{1}{2}\left(p_x^2+p_y^2+p_z^2\right) + yp_x - xp_y - \sum_{n\geq 2} c_n\rho^n P_n\left(\frac{x}{\rho}\right). \tag{13}$$

The Hamiltonians H and K are related by

$$H = K\gamma^2 - \frac{1}{2}(1-\gamma-\mu)^2 - \frac{\mu}{\gamma} - \frac{1-\mu}{1-\gamma}.$$

3.2 Formal Series Solutions

If we skip in (11) the right hand terms (of degree ≥ 2) the solution of the system, restricted to the center manifold (i.e., deleting the exponentially increasing or decreasing terms) is

$$x = \alpha\cos(\omega_1\tau+\phi_1), \qquad y = -\overline{k}\alpha\sin(\omega_1\tau+\phi_1), \qquad z = \beta\cos(\omega_2\tau+\phi_2), \tag{14}$$

with

$$\bar{k} = (\omega_1^2 + 2c_2 + 1)/(2\omega_1).$$

The parameters α, β are arbitrary amplitudes and ϕ_1, ϕ_2 are arbitrary phases. The frequencies in (14) are $\omega_1 = ((2 - c_2 + (9c_2^2 - 8c_2)^{1/2})/2)^{1/2}$, $\omega_2 = c_2^{1/2}$. As (11) is an autonomous system we can take the origin of time such that $\phi_1 = 0$. Furthermore we denote $\omega_1\tau$ and $\omega_2\tau + \phi_2$ as θ_1 and θ_2, respectively. When the right hand terms of (11) are considered we should allow for varying frequencies depending on the amplitudes, according to the Lindstedt–Poincaré method to avoid secular terms. Then θ_1 will be now $d\tau$ and θ_2 will be $f\tau + \phi_2$, where $d = \omega_1 + O(\alpha, \beta)$, $f = \omega_2 + O(\alpha, \beta)$.

We look for solutions in complex exponential form of the type

$$x = \sum x_{i,j,k,m}\alpha^i\beta^j\gamma_1^k\gamma_2^m, \quad y = \sqrt{-1}\sum y_{i,j,k,m}\alpha^i\beta^j\gamma_1^k\gamma_2^m, \quad z = \sum z_{i,j,k,m}\alpha^i\beta^j\gamma_1^k\gamma_2^m, \tag{15}$$

where $\gamma_s = \exp(\sqrt{-1}\theta_s)$, $s = 1, 2$.

Due to the symmetries of the problem the coefficients in (15) satisfy the relations:

1. $i, j \geq 0, i + j \geq 1, |\, k \,| \leq i, |\, m \,| \leq j, i - k \equiv 0 \pmod 2, j - m \equiv 0 \pmod 2$ in all the cases,
2. $x_{i,j,-k,-m} = x_{i,j,k,m}$, $y_{i,j,-k,-m} = -y_{i,j,k,m}$, $z_{i,j,-k,-m} = z_{i,j,k,m}$. Hence it is enough, for the computations, to keep only the terms with $k > 0$ or, if $k = 0$, with $m \geq 0$,
3. For x and y one should have $j \equiv 0 \pmod 2$, and for z one should have $j \equiv 1 \pmod 2$.
4. The Lindstedt–Poncaré method can be normalized in such a way that for $(k, m) = (1, 0)$ the only term appearing in x corresponds to $(i, j) = (1, 0)$. In an analogous way if $(k, m) = (0, 1)$ the only term in z corresponds to $(i, j) = (0, 1)$.
5. In d and f there are only terms of the form $\alpha^i\beta^j$ with i and j even.

The computation of the terms in (15) is done by increasing order of $n = i + j$. For $n = 1$ one has, from (14), the values

$$x_{1,0,1,0} = 1/2, \quad y_{1,0,1,0} = \bar{k}/2, \quad z_{0,1,0,1} = 1/2.$$

Taking into account the properties 1) to 5) one obtains x_n, y_n, z_n (i.e, the terms of total order n in α and β) and, if n is odd, also d_{n-1} and f_{n-1}. In the determination of x_n, y_n, z_n one should impose the conditions $x_{i,j,1,0} = 0$ if $(i, j) \neq (1, 0)$ and $z_{i,j,0,1} = 0$ if $(i, j) \neq (0, 1)$ (see property 4)). This is what allows to obtain d_{n-1} and f_{n-1} for n odd. Details on the recurrences and on the practical implementation concerning the storing of the coefficients can be found in [34].

Finally we make some a priori comments on the running space and CPU time. The programs have been structured in such a way that we can modify a few parameter instructions by setting

$$NG = (j^4 + 7j^3 + 20j^2 + 26j + 6)/6, \quad NG1 = 2j + 1, \quad NG2 = (j^2 + 3j + 2)/2,$$

if the maximal order to run the program is $n = 2j + 1$. One needs $20 + 2n$ vectors of dimension NG and a few ones of dimensions $NG1$, $NG2$. The total space required (in double precision) is $(4j^5 + 56j^4 + 236j^3 + 691j^2 + 881j)/6 + ctant$. If $n = 35$ this is close to 15.5 Mbyte, and it is close to 48 Mbyte if $n = 45$. To keep in mind a simpler expression, the required space up to order n is proportional to $n^4(n + 23)$ for n large. An analysis of the program shows that the CPU time is roughly proportional to n^8.

3.3 Results and Tests

The program described in the above section, has been run for several values of n, both for L_1 and L_2 and for different values of μ. The results below for $\mu = \mu_{\text{Moon}}$ have been run at order 45. The CPU time on an HP 9000/735 computer is 35^m.

The projections on the (X, Y), (X, Z) and (Y, Z) planes of a 2D torus corresponding to $\alpha = 0.15$, $\beta = 0.40$ appear on figures 4 and 5. The expansions above can be used, if $\beta = 0$ or $\alpha = 0$, respectively, to produce the Lyapunov families of "horizontal" and "vertical" orbits, respectively. The figure 6 shows some of them for several values of α (projection (X, Y)) and several values of β (projections (X, Y) and (Y, Z)). In this figure we have used (9) to display the results in synodical coordinates. We note that the "vertical" orbits project on the (X, Y) plane as a curve travelled twice. The (X, Z) projection, not displayed here, looks like an arc of curve travelled twice. The orbits are, approximately, on a vertical cylinder.

For completeness we also display some "halo orbits". These orbits are born from the planar Lyapunov family when they lose stability inside the center manifold of L_2. That is, beyond the initial hyperbolic couple of variables, common to all the orbits in a vicinity of L_2, another couple becomes unstable. This happens when the vertical mode around the planar Lyapunov orbit becomes resonant with the period of the orbit in a 1 to 1 resonance. They are obtained by an expression like (15) but now both angles, $\theta_s, s = 1, 2$ are equal and the amplitudes (α, β) satisfy some relation $g(\alpha, \beta) = 0$, also derived along the Lindstedt–Poincaré procedure (see [28]). The figure 6 shows the (X, Y), (X, Z) and (Y, Z) projections of these orbits for several values of β. The corresponding values of α follow from the $g(\alpha, \beta) = 0$ relation, but in the range shown for β, the amplitude α changes only between 0.1853 and 0.2318 .

The analytical solutions of (8) have been checked against numerical solutions as follows. Given $\rho > 0$ and $\theta \in [0, \pi/2]$, α and β have been obtained by $\alpha = \rho\cos(\theta)$, $\beta = \rho\sin(\theta)$. Then the phases ϕ_1 and ϕ_2 have been taken as multiples of $\pi/4$. Due to the symmetry it is enough to take $\phi_2 < \pi$. From these data we obtain the initial conditions

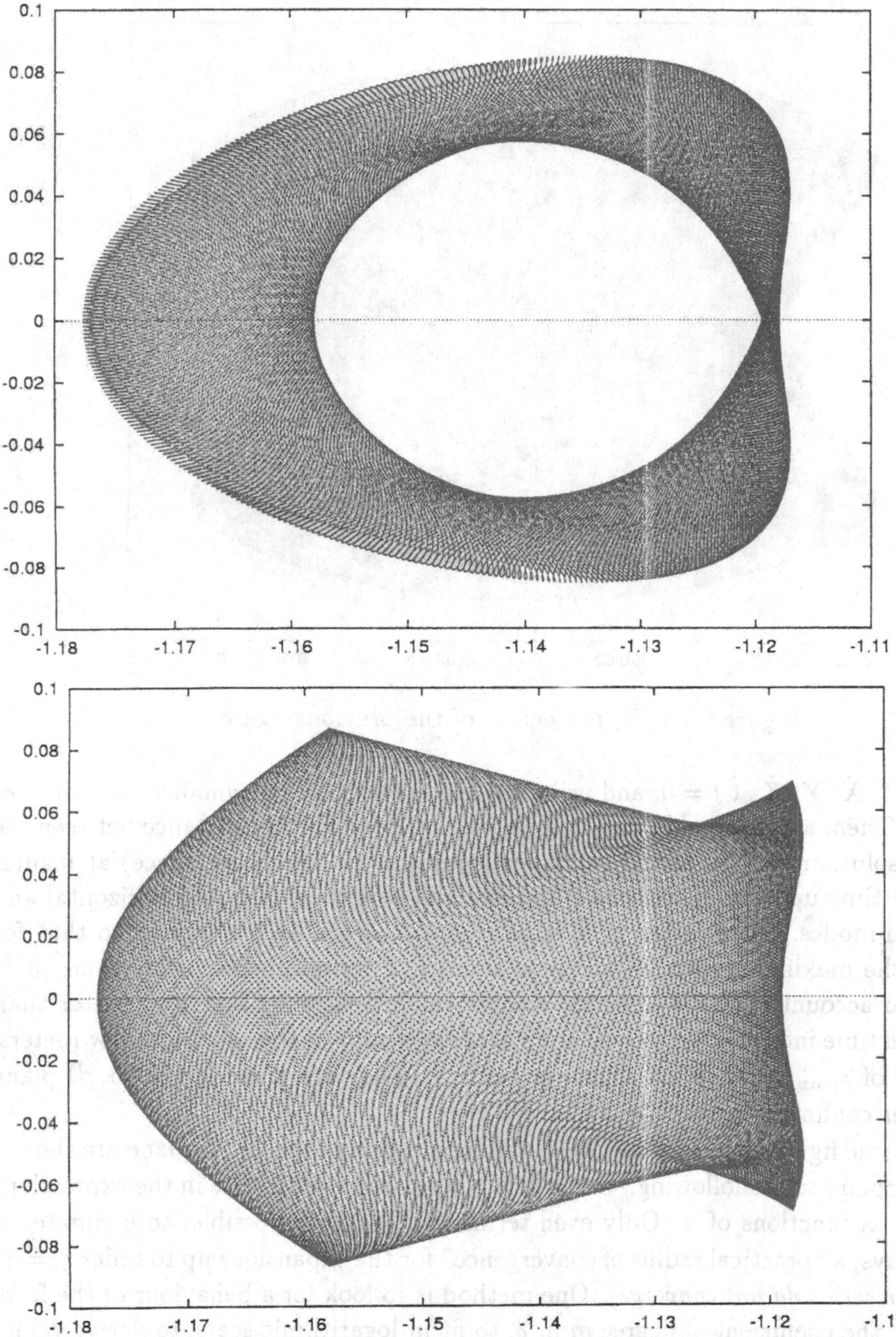

Figure 4: (X,Y) and (Z,Z) projections of a 2D torus corresponding to $\alpha = 0.15$, $\beta = 0.40$, obtained by the Lindstedt–Poincaré method.

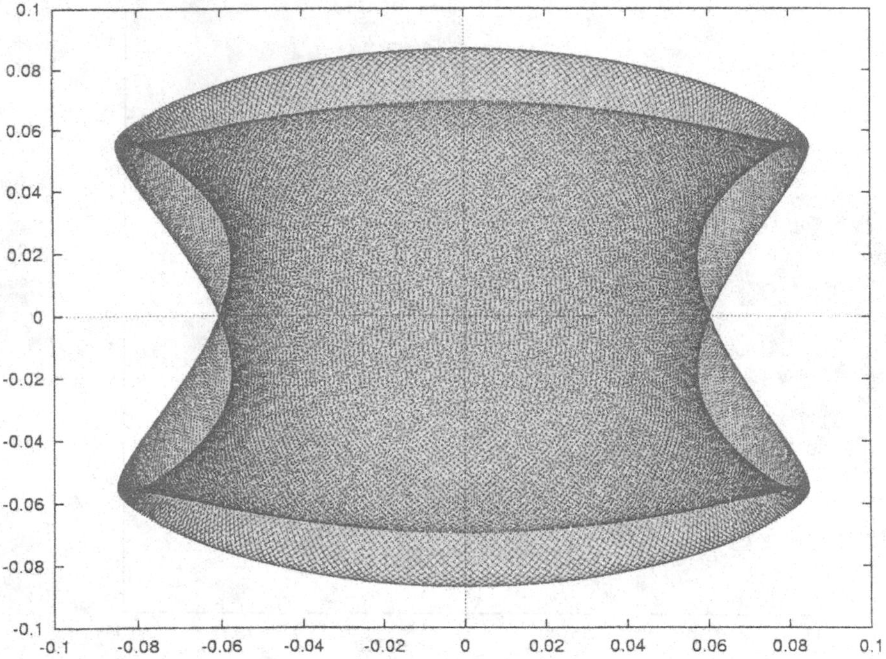

Figure 5: (Y, Z) projection of the previous figure.

for X, Y, Z, $\dot{X}$, $\dot{Y}$, $\dot{Z}$ at $t = 0$, and using (9) the corresponding synodical coordiantes are found. Then, a numerical integration has been started and the distance between the numerical solution and the analytical one is computed (in the phase space) at regular intervals of time up to π. This time is roughly the period of both, the horizontal and the vertical modes. Then, given θ, a value of ρ, ρ_{max}, has been obtained so that for $\rho < \rho_{max}$ the maximum error for all these times and initial phases is less than 10^{-6}. Taking into account that the unstable component increases by a factor greater than 1000 in this time interval, the initial errors are, physically, of the order of a few meters. The values of ρ_{max} versus θ are shown in figure 7. They are given in the (α, β) plane as the inner confining curve.

In the same figure two more curves, confining a region of the (α, β) plane are shown. They correspond to the following. Given θ, it is possible to substitute in the expressions of d and f as functions of ρ. Only even terms appear. It is possible, to estimate, in different ways, a "practical radius of convergence" for the expansions up to order $n = 45$ (despite the series *do not converge*). One method is to look for a behaviour of the form $A \times B^m$ for the coefficient of degree m in ρ, to fit in logarithmic scale, to derive values of B for both d and f, to take the maximum and to use the inverse as an estimate of the radius of convergence. This gives the outer curve in figure 7. Another possibility is to ask for $|t_{44}| < 10^{-3}$, where t_{44} denotes the largest of the last terms in d and f for a given value of ρ. This gives the intermediate curve between the previous two.

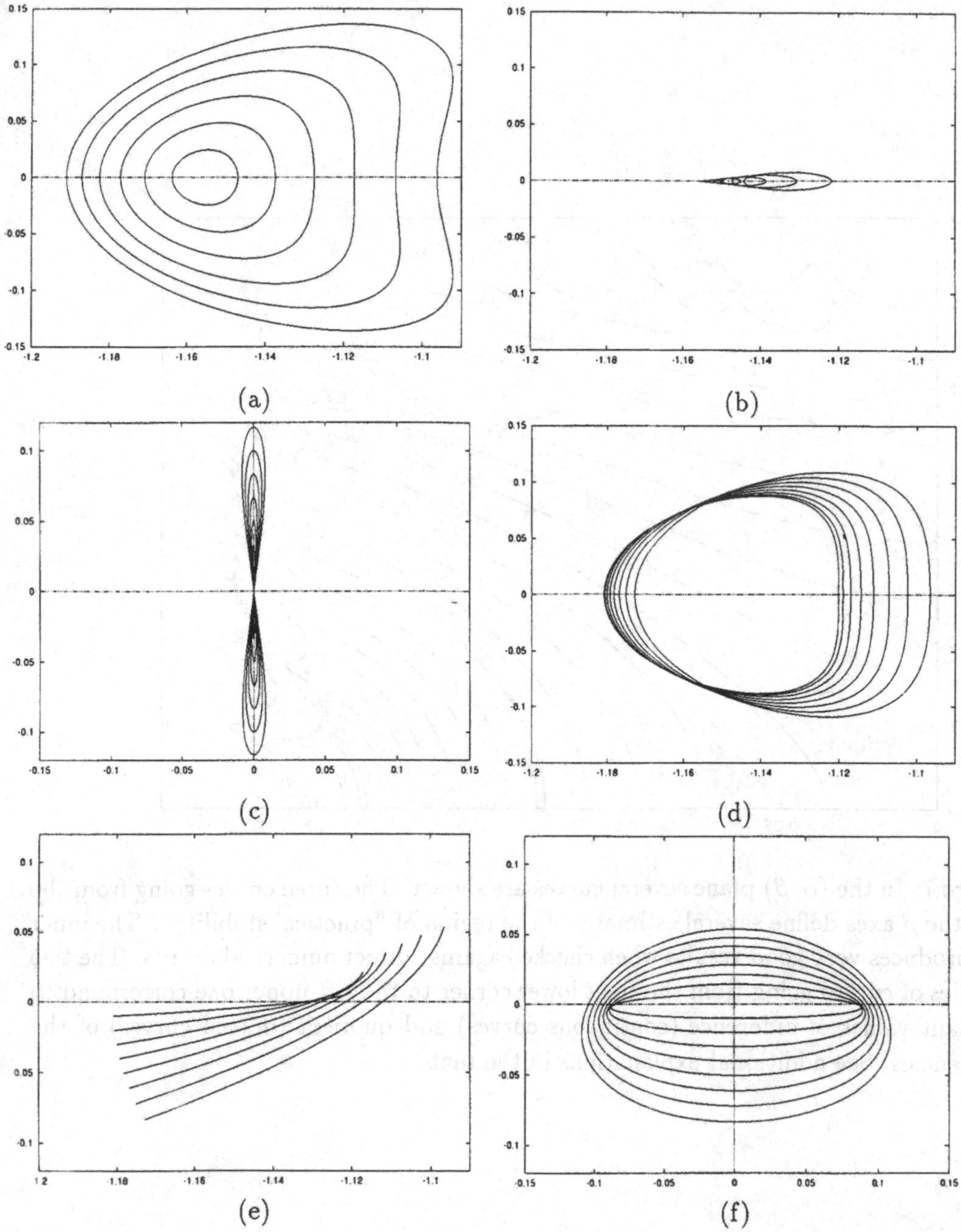

Figure 6: Several projections of periodic orbits obtained using the representation (8): (a) (X, Y) projections of some planar orbits for $\alpha = 0.05, 0.1, 0.15, 0.2, 0.25, 0.3$. (b) and (c) display the (X, Y) and (Y, Z) projections, respectively, for the values of $\beta = 0.1, 0.2, 0.3, 0.4, 0.5, 0.6, 0.7$. (d),(e) and (f) show the (X, Y), (X, Z) and (Y, Z) projections of orbits of the halo family for $\beta = 0.05, 0.1, 0.15, 0.2, 0.25, 0.3, 0.35, 0.4$.

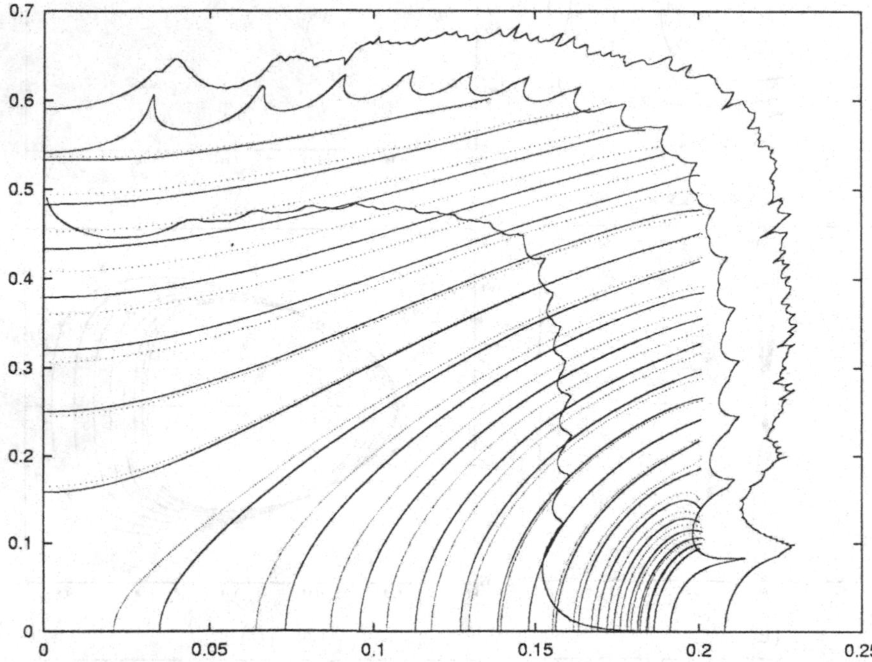

Figure 7: In the (α, β) plane several curves are shown. The three curves going from the α to the β axes define several estimates of the region of "practical stability". The inner one produces very good results when checked against direct numerical results. The two families of curves going from the right lower corner to the left upper one correspond to constant values of difference (continuous curves) and quotient (dotted curves) of the frequencies. See additional explanations in the text.

In the same figure appear two more families of curves. They correspond to constant values of the difference and the quotient of d and f. The values displayed for the difference range from 0.005 to 0.110 with stepsize 0.005 . When the difference increases the curves move from the right lower corner to the left upper one. For the quotient the range goes from 1.0025 to 1.0625 with stepsize 0.0025 .

Further considerations and the applicability of all these families of orbits to spacecraft missions can be found in the references [28], [31],[34] and [38].

3.4 The reduction to the central manifold around L_2

We want to understand how the orbits computed in the previous subsections are organized, to have a global description of a neighborhood of L_2.

First of all we think about skipping the unpleasant unstable terms of the Hamiltonian. This is accomplished by the reduction to the central manifold W^c of dimension 4. We describe how to carry out the computations. After perfoming a linear change of variables going from x, y, p_x, p_y to x_1, x_2, y_1, y_2 and putting $x_3 = z$, $y_3 = p_z$, we have that the Hamiltonian K given by (13) can be written as

$$M = \lambda x_1 y_1 + \frac{1}{2}\omega_1(x_2^2 + y_2^2) + \frac{1}{2}\omega_2(x_3^2 + y_3^2) + \sum_{j\geq 3} M_j(x_1, x_2, x_3, y_1, y_2, y_3),$$

where M_j denotes a homogeneous polynomial of degree j.

Then we pass to an intermediate normal form just trying to cancel in M all the terms such that the total degree in the (x_1, y_1) variables is 1. This is accomplished by making a canonical transformation with a suitable generating function (to be determined). This is possible because there are not problems of small divisors. The denominators appearing in the determination of the generating function (using some complexification to reduce the amount of computation) are of the form

$$(k_1 - l_1)\lambda + \sqrt{-1}(k_2 - l_2)\omega_1 + (k_3 - l_3)\omega_2,$$

with modulus bounded from below by $|\lambda|$. Let $\widetilde{M}$ be this intermediate normal form Hamiltonian. By rearranging terms it can be written as

$$\widetilde{M} = M^0(x_2, x_3, y_2, y_3) + \sum_{j_1+j_2>1} (x_1^{j_1} y_1^{j_2}) M^{j_1, j_2}(x_2, x_3, y_2, y_3).$$

One should have in mind that the new variables are not the sames as the previous ones, despite we keep the same name. The transformation is close to the identity, the closer as the closer is the orbit to L_2.

The fact that small divisors problems do not show up do not means at all that the procedure is convergent. In general one obtains only a C^∞ manifold. The norm of the coefficients of order n increases as $n!$.

It is immediately seen that $I_1 = x_1y_1$ is a first integral. The reduction to the center manifold is obtained by setting $I_1 = 0$. So it remains a two degrees of freedom Hamiltonian, M^0, and furthermore

$$M^0 = \frac{1}{2}\omega_1(x_2^2 + y_2^2) + \frac{1}{2}\omega_2(x_3^2 + y_3^2) + \sum_{j\geq 3} M_j^0(x_2, x_3, y_2, y_3). \quad (16)$$

We point out that if in (16) we put $x_3 = y_3 = 0$ we have an invariant set under the flow associated to M^0. This is now a 1 degree of freedom Hamiltonian. The orbits are simply the Lyapunov periodic orbits around L_1.

By keeping $M^0 = h$ fixed we can use $x_3 = 0$ with $y_3 > 0$ as surface of section. This section, Σ, is not global because it fully contains the corresponding Lyapunov orbit. But we can think that all the points of this orbit (in the boundary of the section) are identified. So we should obtain an S^2 sphere (the classical Hopf fibration) that can be seen as an space of "osculating orbits". For some value of h, h_H say, the level $M^0 = h$ contains the Lyapunov orbit bifurcating to halo orbits. From this value on, the section Σ contains two fixed points associated to the two (symmetrical) halo orbits. It is known that, at least for moderate values of h, they are of elliptic type. At least for values of h not too far from h_H we have in S^2 four fixed points. Two of them correspond to the Lyapunov orbit (hyperbolic) and to the almost vertical periodic orbit (elliptic), respectively. The other two are the halo orbits (elliptic).

A program carrying out this reduction has been developed, implemented and run (up to total order 16). Let ξ and η denote the new variables after skipping the hyperbolic terms in the intermediate normal form. When $M^0(\xi, \eta)$ is available we can do several simulations. The figure 8 shows the result of these simulations for a value of the energy such that the halo orbits are clearly visible. This corresponds to a Poincaré section through $\xi_2 = 0$ (close to $z = 0$ in the initial coordinates). Near the center of the figure one can see a fixed point. It corresponds to a "vertical" periodic orbit. It is surrounded by invariant tori. They belong to the family computed at the beginning of this section, despite the amplitudes are too large and can not be well approximated by the results of the Lindstedt–Poincaré method. The external curve is the Lyapunov planar orbit sitting on this level of the energy. Two other fixed points correspond to the two halo orbits, which are symmetrical the one from the other with respect to $z = 0$) . They are, in turn, surrounded by invariant 2D tori. Between the 2D tori around the vertical orbit and the ones around the halo orbits there is the trace of the stable and unstable manifold of the planar Lyapunov orbit. However, despite the energy is not close to the one of the L_2 point, even the stochastic zone associated to these manifolds is hard to see.

We note that the figure 8 has been displayed with a scale different from the one used in the figure 6. The variables are the ones corresponding to the intermediate normal form. The Moon is located outside the figure, in the negative vertical axis of figure 8.

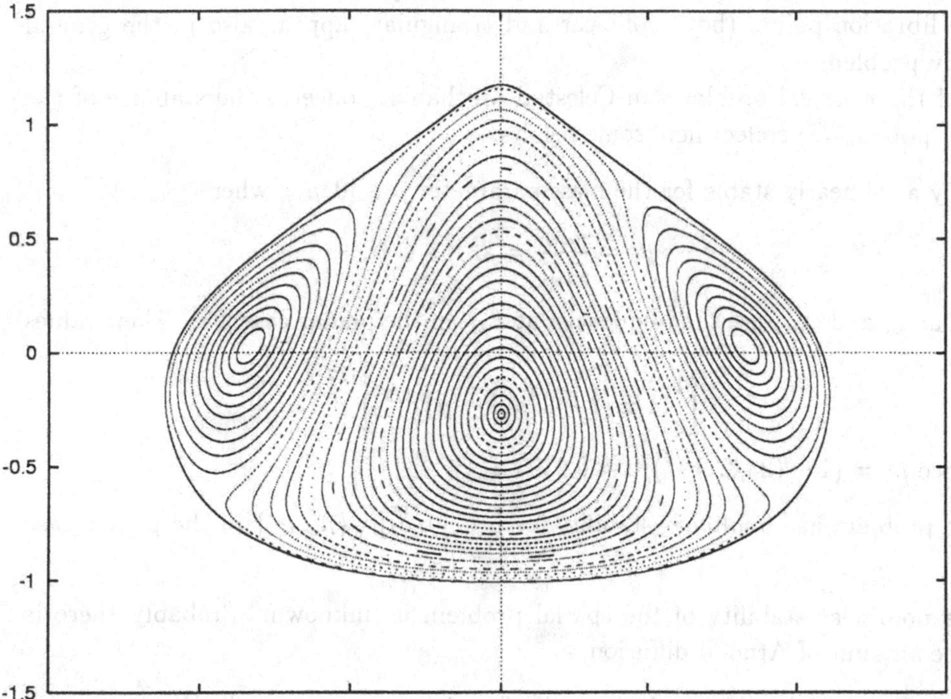

Figure 8: A Poincaré section of the flow of the Hamiltonian reduced to the central manifold of the L_2 point in the Earth-Moon case, for a given level of the energy. See additional explanations in the text.

4 Orbits Near the L_5 Point in the RTBP model. Practical Stability Region Boundaries and Numerical Computation of Unstable Invariant Tori

The RTBP has also libration points sitting outside the line containing the massive bodies. These are the well known Lagrange or triangular points. Using the notation of the previous section they are at rest, in the synodical system, at $(\mu - 1/2, \pm 3^{1/2}/2, 0)$. All these libration points (both collinear and triangular) appear also in the general three body problem.

One of the classical problems in Celestial Mechanics concerns the stability of the triangular points. We collect here some results.

- They are linearly stable for the mass parameter $\mu \in [0, \mu_1]$, where

$$\mu_s = \frac{\omega_{\text{short}}}{\omega_{\text{long}}} = s, \quad s \in \mathbb{N},$$

and ω_{short} and ω_{long} are the frequencies at $L_{4,5}$ in the planar case [98]. Their values are

$$\left[\frac{1}{2}(1 \pm (1 - 27\mu(1-\mu))^{1/2})\right]^{1/2}$$

Hence $\mu_1 = (1 - (23/27)^{1/2})/2 \approx 0.038521$.

- The problem has nonlinear stability for $\mu \in [0, \mu_1] \setminus \{\mu_2, \mu_3\}$ in the planar case [65].

- The nonlinear stability of the spatial problem is unknown. Probably there is some amount of Arnol'd diffusion.

- There is a domain of "practical stability" which is not too small. We summarize how this result is obtained [27], [88].

Let $H = H_2 + H_3 + \ldots$ be the power expansion of the Hamiltonian around $L_{4,5}$, where H_k contains the homogeneous terms of degree k in positions and momenta. By a linear transformation, H_2 can be reduced to

$$-\frac{1}{2}\omega_s(q_1^2 + p_1^2) + \frac{1}{2}\omega_l(q_2^2 + p_2^2) + \frac{1}{2}\omega_z(q_3^2 + p_3^2),$$

where ω_s, ω_l stand for ω_{short}, ω_{long}, and ω_z is the frequency in the vertical direction ($\omega_z = 1$).

A canonical transformation $(q, p) \longrightarrow (Q, P)$, obtained by means of a generating function $G = G_3 + G_4 + \cdots + G_n$, is applied to put H in normal form up to order n. We have the new Hamiltonian

$$\mathcal{H} = N_2 + N_3 + \cdots + N_n + \mathcal{R}_{n+1},$$

where N_k are terms of degree k in the normal form and $N_2 = H_2$. If we only keep $N_2 + N_3 + \cdots + N_n$, the system is integrable. Let

$$I_j = \frac{1}{2}(Q_j^2 + P_j^2),$$

the new momenta. Then, the diffusion of the momenta is due to

$$\dot{I}_j = \{I_j, R_{n+1}\},$$

where $\{\cdot,\cdot\}$ stands for the Poisson bracket. As the remainder $\mathcal{R}_{n+1}$ is of the form $R_{n+1} + R_{n+2} + \cdots$ we introduce some norm $\| R_k \|$. Then one obtains successively bounds for $\| H_k \|, k \geq 2, \| G_k \|, 2 < k \leq n$ (only a finite number of small divisors appears) and for $\| R_k \|, k > n$. The bounds of $\| R_k \|$ are given by means of a recurrence which depends only on the norm of the homogeneous parts of the initial Hamiltonian, $\| H_k \|$, and on the current small divisors which appear till order n.

In a ball of radius ρ in the (Q, P) variables one has

$$|\mathcal{R}_{n+1}| < \sum_{k>n} \rho^k \|R_k\|,$$

where $| \ |$ denotes the sup norm. In a similar way we can bound $|\dot{I}_j|$, the speed of diffusion.

Given T, δ, there exists an initial radius, ρ_0, such that if $(Q, P)_{t=0} \in B_{\rho_0}$ then $(Q, P)_t \in B_{\rho_0(1+\delta)}$ for all $|t| < T$, where B_ρ denotes the ball of radius ρ centered at the origin. We remark that one obtains better results if H_k, N_k, G_k are computed explicitely up to some order for the desired value of μ by means of a symbolic manipulator.

As a result one obtains, in general, Nekhorosev type estimates (i.e., for δ fixed, one has $T \approx \exp(c/\rho^d)$, for some positive constants c, d).

- Numerical simulations ([34], [38]) show a "stable domain" even larger than the one previously described for the planar case [64].

Partial results are known for the elliptic restricted three body problem (i.e., the massive bodies move on elliptic orbits), both concerning local analysis, "practical stability" ([50]) and numerical determination of the "stable domain" ([20], [38]). For the real problems the situation can be more involved. For instance, in the case of the triangular points of the Earth–Moon system, the most important deviation from the RTBP model is due to the influence of the Sun. It has been established that these points are unstable, but regions of practical stability (at least for time intervals very long compared to the duration of space missions) exist at some distance from the triangular points, ranging from 1/4 to 3/4 of the Earth–Moon distance ([38], [96]).

In this section we shall concentrate on the RTBP model for $\mu = 0.0002$. There are several reasons for this choice. From one side it is small enough to consider that perturbation theory can give good approximations. From the other it is big enough so that one needs not a very big amount of time to detect some escapes. Finally it is also close to the actual mass ratio in the Saturn–Titan system.

4.1 Boundaries of the Practical Stability Region

Near the triangular points the system can be considered as three harmonic oscillators, with frequencies at the point equal to ω_{short}, ω_{long} and ω_{vert}, this one being equal to 1. For μ small the values of the ω_{short}, ω_{long} are, respectively, $1 - 27\mu/8 + O(\mu^2)$ and $(27\mu/4)^{1/2} + O(\mu^{3/2})$. Hence, one can also consider, as basic in a neighborhood of L_5, the frequencies ω_{vert}, ω_{long} and $\omega_{\text{vert}} - \omega_{\text{short}}$, whose values are of the order of 1, $\mu^{1/2}$ and μ, respectively.

Provided the frequencies satisfy some conditions (not too close to a strong resonance) there is a "practical stability" region. The previous consideration on the orders of magnitude shows that no strong resonance occurs if μ is small, except the 1 to 1 resonance between the short period and vertical modes. Hence, it takes an extremely large time span to increase the actions by a significant amount. We are interested in the global shape of these regions, because the results of [27] and [88] give rather pessimistic estimates.

To this end the following experiment has been done. Given values of ρ, α and Z, we consider an initial point with synodical coordinates

$$X = \mu + (1+\rho)\cos(2\pi\alpha), \qquad Y = (1+\rho)\sin(2\pi\alpha), \qquad Z,$$

and with zero initial velocity: $\dot{X} = 0, \dot{Y} = 0, \dot{Z} = 0$. The L_5 point corresponds to $\rho = 0$, $\alpha = 1/3$ and $Z = 0$. Then a numerical integration is started up to a given final time. Points close to L_5 just remain moving around it (typically on a 3D torus). But if the initial conditions are sufficiently far from the libration point, the projection of the orbit on the (X, Y) plane crosses the X axis and then it can be considered as escaped from a "large" vicinity of L_5. Due to the shape of some large tori which seem to exist even for large values of Z, it is convenient to take as "escape" criterion from the vicinity of L_5 the condition $Y(t) < Y^*$ for some negative value of Y^*. As a practical value one has taken $Y^* = -0.5$. For the current value of μ we have found orbits lying on tori "around" L_5 such that $Y(t)$ reaches values up to -0.233 . The time span used for the computations is 10,000 times the period of the massive bodies. In the case Saturn–Titan, this means more than 400 years. This can seem a short time interval, but see the comments below.

The figure 9 displays part of the results of this experiment. It gives the projection on the (X, Y) plane of the points which do not escape, according to the previous criterion, starting at the planes of Z value equal to 0.0, 0.3, 0.4, 0.5, 0.6, 0.7 and 0.8 . The ones

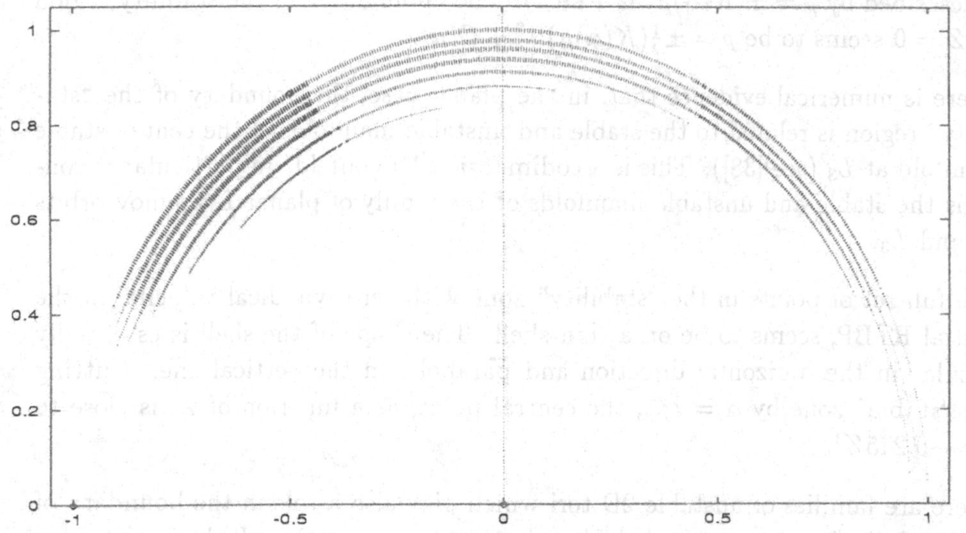

Figure 9: Projections of some sections of the global stability region around L_5 for $\mu = 0.0002$. See additional explanations in the text.

with initial Z equal to 0.1 and 0.2 overlap the projection of the case $Z = 0.0$. The stepsizes in ρ and α have been taken equal to 10^{-5} and 10^{-3}, respectively.

Some remarks must be made concerning the results.

- For small values of μ, as in the present case, the transition from the "stable" zone to instability is quite sharp. This means that increasing the 10,000 revolutions to 100,000, for instance, produces very few losses. On the other hand, most of the points which escape go away from the vicinity of L_5 in a few revolutions. Most of the points which escape after a longer time go away after a time interval of the order of $1/\mu$. This is related to the smallest of the basic frequencies, $\omega_{\text{vert}} - \omega_{\text{short}}$, as said before.

- For the planar case it is seen that the nonescaping points fill an area that we want to identify. To this end take the *zero velocity curve* (z.v.c.) passing through the collinear point L_3, located beyond the larger body at a distance rather close to the distance between the massive bodies. The z.v.c. corresponds to points (X, Y) at which the massless particle at rest has a given value of the energy, in our case the same energy that at L_3. The upper part of this curve (on $Y > 0$) goes, in the angular variable, from $\alpha = 0$ up to a value which tends to $\alpha_{lim} = \cos^{-1}(-(2^{1/2} - 0.5))/\pi \approx 156.09$ degrees, when μ tends to zero. Consider the curve with $\rho = 0$ (that is, at unit distance from the larger body). Then the z.v.c. is located at a distance $O(\mu^{1/2})$ from it. More concretely, there is a function $K(\alpha)$ having a double zero at $\alpha = 0$ and a simple one at $\alpha = \alpha_{lim}$ such that the z.v.c.

is described by $\rho = \pm(K(\alpha)\mu)^{1/2} + O(\mu)$. The "boundary" of the stability region on $Z = 0$ seems to be $\rho = \pm\frac{1}{2}(K(\alpha)\mu)^{1/2} + O(\mu)$.

- There is numerical evidence that, in the planar case, the boundary of the "stability" region is related to the stable and unstable manifolds of the center–stable manifold at L_3 (see [38]). This is a codimension 1 manifold. In particular it contains the stable and unstable manifolds of the family of planar Lyapunov orbits around L_3.

- The full set of points in the "stability" zone with zero synodical velocity, in the spatial RTBP, seems to be on a thin shell. The shape of the shell is essentially circular in the horizontal direction and parabolic in the vertical one. Cutting the "stable" zone by $\alpha = 1/3$, the central point, as a function of Z, is close to $\rho = -0.245Z^2$.

- There are families of unstable 2D tori which play also a role in the boundary of the "stability" region, as we shall show in the next subsection. If the initial value of Z is small these tori reach a vicinity of the L_3 point. Their sections through $Z = 0$ have positive X values. For large initial values of Z, the sections through $Z = 0$ of the 2D tori which seem to be at the "boundary" of the stable region have negative X values.

Summarizing, according to the numerical experiments in the planar case, most of the stable motions take place in tori 2D and they are "bounded" by planar periodic orbits of the Lyapunov family around L_3. In the spatial case, most of the stable motions take place in tori 3D and they are "bounded" by 2D tori associated to the 1 to 1 resonance between the short period and the vertical frequencies. These 2D tori are located at different places depending, mainly, on the initial value of Z. Of course, high order resonances play a role and give rise also to chains of heteroclinic connections taking points away, probably from any neighborhood of any stable point, but this requires a very large amount of time and this mechanism can only be seen for rather large values of μ (say, of the order of 10^{-2}) .

4.2 Unstable 2D Tori: Detection and numerical computation

The figure 10 shows the first 100,000 iterates of the Poincaré section through $Z = 0$ of points starting at rather close points. The first one has been computed with $\rho = -0.1506066340$, $\alpha = 1/3$, $Z = 0.8$ and zero synodical velocity. The second one differs only in the last digit of ρ, which has the value -0.1506066339 . They show, not only the sensitive dependence with respect to the initial conditions, but also the existence of a 2D torus which separates both kinds of motion.

To make last assertion more clear figure 11 shows the first iterates of the Poincaré map, starting at both initial conditions, on the same plot. After passing close to a

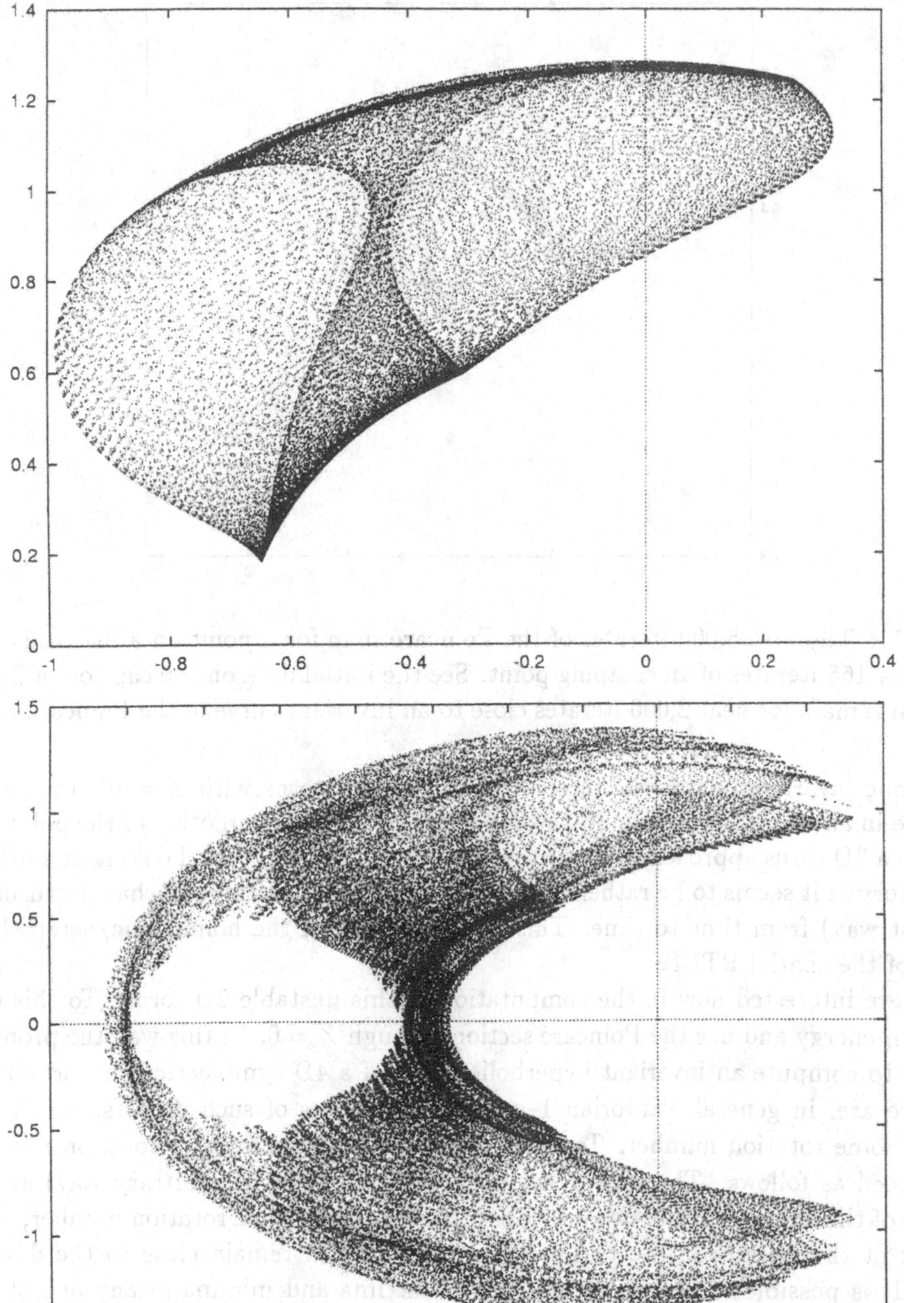

Figure 10: The first 100,000 iterates of the Poincaré map through $Z = 0$ starting with zero synodical velocity at $\alpha = 1/3$, $Z = 0.8$. Top: initial $\rho = -0.1506066340$. Bottom: initial $\rho = -0.1506066339$.

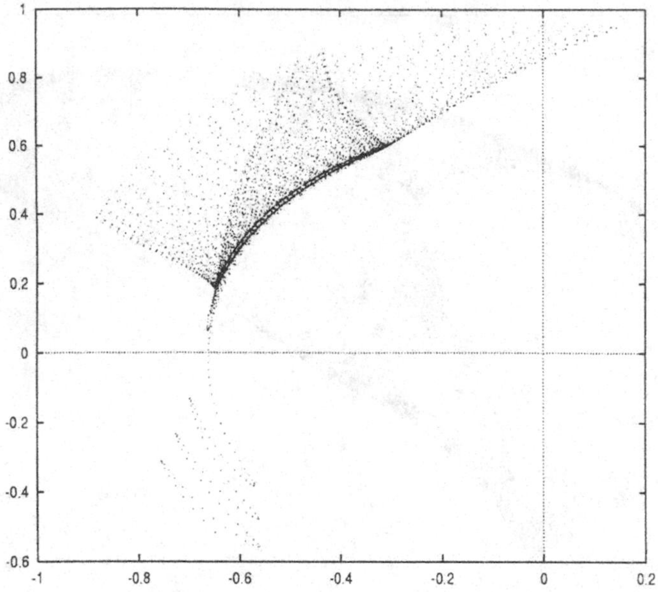

Figure 11: The first 5,000 iterates of the Poincaré map for a point on a 3D torus and the first 4,165 iterates of an escaping point. See the initial data on the caption of figure 7 . Both remain for near 2,000 iterates close to an invariant curve of the Poincaré map.

hyperbolic invariant curve (the intersection of the 2D torus with $Z = 0$), the orbits separate in an exponential way. It is worth to mention that the motion of the point not lying on a 3D torus approaches, in the sequel, several other 2D tori. Looking at suitable time intervals it seems to be rather close to some 3D tori, but this one changes (usually in a fast way) from time to time. This is an evidence of the homoclinic/heteroclinic tangle of the spatial RTBP.

We are interested now in the computation of this unstable 2D torus. To this end we fix an energy and use the Poincaré section through $Z = 0$. In this way the problem reduces to compute an invariant hyperbolic curve of a 4D symplectic transformation. As there are, in general, cantorian 1–parameter families of such objects, we should also fix some rotation number. To select some values of energy and rotation number we proceed as follows. The energy has been choosen (in some arbitrary way) as the average of the energies of the two previous orbits. To have some rotation number, first, we look at the iterates of the two previous orbits which remain close to the desired curve. It is possible to obtain the successive maxima and minima in any one of the current variables ($X, Y, \dot{X}, \dot{Y}$, the value of $\dot{Z}$ being derived from the energy relation (12)) by using interpolation of the different data versus number of iteration. This gives a rotation number $\nu \approx 0.02189307$, that we shall use in the sequel. It is also important the inverse of this number $N = 1/\nu \approx 45.67656$.

To compute a periodic orbit (either stable or not) for a map is, in theory, a simple problem. Assume we denote the map as T and we look for a periodic point of minimal period k. If p is a point in the orbit, we set the equation $T^k p - p = 0$ and we try to solve it by using, for instance, Newton's method. To have success we need some conditions on the properties of the orbit and the initial data should be close enough to p. Furthermore some technical difficulties can occur, for instance if some of the eigenvalues of $D(T^k)(p)$ is very large. In that case a common and very efficient procedure, to solve this difficulty, is based on *parallel shooting* (see, for instance, [97] and also [28]).

The problem is different in the case of an invariant curve. Of course, one can use expansions, as we did in the previous section, if we have some information on a nearby problem, or if the curves are close to some known point. Another possibility is to look for a representation of the coordinates of the curve as a (truncated) Fourier series and to set up a system of equations for the coefficients. This is efficient when the expression of the map is available explicitely. In the present case we want to proceed by direct numerical methods.

As we do not have a return map, as in the case of the periodic orbit, it is possible to *synthetize* it. We look for a fixed point $\mathcal{R}(p) = p$ of the return map $\mathcal{R}$, where $\mathcal{R} = T^N$ with $N = 45.67656$. Certainly we can not compute the power 45.67656 of the Poincaré map, but it is possible to compute the powers in a range $[N_1, N_2]$ containing the desired value of N and compute $\mathcal{R}$, as well as its differential, by interpolation. However the numerical problem has not an unique solution. Indeed, all the points on the curve must be fixed points of the synthesized return map. To determine an unique solution we can fix some of the coordinates of the point p.

In the present problem we have fixed the value of $Y = Y^* = 0.3897977$. This is an arbitrary choice because any choice not too close to the extrema values of Y on the curve could be used, The value that has been taken corresponds to one of the iterates of the Poincaré map for the point sitting on a 3D torus. This iterate will be taken as starting point of the Newton method. Only 3 coordinates $(X, \dot{X}, \dot{Y})$ remain now free, and they are obtained successfully. Note that the Y coordinate of the return map is not asked to coincide with Y^*. This should happen automatically, and is being used as a check of the computations. The Newton process is stopped when the return errors are below 10^{-13}. This is attained in 4 iterates of the procedure. The check of the initial and final values of Y gives also a difference below 10^{-13}. The results are shown in figure 12.

The differential of the return map $D\mathcal{R}$ also provides an estimate of the hyperbolicity of the invariant curve. The dominant eigenvalue is $\lambda_{max} \approx 1.24568$, which implies a dominant eigenvalue for the Poincaré map along the curve $\lambda_{map} \approx 1.004821$. This is a very mild instability which makes the numerical computations much simpler.

We want to remark that it is possible to use a continuation procedure to compute other invariant curves, either by changing the energy, the rotation number or both. When the rotation number takes a rational value (and in the computer approximation

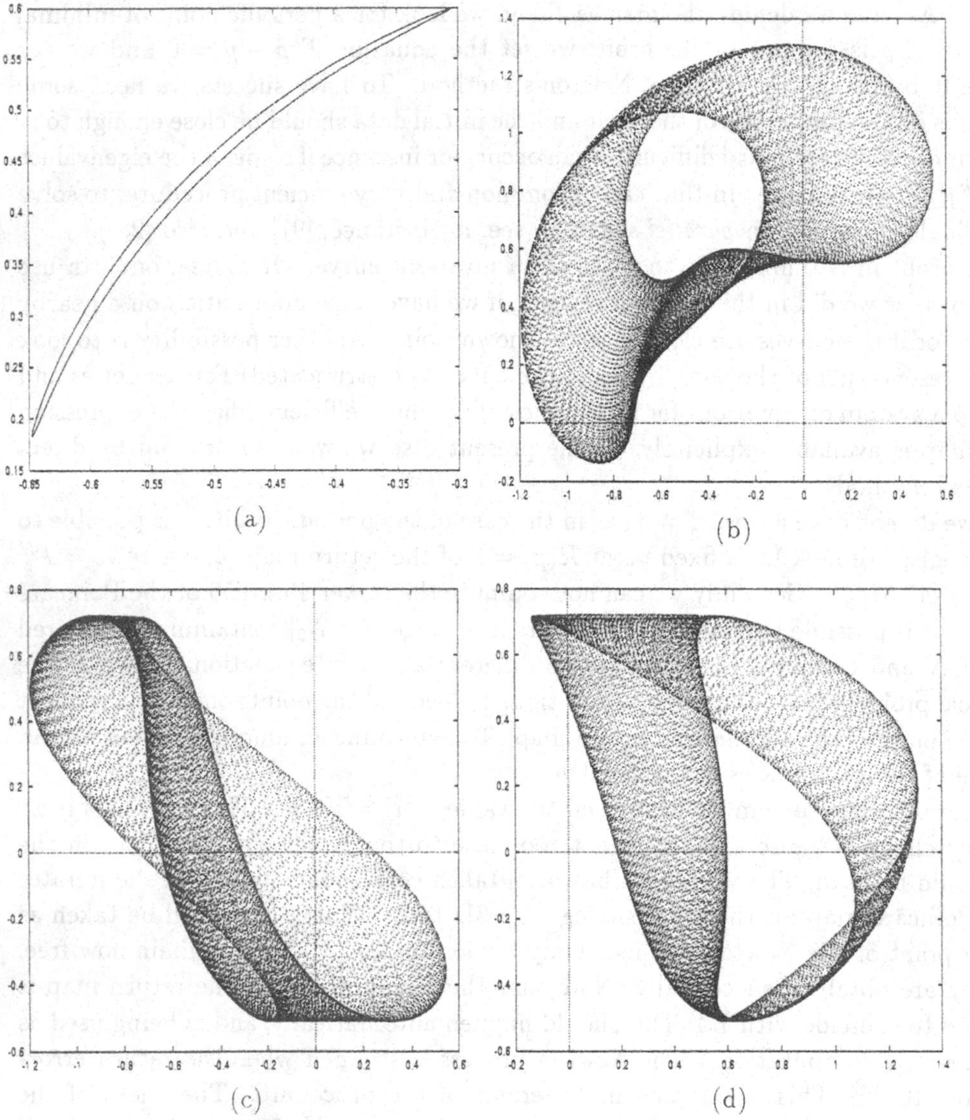

Figure 12: An invariant hyperbolic curve for the Poincaré map through $Z = 0$ of the RTBP as is seen in figure 8 . (a) the (X, Y) projection of the curve. (b), (c) and (d) display the (X, Y), (X, Z) and (Y, Z) of points on the corresponding 2D torus.

this is always the case!), the invariant curve generically do not exists and it is probably replaced by a couple of periodic orbits of elliptic–hyperbolic and hyperbolic–hyperbolic type. Hence, there is no reason at all that any point of these periodic orbits be located on the hyperplane $Y = Y^*$. However, the numerical method can still produce some result without any problem. I tried this both with rotation number $\nu = 1/45$ and $\nu = 1/46$ without trouble. The reason is, obviously, that the return map (T^{45} or T^{46} in this case) has a so slow dynamics that it is below the tolerance of the method (10^{-13}). To see that the method fails in this case it is enough to do all the computations with a higher accuracy.

4.3 A Generalization: Numerical Computation of invariant Tori

The procedure described in the previous subsection can be generalized to the direct numerical computation of invariant tori of flows and maps under reasonable conditions. For simplicity it is sketched here in the case of maps. The case of flows càn be reduced to this one by using a Poincaré section. It is important to mention that only a local section is needed.

First we reformulate the procedure above. Let $\mathbb{T}^1 = \mathbb{R}/\mathbb{Z}$ be the standard 1D torus which is considered as parametrized by values ranging in $[-0.5, 0.5)$. The map on the invariant curve is conjugated to the map $x \longrightarrow x + \nu$ and, hence, after k iterates the current value of x is $x + k\nu \pmod 1$. By taking several values of k (in the concrete example above we have taken k from 38 to 54), one obtains several values of x around 0. By interpolation it is possible to compute the return map $\mathcal{R}$.

In the general case, we look for an m-dimensional torus where the dynamics is conjugated to the map $x \longrightarrow x + \nu$ of the standard torus $\mathbb{T}^m = \mathbb{R}^m/\mathbb{Z}^m$, ν being now in $\mathbb{R}^m$ and satisfying nonresonance conditions: $(k, \nu) \in \mathbb{Z}$ for some $k \in \mathbb{Z}^m$ implies $k = 0$, where $(,)$ denotes the scalar product in $\mathbb{R}^m$. We assume that all the constrains due to first integrals have been taken away and we consider directly the reduced problem. As we are interested on an m-dimensional torus we fix m coordinates of the point p we are looking for. Furthermore we assume that the total dimension n is at least $2m$ and that a variation of the additional $n - m$ coordinates produces a variation of the frequencies ν, provided they exist (probably they do only exist on some cantorian set), which is not degenerate, in the sense that the differential of the frequencies with respect the additional variables has maximal rank. It is also supposed that the remaining $n - 2m$ coordinates, if any, are associated to eigenvalues different from 1. Hence we are assuming, implicitely, that the dynamics along the torus is reducible to constant coefficients (see [49] and [51] in this direction).

Then one considers different iterates of T with number of iterations $k_1, k_2, ..., k_q$ such that on the standard torus the related values of x ($k_1\nu, ..., k_q\nu \pmod 1$)) fall on a prescribed neighborhood of 0. Then, it is possible to synthesize the return map $\mathcal{R}$ as the value at zero of some interpolating function, taking the value $T^{k_j}(p)$ at the point $k_j\nu, j = 1, ..., q$. The condition to be imposed on p is that the $n-m$ free coordinates should satisfy $\mathcal{R}(p) = p$, the remaining m coordinates being used as an additional check. This equation can be solved by Newton's method and the required differential is obtained also by interpolation. Finally it is possible to obtain the eigenvalues of T in the additional $n-2m$ coordinates from the differential of the return map.

This procedure is being tested now in different examples, either conservative or not.

Acknowledgements

The work presented on these lectures has been elaborated along several years. The examples were presented in [94]. The latest results have been supported by DGICYT Grant PB 94–0215 (Spain). Partial support of the EC grant ERBCHRXCT940460, the catalan grant CIRIT 1996S0GR-00105 and the European INTAS grant 93-0339ext is also acknowledged. I thank G. Gómez, À. Jorba and J. Masdemont for many helpful discussions.

References

[1] Abraham, R., Marsden, J.: *Foundations of Mechanics.* Addison-Wesley, 1978.

[2] Arnol'd, V. I.: Proof of a Theorem of A. N. Kolmogorov on the Persistence of Quasi-Periodic Motions under Small Perturbations of the Hamiltonian. *Russ. Math. Surveys,* **18** (1963), 9–36.

[3] Arnol'd, V. I.: Small divisor problems in classical and celestial mechanics. *Russ. Math. Surveys,* **18** (1963), 85–192.

[4] Arnol'd, V. I.: Instability of Dynamical Systems with Several Degrees of Freedom. *Sov. Math. Dokl.,* **5** (1964), 581–585.

[5] Arnold, V.I., Avez, A.: *Ergodic Properties of Classical Mechanics.* Benjamin, N. Y., 1968.

[6] Arnol'd, V. I., Kozlov, V. V., Neishtadt, A. I.: *Dynamical Systems III.* Springer-Verlag, 1988.

[7] Benest, D., Froeschlé, C. (editors): *Modern methods in celestial mechanics,* Editions Frontières, 1990.

[8] Broer H. W., Huitema G. B., Sevryuk M. B.: *Quasi-Periodic Motions in Families of Dynamical Systems: Order amidst Chaos,* Lecture Notes in Math., Vol. 1645, Springer, 1996.

[9] Bruno A.: *Local Methods in Nonlinear Differential Equations.* Springer-Verlag, 1989.

[10] Celletti A., Chierchia L.: Construction of Analytic KAM Surfaces and Effective Stability Bounds. *Commun. Math. Phys.* **118** (1988), 119–161.

[11] Celletti A., Giorgilli A.: On the Stability of the Lagrangian Points in the Spatial Restricted Three Body Problem. *Celestial Mechanics* **50** (1991), 31–58.

[12] Companys, V., Gómez, G., Jorba, À., Masdemont, J., Rodríguez, J., Sim'o, C.: Orbits Near the triangular Libration Points in the Earth-Moon System. Proceedings of the *44th Congress of the International Astronautical Federation, 1993, Graz, Austria*, IAF 1993, Paris.

[13] Companys, V., Gómez, G., Jorba, Á., Masdemont, J., Rodríguez, J., Sim'o, C.: Use of the Earth-Moon Libration Points for Future Missions. AAS 95-404, (1966) 1655-1666.

[14] Delshams, A., Gelfreich, V., Jorba, À., Seara, T., M.: Exponentially Small Splitting of Separatrices under Fast Quasiperiodic Forcing. To appear in *Comm. Math. Phys.*

[15] Delshams, A., Gelfreich, V., Jorba, À., Seara, T., M.: Lower and Upper Bounds for the Splitting of Separatrices for the Pendulum under a Fast Quasiperiodic Forcing. *Electronic Research Anouncements of the AMS*, **3** (1997), 1–10.

[16] Delshams, A., Gutiérrez, P.: Effective Stability and KAM Theory. *J. Diff. Eq.*, **128**, (1996), 415–490.

[17] Delshams, A., Gutiérrez, P.: Estimates on Invariant Tori Near an Elliptic Equilibrium Point of a Hamiltonian System. *J. Diff. Eq.*, **131** (1996), 277–303.

[18] Delshams, A., Seara, T., M.: An Asymptotic Expression for the Splitting of Separatrices of the Rapidly Forced Pendulum. *Commun. Math. Phys.*, **150** (1992), 433–463.

[19] Díez, C., Jorba, À., Simó, C.: A Dynamical Equivalent to the Equilateral Libration Points of the Earth-Moon System. *Cel. Mech.*, **50** (1991), 13–29.

[20] Dvorak, R., Lohinger, E.: Stability zones around the triangular Lagrangian points. In *Predictability, stability and chaos in N-body dynamical systems*, Ed. by A. E. Roy, Plenum Press, 439–446, 1991.

[21] Eliasson, L., H.: Perturbations of Stable Invariant Tori for Hamiltonian Systems. *Ann. Sc. Norm. Super. Pisa, Cl. Sci., Ser. IV*, **15**:1 (1988), 115–147.

[22] Flury, W., Gómez, G., Llibre, J., Martínez, R., Rodríguez, J., Simó, C.: Relative motion near the triangular libration points in the earth-moon system. In *Proceed. of the Optical Interferometry Workshop*, Granada 1987, *ESA-SP* **273**.

[23] Font, J.: *Variedades invariantes y orbitas homoclinicas de L_3 en el Problema Restringido de Tres Cuerpos.* Master Thesis, Univ. Barcelona, 1984.

[24] Fontich, E., Simó, C.: Invariant Manifolds for Near Identity Differentiable Maps and Splitting of Separatrices. *Erg. Th. & Dyn. Systems*, **10** (1990), 319–346.

[25] Fontich, E., Simó, C.: The Splitting of Separatrices for Analytic Diffeomorphisms. *Erg. Th. & Dyn. Systems*, **10** (1990), 295–318.

[26] Giorgilli, A., Galgani, L.: Formal integrals for an autonomous Hamiltonian system near an equilibrium point. *Celest. Mech.* **17** (1978), 267–280.

[27] Giorgilli, A., Delshams, A., Fontich, E., Galgani, L., Simó, C.: Effective Stability for a Hamiltonian System near an Elliptic Equilibrium Point, with an Application to the Restricted Three Body Problem. *J. Diff. Eq.*, **77** (1989), 167–198.

[28] Gómez, G., Llibre, J., Martínez, R., Simó, C.: *Station Keeping of Libration Point Orbits.* ESOC Contract 5648/83/D/JS(SC), Final Report, xii+689 p., 1985.

[29] Gómez, G., Llibre, J., Martínez, R., Simó, C.: Station Keeping of a quasiperiodic halo orbit using invariant manifolds. In *Proceed. 2nd Internat. Symp. on spacecraft flight dynamics*, Darmstadt, october 86, *ESA SP* **225**, 65–70, 1986.

[30] Gómez, G., Llibre, J., Martínez, R., Rodríguez, J., Simó, C.: On the optimal station keeping control of halo orbits. *Acta Astronautica*, **15** (1987), 391–397.

[31] Gómez, G., Llibre, J., Martínez, R., Simó, C.: *Study on Orbits near the Triangular Libration Points in the perturbed Restricted Three-Body Problem.* ESOC Contract 6139/84/D/JS(SC), Final Report, iv+238 p., 1987.

[32] Gómez, G., Jorba, À., Masdemont, J., Simó, C.: Moon's Influence on the Transfer from the Earth to a Halo Orbit around L_1. In A.E. Roy, editor, *Predictability, Stability and Chaos in the N-Body Dynamical Systems*, 283–290, Plenum Press, 1991.

[33] Gómez, G., Jorba, À., Masdemont, J., Simó, C.: Quasiperiodic Orbits as a Substitute of Libration Points in the Solar System. In A.E. Roy, editor, *Predictability, Stability and Chaos in the N-Body Dynamical Systems*, 433–438, Plenum Press, 1991.

[34] Gómez, G., Jorba, À., Masdemont, J., Simó, C.: *Study Refinement of Semi-Analytical Halo Orbit Theory.* ESOC Contract 8625/89/D/MD(SC), Final Report, 213 p., 1991.

[35] Gómez, G., Jorba, À., Masdemont, J., Simó, C.: A Dynamical Systems Approach for the Analysis of the SOHO Mission. In *Proceedings of the Third International Symposium on Spacecraft Flight Dynamics*, 449–456, *ESA SP* **326**, Darmstadt, 1992.

[36] Gómez, G., Jorba, À., Masdemont, J., Simó, C.: A Quasiperiodic Solution as a Substitute of L_4 in the Earth-Moon System. In *Proceedings of the Third International Symposium on Spacecraft Flight Dynamics*, 35–42, *ESA SP* **326**, Darmstadt, 1992.

[37] Gómez, G., Jorba, À., Masdemont, J., Simó, C.: Study of the transfer from the Earth to a Halo orbit around the equilibrium point L_1. *Cel. Mechanics*, **55** (1993), 1–22.

[38] Gómez, G., Jorba, Á., Masdemont, J., Simó, C.: *Study of Poincaré Maps for Orbits Near Lagrangian Points.* ESOC Contract 9711/91/D/ID(SC), Final Report, 276 p., 1993.

[39] Gómez, G., Jorba, À., Masdemont, J., Simó, C.: Study of the transfer between halo orbits in the solar system, *Advances in the Astronautical Sciences*, **84** (1994), 623-638.

[40] Gómez, G., Jorba, À., Masdemont, J., Simó, C.: Practical stability regions related to the Earth-Moon triangular libration points. *Space Flight Dynamics* (1994), 11–17.

[41] Gómez, G., Jorba, À., Masdemont, J., Simó, C.: Mission analysis for orbits in a vicinity of the Earth-Moon triangular libration points. *Space Flight Dynamics* (1994), 18–23.

[42] Gómez, G., Jorba, À., Masdemont, J., Simó, C.: Study of the transfer between halo orbits. Preprint, 1997.

[43] Gómez, G., Masdemont, J., Simó, C.: Lissajous orbits around halo orbits. Preprint, 1997

[44] Henrard, J.: On Brown's Conjecture. *Celest. Mechanics*, **31** (1983), 115–122.

[45] Hofer, H., Zehnder, E.: *Symplectic Invariants and Hamiltonian Dynamics*. Birkhäuser, Basel, 1994.

[46] Holmes, P., Marsden, J.: Melnikov's method and Arnol'd diffusion for perturbations of integrable Hamiltonian systems, *J. Math. Phys.* **23** (1982), 669–675.

[47] Irigoyen, M., Simó, C.: Non-integrability of the $J2$ problem. *Celestial Mechanics*, **55** (1993), 281–287.

[48] Jorba, À., Ramírez-Ros, R., Villanueva, J.: Effective Reducibility of Quasiperiodic Linear Equations close to Constant Coefficients. *SIAM Journal on Mathematical Analysis* **28** (1997), 178–188.

[49] Jorba, À., Simó, C.: On the reducibility of linear differential equations with quasiperiodic coefficients. *J. Diff. Eq.*, **98**, (1992), 111–124.

[50] Jorba, À., Simó, C.: Effective stability for periodically perturbed Hamiltonian systems. In *Hamiltonian Mechanics: Integrability and Chaotic Behavior*, Editor J. Seimenis, Plenum Press, 245-252, 1994.

[51] Jorba, À., Simó, C.: On Quasi-Periodic Perturbations of Elliptic Equilibrium Points. *SIAM Journal on Mathematical Analysis*, **27** (1996), 1704–1737.

[52] Jorba, À., Villanueva, J.: On the Persistence of Lower Dimensional Invariant Tori under Quasiperiodic Perturbations. *Journal of Nonlinear Science* (to appear).

[53] Jorba, À., Villanueva, J.: On the Normal Behaviour of Partially Elliptic Lower Dimensional Tori of Hamiltonian Systems, *Nonlinearity* (to appear).

[54] Jorba, À., Villanueva, J.: Numerical Computation of Normal Forms around some Periodic Orbits of the Restricted Three Body Problem. Preprint (1997).

[55] Kolmogorov, A., N.: The General Theory of Dynamical Systems and Classical Mechanics. *Proc. of the 1954 Intern. Congress Math.*, (1957), 315–333.

[56] Lazutkin, V. F., Simó, C.: Homoclinic orbits in the complex domain. *Intern. Journal on Bifurcation and Chaos* (to appear). (UB preprint No. 203 (1996)).

[57] Llave, R., Rana, D.: Accurate Strategies for Small Denominator problems, *Bull. Amer. Math. Soc.*, **22** (1990), 85–90.

[58] Llave, R., Rana, D.: Accurate Strategies in KAM Problems and their Implementation. In *Computer Aided Proofs in Analysis*, K. Meyer & D. Schmidt (Ed.), The IMA Volumes in Mathematics **28** (1991), 127–146.

[59] Llibre, J., Simó, C.: Homoclinic phenomena in the three body problem. *J. Diff. Eq.*, **37** (1980), 444–465.

[60] Llibre, J., Simó, C.: Oscillatory solutions in the planar restricted three-body problem. *Math. Ann.*, **248** (1980), 153–184.

[61] Llibre, J., Martínez, R., Simó, C.: Transversality of the invariant manifolds associated to the Lyapunov family of periodic orbits near L_2 in the Restricted Three Body Problem. *J. Diff. Eq.*, **58** (1985), 104–156.

[62] Lochak, P., Meunier, C.: *Multiphase Averaging for Classical Systems.* Springer-Verlag, 1988.

[63] MacKay, R. S., Meiss, J. D.: *Hamiltonian Dynamical Systems.* Adam Hilger, 1987.

[64] Mckenzie, R., Szebehely, V.: Non-linear stability motion around the triangular libration points, *Celestial Mechanics* **23**, (1981), 223–229.

[65] Markeev, A. P.: Stability of the triangular Lagrangian solutions of the restricted three body problem in the three dimensional circular case, *Soviet Astronomy* **15**, (1972), 682–686.

[66] Martínez, R., Simó, C.: Blow up of collapsing binaries in the planar three body problem. In C. Albert, editor, *Géométrie sympl. et mécanique*, *Lect. Notes in Math.* **1416**, 255–267, Springer, 1990.

[67] Martínez, R., Simó, C.: A note on the existence of heteroclinic orbits in the planar three body problem. In *Seminar on Dynamical Systems, Euler International Mathematical Institute, St. Petersburg, 1991*, S. Kuksin, V. Lazutkin and J. Pöschel, editors, 129–139, Birkhäuser, 1993.

[68] Meyer, K. R., Hall, G. R.: *Introduction to Hamiltonian Dynamical Systems and the N-Body Problem.* Springer–Verlag, 1992.

[69] Morales, J. J., Simó, C.: Picard-Vessiot theory and Ziglin's theorem. *J. Diff. Eq.*, **107** (1994), 140–162.

[70] Morales, J. J., Simó, C.: Non integrability criteria for Hamiltonians in the case of Lamé normal variational equations. *J. Diff. Eq.* **129** (1996), 111–135.

[71] Morbidelli, A., Giorgilli, A.: Superexponential Stability of KAM Tori, *J. Statist. Phys.* **78** (1995), 1607–1617.

[72] Moser, J.: On Invariant Curves of Area–Preserving Mappings of an Annulus. *Nachr. Akad. Wiss. Göttingen: Math. Phys. Kl. II*, (1962), 1–20.

[73] Moser, J.: Convergent Series Expansions for Quasi-Periodic Motions, *Math. Annalen* **169** (1967), 136–176.

[74] Moser J.: *Lectures on Hamiltonian Systems.* Mem. Am. Math. Soc. 81, 1–60 (1968). Contained in [63].

[75] Moser, J.: Regularization of Kepler's problem and the averaging method on a manifold. *Comm. Pure and Applied Math.* **23** (1970).

[76] Moser, J.: *Stable and Random Motions in Dynamical Systems.* Annals of Mathematics Studies, Princeton Univ. Press, 1973.

[77] Moser, J.: Is the Solar System Stable? *Math. Intelligencer* 1 (1978), 65–71. Contained in [63].

[78] Nekhoroshev, N. N.: An Exponential Estimate of the Time of Stability of Nearly-Integrable Hamiltonian Systems. *Russ. Math. Surveys*, **32** (1977), 1–65.

[79] Perry, A. D., Wiggins, S.: KAM Tori are Very Sticky: Rigorous Lower Bounds on the Time to Move Away from an Invariant Lagrangian Torus with Linear Flow. *Physica* **71 D** (1994), 102–121.

[80] Pöschel, J.: Integrability of Hamiltonian Systems on Cantor Sets, *Comm. Pure Appl. Math.*, **35** (1982), 653–695.

[81] Poincaré, H.: *Les Méthodes Nouvelles de la Mécanique Céleste III.* Gauthier–Villars, 1899.

[82] Rudnev, M., Wiggins, S.: KAM Theory Near Multiplicity one Resonant Surfaces in Perturbations of a-Priori Stable Hamiltonian Systems. Preprint (1996).

[83] Rudnev, M., Wiggins, S.: Existence of Exponentially Small Separatrix Splittings and Homoclinic Connections Between Whiskered Tori in Weakly Hyperbolic Near-Integrable Hamiltonian Systems. Preprint (1997).

[84] Siegel, C. L., Moser, J. K.: *Lectures on Celestial Mechanics.* Springer–Verlag, 1971.

[85] Simó, C.: Homoclinic phenomena and quasi-integrability. In V. Szebehely, editor, *Stability of the Solar System and Its Minor Natural and Artificial Bodies*, 305–316, D. Reidel Pub. Co., Holland, 1985.

[86] Simó, C.: Homoclinic and heteroclinic phenomena in some hamiltonian systems. In K. Meyer and D. Saari, editors, *Hamiltonian Dynamical Systems, Contemporary Math. AMS*, **81** (1988), 193–212.

[87] Simó, C.: Estimates of the error in normal forms of hamiltonian systems. Applications to effective stability. Examples. In A.E. Roy, editor, *Long-term dynamical behaviour of natural and artificial N-body systems*, 481–503, Kluwer, Dordrecht, Holland, 1988.

[88] Simó, C.: Estabilitat de Sistemes Hamiltonians. *Memòries de la Reial Acadèmia de Ciències i Arts de Barcelona*, **48** (1989), 303–348.

[89] Simó, C.: Analytical and numerical computation of invariant manifolds. In [7], 285–330.

[90] Simó, C.: Measuring the lack of integrability in the J_2 problem. In A.E. Roy, editor, *Predictability, Stability and Chaos in the N-Body Dynamical Systems*, 305–309, Plenum Press, 1991.

[91] Simó, C.: Averaging under Fast Quasiperiodic Forcing. In J. Seimenis (Ed.), *Hamiltonian Mechanics: Integrability and Chaotic Behaviour*, Vol **331** of NATO Adv. Sci. Inst. Ser. B Phys., Plenum, New York (1994), 13–34.

[92] Simó, C.: Com entendre el comportament no predictible dels sistemes deterministes. In *Ordre i caos en ecologia*, Publicacions Universitat de Barcelona (1995), 19–54.

[93] Simó, C.: The use of time as a complex variable. *Memòries de la Reial Acadèmia de Ciències i Arts de Barcelona*, **55** (1996), 276–290.

[94] Simó, C.: *Effective computations in Hamiltonian Dynamics.* In Mécanique Céleste, Journée Annuelle de la Société Mathématique de France (1996).

[95] Simó, C.: *An overview on some problems in Celestial Mechanics.* In Summer Course on *Iniciación a los sistemas dinámicos*, organized by the Universidad Complutense de Madrid at El Escorial, July, 1997. Available at `http://www-ma1.upc.es/escorial` .

[96] Simó, C., Gómez, G., Jorba, À., Masdemont, J.: The Bicircular Model near the Triangular Libration Points if the RTBP. In *From Newton to Chaos: Modern techniques for Understanding and Coping with Chaos in N-Body Dynamical Systems*, Ed. A. E. Roy, Plenum Press, 343–370, 1995.

[97] Stoer, J., Bulirsch, R.: *Introduction to Numerical Analysis.* Springer–Verlag, 1983 (second printing).

[98] Szebehely, V.: *Theory of orbits.* Academic Press, 1967.

[99] Whittaker, E. T.: *A Treatise on the Analytical Dynamics of Particles and Rigid Bodies.* Cambridge Univ. Press, 1927.

[100] Wintner, A.: *The analytical foundations of celestial mechanics.* Princeton, 1947.

PART III

STABILITY AND BIFURCATION PROBLEMS

L. Salvadori
University of Trento, Trento, Italy

F. Visentin
University of Naples "Federico II", Naples, Italy

Abstract

The course deals with the stability theory (Parts I,II) and its connection with the bifurcation theory (Part III). Some more recent approaches to stability problems are described, for example by methods which make use of suitable one parameter families of Liapunov functions. Several applications to conservative as well as dissipative mechanical systems with a finite number of degrees of freedom are given. The connection between stability and bifurcation is due to the fact that bifurcation phenomena often occur because of exchange of stability properties under perturbations of the governing equations. Thus stability arguments, in particular appropriate Liapunov functions, may be used not only to analyze the qualitative behavior of the flow near the bifurcating sets, but also to prove the existence of these sets.

PART I

STABILITY PROBLEMS

1. Liapunov stability for ordinary differential equations

Let D be an open set of $\mathbf{R}^n$, $n \geq 1$, which contains the origin 0 and let I be an interval which is unbounded from above. Consider a differential system

$$\dot{x} = f(t, x), \tag{1.1}$$

where $f \in C(I \times D, \mathbf{R}^n)$ is locally Lipschitzian in x, and $f(t, 0) \equiv 0$. Let $\|\cdot\|$ be a norm in $\mathbf{R}^n$ and let ρ be the induced distance. For $a > 0$ denote by $B(a)$ the ball $\{x : \|x\| < a\}$, and by $B[a]$ its closure. Let $\chi = \rho(0, \partial D)$ ($\chi = +\infty$ if $D = \mathbf{R}^n$). Denote by $x(t, t_0, x_0)$ the noncontinuable solution through (t_0, x_0) in $I \times D$, by $J = J(t_0, x_0)$ the interval of existence of $x(t, t_0, x_0)$, and by $J^+ = J^+(t_0, x_0)$ the set $\{t \geq t_0 : t \in J(t_0, x_0)\}$.

*Work supported by the Italian CNR (GNFN) and by MURST (40% funds)

Definition 1.1 *The solution $x \equiv 0$ of (1.1) is said to be:*

(i) *stable if for any $t_0 \in I$ and $\varepsilon \in (0, \chi)$ there exists $\delta = \delta(t_0, \varepsilon) \in (0, \varepsilon)$ such that if $\|x_0\| < \delta$ then $\|x(t, t_0, x_0)\| < \varepsilon$ for $t \geq t_0$;*

(ii) *uniformly stable if property (i) holds with δ independent of t_0;*

(iii) *attractive if for any $t_0 \in I$ there exists a constant $\sigma = \sigma(t_0) \in (0, \chi)$ such that $\|x_0\| < \sigma$ implies $J^+(t_0, x_0) = [t_0, +\infty)$ and $\|x(t, t_0, x_0)\| \to 0$ as $t \to +\infty$, that is*

$$\forall \nu > 0 \; \exists T = T(\nu, t_0, x_0) > 0 \text{ such that } \|x(t, t_0, x_0)\| < \nu \; \forall t \geq t_0 + T;$$

(iv) *uniformly attractive if (iii) holds with σ independent of t_0 and T independent of t_0, x_0;*

(v) *asymptotically stable if it is stable and attractive;*

(vi) *uniformly asymptotically stable if it is uniformly stable and uniformly attractive;*

(vii) *unstable if it is not stable; that is there exist $\eta > 0$ and $t_0 \in I$ such that there exist two sequences $\{x_i\}$ in D, $\|x_i\| \to 0$, $\{t_i\}$ in $J^+(t_0, x_i)$, such that $\|x(t_i, t_0, x_i)\| \geq \eta$ for all $i \in \mathbf{N}$.*

If $f(t + T, x) \equiv f(t, x)$ for some $T > 0$ we may assume $I = \mathbf{R}$. In this case we have $x(t + T, t_0 + T, x_0) \equiv x(t, t_0, x_0)$. As a consequence the stability and the asymptotic stability are always uniform. More particularly when (1.1) is autonomous we use the notation $x(t, x_0)$ for $x(t, 0, x_0)$. In connection with the properties in Definition 1.1 consider now the following functions.

$$\begin{aligned} r(t_0, \varepsilon) &= \sup\{\delta \geq 0 : \|x_0\| \leq \delta \text{ implies } \|x(t, t_0, x_0)\| < \varepsilon \text{ for } t \geq t_0\}; \\ r_u(\varepsilon) &= \inf\{r(t_0, \varepsilon) : t_0 \in I\}; \\ R(t_0) &= \sup\{\eta \geq 0 : \text{ there exist two sequences } \{x_i\} \subset D, \|x_i\| \to 0, \\ &\quad \{t_i\} \subset J^+(t_0, x_i), \text{ such that } \|x(t_i, t_0, x_i)\| \geq \eta \text{ for all } i \in \mathbf{N}\}. \end{aligned}$$

These functions will be called respectively radius of stability, radius of uniform stability, and radius of instability. They reflect in some sense the strength of stability and instability. In particular $x \equiv 0$ is stable if and only if $r(t_0, \varepsilon) > 0$ for any $t_0 \in I$ and $\varepsilon \in (0, \chi)$. In this case the value of r induces a natural ordering for stability. If (1.1) is autonomous $r = r_u$ and R is a constant. Similarly we may define radii for attractivity and uniform attractivity.

Let now $y(t, t_0, x_0)$ be an m–dimensional component of $x(t, t_0, x_0)$ with $m \leq$n. The solution $x \equiv 0$ of (1.1) is said to be stable with respect to y if for any $t_0 \in I$ and $\varepsilon \in (0, \chi)$ there exists $\delta = \delta(t_0, \varepsilon) \in (0, \varepsilon)$ such that if $\|x_0\| < \delta$ then $\|y(t, t_0, x_0)\| < \varepsilon$ for any $t \in J^+(t_0, x_0)$. Analogous definitions occur for partial uniform stability and partial asymptotic stability.

We observe that if $x \equiv 0$ is stable, then for any $t_0 \in I$ and $\varepsilon \in (0, \chi)$ we have $J^+(t_0, x_0) = [t_0, +\infty)$, provided $x_0 \in B(\delta)$. This property is not in general satisfied when the stability occurs only with respect to part of variables.

2. Classical theorems on stability

We recall now some classical theorems of Liapunov direct method in the stability theory. Exhaustive treatments are given in several books. We quote in particular [8,23,32, 41,71,96]. We need some preliminaries. Let J be an open interval and $\varphi : J \to \mathbf{R}$ be a function. The upper right–hand derivative of φ is the function $D^+\varphi : J \to \bar{\mathbf{R}}$ defined by

$$D^+\varphi(t) = \limsup_{h\to 0+} \frac{\varphi(t+h)-\varphi(t)}{h}.$$

The following proposition holds.

Proposition 2.1 *Assume $\varphi : J \to \mathbf{R}$ is continuous. Then: (i) φ is increasing (resp. decreasing) if and only if $D^+\varphi(t) \geq 0$ (resp. $D^+\varphi(t) \leq 0$) for all $t \in J$. (ii) Let $[\alpha, \beta]$ be contained in J. If there exists $k \in \mathbf{R}$ such that $D^+\varphi(t) \geq k$ (resp. $D^+\varphi(t) \leq k$) for all $t \in [\alpha, \beta]$, then $\varphi(\beta) - \varphi(\alpha) \geq k(\beta - \alpha)$ (resp. $\varphi(\beta) - \varphi(\alpha) \leq k(\beta - \alpha)$).*

In the same way we may define the lower right–hand derivative $D_+\varphi$ by

$$D_+\varphi(t) = \liminf_{h\to 0+} \frac{\varphi(t+h)-\varphi(t)}{h}.$$

and the properties given in Proposition 2.1 hold with $D^+\varphi$ replaced by $D_+\varphi$.

Consider now a continuous function $V : I \times D \to \mathbf{R}$ and for any $(t, x) \in I \times D$ consider the noncontinuable solution $x(\tau, t, x)$ of (1.1). We denote by $D^+V : I \times D \to \bar{\mathbf{R}}$ the upper right–hand derivative of V along the solutions of (1.1), that is the function defined by

$$D^+V(t,x) = \limsup_{h\to 0+} \frac{V(t+h, x(t+h,t,x)) - V(t,x)}{h}.$$

Analogously we may define D_+V. Any function $V \in C(I \times D, \mathbf{R})$ which is considered in connection with its derivative D^+V (or D_+V) along the solutions of (1.1) is called a Liapunov function for (1.1). The Liapunov direct method reduces the stability problem to the existence of Liapunov functions V which have along the solutions an appropriate monotonic behavior, then by Proposition 2.1 a convenient derivative D^+V (or D_+V). The method is said to be direct because choosing V in a suitable class of functions the above derivative may be computed without the knowledge of the solutions of (1.1). The following proposition depicts a situation in which this is possible.

Proposition 2.2 *Assume $V \in C(I \times D, \mathbf{R})$ is locally Lipschitzian in x. Then for any $(t, x) \in I \times D$ we have:*

$$D^+V(t,x) = \limsup_{h\to 0+} \frac{V(t+h, x+hf(t,x)) - V(t,x)}{h},$$

$$D_+V(t,x) = \liminf_{h\to 0+} \frac{V(t+h, x+hf(t,x)) - V(t,x)}{h}.$$

If V is C^1 then $D^+V = D_+V = \dot{V}$ with

$$\dot{V} = \frac{\partial V}{\partial t} + \langle \frac{\partial V}{\partial x}, f \rangle.$$

We explicitly point out that the hypothesis in Proposition 2.2 that V is locally Lipschitzian in x is not needed for many of the ensuing general theorems. Only when applications to concrete problems are concerned, it is essential to find the required Liapunov function from a suitable set which allows to obtain the desired information about the derivative D^+V (or D_+V) along the solutions of the given differential equation.

Definition 2.1 *Let $\gamma \in (0,\chi)$. We say that the function $V : I \times D \to \mathbf{R}$ is*

(*i*) *positive (resp. negative) semidefinite on $B[\gamma]$ if $V(t,0) \equiv 0$ and $V(t,x) \geq 0$ (resp. $V(t,x) \leq 0$) in $I \times B[\gamma]$;*

(*ii*) *positive (resp. negative) definite on $B[\gamma]$ if $V(t,0) \equiv 0$ and for any $\alpha \in (0,\gamma)$ there exists $\beta > 0$ such that $t \in I$ and $\|x\| \in [\alpha,\gamma]$ imply $V(t,x) \geq \beta$ (resp. $V(t,x) \leq -\beta$).*

Stating that V is positive definite, i.e. without mentioning $B[\gamma]$, means that V is positive definite on $B[\gamma]$ for some $\gamma \in (0,\chi)$. Similar statements apply to the other properties in Definition 2.1. Let us denote by K the set of the continuous functions c from $\mathbf{R}^+$ into $\mathbf{R}^+$ which are strictly increasing with $c(0) = 0$. One can prove that V is positive (resp. negative) definite on $B[\gamma]$ if $V(t,0) \equiv 0$ and $V(t,x) \geq a(\|x\|)$ (resp. $V(t,x) \leq -a(\|x\|)$) in $I \times B[\gamma]$. Finally, we say that V has the property $P^+(t_0)$ (resp. the property $P^-(t_0)$), $t_0 \in I$, if $V(t,0) \equiv 0$ and for every $\eta \in (0,\chi)$ there exists $x \in B(\eta)$ such that $V(t_0,x) > 0$ (resp. $V(t_0,x) < 0$).

Theorem 2.1 (Liapunov [41]) *Assume there exists for* (1.1) *a function $V \in C(I \times D, \mathbf{R})$ such that: (i) V is positive definite; (ii) D^+V is negative semidefinite. Then $x \equiv 0$ is stable. If in addition (iii) $V(t,x) \to 0$ as $x \to 0$ uniformly in $t \in I$, then $x \equiv 0$ is uniformly stable.*

Actually the addendum in Theorem 2.1 is due to Persidski [63] who first introduced the notion of uniform stability. As an example of the techniques used in Liapunov method, we give the proof of Theorem 2.1. Let $\gamma \in (0,\chi)$ be such that V is positive definite on $B[\gamma]$ and D^+V is negative semidefinite on $B[\gamma]$. For any $\varepsilon \in (0,\gamma)$ there exists then $\beta(\varepsilon) > 0$ such that

$$\|x\| = \varepsilon \Longrightarrow V(t,x) \geq \beta(\varepsilon) > 0.$$

Assume any t_0 in I. Since $V(t_0,\cdot)$ is continuous and $V(t,0) \equiv 0$ for all $t \in I$, there exists $\delta = \delta(t_0,\varepsilon) \in (0,\varepsilon)$ such that $\|x_0\| < \delta$ implies $V(t_0,x_0) < \beta(\varepsilon)$. Assume $x_0 \in B(\delta)$ and consider the noncontinuable solution $x(t,t_0,x_0)$. If there exists $t_1 \in J^+(t_0,x_0)$ such that $\|x(t,t_0,x_0)\| < \varepsilon$ for $t \in [t_0,t_1)$ and $\|x(t_1,t_0,x_0)\| = \varepsilon$, we have $\beta(\varepsilon) > V(t_0,x_0) \geq V(t_1,x(t_1,t_0,x_0)) \geq \beta(\varepsilon)$, then a contradiction. Trivially the above δ is independent of t_0 if $V(t,x) \to 0$ as $x \to 0$ uniformly in $t \in I$. The proof is complete.

Theorem 2.2 (Liapunov [41]) *Assume there exists for* (1.1) *a function $V \in C(I \times D, \mathbf{R})$ such that: (i) V is positive definite; (ii) $V(t,x) \to 0$ as $x \to 0$ uniformly in $t \in I$; (iii) D^+V is negative definite. Then $x \equiv 0$ is uniformly asymptotically stable.*

As a second example of the procedures of the direct method we give a short proof of Theorem 2.2. For a given $L > 0$ let $\gamma \in (0,\chi)$ be such that V is positive definite on $B[\gamma]$, D^+V is negative definite on $B[\gamma]$, and $V(t,x) \leq L$ for $(t,x) \in I \times B[\gamma]$. By virtue of Theorem 2.1 $x \equiv 0$ is uniformly stable and we denote by $\delta(\varepsilon) \in (0,\varepsilon)$ the positive number associated with ε in the definition of uniform stability. Let us prove that $x \equiv 0$ is attractive. Let $\sigma = \delta(\gamma)$, and assume $(t_0,x_0) \in I \times B(\sigma)$. Then we have $x(t,t_0,x_0) \in B(\gamma)$ for all $t \geq t_0$. Clearly the attractivity will be proved if we show that for any $\nu \in (0,\sigma)$ there exists $t^* \geq t_0$ such that $x(t^*,t_0,x_0) \in B(\delta(\nu))$. Indeed this property implies $x(t,t_0,x_0) \in B(\nu)$ for $t \geq t^*$. Suppose then $\delta(\nu) \leq \|x(t,t_0,x_0)\|$ for all $t \geq t_0$. Hence because of (*iii*) we have $V(t,x(t,t_0,x_0)) - V(t_0,x_0) \leq -h(\nu)(t-t_0)$ for some $h(\nu) > 0$ and all $t \geq t_0$, then by virtue of (*i*) a contradiction. It is not difficult to show that the attractivity is uniform by using the property $0 \leq V(t,x) \leq L$ in $I \times B[\gamma]$.

An example of nonuniform asymptotic stability is given by the following theorem.

Theorem 2.3 (Marakhov [44]) *Assume that f is bounded and there exists for* (1.1) *a function $V \in C(I \times D, \mathbf{R})$ such that V is positive definite and D^+V is negative definite. Then $x \equiv 0$ is uniformly asymptotically stable.*

Sufficient conditions for instability are given in each of the three following theorems. The results are obtained by using the technique, illustrated for the proof of Theorem 2.2, in order to show that we may choose x_0 arbitrarily close to $x = 0$ such that $x(t, t_0, x_0)$ cannot remain forever in appropriate regions.

Theorem 2.4 (Liapunov [41]) *Assume there exists for* (1.1) *a function* $V \in C(I \times D, \mathbf{R})$ *such that: (i)* V *has the property* $P^+(t_0)$ *for some* t_0 *in* I*; (ii)* $V(t,x) \to 0$ *as* $x \to 0$ *uniformly in* $t \in I$*; (iii)* D^+V *is positive definite. Then* $x \equiv 0$ *is unstable.*

Theorem 2.5 (Liapunov [41]) *Assume there exists for* (1.1) $\gamma \in (0, \chi)$ *and a function* $V \in C(I \times D, \mathbf{R})$ *such that : (i)* V *has the property* $P^+(t_0)$ *for some* t_0 *in* I*; (ii)* V *is bounded in* $I \times B[\gamma]$*; (iii)* $D^+V(t,x) \geq \lambda V(t,x)$ *for some constant* $\lambda > 0$ *and for all* $(t,x) \in I \times D$*. Then,* $x \equiv 0$ *is unstable.*

Theorem 2.6 (Chetaev [11]) *Assume there exists for* (1.1) $\gamma \in (0, \chi)$, $L > 0$, $t_0 \in I$, *a function* $V \in C(I \times D, \mathbf{R})$*, and an open set* $\Theta \subset B[\gamma]$, $0 \in \partial\Theta$*, such that: (i)* $0 < V(t,x) < L$ *and* $D^+V(t,x) \geq a(V(t,x))$ *for some* $a \in K$ *and for all* $t \geq t_0$ *and* $x \in \Theta$*; (ii)* $V(t,x) = 0$ *for all* $t \geq t_0$ *and* $x \in \partial\Theta \cap B(\gamma)$*. Then* $x \equiv 0$ *is unstable.*

Liapunov Theorems 2.4 and 2.5 are corollaries of Theorem 2.6. Another corollary of Theorem 2.6 is the following which involves t–independent Liapunov functions and is valid for autonomous systems.

Corollary 2.1 *Suppose that system* (1.1) *is autonomous. Assume there exists for* (1.1) $\gamma \in (0, \chi)$*, a functions* $V \in C^1(D, \mathbf{R})$*, and an open set* $\Theta \subset B[\gamma]$, $0 \in \partial\Theta$*, such that: (i)* $V(x) > 0$ *and* $\dot{V}(x) > 0$ *for* x *in* Θ*; (ii)* $V(x) = 0$ *for* $x \in \partial\Theta \cap B(\gamma)$*. Then* $x \equiv 0$ *is unstable.*

In the above theorems the existence of Liapunov functions is assumed. The question arises if such a function exists. In other words given a stability property, can we assert that there exists a Liapunov function which satisfies the conditions in the corresponding theorem? Thus the need to search for converse theorems arises. These theorems are also important in the analysis of the qualitative behavior of the flow of perturbed systems. There have appeared many papers which concern converse theorems. We recall the following results. Persidski [63] has shown that if $f \in C^1$, then the stability of the origin implies the existence of a C^1 function V which satisfies the condition (*i*) and (*ii*) in Theorem 2.1. Kurzweil [33] has shown that if $f \in C^1$, the uniform stability of the origin implies the existence of a C^1 function V which satisfies the condition (*i*), (*ii*), and (*iii*) in Theorem 2.1. Massera [51, 52] was the first to obtain a converse of Liapunov theorem on uniform asymptotic stability. For future reference we state the result.

Theorem 2.7 (Massera [51, 52]) *If* f *is Lipschitzian in* x *uniformly with respect to* t*, and if the null solution of* (1.1) *is uniformly asymptotically stable, then there exists a function* $V : I \times D \to \mathbf{R}$ *which possesses continuous partial derivatives of any order and satisfies the conditions in Theorem* 2.2*. Moreover if* f *is independent of* t *or periodic in* t*, then* V *may be found independent of* t *or periodic in* t *respectively.*

3. Autonomous systems and linear analysis of stability

The results of this section are due to Liapunov [41]. Assume that (1.1) is autonomous and $f \in C^1$. System (1.1) may be written in the form:

$$\dot{x} = Ax + \varphi(x), \tag{3.1}$$

where A is a $n \times n$ constant matrix and $\varphi(x) = o(\|x\|)$ as $x \to 0$. In order to construct appropriate Liapunov functions for (3.1) we will need the following procedure: try to determine a complex constant and an algebraic (nontrivial) form $V(x)$ of degree $m \geq 1$ such that

$$\langle \frac{\partial V}{\partial x}, Ax \rangle = \lambda V.$$

We obtain for the determination of V an homogeneous system of linear algebraic equations in the same number N of the unknowns coefficients of V. Let us denote by $\mathcal{D}_m(\lambda)$ the determinant of the coefficients of this system. In particular we have $\mathcal{D}_1(\lambda) = \det[A - \lambda \mathcal{I}]$ where $\mathcal{I}$ is the $n \times n$ identity. The above form $V(x)$ exists if and only if

$$\mathcal{D}_m(\lambda) = 0. \tag{3.2}$$

The solutions of (3.2) may be expressed in terms of the eigenvalues of A. Precisely the following proposition holds.

Proposition 3.1 *Let $\lambda_1, \lambda_2, \ldots, \lambda_n$ be the eigenvalues of A. The N solutions of* (3.2) *are all obtained by the formula*

$$\lambda = m_1\lambda_1 + m_2\lambda_2 + \ldots + m_n\lambda_n,$$

by assuming for $m_1, m_2, \ldots, m_n$ all the nonnegative integers such that $m_1 + m_2 + \ldots + m_n = m$.

Let U be an m–form, $m \geq 1$, with real coefficients. Consider the equation

$$\langle \frac{\partial V}{\partial x}, Ax \rangle = U(x). \tag{3.3}$$

We may clearly satisfy this equation by assuming for V a form of the same degree m if and only if $\mathcal{D}_m(0) \neq 0$. In this case V is unique (and has real coefficients).

Proposition 3.2 *Assume that the eigenvalues of A have all negative real parts. Assume that U is a definite negative m–form, $m \geq 2$ (even). Then there exists one and only one m–form V which satisfies* (3.3). *Moreover this form is positive definite.*

Proof. Indeed we have $\mathcal{D}_m(0) \neq 0$ and hence there exists a unique m–form V which satisfies (3.3). The null solutions of the linear equation $\dot{x} = Ax$ is asymptotically stable. Along the solution of this equation we have $\dot{V}(x) = U(x)$. Suppose now that $V(y) \leq 0$ for some $y \neq 0$. Hence we would have $V(z) < 0$ for some z and then, since V is a form, $V(x) < 0$ for x arbitrarily close to the origin. By Theorem 2.4 the null solution of $\dot{x} = Ax$ would be unstable, a contradiction. ∎

Proposition 3.3 *Assume that at least one eigenvalue of A has a positive real part and $\mathcal{D}_m(0) \neq 0$. Assume that U is a definite positive m–form. Then there exists one and only one m–form V which satisfies* (3.3) *and $V(x) > 0$ for x arbitrarily close to the origin.*

Proof. The null solution of the linear equation $\dot{x} = Ax$ is unstable. The result relative to the case $\mathcal{D}_m(0) \neq 0$ is obtained by arguments similar to those used in the previous proof and using now Theorem 2.2. ∎

Proposition 3.4 *Assume that at least one eigenvalue of A has a positive real part and $\mathcal{D}_m(0) = 0$. Assume that U is a definite positive m–form. Then there exists one and only one m–form V and a constant $c > 0$ such that*

$$\langle \frac{\partial V}{\partial x}, Ax \rangle = cV(x) + U(x), \tag{3.4}$$

and $V(x) < 0$ for x arbitrarily close to the origin.

Proof. We notice that (3.4) may be written as

$$\langle \frac{\partial V}{\partial x}, A'x \rangle = U(x),$$

where $A' = A - \mathcal{I}c/m$. We choose $c > 0$ so small that $\mathcal{D}_m(c) \neq 0$ and at least one of the eigenvalues of the matrix A' has a positive real part. Then the result is obtained as before by considering now the linear equation $\dot{x} = A'x$ and using again Theorem 2.2. ■

Theorem 3.1 *We have: (i) If all the eigenvalues of A have a negative real part, the null solution of* (3.1) *is asymptotically stable; (ii) if at least one eigenvalue of A has a positive real part, the null solution of* (3.1) *is unstable.*

Proof. To prove *(i)* we consider the function V of Proposition 3.2. The derivative of V along the solutions of (3.1) is

$$\dot{V}_{(3.1)} = U + o(\|x\|^m) \text{ as } x \to 0. \tag{3.5}$$

Then V satisfies all the conditions in Theorem 2.2. Let us prove *(ii)*. If $\mathcal{D}_m(0) \neq 0$ the derivative along the solutions of (3.1) of the function V of Proposition 3.3 is again given by (3.5), but now V satisfies the assumptions in Theorem 2.4 on instability; if $\mathcal{D}_m(0) = 0$ for the function V of Proposition 3.4 we have

$$\dot{V}_{(3.1)} = cV + U + o(\|x\|^m) \text{ as } x \to 0. \tag{3.6}$$

Hence V satisfies the assumptions in Theorem 2.5 on instability. The proof is complete. ■

Thus if the eigenvalues of A have all a negative real part or at least one of them has a positive real part then the stability problem of the null solution of (3.1) is completely solved and the results are the same as those concerning the null solution of the linear equation $\dot{x} = Ax$. In the remaining cases (the so called critical cases), the occurrence of stability or instability depends upon the higher order terms of f and the solution to the stability problem is in general very difficult.

4. Modifications of Liapunov theorems on asymptotic stability and instability

The definitiveness property of D^+V in Theorems 2.2 and 2.4 may be replaced by the weaker assumption that D^+V is only semidefinite provided some additional requirements are satisfied. For periodic or in particular for autonomous systems this is obtained by the following result.

Theorem 4.1 (Barbasin and Krasovskii [2]; Krasovskii [32]). *Assume that $f(t+T, x) \equiv f(t, x)$ for some fixed $T > 0$. Suppose there exist for* (1.1), *$\gamma \in (0, \chi)$ and a function $V \in C(I \times D, \mathbf{R})$, $V(t+T, x) \equiv V(t, x)$ or V time-independent if* (1.1) *is autonomous, such that: (i) $D^+V(t, x) \leq 0$ for every $(t, x) \in I \times B[\gamma]$; (ii) the set $M = \{(t, x) : t \in I, x \in B[\gamma] - 0, D^+V = 0\}$ contains no complete positive semiorbits of* (1.1). *Then: (I) if V is positive definite on $B[\gamma]$, $x \equiv 0$ is uniformly asymptotically stable; (II) if V has the property $P^-(t_0)$ for some t_0 in I, $x \equiv 0$ is unstable.*

Proof. Since we are concerned with a local problem, we may modify f on a complement of some neighborhood of the origin and obtain that the solutions exist for any $t \in \mathbf{R}$. We prove statement *(I)*. It is clear that the null solution is uniformly stable. Assume σ as the number $\delta(\gamma)$ associated to γ in the definition of uniform stability. Hence for any $x_0 \in B(\sigma)$, $t_0 \in \mathbf{R}$, the solution $x(t, t_0, x_0)$ belongs to $B(\gamma)$ for $t \geq 0$. To prove asymptotic stability it is sufficient to show that for any $\nu > 0$ there exists $t^* \geq t_0$ such that $x(t^*, t_0, x_0) \in B(\delta(\nu))$. Assume this is not true. Hence $x(t,t_0,x_0)$ belongs to the compact set $B[\gamma]-B(\delta(\nu))$ for any $t \geq t_0$. Moreover, the function $V(t,x(t,t_0,x_0))$ is nonnegative and decreasing. Then $V(t,x(t,t_0,x_0))$ has a limit $a \geq 0$ as $t \to +\infty$. Consider now the discrete autonomous dynamical system $\pi_{t_0} : \mathbf{Z} \times D \to \mathbf{R}^n$ defined by

$$\pi_{t_0}(n,x) = x(t_0+nT,t_0,x), \text{ for any } x \in D \text{ and } n \in \mathbf{Z}.$$

Denote by $\Lambda^+_{t_0}(x)$ the positive limit set of the orbit $\{\pi_{t_0}(n,x) : n \in \mathbf{Z}\}$, that is

$$\Lambda^+_{t_0}(x) = \{y \in D : \text{ there exists a sequence } \{n_i\} \text{ in } \mathbf{Z} \text{ such that } n_i \to +\infty, \pi_{t_0}(n_i,x) \to y\}.$$

This set is closed and invariant with respect to π_{t_0}. Clearly $\Lambda^+_{t_0}(x_0)$ is nonempty and is contained in $B[\gamma]-B(\delta(\nu))$. Let $y \in \Lambda^+_{t_0}(x_0)$. There exists a divergent sequence $\{n_k\}$ in $\mathbf{Z}$ such that $x(t_0+n_kT,t_0,x_0) \to y$. Taking into account that V is continuous and T-periodic in t, we have

$$\begin{aligned} V(t_0,y) &= \lim_{k\to+\infty} V(t_0,x(t_0+n_kT,t_0,x_0)) \\ &= \lim_{k\to+\infty} V(t_0+n_kT,x(t_0+n_kT,t_0,x_0)) = a. \end{aligned}$$

The π_{t_0}-invariance of $\Lambda^+_{t_0}(x_0)$ assures that $V(t_0,x(t_0+nT,t_0,y)) = a$ for any $n \in \mathbf{Z}$. By virtue of the above periodicity of V and (i), we may conclude that $V(t,x(t,t_0,y)) \equiv a$. Thus $x(t,t_0,y) \in M$ for any t, in contrast with (ii).

The proof in statement (II) is given by absurdo. Since V has the property $P^-(t_0)$ for some t_0 in $\mathbf{R}$, for any $\eta \in (0,\gamma)$ there exists $x_0 \in B(\eta)$, such that $V(t_0,x_0) = V_0 < 0$, Consider the solution $x(t,t_0,x_0)$. Assume that $\|x(t,t_0,x_0)\| < \gamma$ for any $t \geq t_0$. Moreover, since V is periodic in t, one has $V(t,x) \to 0$ as $x \to 0$ uniformly in $t \in \mathbf{R}$. In particular there exists $\lambda > 0$ such that $\|x\| < \lambda \Longrightarrow |V(t,x)| < |V_0|$. Taking into account (ii), we have $V(t,x(t,t_0,x_0)) \leq V_0 < 0$. Then, it has to be $\gamma > \|x(t,t_0,x_0)\| \geq \lambda$ for $t \geq t_0$. Proceeding as for the proof of (I) we get a contradiction. Since x_0 can be chosen arbitrarily close to the origin, (II) follows. ∎

Remark 4.1 *To prove instability in Theorem* 4.1, *hypothesis* (ii) *can be replaced by the following one:*
$(ii)'$ *the set* $M \cap \{(t,x) : V(t,x) < 0\}$ *contains no complete positive semiorbit of* (1.1).

The proofs of the statements (I) and (II) in Theorem 4.1 are not exactly those given by Barbasin and Krasovskii. They are more in the spirit of the so called invariance principle for autonomous systems due to LaSalle ([38,39]). We give a semplified version of this principle (for a more general version see [71]): *Assume that f is independent of t. Let $\Omega \subset D$ be an open set. Let $x(t,x_0)$ be a solution which exists for all $t \geq 0$, and let $\Lambda^+(x_0)$ be its positive limit set:*

$$\Lambda^+(x_0) = \{y \in D : \text{ there exists a sequence } \{t_n\} \text{ in } \mathbf{R} \text{ such that } t_n \to +\infty, x(t_n,x_0) \to y\}.$$

Let $V \in C(D,\mathbf{R})$. Then if $D^+V(x(t,x_0)) \leq 0$ for $t \geq 0$, one has $\Lambda^+(x_0) \cap \Omega \subset M$, where M is the largest invariant set contained in $E = \{x \in \Omega : D^+V(x) = 0\}$.

Dealing with general nonautonomous systems, under the restriction that f is bounded, in the work of Matrosov [54] and in that of Rouche [70] the additional conditions required when D^+V is only semidefinite are expressed by means of a second (scalar or vector respectively) auxiliary function.

Definition 4.1 *Let E be a nonempty subset of D and let $\gamma \in (0,\chi)$.*
(i) *A scalar function $g : I \times D \to \mathbf{R}$ is said to be definitively nonzero with respect to E on $B[\gamma]$ if for each $\eta \in (0,\gamma)$ there exist $\varepsilon > 0$, $\beta > 0$ such that*

$$t \in I, \eta \leq \|x\| \leq \gamma, \text{ and } \rho(x,E) < \varepsilon \Longrightarrow |g(t,x)| > \beta.$$

(ii) *A vector function* $G : I \times D \to \mathbf{R}^k$, $k \geq 1$, *is said to be:*
(1) *uniformly vanishing with respect to* E *on* $B[\gamma]$ *if for every* $\varepsilon > 0$ *there exists* $\delta > 0$ *such that*

$$t \in I, x \in B[\gamma], \text{ and } \rho(x, E) < \delta \Longrightarrow \|G(t, x)\| < \varepsilon;$$

(2) *definitively nonzero with respect to* E *on* $B[\gamma]$ *if for each* $\eta \in (0, \gamma)$ *there exist* $\beta > 0$ *and an open covering* $\{p_1, p_2, \ldots, p_m\}$ *of the set* $\{x \in E : \eta \leq \|x\| \leq \gamma\}$ *such that for every* $i \in \{1, 2, \ldots, m\}$ *there is a component* G_j, $j \in \{1, 2, \ldots k\}$, *of* G *with the property*

$$t \in I, x \in p_i, \text{ and } \eta \leq \|x\| \leq \gamma \Longrightarrow |G_j(t, x)| > \beta.$$

Theorem 4.2 (Matrosov [54]). *Suppose there exist for* (1.1) $\gamma \in (0, \chi)$ *and two functions* $V \in C(I \times D, \mathbf{R})$, $W \in C^1(I \times D, \mathbf{R})$, *such that: (i)* f *is bounded in* $I \times B[\gamma]$; *(ii)* $V(t, x) \to 0$ *as* $x \to 0$ *uniformly in* $t \in I$; *(iii)* $D^+V(t, x) \leq u(x) \leq 0$ *for all* $(t, x) \in I \times B[\gamma]$ *with* $u \in C(D, \mathbf{R}^-)$; *(iv)* W *is bounded in* $I \times B[\gamma]$; *(v)* W *is definitively nonzero with respect to the set* $E = \{x \in B[\gamma] : u(x) = 0\}$ *on* $B[\gamma]$. *Then: (I) if* V *is positive definite,* $x \equiv 0$ *is uniformly asymptotically stable; (II) if* V *has the property* $P^-(t_0)$ *for some* t_0 *in* I, $x \equiv 0$ *is unstable.*

Theorem 4.3 (Rouche [70]). *Suppose there exist for* (1.1) $\gamma \in (0, \chi)$ *and two functions* $V \in C(I \times D, \mathbf{R})$, $W \in C^1(I \times D, \mathbf{R}^k)$, $k \geq 1$, *such that: (i)* f *is bounded in* $I \times B[\gamma]$; *(ii)* $V(t, x) \to 0$ *as* $x \to 0$ *uniformly in* $t \in I$; *(iii)* $D^+V(t, x) \leq u(x) \leq 0$ *for all* $(t, x) \in I \times B[\gamma]$ *where* $u \in C(D, \mathbf{R}^-)$; *(iv)* W *is uniformly vanishing with respect to the set* $E = \{x \in B[\gamma] : u(x) = 0\}$ *on* $B[\gamma]$; *(v)* W *is definitively nonzero with respect to* E *on* $B[\gamma]$. *Then: (I) if* V *is positive definite,* $x \equiv 0$ *is uniformly asymptotically stable; (II) if* V *has the property* $P^-(t_0)$ *for some* t_0 *in* I, $x \equiv 0$ *is unstable.*

The original proofs of Theorems 4.2 and 4.3 are not given here. Instead in the next section we prove, by using a different procedure, two results which are the extension of these theorems to the case that f is not necessarily bounded.

5. One parameter families of Liapunov functions in the analysis of asymptotic stability and instability

The idea of using families of Liapunov functions in the problem of stability has been originated by the remark that any stability concept is a property depending on one or several parameters. Hence, it is natural to associate with each set of values of these parameters a corresponding Liapunov function and then reformulate theorems of the direct method in terms of families of Liapunov functions [77–79]. The difficulty of determining more than one Liapunov function is often balanced by the fact that those functions have less rigid conditions to satisfy. Possible refomulations of Theorems 2.1, 2.2, 2.4 in terms of one parameter families of Liapunov functions are now indicated.

Theorem 5.1 *Assume that for* (1.1) *there exist* $\gamma \in (0, \chi)$ *and a family of functions* $v_\varepsilon \in C(I \times D, \mathbf{R})$, $\varepsilon \in (0, \gamma)$, *such that (i)* $v_\varepsilon(t, 0) \equiv 0$; *(ii)* $t \in I$, $\|x\| = e$ *implies* $v_\varepsilon(t, x) > l_\varepsilon$, $l_\varepsilon > 0$; *(iii)* D^+v_ε *is negative semidefinite on* $B[\gamma]$. *Then* $x \equiv 0$ *is stable.*

Theorem 5.2 *Assume that for* (1.1) *there exist* $\gamma \in (0, \chi)$, *a function* $V \in C(I \times D, \mathbf{R})$ *and a family of functions* $v_\eta \in C(I \times D, \mathbf{R})$, $\eta \in (0, \gamma)$, *such that (i)* $V(t, x) \to 0$ *as* $x \to 0$ *uniformly in* $t \in I$; *(ii)* D^+V *is seminegative definite on* $B[\gamma]$; *(iii)* v_η *is bounded from below in the set* $\{(t, x) \in I \times B[\gamma], \eta \leq \|x\| \leq \gamma\}$; *(iv)* $D^+v_\eta(t, x) < -l_\eta$, $l_\eta > 0$, *in* $\{(t, x) \in I \times B[\gamma], \eta \leq$

$\|x\| \leq \gamma\}$. Then: (I) if V is positive definite, $x \equiv 0$ is asymptotically stable and is uniformly asymptotically stable if v_η is bounded in the set $\{(t,x) \in I \times B[\gamma], \eta \leq \|x\| \leq \gamma\}$; (II) if V has the property $P^-(t_0)$ for some t_0 in I, $x \equiv 0$ is unstable.

Techniques involving families of Liapunov functions have allowed for an extension of the results by Matrosov and Rouche to the case in which the right hand side of (1.1) is not necessarily bounded. The ensuing theorems are corollaries of more general results given in [78,79].

Theorem 5.3 *Suppose there exist for* (1.1) *$\gamma \in (0,\chi)$ and two functions $V \in C(I \times D, \mathbf{R})$, $W \in C^1(I \times D, \mathbf{R})$, such that: (i) $V(t,x) \to 0$ as $x \to 0$; (ii) $D^+V(t,x) \leq u(x) \leq 0$ for all $(t,x) \in I \times B[\gamma]$, with $u \in C(D, \mathbf{R}^-)$; (iii) W and fW are bounded in $I \times B[\gamma]$; (iv) $\dot{W}$ is definitively nonzero with respect to the set $E = \{x \in B[\gamma] : u(x) = 0\}$ on $B[\gamma]$. Then: (I) if V is positive definite, $x \equiv 0$ is uniformly asymptotically stable; (II) if V has the property $P^-(t_0)$ for some t_0 in I, $x \equiv 0$ is unstable.*

Proof. Let $\eta \in (0,\gamma)$ and $C_{\eta\gamma} = \{x \in D, \eta \leq \|x\| \leq \gamma\}$. By virtue of (*iv*), there exist $\varepsilon_\eta > 0$, $\beta_\eta > 0$ such that

$$(t,x) \in I \times C_{\eta\gamma} \text{ and } \rho(x,E) < \varepsilon_\eta \Longrightarrow |\dot{W}(t,x)| > \beta_\eta. \tag{5.1}$$

For fixed $\bar{t} \in I$ consider the sets

$$\begin{aligned} G_\eta^{(1)} &= \{x \in C_{\eta\gamma} : \rho(x,E) < \varepsilon_\eta, (\forall t \in I)\ \dot{W}(\bar{t},x) < -\beta_\eta\}, \\ G_\eta^{(2)} &= \{x \in C_{\eta\gamma} : \rho(x,E) < \varepsilon_\eta, (\forall t \in I)\ \dot{W}(\bar{t},x) > \beta_\eta\}. \end{aligned}$$

From (5.1) it follows that for any $t \in I$ and $x \in G_\eta^{(1)}$ (resp. $x \in G_\eta^{(2)}$) one has $\dot{W}(t,x) < -\beta_\eta$ (resp.$> \beta_\eta$). Define now the two functions $\psi_\eta^{(i)} \in C(C_{\eta\gamma}, \mathbf{R}^+)$, $i = 1,2$, as

$$\begin{aligned} \psi_\eta^{(i)}(x) &= \rho(x,E) && \text{for } x \in G_\eta^{(i)}, \\ \psi_\eta^{(i)}(x) &= \varepsilon_\eta && \text{for } x \in C_{\eta\gamma} - G_\eta^{(i)}. \end{aligned}$$

These functions are both Lipschitzian. Moreover let $\nu_\eta \in C^1(\mathbf{R}^+, [0,1])$ be a function verifying the conditions:

$$\nu_\eta(\tau) = 1 \text{ for } \tau \in [0, \frac{\varepsilon_\eta}{2}], \quad \nu_\eta(\tau) = 0 \text{ for } \tau \in [\varepsilon_\eta, *\infty).$$

It is clear that the two functions $\alpha_\eta^{(i)} \in C(C_{\eta\gamma}, [0,1])$ defined by setting

$$\alpha_\eta^{(i)}(x) = \nu_\eta(\psi_\eta^{(i)}(x)),\ x \in C_{\eta\gamma},$$

are Lipschitzian. Finally, we set

$$h_\eta(t,x) = [\alpha_\eta^{(1)}(x) - \alpha_\eta^{(2)}(x)]W(t,x) \tag{5.2}$$

for any $(t,x) \in I \times C_{\eta\gamma}$. The function h_η is bounded and along the solutions of (1.1) its upper right–hand derivative satisfies the condition

$$D^+h_\eta(t,x) \leq [\alpha_\eta^{(1)}(x) - \alpha_\eta^{(2)}(x)]\dot{W}(t,x) + W(t,x)\mathcal{D}[\alpha_\eta^{(1)}(x) - \alpha_\eta^{(2)}(x)],$$

where $\mathcal{D}$ denotes D^+ or D_+ according to $W(t,x) \geq 0$ or $W(t,x) \leq 0$. By simple computations we obtain:

$$|W(t,x)\mathcal{D}[\alpha_\eta^{(1)}(x) - \alpha_\eta^{(2)}(x)]| \leq K_\eta \|W(t,x)f(t,x)\| \tag{5.3}$$

where K_η is a positive constant. Taking into account (*iii*), we see that the l.h.s. of (5.3) is bounded in $I \times C_{\eta\gamma}$. Furthermore condition (5.1) implies that

$$[\alpha_\eta^{(1)}(x) - \alpha_\eta^{(2)}(x)]\dot{W}(t,x) \leq 0 \text{ in } I \times C_{\eta\gamma}.$$

Hence $D^+ h_\eta$ is bounded from above. Moreover

$$t \in I,\, x \in C_{\eta\gamma}, \text{ and } \rho(x,E) < \frac{\varepsilon_\eta}{2} \Longrightarrow D^+ h_\eta(t,x) < -\beta_\eta. \tag{5.4}$$

We show that we may find a constant $\mu_\eta > 0$ such that one has

$$D^+V(t,x) + \mu_\eta D^+ h_\eta(t,x) \leq -\mu_\eta \beta_\eta \quad \text{for any } (t,x) \in I \times C_{\eta\gamma}.$$

Indeed, if for $(t,x) \in I \times C_{\eta\gamma}$ we have $\rho(x,E) < \frac{1}{2}\varepsilon_\eta$, then (*ii*) and (5.4) imply $D^+ v_\eta(t,x) \leq -\mu_\eta \beta_\eta$. By using again hypothesis (*ii*) we have that there exists A_η such that $(t,x) \in I \times C_{\eta\gamma}$ and $\rho(x,E) \leq \frac{1}{2}\varepsilon_\eta$ imply $D^+V(t,x) < -A_\eta$. As a consequence

$$D^+ v_\eta(t,x) \leq -A_\eta + \mu_\eta B_\eta,$$

where B_η is an upper bound for $D^+ h_\eta$ in $I \times C_{\eta\gamma}$. Hence, we may conclude that, by assuming $\mu_\eta \in (0, A_\eta(\beta_\eta + B_\eta)^{-1})$, $D^+ v_\eta(t,x) \leq -\mu_\eta \beta_\eta$ for all $(t,x) \in I \times C_{\eta\gamma}$. Then the functions V and $v_\eta \in C(I \times C_{\eta\gamma}, \mathbf{R})$ defined by

$$v_\eta(t,x) = V(t,x) + \mu_\eta h_\eta(t,x)$$

satisfy all conditions in Theorem 5.2. Therefore Theorem 5.2 applies and the proof is complete. ∎

Theorem 5.4 *Suppose there exist for* (1.1) $\gamma \in (0,\chi)$ *and two functions* $V \in C(I \times D, \mathbf{R})$, $W \in C^1(I \times D, \mathbf{R}^k)$, $k \geq 1$, *such that:* (*i*) $V(t,x) \to 0$ *as* $x \to 0$ *uniformly in* $t \in I$; (*ii*) $D^+V(t,x) \leq u(x) \leq 0$ *for all* $(t,x) \in I \times B[\gamma]$ *where* $u \in C(D, \mathbf{R}^-)$; (*iii*) W *and* fW_j, $j = 1, \ldots, k$, *are uniformly vanishing with respect to the set* $E = \{x \in B[\gamma] : u(x) = 0\}$ *on* $B[\gamma]$; (*iv*) $\dot{W}$ *is definitively nonzero with respect to* E *on* $B[\gamma]$. *Then:* (*I*) *if* V *is positive definite,* $x \equiv 0$ *is uniformly asymptotically stable;* (*II*) *if* V *has the property* $P^-(t_0)$ *for some* t_0 *in* I, $x \equiv 0$ *is unstable.*

Proof. Let again $\eta \in (0,\gamma)$ and $C_{\eta\gamma} = \{x \in D, \eta \leq \|x\| \leq \gamma\}$. We may assume that the compact set $F_\eta = \{x \in C_{\eta\gamma}, u(x) = 0\}$ is nonempty. Hypothesis (*iv*) implies there exists $\beta_\eta > 0$ and an open covering $\{p_1, p_2, \ldots, p_m\}$ (depending on η) of F_η such that for every $i \in \{1, \ldots, m\}$ there is a component $\dot{W}_j$ of $\dot{W}$ with the property

$$t \in I, x \in C_{\eta\gamma} \cap p_i \Longrightarrow |\dot{W}_j(t,x)| > \beta_\eta. \tag{5.5}$$

Let $L > 0$ be a constant. By virtue of (*iii*), $\{p_1, p_2, \ldots, p_m\}$ may be chosen with the additional property

$$t \in I,\, x \in C_{\eta\gamma} \cap (\cup p_i) \Longrightarrow \|W(t,x)\| < L,\, \|W_j(t,x) f(t,x)\| < L,\, j \in \{1, 2, \ldots, k\}. \tag{5.6}$$

Starting from $\{p_1, p_2, \ldots, p_m\}$ consider now the family $\{\pi_1, \pi_2, \ldots, \pi_k\}$ of sets such that for any fixed j the element π_j is the union of all the sets p_i for which (5.5) holds with this j. Clearly $\{\pi_1, \pi_2, \ldots, \pi_k\}$ is also an open covering of F_η and we have:

$$\begin{aligned} t \in I, x \in C_{\eta\gamma} \cap \pi_j \quad &\Longrightarrow \quad |\dot{W}_j(t,x)| > \beta_\eta, \\ t \in I, x \in C_{\eta\gamma} \cap (\cup \pi_i) \quad &\Longrightarrow \quad \|W(t,x)\| < L,\, \|W_j(t,x) f(t,x)\| < L, \\ &\qquad\quad j \in \{1, 2, \ldots, k\}. \end{aligned}$$

Finally, for $\bar{t} \in I$ fixed and any $j \in \{1, 2, \ldots, k\}$, we set

$$q_j = \pi_j \cap \{x \in B(\gamma), \dot{W}_j(\bar{t}, x) < -\beta_\eta\},$$
$$q_{j+k} = \pi_j \cap \{x \in B(\gamma), \dot{W}_j(\bar{t}, x) > \beta_\eta\}.$$

The sets $\{q_1, q_2, \ldots, q_{2k}\}$ cannot be all empty and they form an open covering of F_η. Moreover, it is easy to see that setting $q = \cup\{q_j, j = 1, 2, \ldots, 2k\}$ the following properties hold:

$$t \in I,\ x \in C_{\eta\gamma} \cap q_j \implies \dot{W}_j(t, x) < -\beta_\eta, \tag{5.7}$$
$$t \in I,\ x \in C_{\eta\gamma} \cap q_{j+k} \implies \dot{W}_j(t, x) > \beta_\eta, \tag{5.8}$$
$$t \in I,\ x \in C_{\eta\gamma} \cap q \implies \|W(t, x)\| < L,\ \|W_j(t, x) f(t, x)\| < L, \tag{5.9}$$

for any $j \in \{1, 2, \ldots, k\}$. Choose now $\varepsilon_\eta \in (0, \rho(F_\eta, \partial q))$ such that the compact set $G_\eta = \{x \in B[\gamma], \rho(x, F_\eta) \le \varepsilon_\eta\}$ is contained in q. Let $\alpha_\eta^{(i)} \in C^1(B(\gamma), [0, 1])$, $i \in \{1, 2, \ldots, 2k\}$, be $2k$ functions verifying the following conditions

(1) the support of $\alpha_\eta^{(i)}$ is in q_i;

(2) $\displaystyle\sum_{i=1}^{2k} \alpha_\eta^{(i)}(x) = 1$ for any $x \in G_\eta$;

(3) $\displaystyle\sum_{i=1}^{2k} \alpha_\eta^{(i)}(x) \le 1$ for any $x \in q - G_\eta$.

Let $h_\eta \in C^1(I \times B(\gamma), \mathbf{R})$ be the function defined by

$$h_\eta(t, x) = \sum_{j=1}^{k} [\alpha_\eta^{(j)}(x) - \alpha_\eta^{(k+j)}(x)] W_j(t, x).$$

Along the solutions of (1.1) we obtain

$$\begin{aligned} \dot{h}_\eta(t, x) &= \sum_{j=1}^{k} [\alpha_\eta^{(j)}(x) - \alpha_\eta^{(k+j)}(x)] \dot{W}_j(t, x) \\ &\quad + \langle f(t, x), grad \sum_{j=1}^{k} [\alpha_\eta^{(j)}(x) - \alpha_\eta^{(k+j)}(x)] W_j(t, x) \rangle. \end{aligned}$$

Hypothesis (*iii*) implies there exists $A_\eta \in (0, \varepsilon_\eta)$ such that

$$\begin{aligned} (t, x) \in I \times C_{\eta\gamma}, \quad & \rho(x, F_\eta) < A_\eta \implies \\ & |f(t, x), grad \sum_{j=1}^{k} [\alpha_\eta^{(j)}(x) - \alpha_\eta^{(k+j)}(x)] W_j(t, x)| < \frac{\beta_\eta}{2}. \end{aligned}$$

Furthermore, from (5.7), (5.8) and the definition of $\alpha_\eta^{(i)}$, $i \in \{1, 2, \ldots, 2k\}$, it follows:

$$(t, x) \in I \times C_{\eta\gamma},\ \ \rho(x, F_\eta) < A_\eta \implies \sum_{j=1}^{k} [\alpha_\eta^{(j)}(x) - \alpha_\eta^{(k+j)}(x)] \dot{W}_j(t, x) < -\beta_\eta.$$

Hence, we have

$$(t, x) \in I \times C_{\eta\gamma}, \quad \rho(x, F_\eta) < A_\eta \implies \dot{h}_\eta(t, x) < -\frac{\beta_\eta}{2}. \tag{5.10}$$

From the same conditions as before, with the addition of (5.9), we have that $\dot{h}_\eta$ admits an upper bound in the set $I \times C_{\eta\gamma}$. On the other hand for some $\lambda_\eta > 0$ we have

$$x \in C_{\eta\gamma}, \quad \rho(x, F_\eta) \geq A_\eta \Longrightarrow u(x) < -\lambda_\eta. \tag{5.11}$$

Taking into account (5.10) and (5.11), and proceeding as in Theorem 5.3, we may prove the existence of two positive constants μ_η, l_η such that

$$(t, x) \in I \times C_{\eta\gamma} \Longrightarrow D^+V(t, x) + \mu_\eta \dot{h}_\eta(t, x) < -l_\eta.$$

Then the functions V and $v_\eta \in C(I \times B(\gamma), \mathbf{R})$ defined by

$$v_\eta(t, x) = V(t, x) + \mu_\eta h_\eta(t, x)$$

satisfy all the hypotheses in Theorem 5.2 and our proof is complete. ∎

Remark 5.1 *In Theorems* 5.3 *and* 5.4 *the instability occurs if more generally* $\dot{W}$ *is supposed to be definitively nonzero on the sets* $E_a = \{x \in B[\gamma] : u(x) = 0, V(x) \leq -a\}$, *for any* $a > 0$ *and (in Theorem* 5.4*)* W *and* Wf *are uniformly vanishing on the same sets.*

We have then indicated a constructive approach to Matrosov idea to use several auxiliary functions in studying asymptotic stability and instability. We have seen that the results by Matrosov and Rouche may be extended to unbounded r.h.s. of the differential equation. The validity of this extension has suggested that suitable one parameter families of Liapunov functions could be used in analyzing the above stability properties for infinite dimensional dynamical systems. Such a procedure in many cases has fruitfully replaced the techniques based on the extension of LaSalle invariance principle to differential equations in abstract spaces (D'Onofrio [15], Marcati [45–48]).

6. Holonomic systems and stability of equilibrium

Let I be an interval in $\mathbf{R}$ unbounded from above and let Ω be an open set in $\mathbf{R}^n$, $n \geq 1$, containing the origin. Let $T \in C^2(\Omega \times \mathbf{R}^n, \mathbf{R})$ and $\Pi \in C^2(\Omega, \mathbf{R})$ be such that $2T(q, v) = \langle v, A(q)v \rangle$, where $A(q)$ is a symmetric and positive definite $n \times n$ matrix. Consider the first order $2n$–dimensional differential system

$$\begin{aligned} \frac{d}{dt}\frac{\partial T}{\partial v} - \frac{\partial T}{\partial q} &= -\lambda(t)\nabla\Pi + \mathcal{D} \\ \frac{dq}{dt} &= v, \end{aligned} \tag{S_λ}$$

where $\lambda \in C^1(I, \mathbf{R})$ and $\mathcal{D} \in C^1(I \times \Omega \times \mathbf{R}^n, \mathbf{R}^n)$ satisfies

$$\langle \mathcal{D}(t, q, v), v \rangle \leq 0 \quad \text{for all} \quad (t, q, v) \in I \times \Omega \times \mathbf{R}^n. \tag{6.1}$$

System (S_λ) may be reduced to the form (1.1) with $x = (q, v)$, $D = \Omega \times \mathbf{R}^n$, f a C^1 function and n replaced by $2n$. In the discussion of (S_λ) we will adopt the notations of Section 1. Furthermore, we denote by $B'(\gamma)$ the ball $\{q \in \mathbf{R}^n : \|q\| < \gamma\}$. Clearly from (6.1) and the continuity of $\mathcal{D}$ it follows $\mathcal{D}(t, q, 0) \equiv 0$. Hence, given $\bar{q} \in \Omega$, (S_λ) admits the solution $q(t) \equiv \bar{q}$, $v(t) \equiv 0$ if and only if $\bar{q}$ is a critical point of Π.

System (S_λ) governs the motion of a mechanical holonomic system $\mathcal{S}$ with n degrees of freedom, whose constraints are time independent and which is subject to the forces $\mathcal{F} = -\lambda\nabla\Pi$

and $\mathcal{D}$. We include the case that $\mathcal{D}$ is nonenergic, that is $\langle \mathcal{D}(t,q,v), v\rangle \equiv 0$, as well as the case that $\mathcal{D}$ is a strictly dissipative force, that is the case where $\langle \mathcal{D}(t,q,v), v\rangle \leq -c(\|v\|)$ for some $c \in K$ and for all $(t,q,v) \in I \times \Omega \times \mathbf{R}^n$. Let H be the total energy $T + \Pi$ of the system. We recall that the derivative along the solutions of (S_λ) of H satisfies the condition $\dot{H}(t,q,v) = \langle \mathcal{D}(t,q,v), v\rangle \leq 0$.

Let $\bar{q}$ be an equilibrium configuration of $\mathcal{S}$, i.e. assume $\nabla\Pi(\bar{q}) = 0$. In this section we analyze the stability of the static solution $q \equiv \bar{q}, v \equiv 0$ of (S_λ). We may assume $\bar{q} = 0$ and $\Pi(0) = 0$. The case of an unbounded λ will be also considered. The results concerning the case $\lambda(t) \equiv 1$ are well known and will be sketched in the next subsection.

6.1. Assume then $\lambda(t) \equiv 1$, that is consider the system (S_1)

$$\begin{aligned} &\frac{d}{dt}\frac{\partial T}{\partial v} - \frac{\partial T}{\partial q} = -\nabla\Pi + \mathcal{D} \\ &\frac{dq}{dt} = v, \end{aligned} \qquad (S_1)$$

Theorem 6.1 (Lagrange–Dirichlet) *The solution $q \equiv 0$, $v \equiv 0$ is uniformly stable if Π has a strict relative minimum at $q = 0$.*

Indeed it is immediate to recognize that the function $V = H$ satisfies all the conditions in Theorem 2.1.

Theorem 6.2 *Suppose the equilibrium position $q = 0$ is isolated. Moreover suppose that $\mathcal{D}$ is completely dissipative and that for some $\gamma > 0$ $\mathcal{D}(t,q,v) \to 0$ as $v \to 0$, uniformly in $(t,q) \in I \times B'[\gamma]$. Then $q \equiv 0$, $v \equiv 0$ is stable if and only if Π has a relative minimum at $q = 0$. Moreover if the above minimum occurs, $q \equiv 0$, $v \equiv 0$ is uniformly asymptotically stable.*

Proof. The result may be obtained as an application of Matrosov theorem 4.2. We may always choose $\gamma > 0$ in such a way that $\mathcal{D}$ is also bounded in $I \times B[\gamma]$ and there are no equilibrium positions in $B'[\gamma] - \{0\}$. Consider the two functions

$$V = H \quad \text{and} \quad W = \langle \frac{\partial T}{\partial v}, \nabla\Pi\rangle.$$

Along the solutions of (S_1) one has

$$\dot{V} = \langle \mathcal{D}(t,q,v), v\rangle \leq -c(\|v\|),\ c \in K.$$

$$\dot{W} = -(\nabla\Pi)^2 + \langle \frac{\partial T}{\partial q}, \nabla\Pi\rangle + \langle \frac{\partial T}{\partial q}, \frac{\partial}{\partial q}\langle \nabla\Pi, v\rangle\rangle + \langle \mathcal{D}, \nabla\Pi\rangle.$$

Hence $\dot{V}$ satisfies the condition *(iii)* in Theorem 4.2. The set E is then

$$E = \{(q,v) \in I \times B[\gamma] : v = 0\},$$

and clearly W and $\dot{W}$ have the required properties. After this the result easily follows. ∎

Suppose now that $\mathcal{D} \equiv 0$ and $n - m$ coordinates $q_{m+1}, \ldots, q_n$, $0 < m < n$, are cyclic, that is T, Π depend on q_j, $j \in \{1, \ldots, m\}$. The determination of the coordinates $q_1, \ldots, q_m$ may be obtained by a differential system which has still the form (S_1):

$$\begin{aligned} &\frac{d}{dt}\frac{\partial \tilde{T}}{\partial w} - \frac{\partial \tilde{T}}{\partial z} = -\nabla\tilde{\Pi}(c,z) + \tilde{\mathcal{D}}(c,z,w) \\ &\frac{dz}{dt} = w, \end{aligned} \qquad (\tilde{S})$$

where $z = (q_1, \ldots, q_m)$, $w = (v_1, \ldots, v_m)$. The functions $\tilde{\Pi}$ and $\tilde{\mathcal{D}}$ involve the $(n-m)$-tpla $c = (c_{m+1}, \ldots, c_n)$ of constants of momenta and $\tilde{T} = \tilde{T}(z, w)$ is a symmetric positive definite quadratic form in w. Moreover $\tilde{\mathcal{D}}$ is linear with respect to w and satisfies $\langle \tilde{\mathcal{D}}, w \rangle \equiv 0$. Then for any fixed c, $(\tilde{S})$ are the equations of motion of a mechanical holonomic system with m degrees of freedom, which is subject to the positional conservative force $-\nabla\tilde{\Pi}$ and the gyroscopic force $\tilde{\mathcal{D}}r$. Assume that for a given value c_0 of c one has $\nabla\tilde{\Pi}(c_0, 0) = 0$, that is for $c = c_0$ system $(\tilde{S})$ admits the solution $z \equiv 0, w \equiv 0$. Let σ be one of the ∞^{n-m} merostatic motions of the original holonomic system corresponding to this solution. By Theorem 6.1 it immediately follows that if $\tilde{\Pi}(c_0, z)$ has a strict relative minimum at $z = 0$, then σ is stable with respect to z, w conditionally to the perturbations for which $c = c_0$ (Routh theorem, [72]). Actually the above minimum of $\tilde{\Pi}(c_0, z)$ is sufficient for σ to be unconditionally stable with respect to z, w [76,77]. See also Pozharitskii [64], Rumiantsev [74], and Karapetyan [26]. An extension to dissipative systems may be found in [77]. A further extension concerning nonautonomous systems was given by Habets and Risito [19].

6.2. We consider now system (S_λ) and assume

(H_1) $\quad \lambda(0) = a > 0$ and $\dot{\lambda}(t) \geq 0$ for all $t \in I$.

Setting $g = \lambda^{\frac{1}{2}}$, and $v = gw$, we have

$$\frac{\partial T}{\partial q}(q, gw) = g^2 \frac{\partial T^\star}{\partial q}(q, w), \quad \frac{\partial T}{\partial v}(q, gw) = g \frac{\partial T^\star}{\partial w}(q, w),$$

where $T^\star(q, w) = T(q, gw)$. Consequently, system (S_λ) in terms of the variables q, w may be written as

$$\begin{aligned} \frac{d}{dt}\frac{\partial T^\star}{\partial w} &= -\frac{\dot{g}}{g}\frac{\partial T^\star}{\partial w} + g\frac{\partial (T^\star - \Pi)}{\partial q} + \frac{1}{g}\mathcal{D}^\star \\ \frac{dq}{dt} &= gw, \end{aligned} \qquad (S_\lambda)'$$

with $\mathcal{D}^\star(t, q, w) = \mathcal{D}(t, q, gw)$. Letting now $V = T^\star + \Pi$, we recognize that along the solutions of $(S_\lambda)'$ one has

$$\dot{V} = -2\frac{\dot{g}}{g}T^\star + \frac{1}{g}\langle \mathcal{D}^\star, w \rangle \leq 0. \tag{6.2}$$

Therefore, we have the following result.

Theorem 6.3 [85] *Under assumption (H_1), if Π has a strict relative minimum at $q = 0$ then the solution $q \equiv 0, v \equiv 0$ of the original system (S_λ) is stable with respect to q and is stable (with respect to both q, v) if λ is bounded.*

Proof. The function V is time independent, $V(0) = 0$, and positive definite. Then, because of (6.2), the solution $q \equiv 0, w \equiv 0$ of $(S_\lambda)'$ is uniformly stable. Hence, for any $\varepsilon > 0$ there exists $\delta^\star > 0$ such that $t_0 \in I$ and $\|(q_0, w_0)\| < \delta^\star$ imply $\|(q(t, t_0, q_0, w_0), w(t, t_0, q_0, w_0))\| < \varepsilon$. Assume the same ε, t_0 and let $\delta = (1 + \frac{1}{a})^{\frac{1}{2}}\delta^\star$. Since $g(t_0) \geq g(0)$, we then have that $\|(q_0, v_0)\| < \delta$ implies $\|(q(t, t_0, q_0, v_0)\| < \varepsilon$ for all $t \in I$. The first part of theorem is proved. If λ is bounded, then for the same $\varepsilon, t_0, q_0, v_0$ we have $\|(q(t, t_0, q_0, v_0), v(t, t_0, q_0, v_0))\| < \varepsilon' = \varepsilon(1 + L)^{\frac{1}{2}}$, were L is an upper bound for λ. ∎

Theorem 6.4 *Assume (H_1). Assume there exists in Ω a fundamental family of open neighborhoods of $q = 0$ such that for any member A of the family one has $\min\{\Pi(q), q \in \partial A\} > 0$. Then the solution $q \equiv 0, v \equiv 0$ of system (S_λ) is stable with respect to q and is stable if λ is bounded.*

We suppose now that the dissipative force is linear in v, that is

$$(H_2) \qquad \mathcal{D}(t,q,v) = \mathcal{B}v,$$

with $\mathcal{B} = \mathcal{B}(t,q)$ a $n \times n$ bounded matrix in $I \times \Omega$. System $(S_\lambda)'$ then becomes

$$\begin{aligned} \frac{d}{dt}\frac{\partial T^*}{\partial w} &= -\frac{\dot g}{g}\frac{\partial T^*}{\partial w} + g\frac{\partial (T^* - \Pi)}{\partial q} + \frac{1}{g}\mathcal{B}w \\ \frac{dq}{dt} &= gw, \end{aligned} \qquad (S_\lambda)''$$

Moreover, (6.2) may be written as

$$\dot V = -2\frac{\dot g}{g}T^* + \frac{1}{g}\langle \mathcal{B}w, w\rangle \le 0. \tag{6.3}$$

In the present case then $\dot V$ is the sum of two quadratic forms in w for which each of them assumes nonpositive values. We will consider the particular case that

$$(H_3) \qquad -2\frac{\dot g}{g}T^* + \frac{1}{g}\langle \mathcal{B}w, w\rangle \le -c(\|w\|), \quad c \in K,$$

for all $(t,q,w) \in I \times \Omega \times \mathbf{R}^n$. In particular (H_3) is satisfied when $\mathcal{B}v$ is a strictly dissipative force, or also when $\frac{\dot\lambda}{\lambda}$ is bounded from below by a positive number. The following result holds.

Theorem 6.5 ([79,85]) *Assume (H_1), (H_2), (H_3) and that $\dot\lambda/\lambda^{\frac{3}{2}}$ is bounded. Moreover, assume that $q = 0$ is an isolated critical point of Π. Then we have:*

(i) if Π has a relative minimum at $q = 0$, then the solution $q \equiv 0, v \equiv 0$ of system (S_λ) is uniformly asymptotically stable with respect to q and is uniformly asymptotically stable when λ is bounded;

(ii) if Π does not have a relative minimum at $q = 0$, then the solution $q \equiv 0, v \equiv 0$ of system (S_λ) is unstable with respect to q.

Proof. Consider the function

$$W = \frac{1}{g}\langle \frac{\partial T^*}{\partial w}, \frac{\partial \Pi}{\partial q}\rangle.$$

Along the solutions of $(S_\lambda)''$ one has

$$\begin{aligned} \dot W \;=\; & \langle -\frac{\partial \Pi}{\partial q}, -2\frac{\dot g}{g^2}\frac{\partial T^*}{\partial w} + \frac{\partial T^*}{\partial q} + \frac{1}{g}\mathcal{B}w\rangle \\ & + \langle \frac{\partial T^*}{\partial w}, \frac{\partial}{\partial q}\langle \frac{\partial \Pi}{\partial q}, w\rangle\rangle - \langle \frac{\partial \Pi}{\partial q}, \frac{\partial \Pi}{\partial q}\rangle. \end{aligned} \tag{6.4}$$

The function V satisfies for system $(S_\lambda)''$ the assumptions (i) and (ii) of Theorem 5.3 with $u(q,w) = -c(\|w\|)$. From (6.4) we see that since $q = 0$ is an isolated critical point of Π there exists $\alpha > 0$ such that the ball $B(\alpha) = \{(q,w) \in \mathbf{R}^n \times \mathbf{R}^n, \|(q,w)\| < \alpha\}$ is contained in $\Omega \times \mathbf{R}^n$ and $\dot W$ is definitively nonzero on the set $E = \{(q,w) \in B[\gamma], u(q,w) = 0\}$. Finally W and Wf^* are bounded on $I \times B(\alpha)$. Here f^* is the right hand side of system $(S_\lambda)''$ written in the vector form (1.1). In conclusion the functions V and W satisfy for system $(S_\lambda)''$ the assumptions (i),(ii),(iii),(iv) of Theorem 5.3. Assume now that Π has a relative minimum at $q = 0$. Since $q = 0$ is an isolated critical point of Π, this minimum is strict. Consequently V is positive definite and then from part (I) of Theorem 5.3 it follows that the solution $q \equiv 0, w \equiv 0$

of system $(S_\lambda)''$ is uniformly asymptotically stable. By trivial arguments similar to those used in the proof of Theorem 6.3, we then recognize that the solution $q \equiv 0, v \equiv 0$ of the original system (S_λ) is uniformly asymptotically stable with respect to q and uniformly asymptotically stable with respect to both q, v when λ is bounded. The first part of the theorem is proved. To prove the second part we observe that if Π does not have a relative minimum at $q = 0$, then in any neighborhood of $q = 0, w = 0$ there are points where $V < 0$. Hence, the solution $q \equiv 0, w \equiv 0$ of system $(S_\lambda)''$ is unstable because of part *(II)* of Theorem 5.3. Let us prove that this solution is unstable with respect to q. From (6.2) it follows that along any solution $(q(t, t_0, q_0, w_0), w(t, t_0, q_0, w_0))$ of $(S_\lambda)''$ we have

$$T^* + \Pi \leq T_0^* + \Pi_0, \tag{6.5}$$

where T_0^*, Π_0, are the initial values of T^*, Π respectively. By using (6.5) one can easily prove that the stability of the solution $q \equiv 0, w \equiv 0$ of system $(S_\lambda)''$ with respect to q would imply its stability with respect to both q, w, which is absurd. Then this solution is unstable with respect to q. From this we easily derive that the solution $q \equiv 0, v \equiv 0$ of the original system (S_λ) is unstable with respect to q. ∎

Remark 6.1 *We notice that by a small modification of Theorem* 5.3 *one can prove that the result concerning the instability in Theorem* 6.5 *more generally holds if the condition that $q = 0$ is an isolated critical point of Π is replaced by the weaker assumption that for some $\varepsilon > 0$ there are no critical points of Π in the set $\{q \in \Omega, \|q\| < \varepsilon, \Pi(q) < 0\}$. By using a known argument* [29] *it follows that under the assumptions (H_1), (H_2), (H_3), and the boundedness of $\dot{\lambda}/\lambda^{\frac{3}{2}}$, if Π is (real) analytic and Π does not have a relative minimum at $q = 0$, then the solution $q \equiv 0, v \equiv 0$ of (S_λ) is unstable with respect to q. Indeed the set of all critical values assumed by Π in a compact set is finite* [90]. *Hence, the continuity of Π implies the existence of $\varepsilon > 0$ such that zero is the only critical value of Π in the set $\{q \in \Omega, \|q\| < \varepsilon\}$. Therefore there are no critical points of Π in the set $\{q \in \Omega, \|q\| < \varepsilon, \Pi(q) < 0\}$.*

The same conclusions hold when $\lambda(t) \equiv 1$ and $\mathcal{D}$ is a completely dissipative force not necessarily linear but satisfying the condition assumed in Theorem 6.2.

Recently there have been several significant improvements and extension of the results in the case of an unbounded $\lambda(t)$. We mention Pucci and Serrin in a series of many papers (see for instance [65–69] and references within). Some of them are concerned with dissipative dynamical system in infinite dimensional space.

6.3. We wish now to give an idea of the type of situations that may occur when λ does not have a fixed sign. For this we simply recall some results given in [83] relative to the case that λ is periodic and the system has only one degree of freedom. Generalizations to holonomic systems with several degrees of freedom may be found in [55]. Consider the second order differential equation

$$\ddot{q} + \dot{q} + \lambda(t)\Pi'(q) = 0,$$

that is the differentials system in the plane

$$\frac{dv}{dt} = -\lambda(t)\Pi'(q) - v$$
$$\frac{dq}{dt} = v. \tag{$\bar{S}_\lambda$}$$

We assume that $\lambda \in C^1(\mathbf{R}, \mathbf{R})$ is periodic and

$$\Pi(q) = bq^m + \sigma(q),$$

with $b \neq 0$, $m \geq 3$, and $\sigma \in C^\infty$ with $\sigma = o(|q|^m)$ as $|q| \to 0$. The following result holds.

Theorem 6.6 ([85]) *Let c be the mean value of λ in a period. Then:*

(I) if $c > 0$, the solution $q \equiv 0, v \equiv 0$ of $(\bar{S}_\lambda)$ is asymptotically stable when m is even and $b > 0$, and is unstable in any other case;

(II) if $c < 0$, the solution $q \equiv 0, v \equiv 0$ of $(\bar{S}_\lambda)$ is asymptotically stable when m is even and $b < 0$, and is unstable in any other case;

(III) if $c = 0$ and λ is not identically zero, then the solution $q \equiv 0, v \equiv 0$ of $(\bar{S}_\lambda)$ is asymptotically stable.

We see that statement (I) agrees with the results in Section 6.2 and shows other situations in which stability properties are preserved when (S_1) is replaced by (S_λ). Statement (II) is equivalent to statement (I). Finally, statement (III) shows that when the mean value of λ in a period is zero and λ is not identically zero, the origin of (S_λ) is always asymptotically stable whereas (I) and (II) imply that if m is odd, or m is even and Π has a strict maximum at $q = 0$, the origin of $(\bar{S}_\lambda)$ is unstable.

7. Converse statements of the Lagrange–Dirichlet theorem for conservative systems

As we have seen, in presence of strictly dissipative forces, and when the equilibrium is isolated, under large regularity assumptions the classic Lagrange–Dirichlet theorem admits a converse statement and all possible cases are covered.

Conversely, when strict dissipations are absent, the problem of stability is in general still open, even if several converse statements of the Lagrange–Dirichlet theorem have been collected. The Liapunov direct method has been fruitfully applied. However, as it was pointed out by Liapunov himself, in the most interesting mechanical cases, in particular the conservative ones, the method does not appear to be exhaustive. Relative to the positional conservative case, different lines of approach in the last few years have been exploited to prove instability when the assumptions in Lagrange–Dirichlet theorem are not satisfied. Essentially they are: (*a*) the use of suitable variational principles in the inspection of the flow near the equilibrium; (*b*) the search for instability results through the existence of nontrivial motions which tend to the equilibrium as $t \to -\infty$; (*c*) algebraic geometry approach. The possibility of stabilizing an unstable equilibrium configuration of a positional conservative system by means of nonenergic forces (which do not contribute to the integral of energy) makes the problem of determining conditions for the inversion of the Lagrange–Dirichlet theorem more difficult in the case of nonpositional conservative systems. At present only few results may be indicated.

7.1. We assume that S is positional conservative. Then the equations of motion of S are written as

$$\begin{aligned} \frac{d}{dt}\frac{\partial T}{\partial v} - \frac{\partial T}{\partial q} &= \nabla\Pi(q) \\ \frac{dq}{dt} &= v, \end{aligned} \tag{7.1}$$

If the potential energy does not have a strict relative minimum at $q = 0$, then the stability problem of the equilibrium $q = 0$ is in general open. The Lagrange–Dirichlet theorem does not admit in general a converse. This is shown by the simple case (Painlevé [60], Wintner [95]) where $n = 1$ and

$$\Pi(q) = \exp(-\frac{1}{q^2})\cos\frac{1}{q}, \quad \text{for } q \neq 0, \quad \text{and} \quad \Pi(0) = 0.$$

In this case Π takes values of both signs in every neigborhood of $q = 0$ and nevertheless $q = 0$ is a stable equilibrium position. This example has suggested the following generalization of the Lagrange–Dirichlet theorem: the equilibrium $q = 0$ is stable if there exists a fundamental family F of open neighborhoods of $q = 0$ such that for each $A \in F$ and $q \in \partial A$ one has $\Pi(q) > 0$. We notice that under general assumptions on T, Π, when $n = 1$ the existence of such a family is also a necessary condition for the stability of the zero solution of (7.1). Unfortunately, when $n \geq 2$ the above generalization of the Lagrange–Dirichlet theorem does not admit in general a converse. Indeed Laloy [36] has proved that if $n = 2$, T is Euclidean and

$$\Pi(q_1, q_2) = \xi(q_1) - \eta(q_2),$$

where

$$\xi(q_1) = \exp(-\frac{1}{q_1^2})\cos\frac{1}{q_1} \quad \text{for } q_1 \neq 0, \quad \text{and} \quad \xi(0) = 0,$$

$$\eta(q_2) = \exp(-\frac{1}{q_2^2})(q_2^2 + \cos\frac{1}{q_2}) \quad \text{for } q_2 \neq 0, \quad \text{and} \quad \eta(0) = 0,$$

then $q = 0$ is a stable equilibrium position while $\Pi(q) < 0$ at every point where $q_1 = q_2 \neq 0$. Therefore, it is natural to search for additional requirements which are able to imply instability when Π does not have a strict relative minimum at $q = 0$, or even more, when the above family F does not exist. We first consider the following conditions on Π:

(A) $\qquad \Pi = \Pi_m + P,$

where $\Pi_m(q)$ is an homogeneous polynomial of degree $m \geq 2$ and

$$\frac{\nabla P(q)}{\|q\|^{m-1}} \to 0 \quad \text{as} \quad q \to 0;$$

(B) the set $\{q : \Pi_m(q) < 0\}$ is nonempty.

In other words, we assume that the potential energy does not have a relative minimum at $q = 0$ and the absence of the minimum is recognizable on the first part of the expansion (A). Classic results state the instability of the equilibrium q=0 in each one of the following cases:

(a) $m = 2$ (Liapunov [41]);

(b) $m \geq 2$ and Π_m is negative definite (Liapunov [41]);

(c) $P = 0$ (Chetaev [12]).

In case (a) at least one eigenvalue of the linear approximation has a positive real part. In case (b) the function $V = -\langle q, \frac{\partial T}{\partial v}\rangle$ has a negative definite derivative along the solutions of (7.1) whereas in every neighborhood of $(0,0)$ there exists $(q, v) \in \Omega \times \mathbf{R}^n$ with $V(q, v) < 0$. Hence all the conditions in Theorem 2.4 are satisfied and the instability of $q = 0$ is proved. Let us consider now the function $V(q, v) = -H\langle q, \frac{\partial T}{\partial v}\rangle$ and the set

$$\Theta = \{(q, v) \in \Omega \times \mathbf{R}^n : \|q\| < \varepsilon, H < 0, \langle q, \frac{\partial T}{\partial v}\rangle > 0\},$$

for some $\varepsilon > 0$. Because of (B) Θ is nonempty and the origin of the phase space is in $\partial\Theta$. In case (c), if ε is sufficiently small, then V and $\dot{V}$ are both positive for $(q,v) \in \Theta$, and $V = 0$ for $(q,v) \in \partial\Theta$. Then the instability of $q = 0$ follows from Corollary 2.1 of Chetaev theorem. Continuing to assume (A) and (B), the above results have had several generalizations.
(1) Palamodov [61] proved that $q = 0$ is unstable if Π is C^2 and $q = 0$ is an isolated critical point for Π_m. The result again follows from Corollary 2.1 of Chetaev theorem with $V = -\langle z, \frac{\partial T}{\partial v}\rangle$, $z = q - \sigma\|q\|^{2-m}\nabla\Pi(q)$, $\sigma > 0$ small, and, for some small $\varepsilon > 0$,

$$\Theta = \{(q,v) \in \Omega \times \mathbf{R}^n : \|q\| < \varepsilon, H < 0, \langle z, \frac{\partial T}{\partial v}\rangle > 0\},$$

(2) Koslov and Palamodov [31] proved that $q = 0$ is unstable if T and Π are both analytic. This result was extended in [30] by Koslov to the case that T and Π are only C^∞. Let (S) be the second order system obtained by setting $v = \dot{q}$ in $(7.1)_1$. In [30],[31] first it is shown that (S) admits a nonzero formal series solution whose general term tends to zero as $t \to +\infty$. In the analytic case this series is actually a solution $q(t)$. In the C^∞ case there exists a solution $q(t)$ (Kuznetsov [35]) which is C^∞ and has such a series as an asymptotic expansion. In both cases we have also $\dot{q}(t) \to 0$ as $t \to +\infty$. Indeed the integral of energy implies that along this solution $\dot{q}(t)$ is bounded in the future and then the positive limit set , say Λ^+, of the solution $q(t)$, $v(t) = \dot{q}(t)$ of (7.1) is nonempty. Since Λ^+ is invariant we see that Λ^+ consists exactly of the origin of the phase space. Then system (7.1) admits a solution $q(t)$, $v(t)$ which tends asymptotically to $(0,0)$ as $t \to +\infty$. Since (S) is reversible, then (7.1) admits the solution $y(t) = q(-t)$, $w(t) = -v(-t)$ which is asymptotic to $(0,0)$ as $t \to -\infty$. This proves the instability of the equilibrium position $q = 0$.
(3) The results in [30,31] have been extended by Moauro and Negrini [56] to the case that T and Π are C^{m+1} (see also[58]). Again the instability is proved by showing the existence of a nontrivial motion which is asymptotic to the equilibrium as $t \to -\infty$. This is obtained by using several transformations to obtain from (7.1) a suitable auxiliary autonomous differential system about which we apply known results concerning the existence of a stable manifold.

A further generalization of the results in [30,31] is obtained in [93] by Taliaferro, by assuming that T and Π are C^2 and

$$\Pi^{((i))}(q) = \Pi_m^{((i))}(q) + O(\|q\|^{m+\varepsilon-i}) \text{ as } q \to 0$$

where $^{(i)}$, $i = 0,1,2$, denotes the Frechet differentiation of order i, $\varepsilon \in (0,1]$ and Π_m is a positively homogeneous C^3 function of degree m. If the set $\{q : \Pi_m(q) < 0\}$ is nonempty, then the equilibrium position $q = 0$ is unstable. The main idea in Taliaferro paper is still the search for a nontrivial motion which is asymptotic to $q = 0$ as $t \to -\infty$. The proof of existence of such a motion is now based on the method of successive approximation.

7.2. The instability Liapunov result in case (b) has been extended by means of techniques of the calculus of variations by Hagedorn (1971, [20]), Liubushin (1980, [40]), and Taliaferro (1980, [91]). In [20] it has been shown that if $\Pi \in C^2(\Omega, \mathbf{R})$ and has a strict relative maximum at $q = 0$, then the equilibrium $q = 0$ is unstable. The proof is obtained by using Jacobi principle which states that if $q(s)$ is a C^2 stationary curve in Ω for the functional

$$\int_0^1 \{[h - \Pi(q(s))]\langle q'(s), A(q(s))q'(s)\rangle\}^{\frac{1}{2}} ds,$$

with $q(0) = 0$, where h is a positive constant, $(')$ is the derivative with respect to s and if $\bar{q}(t) = q(s)$ where

$$t = \alpha(s) = \int_0^s (\langle q'(\xi), A(q(\xi))q'(\xi)\rangle)^{\frac{1}{2}}[h - \Pi(q(\xi))]^{-\frac{1}{2}} d\xi,$$

then $\bar{q}(t)$ is a solution of (7.1) for $0 \le t \le \alpha(1)$ with energy h. One may obtain that $q(0)$ is arbitrarily small by assuming h sufficiently small. The result is achieved by showing that for any $h > 0$ (7.1) has a C^2 stationary curve in Ω with $q(0) = 0$, which comes within any preassigned distance of the boundary of Ω. The existence of such a stationary curve is established by virtue of a theorem of Carathéodory.

Hagedorn result has been generalized in [40] and in [91] to the case in which Π is C^1 and has a local (not necessarily strict) maximum at $q = 0$. Jacobi principle is again used but the existence of an appropriate stationary curve of (7.1) is obtained by means of arguments that allow for the weakening of the assumptions on T, Π.

7.3. We have seen that under very general assumptions the instability of the equilibrium position $q = 0$ follows from the condition that Π does not have a relative minimum and this property is recognizable on the first nonzero part of the expansion of Π. What can be said when Π does not have a minimum and the absence of the minimum is not recognizable on the lowest nonzero form of its expansion? In this context we assume that

$$\Pi = \Pi_2 + \Pi_k + o(\|q\|^k), k \ge 3,$$

where $\Pi_2(q)$ is a semidefinite positive quadratic form and Π_k is an homogeneous polynomial of degree k. Then the equilibrium $q = 0$ is unstable in each one of the two following cases:

(α) Π is analytic, Π does not have a minimum at $q = 0$, and $\frac{\partial^2 \Pi}{\partial q^2}(0)$ is a positive semidefinite matrix with a single zero eigenvalue (Koiter, [27]);

(β) Π is C^∞ and the restriction W of Π_k to the set $\{q : \Pi_2(q) = 0\}$ does not have a minimum at $q = 0$ (Koslov, [30]).

The proof of instability in case (α) is based on Chetaev Theorem 2.6. The set Θ and the function V needed in this case coincide with those considered in case (c)

The instability in case (β) follows from the same arguments used in (2) for the C^∞ case.

In [42] Maffei, Moauro, and Negrini have generalized the instability result in case (β) by assuming that Π is C^m with $m = k+3+\mathrm{I}(k)$, where $\mathrm{I}(k)$ is the integer part of $\frac{(k-3)}{2}$. By means of arguments similar to those employed in [56] and by using a fixed point theorem, the authors are able to prove once more the existence of a motion which is asymptotic to the equilibrium as $t \to -\infty$

An old conjecture is that Lagrange–Dirichlet theorem admits a converse when Π is analytic. Partial results in this direction have been obtained in [22,28,61,92]. Very recently Palamodov [62] has proved the conjecture except for the case in which Π has a nonstrict relative minimum. The proof is based on the construction of appropriate vector fields related to the potential energy and on the search for an appropriate Liapunov function. His analysis has required delicate arguments from algebraic geometry. The open question of the nonstrict minimum has been solved by Laloy and Peiffer [37] in the case $n = 2$ and also T is analytic. The result is an improvement of previous work by Hamel [22] and Silla [89].

PART II

TOTAL STABILITY

1. Total stability for ordinary differential equations

When the dynamics of a physical system are described by differential equations we have to take into account that not only the initial data but also the same differential equations are known within small errors which are completely out of our control. In any concrete problem then, for the observation or the realization of any given evolution, it appears to be necessary to analyze the stable behavior of the corresponding solution with respect not only to the initial data but also to the perturbations of the r.h.s. of the equations of motion. This stability property will be referred to as total stability.

Theoretical as well as applicable problems of total stability have been considered in several papers (see for instance [16,18,34,43,57,82,84,86,87]).

Consider again the differential equation

$$x = f(t, x), \tag{1.1}$$

where we adopt the same notations and assumptions of Section I, 1. In particular $f \in C(I \times D, \mathbf{R}^n)$ is locally Lipschitzian in x, $f(t,0) \equiv 0$, D is an open set which contains the origin, I is an interval not bounded from above, and $\chi = \rho(0, \partial D)$. Consider the vector space

$$\mathcal{F} = \{h \in C(I \times D, \mathbf{R}^n): \text{ locally Lipschitzian in } x \text{ and bounded in } I \times D\},$$

and denote by $|\cdot|$ the C^0 norm in $\mathcal{F}$. For any $h \in \mathcal{F}$ the equation

$$\dot{x} = f(t,x) + h(t,x), \tag{1.2}$$

will be called a perturbed equation of (1.1) and (1.1) will be referred to as the unperturbed equation. Let $x_h(t, t_0, x_0)$ be the noncontinuable solution of (1.2) which satisfies the initial condition (t_0, x_0). The following definition of total stability is equivalent to that given by Dubosin [16].

Definition 1.1 *The solution $x \equiv 0$ of (1.1) is said to be:*

(i) *totally stable if for any $\varepsilon \in (0, \chi)$ and $t_0 \in I$ there exists $\delta_1 = \delta_1(t_0, \varepsilon) \in (0, \varepsilon)$, $\delta_2 = \delta_2(t_0, \varepsilon) > 0$ such that if $\|x_0\| < \delta_1$, $|h| < \delta_2$, then $\|x_h(t, t_0, x_0)\| < \varepsilon$ for $t \geq t_0$;*

(ii) *uniformly totally stable if property (i) holds with δ_1 and δ_2 independent of t_0.*

A relationship between uniform asymptotic stability and uniform total stability is given by the following theorem.

Theorem 1.1 (Malkin, [43]) *Assume f satisfies a Lipschitz condition in x uniformly in t, that is, for every compact subset K of D there exists $L(K) > 0$ such that $\|f(t,x_1) - f(t,x_2)\| \leq L(K)\|x_1 - x_2\|$ for any x_1, x_2 in K. Then, if the null solution of (1.1) is uniformly asymptotically stable, it is uniformly totally stable.*

The proof is easily obtained by appropriate estimations along the solutions of the perturbed equations of the derivative of the Liapunov function of Theorem I,2.7, associated with the uniform asymptotic stability of the null solution of (1.1). Malkin theorem does not admit in general a converse. Hence it leaves open the problem of a characterization of total stability by means of properties concerning the flow of the unperturbed system. This problem has been solved for periodic differential systems for which $f \in C(I \times D, \mathbf{R}^n)$ is locally Lipschitzian in x. Let $U \subset D$ be a compact set such that if $(t_0, x_0) \in I \times U$ then $x(t, t_0, x_0)$ is defined for all $t \geq t_0$. We say that U is quasi-contracting under (1.1) if there exists a compact set $A \subset \text{int}(U)$ and a divergent sequence (t_n), $t_{n+1} > t_n$ such that $\sup\{t_{n+1} - t_n : n \in \mathbf{N}\} < +\infty$ and $x(t_{n+1}, t_n, x_0) \in A$ for all $x_0 \in U$ and $n \in \mathbf{N}$. The following theorem is a generalization of a theorem given by Seibert [87] for autonomous differential systems.

Theorem 1.2 ([82]) *Let $f \in C(I \times D, \mathbf{R}^n)$ be locally Lipschitzian in x and $f(t+T, x) \equiv f(t, x)$ for some fixed $T > 0$. Then the solution $x \equiv 0$ of* (1.1) *is (uniformly) totally stable if and only if it is stable and the origin of $\mathbf{R}^n$ admits a fundamental system of compact neighborhoods which are quasi-contracting under* (1.1).

Taking into account the invariance of volumes for conservative systems, Theorem 1.2 implies that no periodic motions (in particular no equilibrium positions) of conservative systems are totally stable. This lack of total stability raises the following question: in what sense can the motions of conservative systems be observable?

Let $\mathcal{S}$ be a conservative system. A possible explanation of observability may be given by means of the so called *barrier forces*, that is forces neglected in a first approximation but which act on $\mathcal{S}$ and protect $\mathcal{S}$ from the influence of arbitrarily small perturbations of the equations of motion. In this context we mention for example a paper of Cetaev [13], mainly devoted to an isolated equilibrium position in which the potential energy has a strict minimum. The author remarks that in most mechanical systems the conservative schema does not describe accurately the reality because small strictly dissipative forces are ever present. Thus the equilibrium is uniformly asymptotically stable and hence totally stable.

In those cases in which classes of perturbations are of prevalent importance, the observability follows from a stable behavior restricted to these classes. For instance in studying Celestial Mechanics the perturbations which preserve the conservative nature of the system are clearly prevalent. Thus the analysis of conditional total stability, that is total stability with respect to perturbations which satisfy appropriate conditions, arises in the most natural way. This approach has been considered in [57,84]. The results presented here are essentially those appearing in [84].

2. Conditional total stability for ordinary differential equations

Let $\mathcal{E}$ be a normed vector space and denote by $|\cdot|$ its norm. Let $\Lambda \subseteq \mathcal{E}$ be a domain which contains the origin 0 of $\mathcal{E}$ and such that 0 is an accumulation point for Λ. Let $g \in C(I \times D \times \Lambda, \mathbf{R}^n)$ be locally Lipschitzian in x and such that

$$\sup\{\|g(t, x, \lambda) - f(t, x)\|, (t, x) \in I \times D\} \leq a(|\lambda|) \tag{2.1}$$

for any $\lambda \in \Lambda$ and for some $a \in K$. Let $\mathcal{U}$ be the set of all perturbations of f given by

$$\mathcal{U} = \{g(\cdot, \cdot, \lambda) - f : \lambda \in \Lambda\}.$$

For any $\lambda \in \Lambda$ consider the perturbed equation

$$\dot{x} = g(t, x, \lambda), \tag{2.2}$$

and denote by $x_\lambda(t, t_0, x_0)$ the noncontinuable solution of (2.2) which satisfies the initial condition (t_0, x_0). Let $V \in C^1(I \times D, \mathbf{R})$. Often we need to consider the derivative of V along the solutions of the perturbed equation (2.2), $\lambda \in \Lambda$. This derivative will be denoted by $\dot{V}_\lambda$.

Definition 2.1 *The solution $x \equiv 0$ of the unperturbed equation* (1.1) *is said to be:*

(*i*) *$\mathcal{U}$-totally stable if for any $\varepsilon \in (0, \chi)$ and $t_0 \in I$ there exists $\sigma_1 = \sigma_1(t_0, \varepsilon) \in (0, \varepsilon)$, $\sigma_2 = \sigma_2(t_0, \varepsilon) > 0$ such that if $\|x_0\| < \sigma_1$, $\lambda \in \Lambda$, $|\lambda| < \sigma_2$, then $\|x_\lambda(t, t_0, x_0)\| < \varepsilon$ for $t \geq t_0$;*

(*ii*) *$\mathcal{U}$-uniformly totally stable if property (i) holds with σ_1 and σ_2 independent of t_0.*

Assume now in particular that $g(t, 0, \lambda) \equiv 0$ so that $x \equiv 0$ is solution of (2.2) for each $\lambda \in \Lambda$. We point out that the total stability of the zero solution of the unperturbed equation (1.1) is not equivalent to the stability of the zero solution of the perturbed equations. In other words we may have: (*a*) the origin is stable for all perturbed equations and is not $\mathcal{U}$-totally stable; (*b*) the origin is unstable for all perturbed equations (with $\lambda \neq 0$) and is $\mathcal{U}$-totally stable. Later on we will furnish examples which depict both situations. First we state the following proposition.

Proposition 2.1 *Let $g(t, 0, \lambda) = 0$ for all $\lambda \in \Lambda$. Then:*

(*I*) *For any $\lambda \in \Lambda$ let $r(t_0, \varepsilon, \lambda)$ be the radius of stability defined for* (2.1). *Then the zero solution of* (1.1) *is $\mathcal{U}$-totally stable if and only if for any $\varepsilon > 0$ and for any $t_0 \in I$ one has $\liminf_{\lambda \to 0} r(t_0, \varepsilon, \lambda) > 0$.*

(*II*) *For any $\lambda \in \Lambda$ let $r(\varepsilon, \lambda)$ be the radius of uniform stability defined for* (2.1). *Then the zero solution of* (1.1) *is $\mathcal{U}$-uniformly totally stable if and only if for any $\varepsilon > 0$ one has $\liminf_{\lambda \to 0} r(\varepsilon, \lambda) > 0$.*

(*III*) *For any $\lambda \in \Lambda$ let $R(t_0, \lambda)$ be the corresponding radius of instability for* (2.1). *Then $\limsup_{\lambda \to 0} R(t_0, \lambda) > 0$ implies that the zero solution of* (1.1) *is not $\mathcal{U}$-totally stable.*

Examples. Consider in an open bounded subset D of $\mathbf{R}^2$ containing the origin the two systems in $x = (x_1, x_2)$

$$\begin{aligned} \dot{x}_1 &= -x_2 - \lambda^2 x_1 + \lambda x_1(x_1^2 + x_2^2) \\ \dot{x}_2 &= x_1 - \lambda^2 x_2 + \lambda x_2(x_1^2 + x_2^2) \end{aligned} \tag{2.3}$$

$$\begin{aligned} \dot{x}_1 &= -x_2 - x_1(x_1^2 + x_2^2) + \lambda x_1 \\ \dot{x}_2 &= x_1 - x_2(x_1^2 + x_2^2) + \lambda x_2 \end{aligned} \tag{2.4}$$

depending on a parameter $\lambda \in \mathbf{R}^+$. For any $\lambda \in \mathbf{R}^+$ $x \equiv 0$ is solution of systems (2.3) and (2.4). We assume in both cases that the unperturbed system is given by $\lambda = 0$. Let us consider system (2.3) and use the Liapunov function $V(x) = \|x\|^2$. Along the solutions of (2.3) we have $\dot{V}_\lambda(x) = 2\lambda\|x\|^2[\|x\|^2 - \lambda]$. Thus the origin is uniformly asymptotically stable for $\lambda > 0$ and is uniformly stable for $\lambda = 0$. If $\lambda \in (0, \chi)$ the circle $\|x\|^2 = \lambda$ is a trajectory of system (2.3) and hence for $\varepsilon \in (0, \chi)$ the radius $r(\varepsilon, \lambda)$ of stability satisfies $r(\varepsilon, \lambda) < \lambda^{\frac{1}{2}}$. Hence $r(\varepsilon, \lambda) \to 0$ as $\lambda \to 0$. By using Proposition 2.1 we then recognize that the null solution of the unperturbed system is not totally stable with respect to the family of the perturbed systems (2.3), that is it is not $\mathcal{U}$-totally stable with

$$\mathcal{U} = \{h_\lambda : D \to D,\ h_\lambda(x) = \lambda x(\|x\|^2 - \lambda), \lambda \geq 0\}.$$

With regard to system (2.4) we find that the null solution is uniformly asymptotically stable for $\lambda = 0$, and thus unconditionally totally stable by virtue of Theorem 1.1. In particular it is totally stable with respect to the family of perturbed systems (2.4). Let use again the function $V(x) = \|x\|^2$. Along the solutions of (2.4) we have $\dot{V}_\lambda(x) = 2\|x\|^2[\lambda - \|x\|^2]$. Thus the origin is unstable for any $\lambda > 0$. Moreover we notice that given any $\varepsilon \in (0, \chi)$ we have $r(\varepsilon, \lambda) = \varepsilon$ provided λ is sufficiently small. Then, clearly the radius of stability tends to ε as $\lambda \to 0$, in agreement with Proposition 2.1. It is also easy to verify directly that the radius of instability $R(\lambda)$ is equal to $\lambda^{\frac{1}{2}}$ and thus satisfies $R(\lambda) \to 0$ as $\lambda \to 0$.

We notice that the converse of Proposition 2.1,(*III*) does not hold. For instance consider the planar system $\dot{x}_1 = -x_2$, $\dot{x}_2 = -x_1$ in a bounded open set $D \subset \mathbf{R}^2$ which contains the origin and assume for $\mathcal{U}$ a set of perturbations depending on a scalar parameter λ, $\mathcal{U} = \{h_\lambda : D \to D,\ h_\lambda(x) = \lambda x(\|x\|^2 - \lambda)^2, \lambda \geq 0\}$. Then, given any $\lambda > 0$, the origin is unstable for the corresponding perturbed system with a radius of instability which tends to zero as $\lambda \to 0$, while the zero solution of the unperturbed system is stable but not $\mathcal{U}$-totally stable.

3. Holonomic systems and conditional total stability of equilibrium

Consider the mechanical holonomic system $\mathcal{S}$ of Section I, 5.1 with n degrees of freedom, which is subject to the force $-\nabla\Pi$ and the dissipative force $\mathcal{D}$. We assume now that $\mathcal{D}$ is time independent:

$$\begin{aligned} &\frac{d}{dt}\frac{\partial T}{\partial v}(q, v) - \frac{\partial T}{\partial q}(q, v) = -\nabla\Pi(q) + \mathcal{D}(q, v) \\ &\frac{dq}{dt} = v, \end{aligned} \tag{3.1}$$

Since we are concerned with local problems of stability of a static solution, say $q \equiv 0$, $v \equiv 0$, we may restrict the right hand sides of (3.1) to an appropriate neighborhood $\mathcal{N}$ of the origin in the (q, v)-space. Precisely we assume $\mathcal{N} = X \times Y$ where X is a bounded subset of Ω containing the origin of the q-space such that $\bar{X} \subset \Omega$ and Y is a bounded subset of $\mathbf{R}^n$ containing the origin of the v-space. Suppose then $T \in C^2(X \times Y, \mathbf{R})$, $\Pi \in C^2(X, \mathbf{R})$, and $\mathcal{D} \in C^1(X \times Y, \mathbf{R}^n)$. Moreover we suppose that $\mathcal{D}$ satisfies the condition $\langle \mathcal{D}(q, v), v\rangle \leq 0$ for all $(q, v) \in X \times Y$. Once again we include the case that $\mathcal{D}$ is nonenergic as well as the case that $\mathcal{D}$ is a strictly dissipative force. Clearly, system (3.1) may be reduced to the form (1.1) with $x = (q, v)$, $D = X \times Y$. Furthermore, we denote again by $B'(\gamma)$ the ball $\{q \in \mathbf{R}^n : \|q\| < \gamma\}$.

Let $\mathcal{E}_1$ and $\mathcal{E}_2$ be the spaces of bounded functions $C^2(X \times Y, \mathbf{R})$ and $C^2(X, \mathbf{R})$ respectively, both equipped with the norm C^2. Moreover let $\mathcal{E}_3$ be the space of bounded functions $C^1(X \times Y, \mathbf{R}^n)$ equipped with the norm C^1. Consider now the subset Λ_1 of $\mathcal{E}_1$ consisting of all the functions λ_1 such that $\lambda_1(q, v) = \frac{1}{2}\langle v, A_{\lambda_1}(q)v\rangle$ where $A_{\lambda_1}(q)$ is an $n \times n$ symmetric positive definite matrix. Moreover, denote by Λ_3 the subset of $\mathcal{E}_3$ of the functions λ_3 such that $\langle \lambda_3(q, v), v\rangle \leq 0$ for every $(q, v) \in X \times Y$. Finally, let $\Lambda_2 = \mathcal{E}_2$ and $\Lambda = \Lambda_1 \times \Lambda_2 \times \Lambda_3$. For any $\lambda = (\lambda_1, \lambda_2, \lambda_3)$ we assume $|\lambda| = |\lambda_1|_{\mathcal{E}_1} + |\lambda_2|_{\mathcal{E}_2} + |\lambda_3|_{\mathcal{E}_3}$. For any $\lambda = (\lambda_1, \lambda_2, \lambda_3)$ consider the differential system which has the form (3.1) with T,Π,$\mathcal{D}$ replaced by the functions $T+\lambda_1$, $\Pi+\lambda_2$, $\mathcal{D}+\lambda_3$ respectively. This system will be denoted by $\mathcal{S}_\lambda$. It is easy to see that $\mathcal{S}_\lambda$ may assume the form (2.2) with $x=(q,v)$ and g satisfying the condition (2.1). For this system, and with reference to Λ, we will define the set $\mathcal{U}$ as in Section 2. The following theorem shows that under the same hypotheses of the Lagrange–Dirichlet theorem, the zero solution of system (3.1) is totally stable with respect to the above perturbations. The theorem brings out for system (3.1) a property already pointed out in [76] for perturbations depending on a finite number of parameters, and more generally in [77,86].

Theorem 3.1 . *Assume the potential energy has a strict minimum at $q=0$. Then the solution $q\equiv 0$, $v\equiv 0$ of (3.1) is $\mathcal{U}$-totally stable.*

Proof. The total energy H is positive definite. Then, given $\varepsilon\in(0,\chi)$ small and setting $m=\min\{H(q,v),\|(q,v)\|=\varepsilon\}$, one has $m>0$. For each $\lambda\in\Lambda$ let H_λ be the total energy relative to the perturbed system $\mathcal{S}_\lambda$. Along the solutions of $\mathcal{S}_\lambda$ we have $\dot{H}_\lambda(q,v)\le 0$. Moreover there exist $\sigma_1=\sigma_1(\varepsilon)\in(0,\varepsilon)$ and $\sigma_2=\sigma_2(\varepsilon)>0$ such that

$$|H_\lambda(q,v)|<\frac{m}{2}\qquad \text{if } \|(q,v)\|<\sigma_1 \quad\text{and}\quad |\lambda|<\sigma_2,$$

$$H_\lambda(q,v)\ge\frac{m}{2}\qquad \text{if } \|(q,v)\|=\varepsilon \quad\text{and}\quad |\lambda|<\sigma_2.$$

Let $(q_\lambda(t),v_\lambda(t))$ be the noncontinuable solution through (q_0,v_0) with $\|(q_0,v_0)\|<\sigma_1$ and $|\lambda|<\sigma_2$. The existence of $t_1>0$ such that

$$\|(q_\lambda(t),v_\lambda(t))\|<\varepsilon \text{ for each } t\in[0,t_1) \text{ and } \|(q_\lambda(t_1),v_\lambda(t_1))\|=\varepsilon,$$

would imply

$$\frac{m}{2}>|H_\lambda(q_0,v_0)|\ge H_\lambda(q_0,v_0)\ge H_\lambda(q_\lambda(t_1),v_\lambda(t_1))\ge\frac{m}{2},$$

a contradiction. ■

Remark 3.1 . *Let $\lambda\in\Lambda-\{0\}$ be such that $\nabla\Pi_\lambda(0)=0$, i.e. $q=0$ is an equilibrium position of the perturbed system $\mathcal{S}_\lambda$. We emphasize that the total stability of the zero solution of (3.1) does not imply in general that, for λ close to 0, $q\equiv 0$, $v\equiv 0$ is a stable solution of $\mathcal{S}_\lambda$. For example, assume that Π is a positive definite form of (even) degree ≥ 4. Furthermore, assume $\Lambda^*=\{(\lambda_1,\lambda_2,\lambda_3)\in\Lambda:\lambda_1=0,\lambda_2=\mu h(q),\lambda_3=0,\mu\ge 0\}$ where h is a quadratic form whose range contains negative values. In this case, by virtue of Theorem 3.1 the zero solution of the unperturbed system is $\mathcal{U}$-totally stable, whereas for any $\lambda\in\Lambda^*$, $\lambda\ne 0$, the zero solution of $\mathcal{S}_\lambda$ is unstable.*

Theorem 3.2 *Assume that: (i) the potential energy does not have a relative minimum at $q=0$; (ii) there exists $\gamma\in(0,\chi)$ such that there are no equilibrium positions in the set $G=\{q:\|q\|\le\gamma,\Pi(q)<0\}$. Then, the zero solution of (3.1) is not U-totally stable.*

Proof. Consider the subset of Λ

$$\Lambda^*=\{(\lambda_1,\lambda_2,\lambda_3):\lambda_1=0,\lambda_2=0,\lambda_3=\mu\tilde{\mathcal{D}}(q,v),\mu\ge 0\},$$

where $\tilde{\mathcal{D}}\in C(X\times Y,\mathbf{R}^n)$ is a strictly dissipative force. We apply Theorem I, 4.1, assuming $V=H$. By virtue of (i) V has the property $P^-(t_0)$ for any t_0 in $\mathbf{R}$. Moreover, along the solutions of any perturbed equation $\mathcal{S}_{\lambda(\mu)}$ we have

$$\dot{V}_\mu(q,v)=\langle\mathcal{D}_\mu(q,v),v\rangle\le-\mu c(\|v\|), \tag{3.2}$$

where $\dot{V}_\mu = \dot{V}_{\lambda(\mu)}$, $c \in K$, and $\mathcal{D}_\mu = \mathcal{D} + \mu\tilde{\mathcal{D}}$. Hence the set M of Theorem I, 4.1 coincides with the set $\{(q,v) \in B[\gamma] : v = 0\}$. Then, from (*ii*) it follows there are no complete positive semiorbits in $M \cap \{(q,v) \in B[\gamma] : V(q,v) < 0\}$. Hence, by taking into account Remark I, 4.1, the null solution of $\mathcal{S}_{\lambda(\mu)}$ is unstable. Moreover, it is immediate to see that $R(\mu) \geq \gamma$, where γ is the positive constant in (*ii*). By virtue of Proposition 2.1, (*III*) the proof is complete. ■

We clearly have (see Remark I,6.1):

Theorem 3.3 *Assume that* Π *is an analytic function which does not have a relative minimum at* $q = 0$. *Then the zero solution of (3.1) is not* $\mathcal{U}$*-totally stable.*

Returning to consider a C^2 potential energy Π, we wish now to inspect some cases where Π has a nonstrict relative minimum at $q = 0$.

Lemma 3.1 *Assume that* Π *has at* $q = 0$ *a nonisolated zero. For a given* $m \geq 2$ *and for* $\mu \geq 0$ *let*

$$\Pi_\mu(q) = \Pi(q) - \mu\|q\|^m.$$

Assume there exist $\gamma \in (0,\chi)$ *and two sequences* $\{a_i\}$, $\{\mu_i\}$, *with* $a_i \in (0,\gamma)$, $\mu_i > 0$, $a_i \to 0$, $\mu_i \to 0$, *such that for every* $i \in \mathbf{N}$ *and* $\mu = \mu_i$, $\Pi_\mu(q)$ *does not have critical points in the set* $\{q : q \in B'[\gamma] - B'(a_i), \Pi_\mu(q) < 0\}$. *Then the zero solution of (3.1) is not* $\mathcal{U}$*-totally stable.*

Proof. Let $\{q_n\}$ be a sequence of points in $B'(\gamma)$ with $q_n \neq 0$ and $q_n \to 0$. Let Λ^* be the subset of Λ defined by

$$\Lambda^* = \{(\lambda_1, \lambda_2, \lambda_3) : \lambda_1 = 0, \lambda_2 = -\mu\|q\|^m, \lambda_3 = \mu\tilde{D}(q,v), \mu \geq 0\},$$

where $\tilde{D} \in C^1(X \times Y, \mathbf{R}^n)$ is a strictly dissipative force. For each $\mu > 0$ let $\mathcal{S}_{\lambda(\mu)}$ be the perturbed differential system corresponding to this μ in Λ^*. Given any n in $\mathbf{N}$ and $\varepsilon > 0$ there exist $a \in (0,\gamma)$ and $\mu(0,\varepsilon)$ such that $q_n \in B'(\gamma) - B'(a)$ and Π_μ does not have critical points in the set $P = \{q : q \in B'[\gamma] - B'(a), \Pi_\mu(q) < 0\}$. Let $x_0 = (q_n, 0)$ and let $x(t) = (q(t), v(t))$ be the noncontinuable solution of $\mathcal{S}_{\lambda(\mu)}$ such that $x(0) = x_0$. We have $\|x_0\| < \gamma$ and we assume that $x(t)$ exists and satisfies $\|x(t)\| < \gamma$ for all $t \geq 0$. Denote by H_μ the total energy for $\mathcal{S}_{\lambda(\mu)}$ and by $\dot{H}_\mu$ its derivative along the solutions of $\mathcal{S}_{\lambda(\mu)}$. Since $\dot{H}_\mu(q,v) \leq 0$, we have

$$0 \leq \Pi(q(t)) \leq \mu\|q(t)\|^m - \mu\|q_n\|^m.$$

Therefore $\gamma > \|q(t)\| \geq \|q_n\| \geq a$, $x(t) \in B(\gamma) - B(a)$ and $\Pi_\mu(q(t)) < 0$ for all $t \geq 0$. Since Π_μ does not have critical points in P, and $\dot{H}_\mu(q,v) = 0$ if and only if $v = 0$, then, by using again Theorem I, 4.1, we find that the null solution of $\mathcal{S}_{\lambda(\mu)}$ is unstable. Moreover, its radius of instability satisfies the condition $R(\mu) \geq \gamma$. This leads to a contradiction and in view of the arbitrariness of n and ε the proof is complete. ■

Lemma 3.2 *Let us assume that* Π *has a nonstrict relative minimum at* $q = 0$. *Moreover suppose that for some* $\gamma \in (0,\chi)$ *and any* $a \in (0,\gamma)$ *there exists* $k > 0$ *such that*

$$k\Pi(q) - \langle \Pi(q), q \rangle \geq 0, \tag{3.3}$$

in the annulus $B'[\gamma] - B'(a)$. *Then the zero solution of (3.1) is not* $\mathcal{U}$*-totally stable.*

Proof. We may assume $\Pi(q) \geq 0$ in $B'[\gamma]$. For a given $m \geq \max\{2,k\}$ and for $\mu \geq 0$ let

$$\Pi_\mu(q) = \Pi(q) - \mu\|q\|^m.$$

By virtue of Lemma 3.1, the proof will be obtained by showing that for any $\mu > 0$ there are no critical points of Π_μ in the set $\{q : a \le \|q\| \le \gamma,\ \Pi_\mu(q) < 0\}$. For this it is sufficient to prove that if $q \in B'[\gamma] - B'(a)$ satisfies $\langle \nabla\Pi_\mu(q), q\rangle = 0$, then $\Pi_\mu(q) \ge 0$. Since

$$\langle \nabla\Pi_\mu(q), q\rangle = \langle \nabla\Pi(q), q\rangle - m\mu\|q\|^m,$$

it follows that $\langle \nabla\Pi_\mu(q), q\rangle = 0$ implies

$$\Pi_\mu(q) = \Pi(q) - (m)^{-1}\langle \nabla\Pi(q), q\rangle;$$

thus $\Pi_\mu(q) \ge 0$ by virtue of (3.3) and taking into account that $m \ge k$ and $\Pi(q) \ge 0$. The proof is complete. ■

Proposition 3.1 *Suppose that Π has at $q = 0$ a nonisolated zero and*

$$\Pi(q) = \sigma(q)\sum_{j=2}^{h}\Pi_{(j)}(q), \tag{3.4}$$

where $h \ge 2$, $\Pi_{(2)}, \ldots, \Pi_{(h)}$ are forms of degree $2, \ldots, h$ for which

$$\sum_{j=2}^{s}\Pi_{(j)}(q) \ge 0,\ s = 2, \ldots, h, \tag{3.5}$$

and σ is a C^2 function such that $\sigma(q) > 0$ for all $q \ne 0$ in a neighborhood of the origin. Then the zero solution of (3.1) is not $\mathcal{U}$-totally stable.

Proof. Let $\gamma > 0$ be such that $\sigma(q) > 0$ for all $q \in B'(\gamma) - \{0\}$. For any $a \in (0, \gamma)$ there exists $k > h$ such that

$$(k-h)\sigma(q) - \langle \nabla\sigma(q), q\rangle \ge 0, \quad \text{for all } q \in B'[\gamma] - B'(a). \tag{3.6}$$

Then we have:

$$k\Pi(q) - \langle \nabla\Pi(q), q\rangle = \sigma(q)\sum_{j=2}^{h}(k-j)\Pi_{(j)}(q) - \langle \nabla\sigma(q), q\rangle\sum_{j=2}^{h}\Pi_{(j)}(q)$$

$$= [(k-h)\sigma(q) - \langle \nabla\sigma(q), q\rangle]\sum_{j=2}^{h}\Pi_{(j)}(q) + \sigma(q)\sum_{j=2}^{h-1}(h-j)\Pi_{(j)}(q).$$

We easily verify that

$$\sum_{j=2}^{h-1}(h-j)\Pi_{(j)}(q) = \sum_{j=2}^{h-1}\Pi_{(j)}(q) + \sum_{j=2}^{h-2}\Pi_{(j)}(q) + \ldots + \sum_{j=2}^{3}\Pi_{(j)}(q) + \Pi_{(2)}(q).$$

Hence, by virtue of (3.5), (3.6) we get

$$k\Pi(q) - \langle \nabla\Pi(q), q\rangle \ge 0, \quad \text{for all } q \in B'[\gamma] - B'(a).$$

The result follows from Lemma 3.2. ■

Proposition 3.2 *Assume that Π depends only on $r < n$ coordinates, say $q_1, \ldots, q_r$, and is positive definite in these coordinates. Then the solution $q \equiv 0$, $v \equiv 0$ of (3.1) is not $\mathcal{U}$-totally stable.*

Proof. Let $y = (q_1, \ldots, q_r)$ and $z = (q_{r+1}, \ldots, q_n)$. For $\mu \geq 0$ let

$$\Pi_\mu(y, z) = \Pi(y) - \mu\|(y, z)\|^2. \tag{3.7}$$

Let $\gamma \in (0, \chi)$ be such that $\Pi(y) > 0$ for all $y : 0 < \|y\| \leq \gamma$, and assume $\mu > 0$. Clearly $\nabla\Pi_\mu(y, z) = 0$ if and only if

$$\nabla\Pi(y) = 2\mu y \quad \text{and} \quad z = 0.$$

For any $a \in (0, \gamma)$ the critical points $q = (y, z)$ of Π_μ lying in the annulus $B'[\gamma] - B'(a)$ satisfy $\gamma \geq \|y\| \geq a$ and $\|z\| = 0$. Hence if $\mu > 0$ is small enough for each one of these critical points we have $\Pi_\mu(y, z) > 0$. Then the result follows from Lemma 3.1. ∎

PART III

BIFURCATION AND STABILITY

1. Exchange of stability and bifurcation

Consider the dynamics described by a differential equation depending on a parameter.Suppose that at a certain (critical) value of the parameter some invariant set drastically changes its behavior giving rise to new invariant sets. We shall refer to this as a bifurcation phenomenon.

Bifurcation phenomena often occur because of exchange of stability properties of an invariant set under perturbations. This connection between exchange of stability and appearance of bifurcating sets can be exploited in order to obtain existence results and information about the structure of the bifurcating sets. In other words, stability arguments and appropriate Liapunov functions can be used not only for analyzing the qualitative behavior of the flow near the bifurcating sets, but also to prove the existence of these sets. With regard to existence problems, the procedure resulting from the present approach can sometimes fruitfully replace the methods more usually employed, as for example the method involving the implicit function theorem, the Liapunov–Schmidt method and its known variants, and those methods based on topological degree arguments.

Although many of the bifurcation results we will give in the following are valid in the more general context of periodic equations, we will consider only autonomous systems. Precisely, we will be concerned with two kinds of problems:

(*a*) Bifurcation from invariant sets into invariant sets for one–parameter families of autonomous differential equations in $\mathbf{R}^n$, $\dot{u} = f(u, \mu)$, when μ crosses an appropriate value. In this case from suitable stability properties of the unperturbed system one obtain information about existence, structure, and stability properties of the bifurcating sets. Particular interest has the Hopf bifurcation, that is the bifurcation from an equilibrium point into periodic orbits.

(*b*) Generalized Hopf bifurcation. Let us consider in $\mathbf{R}^n$ a differential equation $\dot{u} = f_0(u)$, $f_0(0) = 0$, where f_0 is smooth and the Jacobian matrix $f_0'(0)$ has two purely imaginary eigenvalues $\pm i$, whereas the remaining eigenvalues satisfy a nonresonance condition. For those f close to f_0 (in an appropriate topology) we discuss existence and properties of the periodic orbits of the perturbed equations $\dot{u} = f(u)$ which lie near the origin and have period close to 2π.

The problem of bifurcation has been treated in very many papers and in several books. Some of these papers, in the spirit of our work, will be quoted later. Among the books we mention the books by Andronov et al. [1], Marsden and McCracken [53], Chow and Hale [14].

2. Preliminaries

Consider the differential system

$$\dot{u} = f(u), \tag{2.1}$$

where $f \in C(\mathbf{R}^n, \mathbf{R}^n)$ satisfies conditions that insure global existence and uniqueness of the solutions. We denote by $u(t, u_0)$ the noncontinuable solution of (2.1) through $(0, u_0)$. Let M be a compact set in $\mathbf{R}^n$. The concepts of stability, attractivity, total stability, etc. apply to M, provided the norm $\|u\|$ is replaced by the distance $\rho(u, M)$. For example we say that M is (uniformly) stable if for any $\varepsilon > 0$ there exists $\delta > 0$ such that if $\rho(u_0, M) < \delta$ then

$\rho(u(t, u_0), M) < \varepsilon$ for $t \geq 0$. Also the theorems on stability are easily extended to the stability of compact sets.

For our analysis we need to define the following sets. For each u we denote by $\Lambda^+(u)$ the positive limit set of the orbit passing through u and by $J^+(u)$ the positive first prolongational limit set of u, that is

$$\Lambda^+(u) = \{v \in D : \text{ there exists a sequence } \{t_n\} \text{ in } \mathbf{R} \text{ such that } t_n \to +\infty, u(t_n, u) \to v\};$$
$$J^+(u) = \{v \in D : \text{ there exist sequences } \{t_n\} \text{ in } \mathbf{R} \text{ and } \{u_n\} \text{ in } D \text{ such that } t_n \to +\infty, u_n \to u, u(t_n, u_n) \to v\}.$$

These sets are compact and invariant. The set

$$A^+(M) = \{x \in D : \rho(u(t, u), M) \to 0 \text{ as } t \to +\infty\},$$

is said to be the region of attraction of M.

Definition 2.1 *M is said to be*

(i) *an attractor if $A^+(M)$ is a neighborhood of M;*

(ii) *a uniform attractor if M is an attractor and for any $u_0 \in A^+(M)$ and any neighborhood V of M there exist a neighborhood U of u_0 and $T \in \mathbf{R}$ such that $u(t, U) \in V$ for $t \geq T$.*

Theorem 2.1 *We have:*

(i) *If M is an attractor, then $A^+(M)$ is open.*

(ii) *M is a uniform attractor if and only if the set $\{u \in D : J^+(u) \neq \emptyset, J^+(u) \subset M\}$ is a neighborhood of M (and coincides with $A^+(M)$).*

(iii) *If M is positively invariant and a uniform attractor, then M is asymptotically stable.*

(iv) *If M is asymptotically stable, then M is a uniform attractor.*

Similarly we may define negative attractivity and negative stability properties by analyzing the flow in the past. Thus in particular we may define $\Lambda^-(u)$, $J^-(u)$ and $A^-(M)$. Clearly if $v \in J^+(u)$, then $u \in J^-(v)$. The statements in Theorem 2.1 are immediately extended to negative properties. For instance if M is negatively invariant and a negative uniform attractor, then M is negatively asymptotically stable. This property will also be referred to as complete instability.

For the proof of Theorem 2.1 and more details see for instance Bathia and Szegö [8].

3. Bifurcation for one parameter families of autonomous systems

Consider in $\mathbf{R}^n$ a one parameter family of differential equations

$$\dot{u} = f(u, \mu), \tag{3.1$_\mu$}$$

where $f \in C(\mathbf{R}^n \times \mathbf{R}, \mathbf{R}^n)$ and for each μ satisfies conditions that insure global existence and uniqueness of solutions of $(3.1)_\mu$. We call $(3.1)_0$ the unperturbed equation, and $(3.1)_\mu$ a perturbed equation. Let M_0 be a compact set in $\mathbf{R}^n$ which is invariant with respect to

$(3.1)_0$. For some $\mu_1 > 0$ we consider a family (M_μ), $\mu \in (0, \mu_1)$, of nonempty compact subsets of $\mathbf{R}^n$ which satisfy the two conditions: (*a*) each M_μ is an invariant set for $(3.1)_\mu$; (*b*) $\max\{\rho(u, M_0),\ u \in M_\mu\} \to 0$ as $\mu \to 0$. Then (M_μ) will be called a family continuing M_0 on $(0, \mu_1)$. This continuation will be called proper if in addition $\max\{\rho(u, M_\mu),\ u \in M_0\} \to 0$ as $\mu \to 0$.

Definition 3.1 *Let (M_μ) be a family continuing M_0 on $(0, \mu_1)$. Then $\mu = 0$ will be called a bifurcation value for (M_μ) if there exists $\mu^* \in (0, \mu_1)$ and a second family (M'_μ) continuing M_0 in $(0, \mu^*)$ such that $M_\mu \cap M'_\mu = \emptyset$ for each $\mu \in (0, \mu^*)$. The sets M'_μ will be specified as sets bifurcating from (M_μ), or simply from M_0 if $M_\mu \equiv M_0$.*

Theorem 3.1 ([49]) *Suppose that $f(0, \mu) \equiv 0$ and that the origin is asymptotically stable for $\mu = 0$ and completely unstable for $\mu > 0$. Then there exists $\mu^* > 0$ and a neighborhood $\mathcal{N}$ of the origin such that if $\mu \in (0, \mu^*)$ and M'_μ is the largest invariant compact subset of $\mathcal{N} - \{0\}$, then (M'_μ) is a family of asymptotically stable compact sets bifurcating from $\{0\}$.*

Proof. By virtue of Theorem I, 2.7 there exist $\gamma > 0$ and $V \in C^1(D, \mathbf{R})$ such that

$$a(\|u\|) \le V(u) \le b(\|u\|), \tag{3.2}$$
$$\dot V_0(u) \le -c(\|u\|), \tag{3.3}$$

for some $a, b, c \in K$ and for all u in $B[\gamma]$. Here $\dot V_\mu$ denotes the derivative of V along the solutions of $(3.1)_\mu$. We now determine a neighborhood $\mathcal{N}$ of the origin such that for each μ small $\mathcal{N}$ is asymptotically stable for $(3.1)_\mu$. Choose $\lambda \in (0, a(\gamma))$ and consider the set $\mathcal{N} = \{u : \|u\| \le \gamma,\ V(u) \le \lambda\}$. Clearly, taking into account (3.2), we see that $\mathcal{N}$ contains all the points for which $\|u\| \le b^{-1}(\lambda)$. Thus $\mathcal{N}$ is a compact neighborhood of the origin and is contained in $B(\gamma)$. By (3.3) and continuity arguments we may select $\mu^* > 0$ so that $\dot V_\mu \le -\frac{1}{2}c(b^{-1}(\lambda))$ for all $\mu \in (0, \mu^*)$ and $u \in B[\gamma] - B(b^{-1}(\lambda))$. Hence $\mathcal{N}$ is asymptotically stable and its region of attraction contains a fixed neighborhood $\mathcal{N}^*$ of the origin. Hence by virtue of Theorem 2.1 (*ii*), for each $u \in \mathcal{N}^*$ we have $J^+(u) \ne \emptyset$ and $J^+(u) \subset \mathcal{N}$. Let F_μ be the largest invariant set contained in $\mathcal{N}$. We show now that $A^-(0) \subset F_\mu$. Since $A^-(0)$ is invariant, we need only to prove that $A^-(0) \subset \mathcal{N}$. For, if $u \in A^-(0)$, then there exists $t \in \mathbf{R}^-$ such that $u(t, u) \in \mathcal{N}$ and then $u(-t, u(t, u)) = u \in \mathcal{N}$ by virtue of the positive invariance of $\mathcal{N}$. Set $M'_\mu = F_\mu - A^-(0)$. Clearly M'_μ is the largest invariant compact subset of $\mathcal{N} - \{0\}$. We prove now that M'_μ is a uniform attractor and its region of uniform attraction contains $\mathcal{N}^* - \{0\}$. By virtue of Theorem 2.1 (*iii*) this implies that M'_μ is asymptotically stable. For, assume $u \in \mathcal{N}^* - \{0\}$. We have already seen that $J^+(u) \ne \emptyset$ and $J^+(u) \subset \mathcal{N}$. Moreover, since $J^+(u)$ is invariant we have $J^+(u) \subset F_\mu$. Hence it is sufficient now to prove that $v \in J^+(u)$ implies $v \notin A^-(0)$. Indeed $v \in A^-(0)$ would imply $J^-(v) = \{0\}$. Since $u \in J^-(v)$ we get $u = 0$, a contradiction. Thus M'_μ is a uniform attractor and then asymptotically stable by virtue of Theorem 2.1 (*iii*). Finally it is immediate to see that $M'_\mu \to 0$ as $\mu \to 0$. The proof is then complete. ■

Corollary 3.1 *If $n = 2$ the sets M'_μ of Theorem 3.1 are annular regions between two periodic orbits of $(3.1)_\mu$.*

Related results have been given in [7,50,88]. Consider now the following assumption.

Assumption (A) *Let $f \in C^{k+1}(\mathbf{R}^n \times \mathbf{R}, \mathbf{R}^n)$, $k \ge 3$, and suppose that the Jacobian matrix $D_u f(0, \mu)$ has a couple of complex eigenvalues $\alpha(\mu) \pm i\beta(\mu)$ with $\alpha(0) = 0$, $\alpha'(0) \ne 0$, $\beta(0) \ne 0$, and that the remaining eigenvalues have negative real parts for $\mu = 0$.*

Under Assumption (*A*), system $(3.1)_\mu$ admits a 2–dimensional center manifold S_μ which is tangent to the eigenspace corresponding to the eigenvalues $\alpha(\mu) \pm i\beta(\mu)$. Moreover H_μ is locally invariant and locally attracting (see for instance [14]). Hence, the bifurcating sets (and in particular) all the periodic orbits of $(3.1)_\mu$ close to the origin for small μ are lying on S_μ. Moreover they satisfy a two dimensional system (i.e. the reduction of $(3.1)_\mu$ to S_μ). Hence, an analysis of bifurcation in $\mathbf{R}^2$ gives also information for system $(3.1)_\mu$. This analysis will be accomplished in the next section.

4. Hopf bifurcation on the plane

4.1. Consider in $\mathbf{R}^2$ the one parameter family of differential equations

$$\dot{u} = f(u,\mu), \; f(0,\mu) \equiv 0, \tag{4.1$_\mu$}$$

where $f \in C^{k+1}(\mathbf{R}^2 \times \mathbf{R}, \mathbf{R}^2)$, $k \geq 3$, satisfies conditions that insure global existence of solutions. Denoting by $\alpha(\mu) \pm i\beta(\mu)$ the eigenvalues of the Jacobian matrix $f'_u(0,\mu)$, we assume that $\alpha(0) = 0$, $\alpha'(0) \neq 0$, $\beta(0) = \lambda \neq 0$. Clearly if $\alpha'(0) > 0$ (resp. $\alpha'(0) < 0$) the null solution of $(4.1)_\mu$ is asymptotically stable for $\mu < 0$ (resp. for $\mu > 0$) and completely unstable for $\mu > 0$ (resp. $\mu < 0$).

By means of a suitable change of variables, system $(4.1)_\mu$ may be written as

$$\begin{aligned} \dot{x} &= \alpha(\mu)x - \beta(\mu)y + P(x,y,\mu) \\ \dot{y} &= \alpha(\mu)y + \beta(\mu)x + Q(x,y,\mu), \end{aligned} \tag{4.2$_\mu$}$$

where P, Q are of order ≥ 2 in (x,y) as $(x,y) \to (0,0)$.

If the null solution of $(4.1)_0$ is asymptotically stable for $\mu = 0$, then the conclusions of Corollary 3.1 hold for this system. Thus invariant annular regions will appear near the origin (for $\mu > 0$ or $\mu < 0$ according to $\alpha'(0) > 0$ or $\alpha'(0) < 0$, respectively). On the other hand, in this case we may say more on the bifurcating sets, since Hopf theorem applies.

Theorem 4.1 (Hopf, [25]) *There exist $\varepsilon > 0$, $\sigma > 0$, and a function $\mu \in C^{k-1}([0,\varepsilon), \mathbf{R})$, $\mu(0) = 0$, such that for any $(c,\mu) \in (0,\varepsilon) \times (-\sigma,\sigma)$ the orbit through (c,μ) is closed if and only if $\mu = \mu(c)$.*

We now illustrate the results concerning the stability properties of the bifurcating periodic orbits of $(4.2)_\mu$ given in [59]. First of all, setting $x = r\cos\theta$, $y = r\sin\theta$, system $(4.2)_\mu$ assumes the form

$$\begin{aligned} \dot{r} &= \alpha(\mu)r + \Psi(\theta,r,\mu) \\ \dot{\theta} &= \beta(\mu) + \Theta(\theta,r,\mu), \end{aligned} \tag{4.3$_\mu$}$$

where $\Psi(\theta,r,\mu) = o(r)$, $\Theta(\theta,r,\mu) = O(r)$ as $r \to 0$. Since $\beta(0) > 0$ we see that if $|\mu|$ is small enough and $r(t)$ remains sufficiently small for all $t \geq 0$ the function $\theta(t)$ will increase indefinitely. Since we are interested to the problem of existence and stability of periodic orbits in a neighborhood of the origin, clearly θ can play the role of t. Assuming θ as independent variable, we have for $r(\theta)$ the scalar equation

$$\frac{dr}{d\theta} = R(\theta,r,\mu), \tag{4.4}$$

with $R = (\alpha r + \Psi)(\beta + \Theta)^{-1}$. We may write

$$r(\theta,c,\mu) = u_1(\theta,\mu)c + \varphi(\theta,c,\mu)c^2, \tag{4.5}$$

where $u_1(0,\mu) \equiv 1$, $\varphi(0,c,\mu) \equiv 0$, and $\varphi \in C^k$. By substituting (4.5) into (4.4) and equating the coefficients of the terms with the same powers in c, we obtain in particular

$$\frac{\partial u_1}{\partial \theta} = \frac{\alpha(\mu)}{\beta(\mu)} u_1.$$

Hence

$$u_1(\theta,\mu) = \exp[\frac{\alpha(\mu)}{\beta(\mu)}\theta].$$

Clearly $u_1(\theta,0) \equiv 1$. The problem of determining the periodic orbits of $(4.3)_\mu$ is clearly equivalent to the search of the zeros of the so called displacement function:

$$V(c,\mu) = r(2\pi,c,\mu) - c. \tag{4.6}$$

We give now a short proof of Hopf theorem. From (4.5) and (4.6), we have

$$V(c,\mu) = [u_1(2\pi,\mu) - 1]c + \varphi(2\pi,c,\mu)c^2.$$

Hence

$$V(c,\mu) = c\tilde{V}(c,\mu),$$

where $\tilde{V}(c,\mu) = u_1(2\pi,\mu) - 1 + \varphi(2\pi,c,\mu)c$. The nonzero periodic orbits correspond to the zeros of $\tilde{V}$. Since $\tilde{V}(0,0) = 0$ and

$$\frac{\partial \tilde{V}}{\partial \mu} = [\frac{\partial u_1}{\partial \mu}(2\pi,\mu)]_{(0,0)} = \frac{2\pi}{\lambda}\alpha'(0) \neq 0, \tag{4.7}$$

the assert easily follows from the implicit function theorem.

4.2. Letting $\lambda = \beta(0)$, $X(x,y) = P(x,y,0)$ and $Y(x,y) = Q(x,y,0)$, consider the unperturbed system $(4.2)_0$

$$\begin{aligned} \dot{x} &= -\lambda y + X(x,y) \\ \dot{y} &= \lambda x + Y(x,y), \end{aligned} \tag{4.8}$$

and denote by X_j, Y_j the j-forms of the MacLaurin expansion of X, Y. We now describe an algebraic procedure due to Poincaré. This procedure will be used to obtain complete information on the stability properties of the bifurcating periodic orbits.

(A) Suppose that X, Y are analytic and set

$$F(x,y) = x^2 + y^2 + \sum_{j=3}^{m} F_j(x,y), \tag{4.9}$$

where F_j is a j-form and $m \geq 3$. Furthermore, denote by $[\dot{F}]_j$ the j-form in the expansion of $\dot{F}_{(4.8)}$. The Poincaré procedure consists either in the construction of power series formally satisfying the condition of being integrals of (4.8) or in the determination of m and $F_3, \ldots, F_m$ such that $\dot{F}_{(4.8)}$ is a sign definite function. The results deriving from this procedure are summarized in the following theorem.

Theorem 4.2 *Assume that $F_3, \ldots, F_{m-1}$ satisfy $[\dot{F}]_3 = \ldots = [\dot{F}]_{m-1} = 0$. Then : (a) if m is odd, then there exist one and only one m-form F_m such that $[\dot{F}]_m = 0$; (b) if m is even, then there exists an infinite number of m-forms F_m such that $[\dot{F}]_m = G_m(x^2+y^2)^{\frac{m}{2}}$, where G_m is a constant uniquely determined.*

In both cases the determination of F_m is obtained by solving an algebraic system of linear equations. The right hand sides of these equations involve λ, the coefficients of F_j, X_j, Y_j with $j \in \{2, \ldots, m-1\}$, and, if m is even, the unknown constant G_m. If m is odd, the above equations are independent, while, if m is even, there is exactly one linear relation among the left hand sides of them. It is just this relation (to which the right hand sides must satisfy in order to have compatible equations) that allows to the unique determination of G_m. Actually it can be proved that G_m depends only on F_j, X_j, Y_j with $j \in \{2, \ldots, m-1\}$. Set

$$M = \sup\{m \geq 3, \text{ there exist } F_3, \ldots, F_{m-1} \text{ such that } [\dot{F}]_3 = \ldots = [\dot{F}]_{m-1} = 0\}.$$

Moreover, let G be the constant defined by $G = 0$ if $M = \infty$ and $G = G_M$ if $M \in \mathbf{N}$. The numbers M and G will be respectively referred to as the index and the Poincaré constant of system (4.8). Now we distinguish the two cases:

(*a*) $M = \infty$. In this case the series

$$F = x^2 + y^2 + F_3 + \ldots \tag{4.10}$$

is, at least formally, a first integral of (4.8). It is known that actually (4.8) admits in this case an analitic integral of the form (4.10). By virtue of Liapunov theorem I, 2.1 then the null solution of (4.8) is (nonasymptotically) stable. Moreover $u_l(2\pi) = 0$, $l \geq 2$, and therefore for every small c the orbit of (4.8) is closed.

(*b*) $M \in \mathbf{N}$. In this case the polynomial

$$F = x^2 + y^2 + F_3 + \ldots. + F_M \tag{4.11}$$

satisfies the condition

$$\dot{F}_{(4.8)} = G_M(x^2 + y^2)^{\frac{M}{2}} + o[(x^2 + y^2)^{\frac{M}{2}}]. \tag{4.12}$$

Hence the null solution of (4.8) is asymptotically stable or completely unstable, according to $G_M < 0$ or $G_M > 0$. We prove now that the functions u_l satisfy the conditions $u_l(2\pi) = 0$, $l \in \{2, \ldots, M-2\}$, and $u_{M-1}(2\pi) = g$, where g is a nonzero constant having the same sign of G_M. Indeed, since $r(2\pi, c, 0) - c$ cannot be now identically zero, there exists $s \geq 2$ such that

$$r(2\pi, c, 0) = c + gc^s + o(c^s), \text{ with } g \neq 0. \tag{4.13}$$

The expansion (4.11) in polar coordinates becomes

$$F(\theta, r) = r^2 + r^3\varphi_3(\theta) + \ldots + r^M\varphi_M(\theta),$$

with $\varphi_3, \ldots, \varphi_M$ 2π-periodic functions and

$$r(\theta, c, 0) = c + u_2(\theta, 0)c^2 + o(c^2) \tag{4.14}$$

Then an integration of (4.12) in $[0, 2\pi]$ leads to

$$F(2\pi, r(2\pi, c, 0)) - F(0, c) = \frac{G_M}{\lambda} \int_0^{2\pi} [c^M + o(c^M)] d\theta.$$

Taking into account the periodicity of the functions φ_j we obtain the identity

$$2gc^{s+1} + o(c^{s+1}) = \frac{2G_M\pi}{\lambda} c^M + o(c^M)$$

from which it follows $s = M - 1$ and $g = \frac{2\pi}{\lambda} G_M$. Thus the claimed result holds.

(*B*) Now we consider system (4.8) in the general case $X, Y \in C^{k+1}$.

Definition 4.1 *Let $h \in \{2 \ldots, k\}$ be an integer. The null solution of* (4.8) *is said to be h-asymptotically stable* (*resp. h-completely unstable*) *if* (*i*) *for every* $\xi, \eta \in C^{k+1}(\mathbf{R}^2, \mathbf{R})$, *of order greater than* h *as* $(x, y) \to (0,0)$, *the null solution of the system*

$$\dot{x} = -\lambda y + X_2(x,y) + \ldots + X_h(x,y) + \xi(x,y)$$
$$\dot{y} = \lambda x + Y_2(x,y) + \ldots + Y_h(x,y) + \eta(x,y), \tag{4.15}$$

is asymptotically stable (*resp.completely unstable*) ;

(*ii*) *property* (*i*) *is not satisfied when* h *is replaced by any integer* $m \in \{2, \ldots, h-1\}$.

The 3–asymptotic stability is equivalent to the condition that the origin is a "vague attractor" in the sense of a definition of Ruelle and Takens [73]. It is not difficult to prove the following theorem, by using the properties analyzed in (*A*).

Theorem 4.3 *Let* $h \in \{2, \ldots, k\}$ *be an integer. Then the following propositions are equivalent:*

(*u*) *The null solution of* (4.8) *is* h*–asymptotically stable* (*resp.* h*–completely unstable*).

(*v*) *The index of the system*

$$\dot{x} = -\lambda y + X_2(x,y) + \ldots + X_h(x,y)$$
$$\dot{y} = \lambda x + Y_2(x,y) + \ldots + Y_h(x,y), \tag{4.16}$$

is equal to $h+1$ *and the relative Poincaré constant is* < 0 (*resp.* > 0).

(*w*) *One has*

$$\frac{\partial^j V}{\partial c^j}(0,0) = 0,\ j \in \{1, \ldots, h-1\},\ \frac{\partial^h V}{\partial c^h}(0,0) < 0\ (\textit{resp.} > 0).$$

In addition, if any of propositions (*u*), (*v*), *and* (*w*) *holds,then* h *is odd.*

4.3. We now analyze the relationship for $(4.2)_0$ between attractivity properties of the origin 0 of R^2 for $\mu = 0$ and the family of bifurcating periodic orbits. This family will be denoted by $\{(c, \mu(c)),\ c \in (0, \varepsilon)\}$, where the function $\mu(c)$ introduced in Theorem 4.1 will be called Hopf function for $(4.2)_\mu$. The following theorem holds.

Theorem 4.4 *If the null solution of* $(4.2)_0$ *is* h*–asymptotically stable* (*resp.* h*–completely unstable*) *then the bifurcating periodic orbits of* $(4.2)_\mu$ *are asymptotically stable* (*resp.* h*–completely unstable*).

Proof. Assume $\alpha'(0) > 0$. Analogously we may proceed if $\alpha'(0) < 0$. From the identity $V(c, \mu(c)) \equiv 0$ and the condition $\mu(0) = 0$, $\mu'(0) = 0$ we easily obtain

$$\frac{\partial^{a+1} V}{\partial c^{a+1}}(0,0) = -(s+1)\frac{\partial^2 V}{\partial \mu \partial c}(0,0)\mu^{(s)}(0)$$

for every $s \leq k-1$ such that $\mu'(0) = \ldots. = \mu^{(s-1)}(0) = 0$. Since $\alpha'(0) > 0$ from (4.7) it follows $(\partial^2 V/\partial\mu\partial c)(0,0) > 0$. Then by virtue of Theorem 4.3 we have

$$\mu^{(j)}(0) = 0 \ \text{ for } j \in \{2, \ldots, h-2\} \text{ and } \mu^{(h-1)}(0) > 0\ (\textit{resp.} < 0). \tag{4.17}$$

From (4.17) it follows that μ is strictly increasing (resp. decreasing) in some interval $(0, \varepsilon)$. For each μ there is then only one bifurcating periodic orbit in $(0, \varepsilon)$ and this implies its asymptotic stability by virtue of Corollary 3.1. ∎

Corollary 4.1 *Assume that the r.h.s. of $(4.1)_\mu$ are C^∞ functions in (x, y, μ) and that for $\mu = 0$ they are analytic in x, y. If $\alpha'(0) > 0$ (resp. $\alpha'(0) < 0$), then exactly one of the following possibilities holds:*

(a) *0 is asymptotically stable for $\mu = 0$: the bifurcating periodic orbits are asymptotically stable and occur only for $\mu > 0$ (resp. for $\mu < 0$).*

(b) *0 is completely unstable for $\mu = 0$: the bifurcating periodic orbits are completely unstable and occur only for $\mu < 0$ (resp. for $\mu > 0$).*

(c) *0 is stable but not attracting for $\mu = 0$: the bifurcating periodic orbits have the same property and occur only for $\mu = 0$.*

Proof. Let M, G be the index and the Poincaré constant of (4.8). Clearly, if $M \in \mathbf{N}$, Theorem 4.4 implies all the properties in (a) or in (b) respectively, according to $G < 0$ or $G > 0$. If $M = \infty$, then 0 is stable but not attracting for $\mu = 0$. In this case $\mu(c) \equiv 0$ on some interval $[0, \varepsilon_1)$, with ε_1 sufficiently small and (c) applies. The proof is complete. ■

We conclude by remarking that the condition that the origin is asymptotically stable but not h–asymptotically stable for $\mu = 0$ is not sufficient in general to guarantee that the bifurcating periodic orbits are asymptotically stable. Indeed, consider the system

$$\begin{aligned} \dot{x} &= \mu x - y + x(x^2 + y^2) f(x, y) \\ \dot{y} &= \mu y + x - y(x^2 + y^2) f(x, y), \end{aligned} \tag{4.18}$$

where

$$f(x, y) = \exp[-(x^2 + y^2)]^{-1}\{\sin^2[(x^2 + y^2)^{-2}] + 1\} \text{ for } (x, y) \neq (0, 0),$$
$$f(0, 0) = 0.$$

Since $\alpha(\mu) = \mu$, and $\beta(\mu) = 1$, we have $\alpha(0) = 0$, $\beta(0) = 1$, $\alpha'(0) = 1 > 0$. Hence, the conditions to have Hopf bifurcation are satisfied. Consider the Liapunov function

$$V(x, y) = \frac{1}{2}(x^2 + y^2).$$

Along the solutions of (4.18) one has

$$\dot{V}_\mu = (x^2 + y^2)[\mu - (x^2 + y^2) f(x, y)].$$

Hence we may see that the null solution of (4.18) is asymptotically stable but there does not exist $h > 1$ such that it is h–asymptotically stable. The bifurcating periodic orbits are the circles $x^2 + y^2 = c^2$, and the Hopf function is given by

$$\mu(c) = c^2[\exp(-\frac{1}{c^2})](\sin^2 \frac{1}{c^4} + 1) \text{ for } c \neq 0, \ \mu(0) = 0.$$

By computing $\mu'(c)$ we have for $c \neq 0$

$$\mu'(c) = \frac{2}{c^3}[\exp(-\frac{1}{c^2})][-4 \sin \frac{1}{c^4} \cos \frac{1}{c^4} + c^2(\sin^2 \frac{1}{c^4} + 1)(1 + c^2)].$$

Since we have $\mu(0) = 0$ and $\mu(c) > 0$ for $c > 0$, in any neighborhhood of the origin there are points c in which $\mu'(c) > 0$. Moreover we see that, setting

$$c_n = \sqrt{2}[\pi(4n+1)]^{-\frac{1}{4}},$$

for $n \in \mathbf{N}$ sufficiently large we obtain $\mu'(c_n) < 0$. Hence for any $\varepsilon > 0$, the Hopf function $\mu(c)$ cannot be invertible in $(0, \varepsilon)$. It is easy to prove that the bifurcating periodic orbits cannot be all attracting. Indeed, let c', c'', $0 < c' < c'' < \varepsilon$, such that $\mu(c') = \mu(c'') = \mu$. Denote by γ_1, γ_2 the periodic orbits corresponding to $(c', \mu), (c'', \mu)$ respectively. Let $c^* \in (c', c'')$ be such that $\mu(c^*) \neq \mu$. Thus the orbit through c^* is a nonperiodic orbit lying in the annulus bounded by γ_1 and γ_2. The negative limit set $\Lambda^-(c^*)$ is a periodic orbit of $(4.1)_\mu$ which is attracting in the past and therefore repulsing in the future.

5. Generalized Hopf bifurcation

In this section we analyze the properties of the bifurcating periodic orbits arising in the generalized Hopf bifurcation problem in $\mathbf{R}^n$. As before, the approach is based upon stability properties of the equilibrium point of the unperturbed system. Consider the set of functions $f \in C^\infty(B(a_0), \mathbf{R}^n)$, $a_0 > 0$, $n \geq 2$, with the following metric:

$$|||f||| = \sum_{i=1}^{n} \frac{||f||^{(i)}}{2i(1 + ||f||^{(i)})},$$

where $||f||^{(i)}$ denotes the usual $C^{(i)}$–supremum norm on $B(a_0)$. Let $f_0 \in C^\infty(B(a_0), \mathbf{R}^n)$ and consider the differential equation

$$\dot{u} = f_0(u), \quad f_0(0) = 0. \tag{5.1}$$

Assume the Jacobian matrix $f_0'(0)$ has a purely imaginary pair of eigenvalues $\pm i$ and all other eigenvalues λ_j, $j \in \{1, ..., n-2\}$, satisfy the nonresonance condition $\lambda_j \neq mi$, $m \in \mathbf{Z}$. For those f close to f_0 in the topology defined by the above metric, consider the perturbed equation

$$\dot{u} = f(u), \quad f(0) = 0. \tag{5.2}$$

Again we assume conditions ensuring global existence of solutions. Notice that in $\mathbf{R}^2$, or also in $\mathbf{R}^n$, $n > 2$, when we have only a pair of purely imaginary eigenvalues (giving rise to a 2–dimensional manifold) the Poincaré–Bendixon theory applies. In this case instead, when there are more than two purely imaginary eigenvalues a more difficult situation arises. We will see that in this case the problem may be solved by constructing "quasi–invariant manifolds".

Consider the following properties:

(A) There exists an integer $k > 0$ such that
 (i) there exist $a > 0$, $\delta > 0$ and a neighborhood N of f_0 such that for every $f \in N$ there are at most k nontrivial periodic orbits of (5.2) lying in $B(a)$ whose period is in $[2\pi - \delta, 2\pi + \delta]$;
 (ii) for each integer $j \in \{0, ..., k\}$, for each $a_1 \in (0, a)$, for each $\delta_1 \in (0, \delta)$, and for each neighborhood N_1 of f_0, $N_1 \subset N$, there exists $f \in N_1$ such that (5.2) has exactly j nontrivial periodic orbits lying in $B(a_1)$ whose period is in $[2\pi - \delta_1, 2\pi + \delta_1]$;
 (iii) for any $\bar{a} \in (0, a)$ and $\bar{\delta} \in (0, \delta)$ there exists a neighborhood $\bar{N}$ of f_0, $\bar{N} \subset N$, such that if $f \in \bar{N}$ and if γ is a periodic orbit of (5.2) lying in $B(a)$ whose period is in $[2\pi - \delta, 2\pi + \delta]$, then γ lies in $B(\bar{a})$ with period in $[2\pi - \bar{\delta}, 2\pi + \bar{\delta}]$.

(B) For any neighborhood N of f_0, for any integer $j > 0$, for any $a > 0$, and for any $\delta > 0$, there exists $f \in N$ such that (5.2) has exactly j nontrivial periodic orbits lying in $B(a)$ whose period is in $[2\pi - \delta, 2\pi + \delta]$.

In $\mathbf{R}^2$ Andronov et al. [1] proved that property (A), ((i), (ii)) is a consequence of the origin of (5.1) being h–asymptotically stable or h–completely unstable, where h is an odd integer ≥ 3 and $k = \frac{h-1}{2}$. The problem in $\mathbf{R}^n$ was first considered by Chafee [10]. By using Liapunov–Schmidt method, he obtained a determining equation $\psi(\varepsilon, f) = 0$, where ε is a measure of the amplitude of the bifurcating orbits of (5.1). By assuming that the multiplicity of the zero root of $\psi(\cdot, f)$ is a finite number k, he proved that property (A) holds for this k. In [3] the number k in property (A) has been related to the h–asymptotic stability or the h–complete instability for an auxiliary differential equation (S_h) in $\mathbf{R}^2$. This equation is obtained by an algebraic procedure involving a quasi–invariant manifold. Thus the original problem has been reduced to an analysis of a two dimensional equation. As in [1] we have $k = \frac{h-1}{2}$. The analysis in [3] was completed by observing that when the origin for (S_h) is neither h–asymptotically stable nor h–completely unstable for every $h > 0$ then property (B) holds.

By an appropriate change of coordinates depending on f system (5.2) becomes

$$\begin{aligned} \dot{x} &= \alpha x - \beta y + X(x, y, z, f) \\ \dot{y} &= \alpha y + \beta x + Y(x, y, z, f) \\ \dot{z} &= Az + Z(x, y, z, f). \end{aligned} \tag{5.3}$$

For each f, α and β are constants satisfying $\alpha(f_0) = 0$, $\beta(f_0) = 1$; A is a $(n-2) \times (n-2)$ constant matrix and X, Y, Z are of order ≥ 2 in (x, y, z). The eigenvalues λ_j, $j \in \{1, ..., n-2\}$, of $A_0 \equiv A(f_0)$ satisfy the nonresonance condition $\lambda_j \neq mi$, $m \in \mathbf{Z}$. We denote by $(5.3)_0$ the system (5.3) for $f = f_0$.

Consider now a $(n-2)$–dimensional polynomial of degree h, $h > 0$, given by

$$\Phi^{(h)}(x, y) = \varphi_1(x, y) + \ldots + \varphi_h(x, y), \tag{5.4}$$

where $\varphi_j(x, y)$, $j \in \{1, ..., n-2\}$, is a j–form. We attempt to determine $\varphi_1, \ldots, \varphi_h$ such that along the solutions of $(5.3)_0$ one has

$$\{\frac{d}{dt}[z - \Phi^{(h)}(x, y)]\}_{z=\Phi^{(h)}(x,y)} = o((x^2 + y^2)^{\frac{h}{2}}). \tag{5.5}$$

That is, we have to satisfy the partial differential equation

$$\frac{\partial \varphi_j}{\partial y} x - \frac{\partial \varphi_j}{\partial x} y = A_0 \varphi_j + U_j, \quad j \in \{1, \ldots, n\}, \tag{5.6}$$

where U_j is a $(n-2)$–dimensional polynomial of degree j, depending on $\varphi_1, \ldots, \varphi_{j-1}$. Under the assumption on A_0 ((5.6) admits a unique solution which may be determined recursively, taking into account that $\varphi_1(x, y) = 0$. The 2–dimensional manifold

$$z = \Phi^{(h)}(x, y)$$

is tangent to the eigenspace corresponding to the eigenvalues $\pm i$ at the origin. It will be called a quasi–invariant manifold of order h. Given any $h > 0$, consider the 2–dimensional system

$$\begin{aligned} \dot{x} &= -y + X_0(x, y, \Phi^{(h)}(x, y)) \\ \dot{y} &= x + Y_0(x, y, \Phi^{(h)}(x, y)), \end{aligned} \tag{S_h}$$

where $X_0(x, y, z) = X(x, y, z, f_0)$ and $Y_0(x, y, z) = Y(x, y, z, f_0)$.

Clearly the Poincaré procedure may be applied to system (S_h). Consider the two possible cases:

(*I*) There exists $h > 0$ (and h must be odd) such that the null solution of (S_h) is h–asymptotically stable or h–completely unstable.

(*II*) Case (*I*) does not hold.

We will prove the following theorem.

Theorem 5.1 *In case (I) property (A) holds with* $k = \frac{h-1}{2}$, *in case (II) property (B) holds.*

For the proof we need some preliminaries. By using the transformation of coordinates

$$w = z - \Phi^{(h)}(x, y),$$

system (5.3) may be written as

$$\begin{aligned} \dot{x} &= \alpha x - \beta y + X^{(h)}(x, y, w, f) \\ \dot{y} &= \alpha y + \beta x + Y^{(h)}(x, y, w, f) \\ \dot{w} &= Aw + W^{(h)}(x, y, w, f), \end{aligned} \tag{5.7}$$

where $X^{(h)}(x, y, w, f) = X(x, y, w + \Phi^{(h)}(x, y), f)$ and similarly for $Y^{(h)}$, $W^{(h)}$. For fixed f, the functions $X^{(h)}$, $Y^{(h)}$, $W^{(h)}$ are of order ≥ 2. Set $X_0^{(h)}(x, y, w) = X^{(h)}(x, y, w, f_0)$ and analogously define $Y_0^{(h)}(x, y, w)$, $W_0^{(h)}(x, y, w)$. Clearly we have $W_0^{(h)}(x, y, 0)$ is of order $> h$.

System (S_h) in the new variables becomes

$$\begin{aligned} \dot{x} &= -y + X_0^{(h)}(x, y, 0) \\ \dot{y} &= x + Y_0^{(h)}(x, y, 0). \end{aligned} \tag{S_h}'$$

Now we scale the variable t by replacing t with εt, where ε is now regarded as a parameter (depending on f and the initial conditions (x_0, y_0, w_0)) such that: (1) $\varepsilon = 1$ for $f = f_0$ and $(x_0, y_0, w_0) = (0, 0, 0)$; (2) if $T \geq 0$ is close to 2π, if f is close to f_0, if (x_0, y_0, w_0) is near the origin and the orbit through (x_0, y_0, w_0) is T–periodic, then $\varepsilon = 2\pi/T$. Thus all bifurcating periodic orbits of (5.7) have the same period 2π. System (5.7) assumes the form:

$$\begin{aligned} \dot{x} &= \varepsilon[\alpha x - \beta y + X^{(h)}(x, y, w, f)] \\ \dot{y} &= \varepsilon[\alpha y + \beta x + Y^{(h)}(x, y, w, f)] \\ \dot{w} &= \varepsilon[Aw + W^{(h)}(x, y, w, f)]. \end{aligned} \tag{5.8}$$

We denote by $(5.8)_0$ system (5.8) for $f = f_0$.

Definition 5.1 *A solution* $(x(t), y(t), w(t))$ *of* (5.8) *that exists in* $[0, 2\pi]$ *will be called* $(2\pi, w)$–*solution if* $w(2\pi) = w(0)$.

Clearly every 2π–periodic solution of (5.8) is a $(2\pi, w)$–solution. Then, in order to find the 2π–periodic solutions of (5.8) we need only to inspect the set of $(2\pi, w)$–solutions.

Proposition 5.1 *There exist a neighborhood* N *of* f_0, *and a function* $\tau(x, y, \varepsilon, f)$, $\tau(0, 0, \varepsilon, f) \equiv 0$, *for* ε *close to* 1 *and* $f \in N$ *such that* (1) *for fixed* f, τ *has partial derivatives of any order which are continuous in* (x, y, ε, f), (2) *for every* (x, y, w) *close to the origin,* ε *close to* 1, *and* $f \in N$, *the solution through* (x, y, w) *is a* $(2\pi, w)$–*solution if and only if* $w = \tau(x, y, \varepsilon, f)$.

The proof is based on the variation of parameters formula for the w–equation in (5.8) and on the use of the implicit function theorem, taking into account the hypotheses on A_0.

Proposition 5.2 *The neighborhood N of Proposition 5.1 may be chosen such that if ε is close to 1, $f \in N$, and Γ is a nontrivial periodic 2π-orbit of (5.8) lying in a neighborhood U of the origin, then for any $(x, y, w) \in \Gamma$ we have that the projection of Γ on the (x, y)-plane does not contain $(0,0)$ and intersects any straight line through the origin in exactly two points.*

Proof. Let Γ be any nontrivial 2π-periodic orbit of (5.8) lying in U. For any $(x, y, w) \in \Gamma$, clearly we have $w = \tau(x, y, \varepsilon, f)$. Since $\tau(0,0,\varepsilon, f) = 0$, Γ does not contain $(0,0)$. Now consider the transformation in polar coordinates $x = r\cos\theta$, $y = r\sin\theta$. We need only to show that θ is an increasing function of t along the projection on the (x, y)-plane of any orbit γ. For $r > 0$ along the solutions of (5.8) we have

$$\dot{\theta} = \varepsilon\beta + \Theta(\theta, r, w, \varepsilon, f),$$

where $\Theta(\theta, r, w, \varepsilon, f) = r^{-1}[X^{(h)}(r\cos\theta, r\sin\theta, w, \varepsilon, f)\cos\theta - Y^{(h)}(r\cos\theta, r\sin\theta, w, \varepsilon, f)\sin\theta]$. By substituting $w = \tau(r\cos\theta, r\sin\theta, w, \varepsilon, f)$ in $\Theta(\theta, r, w, \varepsilon, f)$, we easily recognize that it becomes a function of $(\theta, r, \varepsilon, f)$ which tends to zero as $r \to 0$ uniformly in (θ, ε, f). Therefore, since $\beta(f_0) = 1$, we find that if N, U, $|\varepsilon - 1|$ are sufficiently small, then $\dot{\theta}$ is greater than a positive constant along any 2π-periodic orbit of (5.8) lying in U, for any $f \in N$ and ε close to 1. The proof is complete. ■

Proposition 5.2 implies that we may restrict ourselves to the $(2\pi, w)$-solutions through $t_0 = 0$, $x_0 = c$, $y_0 = 0$. Denoting by $(x(t, c, \varepsilon, f), y(t, c, \varepsilon, f), w(t, c, \varepsilon, f))$ such a solution, from the first two equations of (5.8) it follows:

$$x(2\pi, c, \varepsilon, f) = \exp(2\pi\alpha\varepsilon)c\cos 2\pi\beta\varepsilon + \varepsilon\int_0^{2\pi} P(x(t, c, \varepsilon, f), y(t, c, \varepsilon, f), w(t, c, \varepsilon, f), f)dt,$$

$$y(2\pi, c, \varepsilon, f) = \exp(2\pi\alpha\varepsilon)c\sin 2\pi\beta\varepsilon + \varepsilon\int_0^{2\pi} Q(x(t, c, \varepsilon, f), y(t, c, \varepsilon, f), w(t, c, \varepsilon, f), f)dt,$$

where for fixed f, $P(x, y, w, f)$, $Q(x, y, w, f)$ are C^∞ and of order ≥ 2 in (x, y, w). Consider now the continuous function $F(c, \varepsilon, f)$ defined by

$$F(c, \varepsilon, f) = \exp(2\pi\alpha\varepsilon)\sin 2\pi\beta\varepsilon + \varepsilon\int_0^{2\pi} c^{-1}Q(x(t, c, \varepsilon, f), y(t, c, \varepsilon, f), w(t, c, \varepsilon, f), f)dt,$$

$$F(0, \varepsilon, f) = exp(2\pi\alpha\varepsilon)\sin 2\pi\beta\varepsilon.$$

Notice that for each fixed f, F is C^∞ in (c, ε). Consider now the equation

$$F(c, \varepsilon, f) = 0. \tag{5.9}$$

We have

$$F(0, 1, f_0) = 0 \quad \text{and} \quad \frac{\partial F}{\partial \varepsilon}(0, 1, f_0) = 2\pi \neq 0.$$

Hence, by using the implicit function theorem, we have the following proposition.

Proposition 5.3 *There exist $\bar{c} > 0$, $\bar{\eta} > 0$, a neighborhood N of f_0, and a continuous function ε, $\varepsilon(0, f_0) = 1$, defined in $(-\bar{c}, \bar{c}) \times N$ such that: (1) for fixed $f \in N$, ε has partial derivatives of any order which are continuous in (c, f); (2) for every $c \in (-\bar{c}, \bar{c})$, $\varepsilon \in (1 - \bar{\eta}, 1 + \bar{\eta})$, $f \in N$, equation (5.9) holds if and only if $\varepsilon = \varepsilon(c, f)$.*

Due to (5.9) and Proposition 5.3, for any $(c, f) \in (-\bar{c}, \bar{c}) \times N$, a $(2\pi, w)$-solution for which $\varepsilon = \varepsilon(c, f)$, satisfies also the condition $y(2\pi, c, \varepsilon(c, f), f) = 0$. Hence the 2π-periodic orbits will be the solutions for which $x(2\pi, c, \varepsilon(c, f), f) = 0$. Setting $x(t, c, f) = x(t, c, \varepsilon(c, f), f)$,

$y(t,c,f) = y(t,c,\varepsilon(c,f),f)$, $w(t,c,f) = w(t,c,\varepsilon(c,f),f)$, consider the displacement function relative to the $(2\pi, w)$–solutions of (5.8) and $\varepsilon = \varepsilon(c,f)$, defined in $(-\bar{c},\bar{c}) \times N$ as

$$V(c,f) = x(2\pi,c,f) - c.$$

Then the 2π–periodic orbits of (5.8) in our local problem correspond to the zeros of $V(c,f)$. For any integer $h > 0$ consider the MacLaurin expansion in c of the solutions of (5.8) in the form:

$$x(t,c,f) = u_1(t,f)c + u_2(t,f)c^2 + \ldots + u_h(t,f)c^h + o(c^h), \tag{5.10}$$
$$y(t,c,f) = v_1(t,f)c + v_2(t,f)c^2 + \ldots + v_h(t,f)c^h + o(c^h), \tag{5.11}$$
$$w(t,c,f) = w_1(t,f)c + w_2(t,f)c^2 + \ldots + w_h(t,f)c^h + o(c^h), \tag{5.12}$$

where $u_1(0,f) = 1$, $u_j(0,f) = 0$, $j > 1$, $v_j(0,f) = 0$, and $w_j(0,f) = 0$, for $j \geq 0$. Moreover, we may write:

$$\varepsilon(c,f) = \varepsilon_0(f) + \varepsilon_1(f)c + \ldots + \varepsilon_h(f)c^h + o(c^h), \tag{5.13}$$

with $\varepsilon_0(f_0) = 1$. It is $u_1(t,f_0) \equiv 1$. By substituting (5.10) - (5.13) into $(5.8)_0$ we obtain the equation

$$\frac{\partial w_1}{\partial t}(t,f_0) = A_0 w_1(t,f_0),$$

hence $w_1(t,f_0) = [\exp(A_0 t)]w_1(t,f_0)$. Since $w_1(2\pi,f_0) = [\exp(A_0 2\pi)]w_1(2\pi,f_0) = w_1(0,f_0)$, from the hypotheses on A_0, it follows $w_1(t,f_0) \equiv 0$. Taking into account that $W^{(h)}(x,y,0,f_0)$ is of order $> h$ and $w_1(t,f_0) \equiv 0$, we have

$$\frac{\partial w_2}{\partial t}(t,f_0) = A_0 w_2(t,f_0),$$

and, as before, $w_2(t,f_0) \equiv 0$. Analogously we have $w_j(t,f_0) \equiv 0$ for j up to $h-1$. Thus, in order to compute the coefficients $u_j(t,f_0), v_j(t,f_0)$, $j \in \{1,\ldots,h-1\}$, we may put $w = 0$ in the first two equations of $(5.8)_0$. Then we obtain the system

$$\begin{aligned} \dot{r} &= \varepsilon[-y + X_0^{(h)}(x,y,0)] \\ \dot{y} &= \varepsilon + [x + Y_0^{(h)}(x,y,0)]. \end{aligned} \tag{$(S_h)''$}$$

which is exactly $(S_h)'$ with t replaced by εt.

Proposition 5.4 *Assume the null solution of $(S_h)''$ is h–asymptotically stable (resp. h–completely unstable). Then h is odd, and*

$$\frac{\partial^j V}{\partial c^j}(0,f_0) = 0,\ j \in \{1,\ldots,h-1\},\ \frac{\partial^h V}{\partial c^h}(0,f_0) < 0 \ \ (\textit{resp.} > 0). \tag{5.14}$$

Proof. By using the Poincaré procedure, we may construct a polynomial of degree $h+1$ (and h has to be odd and > 1)

$$F(x,y) = x^2 + y^2 + f_3(x,y) + \ldots + f_{h+1}(x,y),$$

which along the solutions of $(S_h)''$ satisfies the condition

$$\dot{F} = \varepsilon G_{h+1}(x^2+y^2)^{\frac{(h+1)}{2}} + o[(x^2+y^2)^{\frac{(h+1)}{2}}],$$

where $G_{h+1} < 0$ (resp. > 0) is a constant. Integrating over $[0, 2\pi]$ along the solutions $\tilde{x}(t,c)$, $\tilde{y}(t,c)$ of $(S_h)''$ with initial conditions $x_0 = c, y_0 = 0$ and $\varepsilon = \varepsilon(c, f_0)$, we obtain

$$F(\tilde{x}(2\pi, c), \tilde{y}(2\pi, c)) - F(c, 0) = 2\pi\varepsilon(c, f_0)G_{h+1}c^{h+1} + o(c^{h+1}). \tag{5.15}$$

Taking into account that $x(t,c,f) - \tilde{x}(t,c) = o(c^h)$, $y(t,c,f) - \tilde{y}(t,c) = o(c^h)$, from (5.10), (5.11), (5.13) and (5.15) we obtain $u_j(2\pi, f_0) = 0$, $j \in \{1, \ldots, h-1\}$, $u_h(2\pi, f_0) = \pi G_{h+1}$. Hence the proof is complete. ∎

Proof of Theorem 5.1. Assume (I). In particular assume that the null solution of (S_h) is h–asymptotically stable. Analogously we may proceed if it is h–completely unstable. We prove that (A) holds. Using Proposition 5.4 we find that $\bar{c} > 0$ and N may be chosen such that $\frac{\partial^h V}{\partial c^h}(c, f) < 0$ for any $(c, f) \in (-\bar{c}, \bar{c}) \times N$. Moreover given $a > 0$ small and N small we can find $\bar{c}$ such that the periodic orbit through $c \in (-\bar{c}, \bar{c})$ lies in $B(a)$. Since the origin in $\mathbf{R}^2$ is a solution of $(S_h)''$ for every f, an application of Rolle theorem assures that $V(c, f)$ has at most $h-1$ zeros different from 0 in $(-\bar{c}, \bar{c})$. Since Proposition 5.2 assures that to each 2π–periodic orbit of (5.8) correspond two zeros of V, we have at most $\frac{h-1}{2}$ nontrivial 2π–periodic orbits lying in $B(a)$. Hence part (i) of property (A) is satisfied with $k = \frac{h-1}{2}$. We now show that (ii) occurs. If $j = 0$ we may choose $f = f_0$. Consider now an integer $j \in \{1, \ldots, k\}$, and consider the perturbation of $(5.8)_0$ given by

$$\begin{aligned}
\dot{x} &= \varepsilon[-y + X_0^{(h)}(x, y, w) + \sum_{i=1}^{j} e_i x(x^2 + y^2)^{k-i}] \\
\dot{y} &= \varepsilon[x + Y_0^{(h)}(x, y, w) + \sum_{i=1}^{j} e_i y(x^2 + y^2)^{k-i}] \\
\dot{w} &= \varepsilon[Aw + W_0^{(h)}(x, y, w)],
\end{aligned} \tag{5.16}$$

where e_i, $i \in \{1, \ldots, j\}$, are constants to be determined. Denote by $V(c, e_1, \ldots, e_j)$ the desplacement function relative to the solutions of (5.16) and by $S(e_1, \ldots, e_j)$ the system of the first two equations of (5.16) for $w = 0$. Clearly

$$V(c, 0, \ldots, 0) = g_0 c^h + o(c^h), \; g_0 < 0.$$

Then, for $c_0 > 0$ and sufficiently small, we have $V(c_0, 0, \ldots, 0) < 0$. There exists $\eta_1 > 0$ such that $V(c_0, e_1, \ldots, e_j) > 0$ for $|e_i| < \eta_1$, $i \in \{1, \ldots, j\}$. Let $e_1 \in (0, \eta_1)$. We have

$$V(c, e_1, 0, \ldots, 0) = g_1 c^{h-2} + o(c^{h-2}), \; g_1 > 0,$$

since the null solution of $S(e_1, 0, ..., 0)$ is $(h-2)$–completely unstable. There exists c_1, $0 < c_1 < c_0$, such that $V(c_1, e_1, 0, \ldots, 0) > 0$. Consequently, there exists $\eta_2 > 0$ such that $V(c_1, e_1, \ldots, e_j) > 0$ for $|e_i| < \eta_2$, $i \in \{1, \ldots, j\}$. Assume now $e_2 \in (0, \eta_2)$. In this case, $S(e_1, e_2, 0, \ldots, 0)$ is $(h-4)$–asymptotically stable. Then, we have $V(c_2, e_1, e_2, 0, \ldots, 0) < 0$ for some c_2, $0 < c_2 < c_1$. Thus we can find $\eta_3 > 0$ such that $V(c_2, e_1, \ldots, e_j) < 0$ for $|e_i| < \eta_3$, $i \in \{1, \ldots, j\}$. Continuing this process, we can find a set of numbers $\bar{c}_1, \bar{c}_2, \ldots, \bar{c}_j$, $c_i < \bar{c}_i < c_{i-1}$ (and then $\bar{c}_i < \bar{c}_{i-1}$), $i \in \{1, \ldots, j\}$, such that $V(\bar{c}_i, e_1, \ldots, e_j) = 0$. Since $c = 0$ is a root of $V(c, e_1, \ldots, e_j)$ of order h-2j and (by virtue of Proposition 5.2) for each positive root of $V(c, e_1, \ldots, e_j)$ we have a negative root, then Rolle theorem assures that $\bar{c}_i$, $i \in \{1, \ldots, j\}$ are the only positive roots of $V(c, e_1 \ldots, e_j)$. Moreover, we can obtain that the $\bar{c}_i$, $i \in \{1, \ldots, j\}$, can be made as close to $c = 0$ as we desire, by choosing c_0 sufficiently small. This completes the proof of part (ii). To prove (iii) of property (A), we observe that for $\bar{c} > 0$ sufficiently small and $c \in [0, \bar{c}]$ we have $|V(c, f)| \geq \lambda c^n$, for some $\lambda > 0$. Because of the continuity of $V(c, f)$ in (c, f_0) we have for each

$c' \in [0, \bar{c}]$ there exists a neighborhood $N_{c'}$ of f_0 contained in N such that all the roots of $V(c, f)$ lying in $[0, \bar{c}]$ lie in $[0, c']$. From this property $(A)(iii)$ follows.

Assume now (II). That is, assume that for every $h > 0$ the null solution of (S_h) is neither h-asymptotically stable nor h-completely unstable. For any positive integer j assume $h = 2j+1$ and consider the perturbation of $(5.8)_0$ given by

$$\begin{aligned} \dot{x} &= \varepsilon[-y + X_0^{(h)}(x, y, w) + bx(x^2 + y^2)^{\frac{(h-1)}{2}}] \\ \dot{y} &= \varepsilon[x + Y_0^{(h)}(x, y, w) + by(x^2 + y^2)^{\frac{(h-1)}{2}}] \\ \dot{w} &= \varepsilon[Aw + W_0^{(h)}(x, y, w)], \end{aligned} \tag{5.17}$$

where b is a constant. Then, we have that for the corresponding reduced system $(S_h)''$ the origin is either h-asymptotically stable or h-completely unstable according to $b < 0$ or $b > 0$ respectively. Thus, we have reduced the problem to the previous case. Since j and b are arbitrary, property (B) holds, completing the proof of the theorem. ∎

We conclude by remarking that some of the result of this part have been extended to the periodic case [5,6,83,94].

References

[1] Andronov A., Leontovich E., Gordon I., Mayer A., "Theory of Bifurcation of Dynamical Systems in the Plane", Israel Program of Scientific Translations, Jerusalem, 1971.

[2] Barbashin E.A., Krasovskii N.N., *On the stability of motion in the large*, Doklady Akad. Nauk SSSR, **86** (3), 1952.

[3] Bernfeld S.R., Negrini P., Salvadori L., *Quasi-invariant manifolds, stability and generalized Hopf bifurcation*, Ann. Mat Pura Appl., **130**, 105-119, 1982.

[4] Bernfeld S.R., Salvadori L., *Generalized Hopf bifurcation and h-asymptotic stability*, Nonlinear Analysis: TMA, **4** (6), 1091-1107, 1980.

[5] Bernfeld S.R., Salvadori L., Visentin F., *Hopf bifurcation and related stability problems for periodic differential systems*, J. Math. Anal. Appl., **116** (2), 427-438, 1986.

[6] Bernfeld S.R., Salvadori L., Visentin F., *Bifurcation for periodic differential equations at resonance, Differential and Integral Equations*, **3** (1), 1-12, 1990.

[7] Bertotti M.L., Moauro V., *Bifurcation and total stability*, Sem. Mat. Univ. Padova, **71**, 131-139, 1984.

[8] Bhatia N.P., Szegö G.P., "Dynamical Sistems: Stability Theory and Applications", Springer, 1967.

[9] Chafee N., *The bifurcation of one or more closed orbit from an equilibrium point of an autonomous differential system*, J. Differential Equations., **4**, 661-679, 1968.

[10] Chafee N., *Generalized Hopf bifurcation and perturbations in a full neighborhoods of a given vector field*, Indiana Univ. Math. J., **27**, 173-194, 1978.

[11] Chetaev A.G., *A theorem on instability* (Russian), Dokl. Akad. Nauk. SSSR, **1**, 529-531, 1934.

[12] Chetaev N.G., *On the instability of equilibrium in certain cases when the force function is not a maximum*, Akad. Nauk SSSR Prikl. Math. Mekh., **16**, 89-93, 1952.

[13] Chetaev N.G., *Note on a classical Hamiltonian theory*, PMM, **24** (1), 1960.

[14] Chow S.N., Hale J.K., "Methods of Bifurcation Theory", Springer, 1982.

[15] D'Onofrio B.M., *The stability problem for some nonlinear evolution equation*, Boll. Un. Mat. Ital., **17B**, 425-439, 1980.

[16] Dubosin G.N., *On the problem of stability of a motion under constanctly acting perturbations*, Trudy Gos. Astron. Inst. Sternberg, **14** (1), 1940.

[17] Furta S.D., *On the asymptotic solutions of the equations of motions of mechanical systems*, P.M.M., **50**, 726-731, 1986

[18] Gorsin S., *On the stability of motion under constanctly acting perturbations*, Izv. Akad. Nauk Kazakh. SSR **56**, Ser. Mat. Mekh. **2**, 1948.

[19] Habets P., Risito C., *Stability criteria for systems with first integrals, generalizing theorems of Routh and Salvadori*, Equations différentielles et fonctionnelles non linéaires, éd. P. Janssens, J. Mawhin et N. Rouche, Hermann, Paris, 570-580, 1973.

[20] Hagedorn P., *Die Umkehrung der Stabilitätssätze von Lagrange-Dirichlet und Routh*, Arch. Rational Mech. Anal., **42**, 281-316, 1971.

[21] Hagedorn P., *Über die Instabilität konservativer Systeme mit gyroskopischen Kräften*, Arch. Rational Mech. Anal.,**58**, 1-9, 1975.

[22] Hamel G., *Über die Instabilität der Gleichgewlchtslage eines Systems von zwei Freihitsgraden*, Math. Ann., **57**, 541-553, 1904.

[23] Hahn W., "Stability of Motion", Springer-Verlag, 1967.

[24] Hahn W., *On Salvadori's one-parameter families of Liapunov functions*, Ricerche di Matematica, **XX**, 193-197, 1971.

[25] Hopf E. *Abzweigung einer periodischen Läsung von einer stationären Läsung eines Differential systems*, Ber. Math. Phys. Schsische Akad. Wiss. Leipzig **94**, 1-22, 1942.

[26] Karapetyan A.V., *The Routh theorem and its extensions*, Colloquia Mathematica, "Qualitative Theory of Differential Equations", Szeged, 271-290, 1988.

[27] Koiter W.T., *On the instability of equilibrium in the absence of a minimum of the potential energy*, Nederl. Akad. Wetensch. Proc. ser. B, **68**, 107-113, 1965.

[28] Koslov V.V., *The instability of equilibrium in a potential field*, Uspekhi Mat. Nauk, **36** (1), 209-210, 1981.

[29] Koslov V.V., *Equilibrium instability in a potential field, taking account of a viscous friction*, P.M.M., U.S.S.R., **45**, 417-418, 1982.

[30] Koslov V.V., *Asymptotic motions and the inversion of the Lagrange-Dirichlet theorem*, P.M.M., U.S.S.R., **50**, 719-725, 1986.

[31] Koslov V.V., Palamodov V.P., *On asymptotic solutions of the equations of classical mechanics*, Soviet Math. Dokl., **25**, 335-339, 1982.

[32] Krasovskii N.N., "Problems of the Theory of Stability of Motion", Stanford Univ. Press, Stanford, California, 1963; traslation of the Russian edition, Moskow, 1959.

[33] Kurzweil J., *Reversibility of Liapunov's first theorem on stability of motion* (Russian), Czechoslovak Math. J., **5**, 382-398, 1955.

[34] Kuzmin P.A., *Stability with parametric disturbance*, PMM, **21** (1), 1957.

[35] Kuznetsov A.N., *Differenziable solutions to degenerate systems of ordinary equations*, Funk. Analiz Pril., **6**, 41-51, Functional Analysis and its Appl., **6** (2), 119-127, 1972.

[36] Laloy M., *On equilibrium instability for conservative and partially dissipative mechanical systems*, Int. J. Non–linear Mech., **2**, 295-301, 1976.

[37] Laloy M., Peiffer K., *On the instability of equilibrium when the potential has a non–strict local minimum*, Arch. Rational Mech. Anal., 213-222, **78**, 1982.

[38] La Salle J.P., *Asymptotic stability criteria*, Proc. Symp. Appl Math., **13**, 299-307, 1962.

[39] LaSalle J. P., *Stability theory for ordinary differential equations*, J. Differential Equations, **4**, 57-65, 1968.

[40] Liubushin E., *On instability of equilibrium when the force function is not a maximum*, J. Appl. Math. Mech., **44**, 158-162, 1980.

[41] Liapunov, M.A., "Problème général de la stabilité du mouvement", Photo–reproduction in Annals of Mathematics, Studies n 17, Princeton University Press, Princeton, 1949, of the 1907 French translation of the fundamental Russian paper of Lyapunov published in Comm. Soc. Math., Kharkov, 1892.

[42] Maffei C., Moauro V., Negrini P., *On the inversion of the Dirichlet–Lagrange theorem in a case of nonhomogeneous potential*, Differential and Integral Equations, **4** (4), 767-782, 1991.

[43] Malkin I.G., *Stability in the case of constantly acting disturbances*, PMM, **8**, 1944.

[44] Marakhov M., *On a theorem of stability*, Izv. Fiz. Mat. Obs. Kazan. Univ., **12** (3), 171-174 (Russian), 1940.

[45] Marcati P., *Stability analysis of abstract hyperbolic equations using families of Liapunov functions*, Equadiff 82, Lect. Notes in Math. 1017, Springer–Verlag, 1983.

[46] Marcati P., *Decay and stability for nonlinear hyperbolic equations*, J. Differential Equations, **55**, 30-58, 1984

[47] Marcati P., *Stability for second order abstract evolution equations*, Nonlinear Analysis: TMA, **8**, 237-252, 1984.

[48] Marcati P., *Abstract stability theory and applications to hyperbolic equations with time dependent dissipative force fields*, Computers and Maths with Appl., **12A**, 541-550, 1986.

[49] Marchetti F., Negrini P., Salvadori L., Scalia M., *Liapunov direct method in approaching bifurcation problems*, Ann. Mat. Pura Appl., **108** (4), 211-226, 1976.

[50] Marchetti F., *Some stability problems from a topological viewpoint*, Acc. Naz. Lincei Sci. Fis. Mat., **60** (6), 733-742, 1976.

[51] Massera J.L., *On Liapunov conditions on stability*, Ann. Mat., **50**, 705-721, 1949.

[52] Massera J.L., ***Contributions to stability theory***, Ann. Mat., **64**, 1956, 182-206, ***Erratum***, Ann. Mat., **68**, 202, 1958.

[53] Marsden J , Mc Cracken M.F., "The Hopf Bifurcation and its Applications", Springer, 1976.

[54] Matrosov V.M., ***On the stability of motion***, PMM, **26**, 1962, 885-895.

[55] Moauro V., ***A stability problem for holonomic systems***, Ann. Mat. Pura Appl., (4), **139**, 227-236, 1985.

[56] Moauro V , Negrini P., ***On the inversion of Lagrange-Dirichlet theorem***, Differential and Integral Equations, **2** (4), 471-478, 1989.

[57] Moauro V., Salvadori L, Scalia M., ***Total stability and classical Hamiltonian theory***, in Nonlinear Equations in Abstract Spaces (V. Lakshmikantham ed.), Academic Press, 149-159, 1978.

[58] Negrini P., ***On the inversion of Lagrange-Dirichlet theorem***, Resenahas, IME-USP, **2** (1), 83-114, 1995.

[59] Negrini P., Salvadori L., ***Attractivity and Hopf bifurcation***, Nonlinear Analysis: TMA, **3** (1), 87-99, 1979.

[60] Painlevé P., ***Sur la stabilité de l'équilibre***, C.R. Acad. Sci. Paris, sér. A-B, **138**, 1555-1557, 1904

[61] Palamodov V.P., ***Stability of equilibrium in a potential field***, Functional Anal. Appl., **11**, 277-289, 1978.

[62] Palamodov V.P., ***Stability of motion and algebraic geometry***, Amer. Math. Soc. Trans., **168** (2), 5-19, 1995.

[63] Persidski S.K., ***On the stability of motion in first approximation***, Mat. Sb, **40**, 284-293, 1933.

[64] Pozharitskii G.K., ***On the construction of Liapunov functions from the integrals of the equations of the perturbed motion***, PMM, USSR, **22**, 145-154, 1958.

[65] Pucci P., Serrin J., ***Asymptotic stability for ordinary differential systems with time dependent restoring potentials***, Arch. Rat. Mech. Anal., **132**, 207-232, 1995.

[66] Pucci P., Serrin J., ***Precise damping conditions for global asymptotic stability for nonlinear second order systems***, Acta Math., **170**, 275-307, 1993.

[67] Pucci P., Serrin J., ***Precise damping conditions for global asymptotic stability for nonlinear second order systems II***, J. Differential Equations, **113**, 505-534, 1994.

[68] Pucci P., Serrin J., ***Remarks on Liapunov stability***, Differential and Integral Equations, **8** (6), 1265-1278, 1995.

[69] Pucci P., Serrin J., ***Asymptotic stability for non-autonomous dissipative wave systems***, Comm. Pure Appl. Math., **49**, 177-216, 1996.

[70] Rouche N., ***On the stability of motion***, Int. J. Non-Lin. Mech., **3**, 295-306, 1968.

[71] Rouche N., Habets P. and Laloy M., "Stability Theory by Liapunov's Direct Method", Springer Verlag, New York, 1977.

[72] Routh E.J., "The Advanced Part of a Treatise on the Dynamics of a System of Rigid Bodies", 1st Ed London: MacMillan, 1860.

[73] Ruelle J.E. and Takens F, *On the nature of turbolence*, Comm. Math. Phys., **20**, 167-192, 1971.

[74] Rumiantsev V.V., *On the stability of steady state motions*, Prikl. Mat. Meh., **32**, 504-508, J. Appl. Math. Mech., **32**, 517-521, 1968.

[75] Rumiantsev V.V. and Oziraner A.S., "Stability and Stabilization of Motion With Respect to Part of Variables" (in Russian), Moskow, 1987.

[76] Salvadori L., *Un'osservazione su un criterio di stabilità del Routh*, Rend, Acc. Sci. Fis.Mat.Nat., Napoli, **XX**, 269-272, 1953.

[77] Salvadori L., *Sulla stabilità del movimento*, Le Matematiche, **XXIV**, 218-239, 1969

[78] Salvadori L., *Una generalizzazione di alcuni teoremi di Matrosov*, Ann. Mat. Pura Appl., Serie IV, **84**, 83-94, 1970.

[79] Salvadori L., *Famiglie ad un parametro di funzioni di Liapunov nello studio della stabilità*, Symposia Mathematica, **6**, 309-330, 1971.

[80] Salvadori L., *An approach to bifurcation via stability theory*, J. Math. Phis. Sci. **18** (1), 99-110, 1984.

[81] Salvadori L., *Stability problems for holonomic mechanical systems*, in "La Mécanique analytique de Lagrange et son héritage". Atti dell'Accademia delle Scienze di Torino, **126** (suppl. 2), 151-168, 1992.

[82] Salvadori L., Schiaffinò A., *On the problem of total stability*, Nonlinear Analysis: TMA, **1** (3), 1977.

[83] Salvadori L., Visentin F., *Sul problema della biforcazione generalizzata di Hopf per sistemi periodici*, Rend. Sem. Mat. Univ. Padova, **68**, 129-147, 1982.

[84] Salvadori L., Visentin F., *Sulla stabilità totale condizionata dell'equilibrio nella meccanica dei sistemi olonomi*, Rend. Mat., Serie VII, **12**, 475-495, 1992.

[85] Salvadori L., Visentin F., *Time-dependent forces and stability of equilibrium for holonomic systems*, Ricerche di Matematica, **XLI**, 287-275, 1992.

[86] Savchenko A.I., *On the stability of motions of conservative mechanical systems under continually-acting perturbations*, PMM, **38**, 240-245, 1974.

[87] Seibert P., *Estabilidad bajo perturbaciones sostenidas y su generalizacion en flujos continuos*, Acta Mexicana Cienc. y Tecnol., **11** (3), 1968.

[88] Seibert P. , Florio J.S., *On the foundations of bifurcation theory*, Nonlinear Anal.: TMA, **22** (8), 927-944, 1944.

[89] Silla L., *Sulla instabilità dell'equilibrio di un sistema materiale in posizioni non isolate*, Rend. Accad. Lincei, **17**, 347-355, 1908.

[90] Soucek J. , Soucek V., *Morse-Sard theorem for real analytic functions*, Comment. Math Universitatis Carolinae, **13**, 45-51, 1972.

[91] Taliaferro S.D., *An inversion of the Lagrange–Dirichlet stability theorem*, Arch. Rational Mech. Anal., **73**, 183-190, 1980.

[92] Taliaferro S.D., *Stability for two dimensional analytic potentials*, J. Differential Equations, **35**, 248-265, 1980.

[93] Taliaferro S.D., *Instability of an equilibrium in a potential field*, Arch. Rational Mech. Anal., **109**, 183-194, 1990.

[94] Visentin F., *Stability properties of bifurcating periodic solutions of periodic systems*, in Differential Equations: Stability and Control, Elaydi Ed., Lecture Notes in Pure and Applied Mathematics, Marcel Dekker, **127**, 513-520, 1990.

[95] Wintner A., "The analytical foundation of celestial mechanics", Univ.Press, Princeton, 1941.

[96] Yoshizawa T.,"Stability theory by Liapunov's second method", The Math. Soc. of Japan, Tokio, 1966.

PART IV

INVARIANT SETS OF MECHANICAL SYSTEMS

A.V. Karapetyan
Russian Academy of Sciences, Moscow, Russia

ABSTRACT

There are discussed problems of the existence and stability of invariant sets of mechanical systems (in particular, steady motions and relative equilibria).

The existence problem of stable motions (zero-dimensional invariant sets) firstly was investigated in the [1]. Really, the famous Routh theory [1]-[15] gives stability conditions of steady motions of conservative mechanical systems with first integrals as well as construction method of such steady motions. This method was spreaded for the construction problem of steady motions not only stable [2], [7] and for dissipative system with first integrals [12]-[15].

Moreover the Routh theory was modified to the existence and stability problems of invariant sets of dynamical systems with an ungrowing function and first integrals [7], [14] (in particular, of conservative and dissipative mechanical systems with symmetry [16]-[19]).

Chapter 1

Introduction.

1.1 Mechanical systems with cyclic coordinates.

Let us consider a mechanical systems with n degrees of freedom discribed by generalized coordinates $\mathbf{q} \in \mathbf{R}^n$ and velocities $\dot{\mathbf{q}} \in \mathbf{R}^n$ $\left(\dot{\mathbf{q}} = \dfrac{d\mathbf{q}}{dt}, \quad t \in [0, +\infty)\right)$. Let $T = \dfrac{1}{2}(A(\mathbf{q})\dot{\mathbf{q}}, \dot{\mathbf{q}})$ be kinetic energy, $V(\mathbf{q})$ a potential energy and $\mathbf{Q}(\mathbf{q}, \dot{\mathbf{q}})$ nonconservative generalized forces ($A(\mathbf{q}) \in \mathcal{C}^2$ is a positive definite symmetric $n \times n$-matrix, $V(\mathbf{q}) \in \mathcal{C}^2 : \mathbf{R}^n \to \mathbf{R}$, $\mathbf{Q}(\mathbf{q}, \dot{\mathbf{q}}) \in \mathcal{C}^1 : \mathbf{R}^n \times \mathbf{R}^n \to \mathbf{R}^n$). If $\mathbf{Q}(\mathbf{q}, \mathbf{0}) \equiv \mathbf{0}$, $(\mathbf{Q}, \dot{\mathbf{q}}) \leq 0$ then the generalized forces $\mathbf{Q}$ are dissipative. Here and further $(.,.)$ is a scalar product.

Equations of motion of the system can be written in the Lagrange form

$$\frac{d}{dt}\frac{\partial L}{\partial \dot{\mathbf{q}}} = \frac{\partial L}{\partial \mathbf{q}} + \mathbf{Q}, \tag{1.1}$$

where $L = T - V$ is the Lagrangian of the system. If $\mathbf{Q}$ are dissipative forces then the equations (1.1) admit the energy equation

$$\frac{dH}{dt} = (\mathbf{Q}, \dot{\mathbf{q}}) \leq 0, \tag{1.2}$$

where $H = T + V$ is the total mechanical energy of the system. In this case H is an ungrowing function. If $(\mathbf{Q}, \dot{\mathbf{q}}) \equiv 0$ (in particular, $\mathbf{Q} \equiv \mathbf{0}$) then equations (1.1) admit the energy integral

$$H = h \equiv \text{const}. \tag{1.3}$$

In this case the mechanical system is conservative.

Assume that the kinetic energy, the potential energy and generalized dissipative forces do not depend on some generalized coordinates $\mathbf{s} \in \mathbf{R}^k$ and generalized dissipative forces corresponding to these coordinates are absent

$$T = T(\mathbf{r}, \dot{\mathbf{r}}, \dot{\mathbf{s}}), \quad V = V(\mathbf{r}), \quad \mathbf{Q} = \mathbf{Q}^{(m)}(\mathbf{r}, \dot{\mathbf{r}}) \in \mathbf{R}^m \tag{1.4}$$

($\mathbf{r} \in \mathbf{R}^m$, $\mathbf{s} \in \mathbf{R}^k$, $m + k = n$). Coordinates $\mathbf{r}$ and $\mathbf{s}$ are essential and cyclic coordinates respectively.

The Lagrange equations (1.1) of the system with cyclic coordinates have the following form

$$\frac{d}{dt}\frac{\partial T}{\partial \dot{\mathbf{r}}} = \frac{\partial T}{\partial \mathbf{r}} - \frac{\partial V}{\partial \mathbf{r}} + \mathbf{Q}^{(m)}; \quad \frac{d}{dt}\frac{\partial T}{\partial \dot{\mathbf{s}}} = \mathbf{0}. \tag{1.5}$$

Obviously, equations (1.5) admit k first integrals

$$\mathbf{p} = \frac{\partial T}{\partial \dot{\mathbf{s}}} = \mathbf{c} = \text{const} \quad (\mathbf{c} \in \mathbf{R}^k). \tag{1.6}$$

Let us separate $n \times n$-matrix $\mathbf{A}$ to $m \times m$-matrix $\mathbf{A}_{mm}$, $m \times k$-matrix $\mathbf{A}_{mk}$, $k \times m$-matrix $\mathbf{A}_{km} = \mathbf{A}_{mk}^{\mathrm{T}}$ and $k \times k$-matrix $\mathbf{A}_{kk}$ (symbol T denotes the transposition) such that

$$T = \frac{1}{2}\left[(\mathbf{A}_{mm}\dot{\mathbf{r}}, \dot{\mathbf{r}}) + (\mathbf{A}_{mk}\dot{\mathbf{s}}, \dot{\mathbf{r}}) + (\mathbf{A}_{km}\dot{\mathbf{r}}, \dot{\mathbf{s}}) + (\mathbf{A}_{kk}\dot{\mathbf{s}}, \dot{\mathbf{s}})\right]. \tag{1.7}$$

Then the first integrals (1.6) can be written in the form

$$\mathbf{p} \equiv \mathbf{A}_{km}\dot{\mathbf{r}} + \mathbf{A}_{kk}\dot{\mathbf{s}} = \mathbf{c}. \tag{1.8}$$

Introduce the Routh function

$$R = R(\mathbf{r}, \dot{\mathbf{r}}, \mathbf{c}) = (L - (\dot{\mathbf{s}}, \mathbf{c}))_* = R_2 + R_1 + R_0,$$
$$R_2 = \frac{1}{2}(\mathbf{M}(\mathbf{r})\dot{\mathbf{r}}, \dot{\mathbf{r}}), \quad R_1 = (\mathbf{g}_c(\mathbf{r}), \dot{\mathbf{r}}), \quad R_0 = -V_c(\mathbf{r});$$

$$\mathbf{M} = \mathbf{A}_{mm} - \mathbf{A}_{mk}\mathbf{A}_{kk}^{-1}\mathbf{A}_{km}, \quad \mathbf{g}_c = \mathbf{A}_{mk}\mathbf{A}_{kk}^{-1}\mathbf{c}, \quad V_c = V + \frac{1}{2}\left(\mathbf{A}_{kk}^{-1}\mathbf{c}, \mathbf{c}\right).$$

The subscript star means that expression $(\cdot)_*$ was obtained by eliminating quantities $\dot{\mathbf{s}}$ from expression $(\cdot)$ using formula

$$\dot{\mathbf{s}} = \mathbf{A}_{kk}^{-1}(\mathbf{c} - \mathbf{A}_{km}\dot{\mathbf{r}}) \tag{1.9}$$

followed relation (1.8).

Thus equations (1.5) can be represented in the form

$$\frac{d}{dt}\frac{\partial R}{\partial \dot{\mathbf{r}}} = \frac{\partial R}{\partial \mathbf{r}} + \mathbf{Q}^{(m)}, \tag{1.10}$$

$$\dot{\mathbf{c}} = \mathbf{0}, \quad \dot{\mathbf{s}} = -\frac{\partial R}{\partial \mathbf{c}} \tag{1.11}$$

(the Routh equations). Equations (1.10) can be considered independently on equations (1.11) and represented in the form

$$\frac{d}{dt}\frac{\partial R_2}{\partial \dot{\mathbf{r}}} = \frac{\partial R_2}{\partial \mathbf{r}} + \mathbf{G}_c\dot{\mathbf{r}} - \frac{\partial V_c}{\partial \mathbf{r}} + \mathbf{Q}^{(m)}, \tag{1.12}$$

$$\left(\mathbf{G}_c = \left(\frac{\partial \mathbf{g}_c}{\partial \mathbf{r}}\right)^{\mathrm{T}} - \left(\frac{\partial \mathbf{g}_c}{\partial \mathbf{r}}\right), \quad \mathbf{G}_c^{\mathrm{T}} = -\mathbf{G}_c \right).$$

Equations (1.12) can be considered as equations of motion in the Lagrange form of some mechanical system with m degrees of freedom. Variables $\mathbf{r}$ are generalized coordinates of this system, R_2 is the kinetic energy, V_c is its potential energy, $\mathbf{G}_c\dot{\mathbf{r}}$ and $\mathbf{Q}^{(m)}$ are generalized gyroscopic and dissipative forces respectively. Such system is the reduced system and V_c is the reduced potential [2]. The reduced system admits an ungrowing function $H_* \equiv H_c = R_2 + V_c$, because $\dfrac{dH_c}{dt} = (\mathbf{Q}^{(m)}, \dot{\mathbf{r}}) \leq 0$, or, in the conservative case, when $(\mathbf{Q}^{(m)}, \dot{\mathbf{r}}) \equiv 0$, the Jacobi first integral $H_c = \mathrm{const}$

1.2 Steady motions of mechanical systems with cyclic coordinates.

The reduced system can stay in equilibria positions

$$\mathbf{r} = \mathbf{r}_c^0, \quad \dot{\mathbf{r}} = \mathbf{0} \tag{1.13}$$

if $\mathbf{r}_c^0$ satisfy equations

$$\frac{\partial V_c}{\partial \mathbf{r}} = \mathbf{0}. \tag{1.14}$$

The equilibria (1.13) of the reduced system correspond to steady motions of the original system with cyclic coordinates (see (1.9))

$$\mathbf{r} = \mathbf{r}_c^0, \quad \dot{\mathbf{r}} = \mathbf{0}, \quad \mathbf{s} = \mathbf{s}_c^0(t) = \dot{\mathbf{s}}_c^0 t + \mathbf{s}^0, \quad \dot{\mathbf{s}} = \dot{\mathbf{s}}_0 = \mathbf{A}_{kk}^{-1}(\mathbf{r}_c^0)\mathbf{c}, \tag{1.15}$$

($\mathbf{s}^0 \in \mathbf{R}^k$). Thus we have

Theorem 1.1. *If the reduced potential* $V_c(\mathbf{r})$ *takes a stationary value* $\mathbf{r}_c^0$ *then relations* (1.15) *represent some steady motion of the system.*

Theorem 1.1 was done by E.J.Routh [1],[2]. Note that steady motions (1.15) can be called stationary motions because they give stationary value to the total mechanical energy H on conditions (1.6).

Remark 1.1. The point $\mathbf{r}_c^0$ and the corresponding to it steady motion (1.15) depend on parameters $\mathbf{c}$. It means that steady motions (1.15) form some family.

Remark 1.2. The reduced potential V_c even for fixed values of parameters $\mathbf{c}$ can take stationary value not only in the point $\mathbf{r}_c^0$, but, generally speaking, in some other points $\mathbf{r}_c^1$, $\mathbf{r}_c^2$, ... (system (1.14) can have a number solutions). These points and the corresponding steady motions also depend on parameters $\mathbf{c}$ and form corresponding families.

Remark 1.3. All steady motions of the system can be represented in the space $\mathbf{R}^m \times \mathbf{R}^k$ ($\mathbf{r} \in \mathbf{R}^m$, $\mathbf{c} \in \mathbf{R}^k$) as surfaces $\mathbf{r} = \mathbf{r}^0(\mathbf{c})$, $\mathbf{r}^1(\mathbf{c})$, $\mathbf{r}^2(\mathbf{c})$, ... These surfaces can have (for specials values $\mathbf{c}^*$) common points, which are points of bifurcation (by Poincare). Note that if $\mathbf{r}^i(\mathbf{c}^*) = \mathbf{r}^j(\mathbf{c}^*) = \mathbf{r}^*$ $(i \neq j)$ then

$$\det\left[\frac{\partial^2 V_{c^*}}{\partial \mathbf{r}^2}\right]_{\mathbf{r}=\mathbf{r}^*} = 0.$$

Theorem 1.2. *If the reduced potential takes a local strict minimum in the point* $\mathbf{r}_c^0$ *then the steady motion* (1.15) *is stable with respect to variables* $\mathbf{r}$, $\dot{\mathbf{r}}$, $\dot{\mathbf{s}}$.

Theorem 1.2 was proved by L.Salvadori in the [8] (for conservative systems; see also [9]-[11]) and in the [12],[13] (for dissipative systems).

Note that a steady motion (1.15) is always unstable with respect to cyclic coordinates $\mathbf{s}$, because for perturbed motions ($\mathbf{r} = \mathbf{r}(t)$, $\mathbf{s} = \mathbf{s}(t)$) in general $\dot{\mathbf{s}}(0) \neq \dot{\mathbf{s}}_c^0$.

Remark 1.4. If

$$\det\left[\frac{\partial^2 V_c}{\partial \mathbf{r}^2}\right]_0 = \det\left[\frac{\partial^2 V_c}{\partial \mathbf{r}^2}\right]_{\mathbf{r}=\mathbf{r}_c^0} \neq 0$$

then the number of negative eigenvalues of matrix $\left[\frac{\partial^2 V_c}{\partial \mathbf{r}^2}\right]_0$ is ind $\delta^2 V_c$ or Poincare degree of instability of corresponding steady motion (1.15). If the Poincare degree of instability of steady motion (1.15) is equal to zero, then all eigenvalues of matrix $\left[\frac{\partial^2 V_c}{\partial \mathbf{r}^2}\right]_0$ are positive and therefore the reduced potential V_c has a local strict minimum in the point $\mathbf{r}_c^0$. If the Poincare degree of instabilty of steady motion (1.15) is odd then $\det\left[\frac{\partial^2 V_c}{\partial \mathbf{r}^2}\right]_0 < 0$.

Theorem 1.3. *If the Poincare degree of instability of steady motion* (1.15) *is odd then this steady motion is unstable.*

Theorem 1.3 was proved by Lord Kelvin [20] (for linear systems) and N.G.Chetaev [21] (in general case).

Theorems 1.1–1.3 are correct as for dissipative as well for conservative mechanical systems. The next two theorems are correct only for dissipative mechanical systems.

Remark 1.5. If dissipative forces $\mathbf{Q}^{(m)}$ are such that $(\mathbf{Q}^{(m)}, \dot{\mathbf{r}}) = 0$ only for $\dot{\mathbf{r}} = \mathbf{0}$ then function H_c preserves its initial value only in equilibria of the reduced system. It means that the total mechanical energy H of the original mechanical system with cyclic coordinates preserves its initial value only at steady motions. In this case we will say that forces $\mathbf{Q}^{(m)}$ have the total dissipation with respect to essential generalized velocities.

Theorem 1.4. *If the reduced potential takes a local strict minimum for fixed values* $\mathbf{c}^0$ *of parameters* $\mathbf{c}$ *in the point* $\mathbf{r}^0$ $(\mathbf{c}^0)$, *this point is isolated from other stationary points* $\mathbf{r}^1$, $\mathbf{r}^2$,... *of the reduced potential (if these points exist at all), and dissipative forces have the total dissipation with respect to essential generalized velocities then the corresponding to these parameters* $\mathbf{c}^0$ *steady motion* (1.15) *is stable and every perturbed motion fairly close to the unperturbed tends to some steady motion* (1.15) *corresponding to perturbed values of parameters* $\mathbf{c}$ *as* $t \to +\infty$.

Theorem 1.5. *If the reduced potential takes a stationary value, that is not even nonstrict minimum, for fixed values* $\mathbf{c}^0$ *of parameters* $\mathbf{c}$ *in the point* $\mathbf{r}^0$ $(\mathbf{c}^0)$, *this point is isolated from other stationary points* $\mathbf{r}^1, \mathbf{r}^2, \ldots$ *of the reduced potential (if these points exist at all) and dissipative forces have the total dissipation with respect to essential generalized velocities then the corresponding to these parameters* $\mathbf{c}^0$ *steady motion* (1.15) *is unstable.*

Theorem 1.4 and 1.5 were proved by V.V.Rumyantsev [11] and L.Salvadori [12] (see also [20]-[22]).

Remark 1.6. If the Poincare degree of instability of a steady motion is even then this steady motion may be stable (gyroscopic stabilization [20], [22]) without dissipative forces with total dissipation and unstable with the latters. So the gyroscopic stability is nonsecular as it is disturbed by dissipative forces. If the Poincare degree of instability is zero then the steady motion is stable in the secular sense (see theorems 1.2–1.5).

1.3 Relative equilibria of mechanical systems with cyclic coordinates.

Let us consider a mechanical systems with cyclic coordinates (see (1.4)) and assume that $\dot{\mathbf{s}} \equiv \boldsymbol{\omega} = \text{const}$ for all motions of the system ($\boldsymbol{\omega} \in \mathbf{R}^k$). It means that there are some control forces $\mathbf{Q}_\omega^{(k)}(\mathbf{r}, \dot{\mathbf{r}}) \in \mathcal{C} : \mathbf{R}^m \times \mathbf{R}^m \to \mathbf{R}^k$ preserving initial values $\boldsymbol{\omega}$ of cyclic velocities $\dot{\mathbf{s}}$. These forces can be defined by relations

$$\left(\frac{d}{dt}\frac{\partial T}{\partial \dot{\mathbf{s}}}\right)_{\dot{\mathbf{s}} \equiv \omega} = \mathbf{Q}_\omega^{(k)}(\mathbf{r}, \dot{\mathbf{r}}), \tag{1.16}$$

while essential variables $\mathbf{r}$ and $\dot{\mathbf{r}}$ satisfy equations

$$\frac{d}{dt}\frac{\partial \Lambda}{\partial \dot{\mathbf{r}}} = \frac{\partial \Lambda}{\partial \mathbf{r}} + \mathbf{Q}^{(m)} \tag{1.17}$$

(see (1.1)). Here

$$\Lambda = (L)_{\dot{\mathbf{s}} \equiv \omega} = \Lambda_2 + \Lambda_1 + \Lambda_0,$$

$$\Lambda_2 = \frac{1}{2}(\mathbf{A}_{mm}(\mathbf{r})\dot{\mathbf{r}}, \dot{\mathbf{r}}), \quad \Lambda_1 = (\mathbf{g}_\omega(\mathbf{r}), \dot{\mathbf{r}}), \quad \Lambda_0 = -V_\omega(\mathbf{r}),$$

$$\mathbf{g}_\omega(\mathbf{r}) = \mathbf{A}_{mk}(\mathbf{r})\boldsymbol{\omega}, \quad V_\omega(\mathbf{r}) = V(\mathbf{r}) - \frac{1}{2}(\mathbf{A}_{kk}(\mathbf{r})\boldsymbol{\omega}, \boldsymbol{\omega}).$$

Thus equations (1.16) can be considered as equations of motion in the Lagrange form of some mechanical system with m degrees of freedom. Variables $\mathbf{r}$ are generalized coordinates of this system, Λ_2 is its kinetic energy, V_ω is its potential energy, $\mathbf{G}_\omega \dot{\mathbf{r}}$ and $\mathbf{Q}^{(m)}$ are generalized gyroscopic and dissipative forces. Such system is the restricted system and V_ω is the changed potential

[3],[10],[23]–[26]. The restricted system admit an ungrowing function $H_\omega = \Lambda_2 + V_\omega$ or, in the conservative case, when $\left(\mathbf{Q}^{(m)}, \dot{\mathbf{r}}\right) = 0$ the Jacobi first integral $H_\omega =$ const.

Evidently, equations (1.17) of the restricted system have the same structure as equations (1.12) of the reduced system. There is only one difference: parameters ω are not perturbed in the original system, while parameters $\mathbf{c}$ can perturb in the original system with cyclic coordinates.

The restricted system (1.17) can stay in equilibria positions

$$\mathbf{r} = \mathbf{r}^0_\omega, \quad \dot{\mathbf{r}} = 0 \tag{1.18}$$

if $\mathbf{r}^0_\omega$ satisfy equations

$$\frac{\partial V_\omega}{\partial \mathbf{r}} = 0. \tag{1.19}$$

These equilibria of the restricted system (1.17) correspond to relative equilibria of the original restricted system with control forces

$$\mathbf{r} = \mathbf{r}^0_\omega, \quad \dot{\mathbf{r}} = 0, \quad \mathbf{s} = \omega t + \mathbf{s}^0, \quad \dot{\mathbf{s}} = \omega \quad (\mathbf{s}^0 \in \mathbf{R}^k) \tag{1.20}$$

Similarly to the previous section we have

Theorem 1.6. *If the changed potential $V_\omega(\mathbf{r})$ takes a stationary value in some point $\mathbf{r}^0_\omega$ then relations* (1.20) *represent some relative equilibria of the system.*

Theorem 1.7. *If the changed potential takes a local strict minimum in the point $\mathbf{r}^0_\omega$ then relative equilibrium* (1.20) *is stable.*

Theorem 1.8. *If the Poincare degree of instability of relative equilibrium* (1.20) *is odd then this relative equilibrium is unstable.*

Theorem 1.9. *If the changed potential takes a local strict minimum in the point $\mathbf{r}^0_\omega$, this point is isolated from other stationary points $\mathbf{r}^1_\omega$, $\mathbf{r}^2_\omega$,... of the changed potential (if these points exist at all) and dissipative forces have the total dissipation with respect to essential generalized velocities then the relative equilibrium* (1.20) *is asymptotically stable.*

Theorem 1.10. *If the changed potential takes a stationary value, that is not even nonstrict minimum, in the point $\mathbf{r}^0_\omega$, this point is isolated from other stationary points $\mathbf{r}^1_\omega$, $\mathbf{r}^2_\omega$,... of the changed potential (if these points exist at all) and dissipative forces have the total dissipation with respect to essential generalized velocities then the relative equilibrium is unstable.*

Note that stability, asymptotic stability and instability in theorems 1.7–1.10 one has to understand with respect to variables $\mathbf{r}$ and $\dot{\mathbf{r}}$. Evidently remarks 1.1–1.6 are spreaded to relative equilibria too.

1.4 The correspondence of relative equilibria with steady motions and the relationship between their stability conditions.

The reduced potential V_c differs from the changed potential V_ω. Neverless the problem of determining the steady motions (see (1.14)) and relative equilibria (see (1.19)) of systems with cyclic coordinates are in a sense equivalent. The rigorous formulation is as follows:

Theorem 1.11. *For any set* $\mathbf{c}$ *of constants of the cyclic integrals (for any set* $\boldsymbol{\omega}$ *of constants of the cyclic velocities) there exists a set of constants* $\boldsymbol{\omega}$ *(a set of constants* $\mathbf{c}$*) such that a solution of system* (1.19) *coincides with that of system* (1.14).

Theorem 1.11 was proved by M. Pascal [23] and S. Ja. Stepanov [24] (see also [3], [10], [25]).

Note that the correspondence of constants $\mathbf{c}$ ($\boldsymbol{\omega}$) and $\boldsymbol{\omega}$ ($\mathbf{c}$) is given by the relation (see (1.19))

$$\mathbf{c} = \mathbf{A}_{kk}(\mathbf{r}_\omega^0)\boldsymbol{\omega} \quad \left(\boldsymbol{\omega} = \mathbf{A}_{kk}^{-1}(\mathbf{r}_c^0)\mathbf{c}\right) \tag{1.21}$$

Thus on conditions (1.21) $\mathbf{r}_c^0 = \mathbf{r}_\omega^0$ ($\mathbf{r}_\omega^0 = \mathbf{r}_c^0$), where $\mathbf{r}_c^0$ and $\mathbf{r}_\omega^0$ are solutions of equations (1.14) and (1.19) respectively.

The analogous correspondence in regard to the stability of steady motions and relative equilibria is not complete. In rigorous terms:

Theorem 1.12. *If the changed potential* V_ω *has a strict local minimum at point* $\mathbf{r}_\omega^0$ *then the reduced potential* V_c *has a strict local minimum at point* $\mathbf{r}_c^0 = \mathbf{r}_\omega^0$ *on condition* (1.21).

Theorem 1.12 was proved by M. Pascal [23] and S. Ja. Stepanov [24] (see also [3],[10],[25]).

Thus if a relative equilibrium is stable in the secular sense then the corresponding steady motion is stable in the same sense too. However in general case, a steady motion can be stable in the secular sense even the corresponding relative equilibrium is unstable.

Let us say that a steady motion (relative equilibrium) is trivial if $\mathbf{r}_c^0$ ($\mathbf{r}_\omega^0$) is independent of $\mathbf{c}$ ($\boldsymbol{\omega}$). Note that trivial steady motions correspond to trivial relative equilibria and vice versa.

Theorem 1.13. *Degrees of instability of trival steady motions and corresponding relative equilibria coincide.*

Theorem 1.13 was proved by A. V. Karapetyan an S. Ja. Stepanov [26].

Corollary. *Conditions for secular stability of non-degenerate* $\left(\det\left[\dfrac{\partial^2 V_{c,\omega}}{\partial \mathbf{r}^2}\right] \neq 0\right)$ *trivial steady motions and corresponding trivial relative equilibria coincide.*

Chapter 2

Invariant sets of dynamical systems with first integrals.

2.1 Preliminary information.

Results given in Introduction are not invariant respectively to a choice of generalized coordinates. Moreover, they deal with steady motion and relative equilibria which represent zero-dimensional invariant sets only. In this chapter we will consider general dynamical systems (in particular, mechanical) described by ordinary differential equations

$$\dot{\mathbf{x}} = \mathbf{f}(\mathbf{x}). \qquad (2.1)$$

Here $\mathbf{x} \in \mathbf{X} \subseteq \mathbf{R}^n$ is n-vector of phasic variables, $\dot{\mathbf{x}} = \frac{d\mathbf{x}}{dt}$, $t \in [0; +\infty)$, $\mathbf{f}(\mathbf{x}) \in \mathcal{C}^1 : \mathbf{X} \to \mathbf{R}^n$.

Let $\mathbf{x}(t; \mathbf{x}^0)$ denote a solution of equations (2.1) with initial conditions $\mathbf{x}(0; \mathbf{x}^0) = \mathbf{x}^0 \in \mathbf{X}$ ($t = 0$ can be chosen as the initial time without loss of generality because equations (2.1) do not depend on time explicitly). This solution corresponds to some motion of the dynamical system, so we will call it motion $\mathbf{x}(t; \mathbf{x}^0)$ of the system.

Definition 2.1. A set $\mathbf{X}_0 \subset \mathbf{X}$ is an invariant set of the system if $\mathbf{x}(t; \mathbf{x}^0) \in \mathbf{X}_0$ for all $t \geq 0$ and arbitrary $\mathbf{x}^0 \in \mathbf{X}_0$.

Remark 2.1. If $\dim \mathbf{X}_0 = 0$ then $\mathbf{X}_0 = \{\mathbf{x}^0\}$, where $\mathbf{x}^0 \in \mathbf{X}$ is an invariant point of equations (2.1) ($\mathbf{f}(\mathbf{x}^0) = \mathbf{0}$), so $\mathbf{x}(t; \mathbf{x}^0) \equiv \mathbf{x}^0$ is a steady motion of the system.

Remark 2.2. If $\dim \mathbf{X}_0 = d > 0$ ($d < n$) then $\mathbf{X}_0 = \{\mathbf{x}^0 \in \mathbf{X} : \varphi_0(\mathbf{x}) = \mathbf{0}\}$ where $\varphi(\mathbf{x}) \in \mathcal{C}^1 : \mathbf{X} \to \mathbf{R}^{n-d}$. Motions of the system on an invariant set $\mathbf{X}_0$ satisfy equations

$$\dot{\mathbf{x}} = \mathbf{f}(\mathbf{x})|_{\varphi_0(\mathbf{x})=0}$$

and depend on time.

Remark 2.3. [7] If a set $\{\mathbf{x} \in \mathbf{X} : \varphi_0(\mathbf{x}) = \mathbf{0}\}$ is an invariant set of the system then

$$\left(\frac{\partial\varphi_0}{\partial\mathbf{x}}\,\mathbf{f}\right)_{\varphi_0=0} = \mathbf{0}.$$

If there are fulfilled the latter relation and

$$\operatorname{rank}\left(\frac{\partial\varphi_0}{\partial\mathbf{x}}\,\mathbf{f}\right)_{\varphi_0=0} = n - d$$

then a set $\{\mathbf{x} \in \mathbf{X} : \varphi_0(\mathbf{x}) = \mathbf{0}\}$ is an invariant set of the system.

Definition 2.2. Function $U(\mathbf{x}) \in \mathcal{C}^1 : \mathbf{X} \to \mathbf{R}$ is an ungrowing function of the system if $\frac{dU}{dt} = (\operatorname{grad} U, \mathbf{f}) \leq 0$ for all $\mathbf{x} \in \mathbf{X}$. Function $U(\mathbf{x}) \in \mathcal{C}^1 : \mathbf{X} \to \mathbf{R}$ is a first integral of the system if $\frac{dU}{dt} = (\operatorname{grad} U, \mathbf{f}) \equiv 0$.

Definition 2.3. A compact invariant set $\mathbf{X}_0$ of the system is stable if for any small positive number ε there exists a positive number δ such that $\operatorname{dist}(\mathbf{x}(t; \mathbf{x}^*); \mathbf{X}_0) < \varepsilon$ for all $t \geq 0$ and arbitrary $\mathbf{x}^* \in \mathbf{X}$ with $\operatorname{dist}(\mathbf{x}^*; \mathbf{X}_0) < \delta$; in particular, a steady motion $\mathbf{x}(t; \mathbf{x}^0) \equiv \mathbf{x}^0$ is stable if for any small positive number ε there exists a positive number δ such that $\| \mathbf{x}(t; \mathbf{x}^*) - \mathbf{x}^0 \| < \varepsilon$ for all $t \geq 0$ and an arbitrary $\mathbf{x}^* \in \mathbf{X}$ with $\| \mathbf{x}^* - \mathbf{x}^0 \| < \delta$; otherwise $\mathbf{X}_0$ and $\mathbf{x}^0$ are unstable invariant set and steady motion respectively.

Here and further $\operatorname{dist}(\mathbf{x}; \mathbf{X}_0) = \min_{\mathbf{y} \in \mathbf{X}_0} \| \mathbf{x} - \mathbf{y} \|$; $\| \mathbf{x} \|$ is a norm of vector $\mathbf{x}$.

Definition 2.4. A motion $\mathbf{x}(t; \mathbf{x}^0)$ of the system is stable with respect to functions $\mathbf{y}(\mathbf{x}) \in \mathcal{C} : \mathbf{X} \to \mathbf{R}^m$ $(m < n)$ if for any positive number ε there exists a positive number δ such that $\| \mathbf{y}(\mathbf{x}(t; \mathbf{x}^*)) - \mathbf{y}(\mathbf{x}(t; \mathbf{x}^0)) \| < \varepsilon$ for all $t \geq 0$ and an arbitrary $\mathbf{x}^* \in \mathbf{X}$ with $\| \mathbf{x}^* - \mathbf{x}^0 \| < \delta$; otherwise a motion $\mathbf{x}(t; \mathbf{x}^0)$ is unstable with respect to functions $\mathbf{y}(\mathbf{x})$.

Definition 2.5. A compact invariant set $\mathbf{X}_0$ of the system is asymptotically stable if it is stable and $\lim_{t \to +\infty} \operatorname{dist}(\mathbf{x}(t; \mathbf{x}^*); \mathbf{X}_0) = 0$, where $\operatorname{dist}(\mathbf{x}^*; \mathbf{X}_0) < \delta$; in particular, a motion $\mathbf{x}(t; \mathbf{x}^0)$ of the system is asymptotically stable with respect to functions $\mathbf{y}(\mathbf{x})$ if it is stable with respect to these functions and $\lim_{t \to +\infty} \| \mathbf{y}(\mathbf{x}(t; \mathbf{x}^*)) - \mathbf{y}(\mathbf{x}(t; \mathbf{x}^0)) \| = 0$, where $\| \mathbf{x}^* - \mathbf{x}^0 \| < \delta$ (δ is some positive number).

2.2 Invariant sets.

Consider dynamical system (2.1) and assume that it admits an ungrowing function $U_0(\mathbf{x}) \in \mathcal{C}^2 : \mathbf{X} \to \mathbf{R}$ and k first integrals $\mathbf{U}(\mathbf{x}) \in \mathcal{C}^2 : \mathbf{X} \to \mathbf{R}^k$, $(k < n)$.

We will say that a function $U_0(\mathbf{x})$ takes a non-degenerate stationary value on some set $\mathbf{X}_0 \subset \mathbf{X}$ for fixed values of first integrals $\mathbf{U}(\mathbf{x}) = \mathbf{c}$ $(\mathbf{c} \in \mathbf{R}^k)$ if $\mathbf{X}_0$ is the maximal connected subset of set $\{\mathbf{x} \in \mathbf{X} : \delta U_0|_{\mathbf{U}=\mathbf{c}} = 0\}$ and $\delta^2 U_0|_{\mathbf{U}=\mathbf{c}} \neq 0$ for $\mathbf{x} \in \mathbf{X}_0$.

Theorem 2.1.[14] *If an ungrowing function $U_0(\mathbf{x})$ of the system takes a non-degenerate stationary value on some set $\mathbf{X}_0$ for fixed values of first integrals $\mathbf{U}(\mathbf{x}) = \mathbf{c}$ of the system then $\mathbf{X}_0$ is an invariant set of this system.*

Proof. Let function $U_0(\mathbf{x})$ takes a stationary value on a set $\mathbf{X}_0$ for fixed values $\mathbf{c}^0$ of first integrals $\mathbf{U}(\mathbf{x}) = \mathbf{c}$. Then this set is defined by relations

$$\frac{\partial W}{\partial \mathbf{x}} \equiv \frac{\partial U_0}{\partial \mathbf{x}} + \frac{\partial(\boldsymbol{\lambda}, \mathbf{U})}{\partial \mathbf{x}} = 0, \tag{2.2}$$

$$\frac{\partial W}{\partial \boldsymbol{\lambda}} \equiv \mathbf{U} - \mathbf{c}^0 = 0, \tag{2.3}$$

where $W = U_0 + (\boldsymbol{\lambda}, (\mathbf{U} - \mathbf{c}^0)) = W(\mathbf{x}; \boldsymbol{\lambda})$, $\boldsymbol{\lambda} \in \mathbf{R}^k$ is a vector of unknown constants (Lagrange's multipliers).

Consider a scalar product of the left part of (2.2) with the right part of (2.1) and take relations $\left(\frac{\partial \mathbf{U}}{\partial \mathbf{x}}, \mathbf{f}\right) \equiv 0$ into account ($\mathbf{U} = \mathbf{c}$ are first integrals): $(\operatorname{grad} U_0, \mathbf{f})_{\mathbf{X}_0} = 0$. It means that $(\operatorname{grad} U_0, \mathbf{f})$ takes a maximal value on the set $\mathbf{X}_0$ ($(\operatorname{grad} U_0, \mathbf{f}) \leq 0$). So

$$\frac{\partial}{\partial \mathbf{x}}\left(\frac{\partial U_0}{\partial \mathbf{x}}, \mathbf{f}\right) \equiv \frac{\partial^2 U_0}{\partial \mathbf{x}^2}\mathbf{f} + \frac{\partial \mathbf{f}}{\partial \mathbf{x}}\frac{\partial U_0}{\partial \mathbf{x}} = 0 \tag{2.4}$$

on the set $\mathbf{X}_0$.

On other side $(\operatorname{grad} \mathbf{U}, \mathbf{f}) \equiv 0$. So

$$\frac{\partial}{\partial \mathbf{x}}\left(\frac{\partial \mathbf{U}}{\partial \mathbf{x}}, \mathbf{f}\right) \equiv \frac{\partial^2 \mathbf{U}}{\partial \mathbf{x}^2}\mathbf{f} + \frac{\partial \mathbf{f}}{\partial \mathbf{x}}\frac{\partial \mathbf{U}}{\partial \mathbf{x}} = 0 \tag{2.5}$$

for all $\mathbf{x} \in \mathbf{X}$. Thus (see (2.4) and (2.5))

$$\left(\frac{\partial}{\partial \mathbf{x}}\left(\frac{\partial W}{\partial \mathbf{x}}\right), \mathbf{f}\right) \equiv \frac{\partial^2 U_0}{\partial \mathbf{x}^2}\mathbf{f} + \frac{\partial^2(\boldsymbol{\lambda}, \mathbf{U})}{\partial \mathbf{x}^2}\mathbf{f} = -\frac{\partial \mathbf{f}}{\partial \mathbf{x}}\frac{\partial W}{\partial \mathbf{x}} = 0$$

on the set $\mathbf{X}_0$.

Taking the identity $(\operatorname{grad} \mathbf{U}, \mathbf{f}) \equiv 0$ and the condition $\delta^2 U_0|_{\mathbf{U}=\mathbf{c}^0} \neq 0$ for $\mathbf{x} \in \mathbf{X}$ into account we conclude that the set $\mathbf{X}_0$ ((2.2),(2.3)) is an invariant set of the system according to remark 2.3.

Corollary 2.1. If the first integral $U_0(\mathbf{x})$ takes a non-degenerate stationary value on some set $\mathbf{X}_0 \subset \mathbf{X}$ for fixed values of other first integrals then $\mathbf{X}_0$ is an invariant set of the system.

Proof. If $U_0(\mathbf{x})$ is the first integral then $(\operatorname{grad} U_0, \mathbf{f}) \equiv 0$ and therefore $U_0(\mathbf{x})$ is an ungrowing function. So one can use theorem 2.1 (see also [7]).

Remark 2.4. Motions of the system belonging on the set $\mathbf{X}_0$ depend on time and coincide with steady motions of the system if $\dim \mathbf{X}_0 = 0$ only. Nevertheless we can consider these motions as stationary because they give a stationary value to an ungrowing function on fixed levels of first integrals.

Remark 2.5. An invariant set $\mathbf{X}_0$ giving to function U_0 a stationary value for fixed constants of first integrals $\mathbf{U} = \mathbf{c}$ depends on these constants. It means that invariant set $\mathbf{X}_0$ forms some family of invariant sets $\mathbf{X}_0(\mathbf{c})$. Evidently, function U_0 does not decrease on all sets of this family, because $(\operatorname{grad} U_0, \mathbf{f}) = 0$ for all $\mathbf{x} \in \mathbf{X}_0(\mathbf{c})$.

Remark 2.6. Function $U_0(\mathbf{x})$ (even for fixed values of first integrals $\mathbf{U}(\mathbf{x}) = \mathbf{c}$) can take stationary value not only on the set $\mathbf{X}_0$ but, generally speaking, on other sets $\mathbf{X}_1, \mathbf{X}_2, \ldots$ too (system (2.2), (2.3) can have several solutions). Sets $\mathbf{X}_1, \mathbf{X}_2, \ldots$ correspond to the same values of the first integrals (by assumption), but, in general, to different values of this function. These sets also depend on constants $\mathbf{c}$ and form some families of invariant sets $\mathbf{X}_1(\mathbf{c})$, $\mathbf{X}_2(\mathbf{c})$,... Evidently, $(\operatorname{grad} U_0, \mathbf{f}) = 0$ for all $\mathbf{x} \in \mathbf{X}_0 \cup \mathbf{X}_1 \cup \mathbf{X}_2 \ldots$

Remark 2.7. Families of invariant sets $\mathbf{X}_0(\mathbf{c})$, $\mathbf{X}_1(\mathbf{c})$, $\mathbf{X}_2(\mathbf{c})$,... can have different dimension and (for special values $\mathbf{c}^*$) common points. Such points are points of bifurcation (by Poincare). The union of these families $\mathbf{X}(\mathbf{c}) = \mathbf{X}_0(\mathbf{c}) \cup \mathbf{X}_1(\mathbf{c}) \cup \mathbf{X}_2(\mathbf{c})\ldots$ represents in the space $\mathbf{X} \times \mathbf{R}^k$ $(\mathbf{x} \in \mathbf{X}, \quad \mathbf{c} \in \mathbf{R}^k)$ some self-intersecting surface (Poincare-diagram).

2.3 Stability of invariant sets.

Theorem 2.2. [14] *If an ungrowing function $U_0(\mathbf{x})$ takes a local strict minimum for fixed values $\mathbf{c}^0$ of first integrals $\mathbf{U}(\mathbf{x}) = \mathbf{c}$ on a compact set $\mathbf{X}_0(\mathbf{c}^0)$ then $\mathbf{X}_0(\mathbf{c}^0)$ is a stable invariant set of the system and every stationary motion $\mathbf{x}(t; \mathbf{x}^0)$ $(\mathbf{x}^0 \in \mathbf{X}_0(\mathbf{c}^0))$ is stable with respect to* $\operatorname{dist}(\mathbf{x}; \mathbf{X}_0(\mathbf{c}^0))$.

Proof. Note that there is no non-degeneration condition of function U_0 on the set $\mathbf{X}_0(\mathbf{c}^0)$ in the formulation of the theorem 2.2. So firstly we have to prove that the set $\mathbf{X}_0(\mathbf{c}^0)$ giving the minimum to an ungrowing function is an invariant set.

Assume that function U_0 takes minimal value $m_0(\mathbf{c}^0)$ on the set $\mathbf{X}_0(\mathbf{c}^0)$. Consider some motion $\mathbf{x}(t; \mathbf{x}^0)$ of the system with initial conditions $\mathbf{x}(0; \mathbf{x}^0) = \mathbf{x}^0 \in \mathbf{X}_0(\mathbf{c}^0)$ (evidently $\mathbf{U}(\mathbf{x}^0) = \mathbf{c}^0$). Then

$$U_0(\mathbf{x}(t; \mathbf{x}^0)) \leq m_0(\mathbf{c}^0), \tag{2.6}$$

because $U_0(\mathbf{x})$ is an ungrowing function. On the other side value $m_0(\mathbf{c}^0)$ corresponds to a strict minimum of function U_0 for $\mathbf{U}(\mathbf{x}) = \mathbf{c}^0$. Therefore

$$U_0(\mathbf{x}(t; \mathbf{x}^0)) \geq m_0(\mathbf{c}^0). \tag{2.7}$$

Thus (see (2.6) and (2.7)) $U_0(\mathbf{x}(t; \mathbf{x}^0)) \equiv m_0(\mathbf{c}^0)$ and $\mathbf{x}(t; \mathbf{x}^0) \in \mathbf{X}_0(\mathbf{c}^0)$ because namely the set $\mathbf{X}(\mathbf{c}^0)$ gives the value $m_0(\mathbf{c}^0)$ to function $U_0(\mathbf{x})$ for $\mathbf{U}(\mathbf{x}) = \mathbf{c}^0$. It means that $\mathbf{X}_0(\mathbf{c}^0)$ is an invariant set of the system.

Consider a set

$$\{\mathbf{x} \in \mathbf{X} : \operatorname{dist}(\mathbf{x}; \mathbf{X}_0(\mathbf{c}^0)) = \varepsilon\} \subset \mathbf{X} \tag{2.8}$$

for some small positive number ε. Set $\mathbf{X}_0(\mathbf{c}^0)$ is a compact, so set (2.8) is a compact too and function $U_0(\mathbf{x}) - m_0(\mathbf{c}^0)$ is always bounded from below on this set. It means that there exists a positive number σ_1 such that $U_0(\mathbf{x}) - m_0(\mathbf{c}^0) > -\sigma_1$ on set (2.8). If variables $\mathbf{x}$ verify relations

$$\mathbf{U}(\mathbf{x}) = \mathbf{c}^0 \tag{2.9}$$

then function $U_0(\mathbf{x}) - m_0(\mathbf{c}^0) \geq \sigma_2 > 0$ on set (2.8) because function $U_0(\mathbf{x})$ on condition (2.9) takes a strict minimal value $m_0(\mathbf{c}^0)$ on set $\mathbf{X}_0(\mathbf{c}^0)$. By continuity there exist positive numbers σ_3 and σ_4 such that inequality $\| \mathbf{U}(\mathbf{x}) - \mathbf{c}_0 \| < \sigma_3$ implies inequality $U_0(\mathbf{x}) - m_0(\mathbf{c}^0) > \sigma_4$. Then, choosing a positive number $\mu < \dfrac{\sigma_3}{\sigma_4}$ we have that function

$$W = \mu(U_0(\mathbf{x}) - m_0(\mathbf{c}^0)) + \| \mathbf{U}(\mathbf{x}) - \mathbf{c}^0 \|$$

is bounded from below by a positive number σ ($\sigma < \min(\sigma_3 - \mu\sigma_1; \mu\sigma_4)$) on set (2.8). For this number σ one can find a positive number δ such that domain

$$\operatorname{dist}(\mathbf{x}, \mathbf{X}_0(\mathbf{c}^0)) < \delta \tag{2.10}$$

is completely inside domain $W(\mathbf{x}) < \sigma$ that is completely inside domain

$$\operatorname{dist}(\mathbf{x}, \mathbf{X}_0(\mathbf{c}^0)) < \varepsilon \tag{2.11}$$

Function $W(\mathbf{x})$ does not increase in this domain, because

$$\frac{dW}{dt} = \mu \frac{dU_0}{dt} = \mu(\operatorname{grad} U_0, \mathbf{f}) \leq 0.$$

Thus every perturbed motion of the sistem with initial conditions from domain (2.10) does not leave domain $W < \sigma$ and therefore domain (2.11). So $\mathbf{X}_0(\mathbf{c}^0)$ is a stable invariant set and every motion $\mathbf{x}(t, \mathbf{x}^0) \subset \mathbf{X}_0(\mathbf{c}^0)$ is stable with respect to $\operatorname{dist}(\mathbf{x}, \mathbf{X}_0(\mathbf{c}^0))$.

Corollary 2.2. If the first integral $U_0(\mathbf{x})$ takes a local strict minimum or maximum for fixed values $\mathbf{c}^0$ of other first integrals $\mathbf{U}(\mathbf{x}) = \mathbf{c}$ on a compact set $\mathbf{X}_0(\mathbf{c}^0)$ then $\mathbf{X}_0(\mathbf{c}^0)$ is a stable invariant set of the system and every stationary motion $\mathbf{x}(t, \mathbf{x}^0)$ ($\mathbf{x}^0 \in \mathbf{X}_0(\mathbf{c}^0)$) is stable with respect to $\operatorname{dist}(\mathbf{x}, \mathbf{X}_0(\mathbf{c}^0))$.

Proof. If $U_0(\mathbf{x})$ is the first integral then $U_0(\mathbf{x})$ and $-U_0(\mathbf{x})$ are ungrowing functions. Under conditions of corollary 2.2 one of these functions takes a local strict minimum on set $\mathbf{X}_0(\mathbf{c}^0)$. So one can use theorem 2.2.

Note that given proof of theorem 2.2 is some modification of Salvadori's proof of theorem 1.2 in dissipative case (see [12],[13]).

2.4 The Poincare degree of instability.

Function $U_0(\mathbf{x})$ has a local strict minimum [has no even nonstrict minimum] for fixed values $\mathbf{c}^0$ of first integrals $\mathbf{U}(\mathbf{x}) = \mathbf{c}$ on set $\mathbf{X}_0(\mathbf{c}^0) = \{\mathbf{x} \in \mathbf{X} : \varphi_0(\mathbf{x}) =$

$\mathbf{0}\}(\varphi_0(\mathbf{x}) \in C^1 : \mathbf{X} \to \mathbf{R}^{n-d}; d = \dim \mathbf{X}_0(\mathbf{c}^0))$ if the second variation of function $W = U_0 + (\boldsymbol{\lambda}, (\mathbf{U} - \mathbf{c}^0))$ is positive definite with respect to derivations $\xi = \left(\frac{\partial \varphi_0}{\partial \mathbf{x}}\right)_{\varphi_0 = 0} \delta \mathbf{x}$ from this set [can take negative values in some neighborhood of this set] on linear manifold $\delta \mathbf{U} = \mathbf{0}$:

$$\delta^2 W(\mathbf{X}_0(\mathbf{c}^0); \boldsymbol{\lambda}^0(\mathbf{c}^0))\Big|_{\delta \mathbf{U}(\mathbf{X}_0(\mathbf{c}^0)) = \mathbf{0}} > 0 \quad \forall \xi : 0 < \| \xi \| < \delta$$

$$\left[\exists \xi \quad (0 < \| \xi \| < \delta) : \delta^2 W(\mathbf{X}_0(\mathbf{c}^0); \boldsymbol{\lambda}^0(\mathbf{c}^0))\Big|_{\delta \mathbf{U}(\mathbf{X}_0(\mathbf{c}^0)) = \mathbf{0}} < 0\right]$$

(δ is some positive number). This second variation can be represented in the form $\frac{1}{2}(\mathbf{Q}_0 \boldsymbol{\eta}, \boldsymbol{\eta})$, where $\boldsymbol{\eta} \in \mathbf{R}^{n-d-k}$ and $\mathbf{Q}_0$ is $(n-d-k) \times (n-d-k)$ - matrix ($n = \dim \mathbf{x}, d = \dim \mathbf{X}_0(\mathbf{c}^0), k = \dim \mathbf{U}$). The index of this quadratic form is the Poincare degree of instability: if the Poincare degree of instability is [is not] zero, then function $U_0(\mathbf{x})$ takes a local strict minimum [has no even nonstrict minimum] for fixed values $\mathbf{c}^0$ of first integrals $\mathbf{U}(\mathbf{x}) = \mathbf{c}$ on set $\mathbf{X}_0(\mathbf{c}^0)$.

Consider the case when $\dim \mathbf{X}_0 = 0 : \mathbf{X}_0(\mathbf{c}) = \{\mathbf{x}^0(\mathbf{c})\}$, where $\mathbf{x}^0(\mathbf{c})$ is a family of steady motions of the system $(\mathbf{f}(\mathbf{x}^0(\mathbf{c})) \equiv \mathbf{0})$. Let $\xi = \delta \mathbf{x} = \mathbf{x} - \mathbf{x}^0(\mathbf{c}^0)$, then the linearized system of perturbed motion has the form

$$\dot{\xi} = \left(\frac{\partial \mathbf{f}}{\partial \mathbf{x}}\right)_0 \xi \tag{2.12}$$

The second variation of function W and linear manifold $\delta \mathbf{U} = \mathbf{0}$ have the forms

$$\frac{1}{2}\left(\left(\frac{\partial^2 W}{\partial \mathbf{x}^2}\right)_0 \xi, \xi\right), \tag{2.13}$$

$$\left(\frac{\partial \mathbf{U}}{\partial \mathbf{x}}\right)_0 \xi = \mathbf{0} \tag{2.14}$$

recpectively. Here and further $(\cdot)_0 = (\cdot)_{x = x^0(c^0)}$.

Remark 2.8. Quadratic form (2.13) is an ungrowing function and linear forms (2.14) are first integrals of linear system (2.12).

Linear system (2.12) and quadratic form (2.13) on linear manifold (2.14) can be represented in the forms

$$\dot{\boldsymbol{\eta}} = \mathbf{P}_0 \boldsymbol{\eta}, \tag{2.15}$$

$$\frac{1}{2}(\mathbf{Q}_0 \boldsymbol{\eta}, \boldsymbol{\eta}) \tag{2.16}$$

respectively ($\dim \boldsymbol{\eta} = \dim \boldsymbol{\xi} - \dim \mathbf{U} = n - k$). Quadratic form (2.16) is an ungrowing function of system (2.15) : $(\mathbf{Q}_0 \mathbf{P}_0 \boldsymbol{\eta}, \boldsymbol{\eta}) \leq 0$. The latter inequality implies

$$(\mathbf{Q}_0(-\mathbf{P}_0)\boldsymbol{\eta}, \boldsymbol{\eta}) \geq 0, \tag{2.17}$$

Assume that $\det \mathbf{P}_0 \neq 0, \det \mathbf{Q}_0 < 0$, then $\det \mathbf{Q}_0(-\mathbf{P}_0) > 0$ (see (2.17)) and therefore $\det(-\mathbf{P}_0) < 0$. Thus the characteristic equation ($\mathbf{I}$ is the unit matrix)

$$\chi(æ) = \det(\mathbf{I}æ - \mathbf{P}_0) = 0 \tag{2.18}$$

of system (2.15) has a positive root, because

$$\chi(0) = \det(-\mathbf{P}_0) < 0, \quad \chi(+\infty) = +\infty.$$

It means that the characteristic equation $æ^k\chi(æ) = 0$ of system (2.12) has a positive root.

Theorem 2.3. [27],[28] *If the Poincare degree of instability of a steady motion* $\mathbf{x}^0(\mathbf{c}^0)$ *is odd and* rank $\left(\frac{\partial \mathbf{f}}{\partial \mathbf{x}}\right)_0 = n - k$ *then this steady motion is unstable.*

Proof. Under condition of theorem 2.3 $\det \mathbf{P}_0 \neq 0$ because rank $\left(\frac{\partial \mathbf{f}}{\partial \mathbf{x}}\right)_0 =$ rank $\mathbf{P}_0$, and $\det \mathbf{Q}_0 < 0$ because the Poincare degree of instability is odd. Therefore the characteristic equation of the linearized system of perturbed motion has a positive root. Thus the unperturbed motion is unstable according to the Lyapunov theorem on instability by the first aproximation [5].

Remark 2.9. The formulation of theorem 2.3 does not depend on the condition is $U_0(\mathbf{x})$ an ungrowing function or the first integral.

2.5 Stability and instability of invariant sets in the case when $U_0(\mathbf{x})$ is an ungrowing function.

Theorems 2.1–2.3 are correct in the case when $U_0(\mathbf{x})$ is the first integral (see corollaries 2.1,2.2 and remark 2.9). The next two theorems are correct only in the case when $U_0(\mathbf{x})$ is an ungrowing function.

Theorem 2.4. [14] *If an ungrowing function* $U_0(\mathbf{x})$ *takes a local strict minimum for fixed values* $\mathbf{c}^0$ *of the first integrals* $\mathbf{U}(\mathbf{x}) = \mathbf{c}$ *on a compact set* $\mathbf{X}_0(\mathbf{c}^0)$ *and preserves its initial value in some neighborhood of this set only on the family* $\mathbf{X}_0(\mathbf{c})$ *then* $\mathbf{X}_0(\mathbf{c}^0)$ *is a stable invariant set and every pertubed motion of system fairly close to the invariant set* $\mathbf{X}_0(\mathbf{c}^0)$ *asymptotically tends to some set* $\mathbf{X}_0(\mathbf{c})$ *corresponding to perturbed values of constants* $\mathbf{c}$ *as* $t \to +\infty$.

Proof. Note that on the additional condition of theorem 2.4 (by comparison with theorem 2.2) set $\mathbf{X}_0(\mathbf{c}^0)$ is isolated from other invariant sets $\mathbf{X}_1, \mathbf{X}_2, \ldots$ if they exist at all.

As before, assume that function $U_0(\mathbf{x})$ takes a minimal value $m_0(\mathbf{c}^0)$ on set $\mathbf{X}_0(\mathbf{c}^0)$. According to theorem 2.2 $\mathbf{X}_0(\mathbf{c}^0)$ is a stable invariant set and every motion $\mathbf{x}(t, \mathbf{x}^0) \subset \mathbf{X}_0(\mathbf{c}^0)$ is stable with respect to $\mathrm{dist}\,(\mathbf{x}, \mathbf{X}_0(\mathbf{c}^0))$. So every perturbed motion $\mathbf{x}(t)$ is inside domain (2.11) for every small positive number ε if at the initial time inequality $\mathrm{dist}(\mathbf{x}(0), \mathbf{X}_0(\mathbf{c}^0)) < \delta$ is fulfilled (see (2.10)).

Function $U_0(\mathbf{x})$ preserves its initial value in some neighborhood of set $\mathbf{X}_0(\mathbf{c}^0)$ only on the family $\mathbf{X}_0(\mathbf{c})$. Thus one can choose a positive number ε such sufficiently small that domain (2.11) would not contain points of other invariant sets $\mathbf{X}_1, \mathbf{X}_2, \ldots$ if these sets exist at all (as it was noted above, function $\frac{dU_0}{dt} \equiv (\operatorname{grad} U_0, \mathbf{f})$ vanishes on all these sets). Function $U_0(\mathbf{x}(t))$ has to tend (as ungrowing in domain (2.11)) to some limit u_0 being at all time not less than it:

$$U_0(\mathbf{x}(t)) \geq u_0 \tag{2.19}$$

Assume that some perturbed motion $\mathbf{x}(t)$ does not approach some set $\mathbf{X}_0(\mathbf{c})$ corresponding to perturbed values of first integrals $\mathbf{U}(\mathbf{x}) = \mathbf{c}$. Then there exists a sequence of points

$$\mathbf{x}^p = \mathbf{x}(p\tau) \quad (p = p_1, p_2, \ldots; \quad p_1 < p_2 < \ldots; \tau = \text{const} > 0) \tag{2.20}$$

such that $\operatorname{dist}(\mathbf{x}^p, \mathbf{X}_0(\mathbf{c}^0)) \geq \gamma$, where γ is some positive finite number. Evidently in the bounded domain (2.11) one can choose a subsequence

$$\mathbf{x}^q = \mathbf{x}(q\tau) \quad (q = p_{q_1}, p_{q_2}, \ldots; \quad q_1 < q_2 < \ldots) \tag{2.21}$$

convergent to some point $\mathbf{x}^*$ with (by continuity)

$$U_0(\mathbf{x}^*) = u_0, \qquad \operatorname{dist}(\mathbf{x}^*, \mathbf{X}_0(\mathbf{c}^0)) \geq \gamma. \tag{2.22}$$

Consider now motions $\mathbf{x}^*(t) \equiv \mathbf{x}(t, \mathbf{x}^*)$ and $\mathbf{x}^q(t) \equiv \mathbf{x}(t, \mathbf{x}^q)$ beginning at the initial time in points $\mathbf{x}^*$ and $\mathbf{x}^q$ correspondently. Since $\mathbf{x}^*$ satisfy relations (2.22) and $\frac{dU_0}{dt} = 0$ only for $\mathbf{x} \in \mathbf{X}_0(\mathbf{c})$ (in domain (2.11)) then there exists a moment t_* such that

$$U_0(\mathbf{x}^*(t_*)) = u_* < u_0 \tag{2.23}$$

Sequence $\{\mathbf{x}^q\}$ converges to point $\mathbf{x}^*$, so (by the continuous dependence of solutions of differential equations on initial conditions on a finite time interval)

$$\| \mathbf{x}^*(t_*) - \mathbf{x}^q(t_*) \| < \alpha \qquad \forall q > q_*(\alpha)$$

for every positive number α. Then (by continuity)

$$\left| U_0\big(\mathbf{x}^q(t_*)\big) - U_0\big(\mathbf{x}^*(t_*)\big) \right| < \beta \qquad \forall q > q_*(\alpha) = q_*(\alpha(\beta)) = q^*(\beta)$$

for every given number β. Choosing $\beta < u_0 - u_*$ we have that $U_0(\mathbf{x}^q(t_*)) < u_* + \beta < u_0$. Using the group property of solutions of autonomic differintial equations $\mathbf{x}^q(t_*) = \mathbf{x}(t_* + q\tau)$ we can write the latter inequality in the form

$$U_0(\mathbf{x}(t_* + q\tau)) < u_0 \qquad \forall q > q^* \tag{2.24}$$

that is contradiction with relation (2.19).

Thus every perturbed motion $\mathbf{x}(t)$ fairly close to the invatiant set $\mathbf{X}_0(\mathbf{c}^0)$ (see (2.11)) approaches one of invariant sets $\mathbf{X}_0(\mathbf{c})$ as $t \to +\infty$.

Theorem 2.5. [14] *If an ungrowing function $U_0(\mathbf{x})$ takes a non-degenerate stationary value, that is not even nonstrict minimum, for fixed value $\mathbf{c}^0$ of first integral $\mathbf{U}(\mathbf{x}) = \mathbf{c}$ on some compact set $\mathbf{X}_0(\mathbf{c}^0)$ and preserves its initial value in some neighborhood of this set only on the family $\mathbf{X}_0(\mathbf{c})$ then $\mathbf{X}_0(\mathbf{c}^0)$ is an unstable invariant set of the system.*

Proof. Note that $\mathbf{X}_0(\mathbf{c}^0)$ is an invariant set according to theorem 2.1. Assume that function $U_0(\mathbf{x})$ takes a stationary value $m_0(\mathbf{c}^0)$ on set $\mathbf{X}_0(\mathbf{c}^0)$ for $\mathbf{U}(\mathbf{x}) = \mathbf{c}^0$ and this value is not even nonstrict minimum. Then function $U_0(\mathbf{x}) - m_0(\mathbf{c}^0)$ can take negative values on set (2.9).

Consider some perturbed motion $\mathbf{x}(t)$ with initial conditions satisfying relations

$$U_0(\mathbf{x}(0)) < m_0(\mathbf{c}^0), \qquad \mathrm{dist}(\mathbf{x}(0), \mathbf{X}_0(\mathbf{c}^0)) < \delta, \qquad \mathbf{U}(\mathbf{x}(0)) = \mathbf{c}^0$$

where δ is any small positive number. Then

$$\mathrm{dist}(\mathbf{x}(0), \mathbf{X}_0(\mathbf{c}^0)) > 0 \tag{2.25}$$

because otherwise $\mathbf{x}(0) \in \mathbf{X}_0(\mathbf{c}^0)$ and $U_0(\mathbf{x}(0)) = m_0(\mathbf{c}^0)$.

Assume that at all time motion $\mathbf{x}(t)$ belongs to domain (2.11), where ε is some sufficiently small positive number. As before we choose ε such small that domain (2.11) does not contain points of other invariant sets $\mathbf{X}_1, \mathbf{X}_2, \ldots$ if they exist at all. Then $\dfrac{dU_0(\mathbf{x}(t))}{dt} < 0$.

Function $U_0(\mathbf{x}(t))$ is bounded in domain (2.11) and has to tend to some limit being at all time not less than it (see (2.19)). As before (since $\mathbf{x}(t) \notin \mathbf{X}_0(\mathbf{c}^0)$, see (2.25)) there exists a sequence of points like (2.20) with $\mathrm{dist}(\mathbf{x}^p, \mathbf{X}_0(\mathbf{c}^0)) \geq \gamma$, where γ is some finite positive number. In the bounded domain one can choose a subsequence like (2.21) convergent to some point $\mathbf{x}^*$ with

$$U_0(\mathbf{x}^*) = u_0, \qquad \mathrm{dist}(\mathbf{x}^*, \mathbf{X}_0(\mathbf{c}^0)) \geq \gamma$$

Consider now motions $\mathbf{x}^*(t)$ and $\mathbf{x}^q(t)$ beginning at the initial time in points $\mathbf{x}^*$ and $\mathbf{x}^q$ correspondingly. Evidently ($\mathbf{x}^* \notin \mathbf{X}_0(\mathbf{c}^0)$) there exists a moment t_* such that condition (2.23) is fulfilled. Then repeating the given above procedure (see the proof of theorem 2.4) we get inequality (2.24) that is in contradiction with relation (2.19).

Thus the assumption that motion $\mathbf{x}(t)$ at all time belongs to domain (2.11) is not correct. It meant that invariant set $\mathbf{X}_0(\mathbf{c}^0)$ is unstable.

Remark 2.10. It follows from given proof that any motion $\mathbf{x}^0(t) \subset \mathbf{X}_0(\mathbf{c}^0)$ is unstable with respect to $\mathrm{dist}(\mathbf{x}, \mathbf{X}_0(\mathbf{c}^0))$ (under conditions of theorem 2.5) if there exists some moment t_0 such that domain $\left\|\mathbf{x} - \mathbf{x}^0(t_0)\right\| < \rho$ contains points $\mathbf{x}$ with $U_0(\mathbf{x}) - m_0(\mathbf{c}^0) < 0$ (for any small positive number ρ). In particular, all motions $\mathbf{x}^0 \subset \mathbf{X}_0(\mathbf{c}^0)$ are unstable with respect to $\mathrm{dist}(\mathbf{x}, \mathbf{X}_0(\mathbf{c}^0))$ if an ungrowing function has a local strict maximum on set $\mathbf{X}_0(\mathbf{c}^0)$ for fixed values $\mathbf{c}^0$ of first integral and preserves its initial value in some neighborhood of this set only on the family $\mathbf{X}_0(\mathbf{c})$.

Note that given proofs of theorems 2.4 and 2.5 are some modifications of the Barbashin-Krasovskii and Krasovskii theorems on asymptotic stability and instability [22].

Chapter 3

Invariant sets of mechanical systems with symmetry.

3.1 Invariant sets of stationary motions.

In mechanical systems an ungrowing function and first integral have usually some specific form (see, for example, the Introduction). In this chapter we will consider a dissipative (in particular, conservative) mechanical system with k - parametrical symmetry group. Let $\mathbf{v} \in \mathbf{R}^n$ and $\mathbf{r} \in \mathbf{M} \subseteq \mathbf{R}^m$ be quasivelocities (in particular, moments or generalized velocities) and coordinates of this system respectively. $\mathbf{M}$ is configuration space of essential coordinates; $\dim \mathbf{M} \leq n$.

The total mechanical energy can be represented in the form

$$U_0(\mathbf{v};\mathbf{r}) = \frac{1}{2}(\mathbf{A}(\mathbf{r})\mathbf{v}, \mathbf{v}) + V(\mathbf{r}) \qquad (3.1)$$

Here $\mathbf{A}(\mathbf{r})$ is positive definite $n \times n$ - matrix ($\mathbf{A}(\mathbf{r}) \in \mathcal{C}^2$) of a kinetic energy, $V(\mathbf{r}) \in \mathcal{C}^2 : \mathbf{M} \to \mathbf{R}$ is a potential energy of a system. Evidently, the total mechanical energy is an growing function (in conservative case it is the first integral).

The first integrals corresponding the symmetry group can be represented in the form

$$U(\mathbf{v};\mathbf{r}) = \mathbf{B}^T(\mathbf{r})\mathbf{v} = \mathbf{c} = \text{const} \quad (\mathbf{c} \in \mathbf{R}^k) \qquad (3.2)$$

Here $\mathbf{B}(\mathbf{r})$ is $n \times k$ - matrix of Noether's integrals ($\mathbf{B}(\mathbf{r}) \in \mathcal{C}^2; \operatorname{rank} \mathbf{B} = k$).

Taking a special form of ungrowing function (3.1) and first integrals (3.2) into account one can solve the problem on stationary values of total mechanical energy on fixed levels of Noether's integrals in two steps. Really, function (3.1) of variables $\mathbf{v}$ and $\mathbf{r}$ on conditions (3.2) takes a stationary value with respect to these variables if and only if it takes a minimal value with respect quasi-velocities $\mathbf{v}$ on conditions (3.2) and this minimum, depending on coordinates $\mathbf{r}$, takes a stationary value with respect to these coordinates on conditions $\mathbf{r} \in \mathbf{M}$.

Let us find $\min\limits_{\mathbf{v}} U_0\Big|_{\mathbf{U}=\mathbf{c}} = V_c(\mathbf{r})$. Consider function $W = U_0 - (\boldsymbol{\lambda}, (\mathbf{U}-\mathbf{c}))$, where $\boldsymbol{\lambda} \in \mathbf{R}^k$ is a vector of unknown Lagrange's multipliers, and write conditions of its minimum with respect to variables $\mathbf{v}$ for fixed constants $\mathbf{c}$:

$$\frac{\partial W}{\partial \mathbf{v}} = \mathbf{A}\mathbf{v} - \mathbf{B}\boldsymbol{\lambda} = 0, \qquad \frac{\partial W}{\partial \boldsymbol{\lambda}} - \left(\mathbf{B}^T\mathbf{v} - \mathbf{c}\right) = 0 \tag{3.3}$$

Solving system (3.3) we find

$$\mathbf{v} = \mathbf{A}^{-1}\mathbf{B}\left(\mathbf{B}^T\mathbf{A}^{-1}\mathbf{B}\right)^{-1}\mathbf{c} = \mathbf{v}_c(\mathbf{r}) \tag{3.4}$$

$$V_c(\mathbf{r}) = V + \frac{1}{2}\left(\left(\mathbf{B}^T\mathbf{A}^{-1}\mathbf{B}\right)^{-1}\mathbf{c}, \mathbf{c}\right) \tag{3.5}$$

Function $V_c(\mathbf{r})$ is a effective potential [16]–[19]. Evidently it depends on variables $\mathbf{r} \in \mathbf{M}$ and parameters $\mathbf{c} \in \mathbf{R}^k$. Note that if $\mathbf{r}$ are essential generalized (i.e. independent) coordinates and $\mathbf{v} = (\dot{\mathbf{r}}, \dot{\mathbf{s}})$, where $\mathbf{s}$ are cyclic generalized coordinates, then the effective potential (3.5) coincides with the reduced potential given in the introduction.

Theorem 3.1. [18],[19] *If the effective potential takes a non-degenerate stationary value on some set* $\mathbf{M}_0 \subset \mathbf{M}$ *then* $\mathbf{M}_0$ *is an invariant set in the configuration space* $\mathbf{M}$ *and*

$$\mathbf{N}_0 = \mathbf{M}_0 \times \{\mathbf{v} \in \mathbf{R}^n : \mathbf{v} = \mathbf{v}_c(\mathbf{r}), \mathbf{r} \in \mathbf{M}_0\} \subset \mathbf{M} \times \mathbf{R}^n \tag{3.6}$$

is an invariant set of the system in the phasic space $\mathbf{M} \times \mathbf{R}^n$.

Proof. Under conditions of theorem 3.1 ungrowing function 3.1 takes a non-degenerate stationary value on fixed levels of first integral (3.2) on set $\mathbf{N}_0$ and theorem 3.1 follows from theorem 2.1.

Remark 3.1. Set $\mathbf{M}_0 \subset \mathbf{M}$ and corresponding to it set $\mathbf{N}_0 \subset \mathbf{M} \times \mathbf{R}^n$ depend on parameters $\mathbf{c}$. The effective potential can take stationary values even for fixed parameters $\mathbf{c}$ not only on set $\mathbf{M}_0$, but also on some other sets $\mathbf{M}_1, \mathbf{M}_2, \ldots$. These sets and corresponding to them sets $\mathbf{N}_1, \mathbf{N}_2, \ldots$ depend on parameters $\mathbf{c}$ too. The total mechanical energy does not decrease on set $\mathbf{N} = \mathbf{N}_0 \cup \mathbf{N}_1 \cup \mathbf{N}_2 \ldots$. Motions of the system $\mathbf{r} = \mathbf{r}(t), \mathbf{v} = \mathbf{v}(t)$ belonging to set $\mathbf{N}$ may be considered as stationary motions because they give stationary values to the ungrowing function on fixed levels of the first integrals.

Remark 3.2. If $\mathbf{M} = \mathbf{R}^m$ then we have to solve system $\dfrac{\partial V_c}{\partial \mathbf{r}} = \mathbf{0}$ to find invariant sets. If $\mathbf{M} = \{\mathbf{r} \in \mathbf{R}^m : \boldsymbol{\psi}(\mathbf{r}) = \mathbf{0}\}$, where $\boldsymbol{\psi}(\mathbf{r}) \in \mathcal{C}^1 : \mathbf{R}^m \to \mathbf{R}^{m-\mu}$ $(\mu = \dim \mathbf{M})$, then we have to solve system $\dfrac{\partial W_c}{\partial \mathbf{r}} = \mathbf{0}, \boldsymbol{\psi}(\mathbf{r}) = \mathbf{0}$, where $W_c(\mathbf{r}) = V_c + (\boldsymbol{\nu}, \boldsymbol{\psi})$ and $\boldsymbol{\nu}$ is μ-vector of unknown Lagrange's multipliers.

3.2 Stability of invariant sets of stationary motions.

Theorem 3.2. [18],[19] *If the effective potential takes a local strict minimum for fixed values* $\mathbf{c}^0$ *of parameters* $\mathbf{c}$ *on some compact set* $\mathbf{M}_0(\mathbf{c}^0) \subset \mathbf{M}$ *then* $\mathbf{M}_0(\mathbf{c}^0)$ *is a stable invariant set in the configuration space and* $\mathbf{N}_0(\mathbf{c}^0)$ *is a stable invariant set of the system.*

Proof. Theorem 3.2 follows from theorem 2.2 because compact set $\mathbf{N}_0(\mathbf{c}^0)$ gives to ungrowing function (3.1) a local strict minimum for fixed values $\mathbf{c}^0$ of first integrals (3.2).

Consider the case when $\dim \mathbf{M}_0 = 0$, so $\mathbf{M}_0(\mathbf{c}) = \{\mathbf{r}^0(\mathbf{c})\}$. Then (see (3.6)) $\dim \mathbf{N}_0 = 0$, so $\mathbf{N}_0(\mathbf{c}) = \{\mathbf{r}^0(\mathbf{c}); \mathbf{v}^0(\mathbf{c})\}$, where $\mathbf{v}^0(\mathbf{c}) = \mathbf{v}_c^0(\mathbf{r}^0(\mathbf{c}))$ (see (3.4)). It means that the system admits steady motions $\mathbf{r} = \mathbf{r}^0; \mathbf{v} = \mathbf{v}^0$. In particular, if $\mathbf{v} = (\dot{\mathbf{r}}, \dot{\mathbf{s}})$, where $\mathbf{r}$ and $\mathbf{s}$ are essential and cyclic generalized coordinates respectively, then $\mathbf{v}^0 = (0, \dot{\mathbf{s}}^0)$ and steady motions $\mathbf{r} = \mathbf{r}^0; \mathbf{v} = \mathbf{v}^0$ are the same that were considered in the introduction.

Equations of motion of a system discribed by phasic variables $\mathbf{r}$ and $\mathbf{v}$ may be represented in the form

$$\dot{\mathbf{r}} = \mathbf{F}(\mathbf{r})\mathbf{v}; \quad \dot{\mathbf{v}} = \mathbf{f}(\mathbf{r}; \mathbf{v}) \tag{3.7}$$

Here $\mathbf{F}(\mathbf{r}) \in \mathcal{C}^1$ is $m \times n$ - matrix and $\mathbf{f}(\mathbf{r}; \mathbf{v}) \in \mathcal{C}^1 : \mathbf{M} \times \mathbf{R}^n \to \mathbf{R}^n$ with $\mathbf{F}(\mathbf{r}^0)\mathbf{v}^0 = 0$ and $\mathbf{f}(\mathbf{r}^0; \mathbf{v}^0) = 0$ because $\{\mathbf{r}^0; \mathbf{v}^0\}$ is a steady motion of system (3.7) by assumption.

Let $\mathbf{x} = \mathbf{r} - \mathbf{r}^0, \mathbf{y} = \mathbf{v} - \mathbf{v}^0$, then the linearized system of perturbed motion of the system has the form

$$\begin{pmatrix} \dot{\mathbf{x}} \\ \dot{\mathbf{y}} \end{pmatrix} = \mathbf{S}_0 \begin{pmatrix} \mathbf{x} \\ \mathbf{y} \end{pmatrix} \tag{3.8}$$

Here $\mathbf{S}_0$ is a $(m+n) \times (m+n)$ - matrix of the form

$$\mathbf{S}_0 = \begin{pmatrix} \dfrac{\partial \mathbf{F}\mathbf{v}}{\partial \mathbf{r}} & \mathbf{F} \\ \dfrac{\partial \mathbf{f}}{\partial \mathbf{r}} & \dfrac{\partial \mathbf{f}}{\partial \mathbf{v}} \end{pmatrix}_0 \tag{3.9}$$

According to theorem 2.3 we have

Theorem 3.3. *If the Poincare degree of instability of a steady motion is odd and* $\operatorname{rank} \mathbf{S}_0 = n - k + \mu$ *then this steady motion is unstable.*

Note that if $\mathbf{M} = \mathbf{R}^m$ then $\mu = m$ (see remark 3.2).

Theorems 3.1–3.3 are correct as well for dissipative systems as conservative ones. The next two theorems are correct only for dissipative systems.

Theorem 3.4. [18],[19] *If the effective potential takes a local strict minimum for fixed values* $\mathbf{c}^0$ *of parameters* $\mathbf{c}$ *on some compact set* $\mathbf{M}_0(\mathbf{c}^0) \subset \mathbf{M}$ *and the total mechanical energy preserves its initial value only on the family* $\mathbf{N}_0(\mathbf{c})$ *of invariant sets (in some neighborhood of set* $\mathbf{N}_0(\mathbf{c}^0)$*) then* $\mathbf{N}_0(\mathbf{c}^0)$ *is a stable invariant set and* *every perturbed motion of the system fairly close to set* $\mathbf{N}_0(\mathbf{c}^0)$ *tends to some set* $\mathbf{N}_0(\mathbf{c})$ *corresponding to perturbed values of parameters* $\mathbf{c}$ *as* $t \to +\infty$.

Theorem 3.5. [18],[19] *If the effective potential takes a non-degenerate stationary value, that is not even nonstrict minimum, for fixed values* $\mathbf{c}^0$ *of parameters* $\mathbf{c}$ *on a compact set* $\mathbf{M}_0(\mathbf{c}^0) \subset \mathbf{M}$ *and the total mechanical energy preserves its initial value only on the family* $\mathbf{N}_0(\mathbf{c})$ *of invariant sets (in some neighborhood of set* $\mathbf{N}_0(\mathbf{c}^0)$*) then* $\mathbf{N}_0(\mathbf{c}^0)$ *is an unstable invariant set.*

Theorem 3.4 and 3.5 follow from theorems 2.4 and 2.5 respectively.

Evidently, if the total mechanical energy preserves its initial value only on set $\mathbf{N} = \mathbf{N}_0 \cup \mathbf{N}_1 \cup \mathbf{N}_2 \ldots$ then the additional condition of theorems 3.4 and 3.5 (except conditions of minimal or nonminimal stationary values) are fulfilled for isolated (for fixed values of parameters $\mathbf{c}$) invariant sets.

Remark 3.3. Invariant sets $\mathbf{N}_i(\mathbf{c}^0)$ and $\mathbf{N}_j(\mathbf{c}^0)$ $(i \neq j)$ are isolated one from another if and only if sets $\mathbf{M}_i(\mathbf{c}^0)$ and $\mathbf{M}_j(\mathbf{c}^0)$ are isolated one form another (see (3.6)).

Remark 3.4. The index of the second variation of the total mechanical energy on the linear manifold $\delta\mathbf{U} = \mathbf{0}$ is equal to the index of second variation of the effective potential. So the Poincare degree of instability of invariant set $\mathbf{N}_0(\mathbf{c}^0)$ is equal to $\operatorname{ind} \delta^2 V_c^0(\mathbf{M}_0(\mathbf{c}^0))$ $(\mathbf{r} \in \mathbf{M})$.

Remark 3.5. Invariant sets $\mathbf{N}_0(\mathbf{c}), \mathbf{N}_1(\mathbf{c}), \mathbf{N}_2(\mathbf{c}), \ldots$ can be represented in the space $\mathbf{R}^n \times \mathbf{M} \times \mathbf{R}^k$ $(\mathbf{v} \in \mathbf{R}^n, \mathbf{r} \in \mathbf{M}, \mathbf{c} \in \mathbf{R}^k)$ as some surfaces. Invariant (in the configuration space) sets $\mathbf{M}_0(\mathbf{c}), \mathbf{M}_1(\mathbf{c}), \mathbf{M}_2(\mathbf{c}), \ldots$ can be represented in the space $\mathbf{M} \times \mathbf{R}^k$ $(\mathbf{r} \in \mathbf{M}, \mathbf{c} \in \mathbf{R}^k)$ as some surfaces too. According to remarks 3.3 and 3.4 the latter surfaces can be considered as Poincare - diagrams.

Remark 3.6. According to the definition of the effective potential $V_c(\mathbf{r}) \leq U_0(\mathbf{v}; \mathbf{r}) \leq h$, where h is initial value of the total mechanical energy. So inequality $V_c(\mathbf{r}) \leq h$ defines in the configuration space domains of possible motions for given values $\mathbf{c}$ of the first integrals and the initial value h of the total mechanical energy. The topological type of these domains changes if $(\mathbf{c}; h) \in \Sigma$, where $\Sigma = \Sigma_0 \cup \Sigma_1 \cup \Sigma_2 \ldots$,

$$\Sigma_s = \{(\mathbf{c}; h) \in \mathbf{R}^k \times \mathbf{R} : h = h_s(\mathbf{c}) = V_c(\mathbf{M}_s(\mathbf{c}))\} \qquad (s = 0, 1, 2, \ldots)$$

For conservative systems with symmetry [16],[17] all points of the space $(\mathbf{c}; h)$ are invariant. For dissipative systems there are two possibilities [19]. If a point $\mathbf{P}(\mathbf{c}_p; h_p) \in \Sigma$ then it is invariant point in this space. If a point $\mathbf{Q}(\mathbf{c}_q; h_q) \notin \Sigma$ then it can evaluate along the straight line $\mathbf{c} = \mathbf{c}_q$ asymptotically tending to some point $\overline{\mathbf{Q}}(\mathbf{c}_q; \overline{h}_q)$ with $\overline{h}_q < h_q$ and it necessary evaluates if $\dfrac{dU_0}{dt} = 0$ only for $(\mathbf{v}; \mathbf{r}) \in \mathbf{N}$. Moreover, in the latter case $\overline{\mathbf{Q}} \in \Sigma$. Surface Σ in the space $(\mathbf{c}; h)$ represents Smale - diagram.

3.3 Invariant sets of relative equilibria and their stability.

Let us assume that the system is subjected to control forces which assure the relations

$$(\mathbf{B}^T\mathbf{A}^{-1}\mathbf{B})^{-1}\mathbf{B}\mathbf{v} = \omega = \text{const} \quad (\omega \in \mathbf{R}^k) \tag{3.10}$$

for all motions of the system, not only for steady motions considered in the previous section of this chapter.

Introduce the changed potential of the system restricted by relations (3.10):

$$V_\omega = V - \frac{1}{2}(\mathbf{B}^T\mathbf{A}^{-1}\mathbf{B}\omega, \omega) \tag{3.11}$$

Similarly to the previous results we have

Theorem 3.6. *If the changed potential takes a non-degenerate stationary value on some set* $\mathbf{M}_0 \subset \mathbf{M}$ *then* $\mathbf{M}_0$ *is an invariant set in the configuration space* $\mathbf{M}$ *and*

$$\mathbf{N}_0 = \mathbf{M}_0 \times \{\mathbf{v} \in \mathbf{R}^n : \mathbf{v} = \mathbf{v}_\omega(\mathbf{r}), \mathbf{r} \in \mathbf{M}_0\} \subset \mathbf{M} \times \mathbf{R}^n \tag{3.12}$$

where $\mathbf{v}_\omega(\mathbf{r}) = \mathbf{A}^{-1}\mathbf{B}\omega$, *is an invariant set of the restricted system in the phasic space.*

Evidently, set $\mathbf{M}_0$ as well as set $\mathbf{N}_0$ depend on parameters ω : $\mathbf{M}_0 = \mathbf{M}_0(\omega), \mathbf{N}_0 = \mathbf{N}_0(\omega)$. Similarly to the section 1.3 we will call motions of restricted system belonging to invariant set as relative equilibria. See also remark 3.2 with replasement V_c to V_ω.

Theorem 3.7. *If the changed potential* $V_\omega(\mathbf{r})$ *takes a local strict minimum on some compact set* $\mathbf{M}_0(\omega) \subset \mathbf{M}$ *then* $\mathbf{M}_0(\omega)$ *is a stable invariant set in the configuration space and* $\mathbf{N}_0(\omega)$ *is a stable invariant set of the restricted system.*

Theorem 3.8. *If* $\dim \mathbf{M}_0 = 0$ $(\mathbf{M}_0(\omega) = \{\mathbf{r}^0(\omega)\})$, *the Poincare degree of instability of relative equilibria* $\{\mathbf{r} = \mathbf{r}^0(\omega), \mathbf{v} = \mathbf{v}^0(\omega) = \mathbf{v}_\omega(\mathbf{r}^0)\}$ *is odd and rank of* $(n-k+m) \times (n-k+m)$ *- matrix of the linearized system of perturbed motions of the restricted system is equal to* $n-k+\mu$ *then this relative equilibrium is unstable.*

Theorem 3.9. *If the changed potential* $V_\omega(\mathbf{r})$ *takes a local strict minimum on some compact set* $\mathbf{M}_0(\omega) \subset \mathbf{M}$ *and the total mechanical energy of the restricted system preserves its initial value only on set* $\mathbf{N}_0(\omega)$ *(in some neighborhood of this set) then* $\mathbf{N}_0(\omega)$ *is an asymptotically stable invariant set of the restricted system.*

Theorem 3.10. *If the changed potential* $V_\omega(\mathbf{r})$ *takes a non-degenerate stationary value, that is not even nonstrict minimum, on some compact set* $\mathbf{M}_0(\omega) \subset \mathbf{M}$ *and the total mechanical energy of the restricted system preserves its initial value only on set* $\mathbf{N}_0(\omega)$ *(in some neighborhood of this set) then* $\mathbf{N}_0(\omega)$ *is an unstable invariant set of the restricted system.*

3.4 The correspondence of invariant sets of relative equilibria and stationary motions and relationship between their stability conditions.

Consider a general case, when $\mathbf{M} = \{\mathbf{r} \in \mathbf{R}^m : \boldsymbol{\psi}(\mathbf{r}) = 0\}$ (see remark 3.2) then sets $\mathbf{M}_0(\mathbf{c})$ and $\mathbf{M}_0(\boldsymbol{\omega})$ are determined from equations

$$\frac{\partial W_c}{\partial \mathbf{r}} = 0, \quad \boldsymbol{\psi}(\mathbf{r}) = 0, \qquad (W_c = V_c + (\boldsymbol{\nu}, \boldsymbol{\psi})) \tag{3.13}$$

$$\frac{\partial W_\omega}{\partial \mathbf{r}} = 0, \quad \boldsymbol{\psi}(\mathbf{r}) = 0, \qquad (W_\omega = V_\omega + (\boldsymbol{\nu}, \boldsymbol{\psi})) \tag{3.14}$$

respectively (in both cases $\boldsymbol{\nu}$ is a μ - dimensional vector of unknown Lagrange's multipliers; $\mu = \dim \mathbf{M}$).

Theorem 3.11. [29] *For any set* $\mathbf{c}$ *of constants of Noether's integrals* (3.2) *(for any set* $\boldsymbol{\omega}$ *of constants of restrictions* (3.10) *) there exists a set of constants* $\boldsymbol{\omega}$ *(a set of constants* $\mathbf{c}$*) such that the solution of system* (3.14) *coincides with that of system* (3.13)*:* $\mathbf{M}_0(\boldsymbol{\omega}) = \mathbf{M}_0(\mathbf{c})$.

Proof. Let $\mathbf{M}_0(\mathbf{c})$ and $\boldsymbol{\nu}^0(\mathbf{c})$ be a solution of system (3.13):

$$\frac{\partial V}{\partial \mathbf{r}} + \frac{\partial C}{\partial \mathbf{r}} + \frac{\partial(\boldsymbol{\psi}, \boldsymbol{\nu})}{\partial \mathbf{r}} = 0, \qquad \boldsymbol{\psi} = 0 \tag{3.15}$$

for $\mathbf{r} \in \mathbf{M}_0(\mathbf{c})$ and $\boldsymbol{\nu} = \boldsymbol{\nu}^0(\mathbf{c})$. Here $C = \frac{1}{2}(\mathbf{D}^{-1}\mathbf{c}, \mathbf{c})$, $\mathbf{D} = \mathbf{B}^T\mathbf{A}^{-1}\mathbf{B}$. Put

$$\boldsymbol{\omega} = (\mathbf{B}^T\mathbf{A}^{-1}\mathbf{B})_0\mathbf{c} \tag{3.16}$$

Consider now system (3.14) from which $\mathbf{M}_0(\boldsymbol{\omega})$ and $\boldsymbol{\nu}^0(\boldsymbol{\omega})$ are determined:

$$\frac{\partial V}{\partial \mathbf{r}} - \frac{\partial \Omega}{\partial \mathbf{r}} + \frac{\partial(\boldsymbol{\psi}, \boldsymbol{\nu})}{\partial \mathbf{r}} = 0, \qquad \boldsymbol{\psi} = 0 \tag{3.17}$$

Here $\Omega = \frac{1}{2}(\mathbf{D}\boldsymbol{\omega}, \boldsymbol{\omega})$. Taking the obvious identity $\dfrac{\partial(\mathbf{D}\mathbf{D}^{-1})}{\partial \mathbf{r}} \equiv \mathbf{0}$ into consideration one can express the second term of first equation in system (3.17) as

$$-\frac{\partial \Omega}{\partial \mathbf{r}} = \mathbf{D}\frac{\partial \Omega^{-1}}{\partial \mathbf{r}}\mathbf{D}$$

Using formula (3.16) we conclude that this system is satisfied by $\mathbf{M}_0(\mathbf{c})$ and $\boldsymbol{\nu}^0(\mathbf{c})$. Consequently, $\mathbf{M}_0(\boldsymbol{\omega}) = \mathbf{M}_0(\mathbf{c})$ and $\boldsymbol{\nu}^0(\boldsymbol{\omega}) = \boldsymbol{\nu}^0(\mathbf{c})$ (on condition that (3.16) holds). One proves similarly that if $\mathbf{M}_0(\boldsymbol{\omega})$ and $\boldsymbol{\nu}^0(\boldsymbol{\omega})$ are solutions of system (3.17) then $\mathbf{M}_0(\mathbf{c}) = \mathbf{M}_0(\boldsymbol{\omega})$ and $\boldsymbol{\nu}^0(\mathbf{c}) = \boldsymbol{\nu}^0(\boldsymbol{\omega})$, where $\mathbf{M}_0(\mathbf{c})$ and $\boldsymbol{\nu}^0(\mathbf{c})$ are solutions of system (3.15) on condition that

$$\mathbf{c} = (\mathbf{B}^T\mathbf{A}^{-1}\mathbf{B})_0\boldsymbol{\omega} \tag{3.18}$$

Theorem 3.12. [29] *If the changed potential V_ω has a strict local minimum on set $\mathbf{M}_0(\omega)$ then the effective potential V_c has a strict local minimum on set $\mathbf{M}_0(\mathbf{c})$ on conditions that $\mathbf{c}$ and $\boldsymbol{\omega}$ satisfy relationship* (3.16) *(or* (3.18)*).*

Proof. Suppouse that condition (3.16) (or (3.18)) is satisfied. Then $\mathbf{M}_0(\mathbf{c}) = \mathbf{M}_0(\omega) = \mathbf{M}_0$. In this case

$$(V_c(\mathbf{r}) - V_c(\mathbf{M}_0)) - (V_\omega(\mathbf{r}) - V_\omega(\mathbf{M}_0)) = C - C_0 + \Omega - \Omega_0 =$$

$$= \frac{1}{2}\Big((\mathbf{D}^{-1} - \mathbf{D}_0^{-1})\mathbf{c}, \mathbf{c}\Big) + \frac{1}{2}\Big((\mathbf{D} - \mathbf{D}_0)\boldsymbol{\omega}, \boldsymbol{\omega}\Big) =$$

$$= \frac{1}{2}\Big((\mathbf{D}_0\mathbf{D}^{-1}\mathbf{D}_0 - 2\mathbf{D}_0 + \mathbf{D})\boldsymbol{\omega}, \boldsymbol{\omega}\Big) =$$

$$= \frac{1}{2}\Big((\mathbf{D}^{-1}(\mathbf{D}_0 - \mathbf{D}))\boldsymbol{\omega}, (\mathbf{D}_0 - \mathbf{D})\boldsymbol{\omega}\Big) \geq 0$$

Therefore $(V_c(\mathbf{r}) - V_c(\mathbf{M}_0)) \geq (V_\omega(\mathbf{r}) - V_\omega(\mathbf{M}_0))$, so if $V_\omega(\mathbf{r}) > V_\omega(\mathbf{M}_0)$ then $V_c(\mathbf{r}) > V_c(\mathbf{M}_0)$.

Corollary 3.1. If an invariant set of relative equilibria of the restricted system is stable in the secular sense then the corresponding invariant set of stationary motions of the free system is stable in the secular sense.

Note that the convert is not necessary true: invariant sets of stationary motions can be stable in the secular sense even if the corresponding invariant sets of relative equilibria are unstable.

Let us call invariant set $\mathbf{M}_0(\mathbf{c})$ ($\mathbf{M}_0(\omega)$) trivial if $\mathbf{M}_0(\mathbf{c})$ ($\mathbf{M}_0(\omega)$) is independent on $\mathbf{c}$ (on $\boldsymbol{\omega}$). Obviously (see (3.15) and (3.17)) trivial invariant sets $\mathbf{M}_0(\mathbf{c})$ and $\mathbf{M}_0(\omega)$ satisfy relations

$$\frac{\partial V}{\partial \mathbf{r}} + \frac{\partial(\boldsymbol{\psi}, \boldsymbol{\nu})}{\partial \mathbf{r}} = 0, \qquad \boldsymbol{\psi} = 0 \tag{3.19}$$

that is, they always coincide: $\mathbf{M}_0(\mathbf{c}) = \mathbf{M}_0(\omega) = \mathbf{M}_0$.

When that happens one has

$$\frac{\partial C}{\partial \mathbf{r}} = 0, \qquad \frac{\partial \Omega}{\partial \mathbf{r}} = 0, \qquad (\mathbf{r} \in \mathbf{M}_0) \tag{3.20}$$

identically with respect to $\mathbf{c}$ and $\boldsymbol{\omega}$ correspondingly.

Theorem 3.13. [29] *The degrees of instability for trivial invariant sets of stationary motions and relative equilibria always coincide.*

The proof follows in an obvious manner from (3.18)-(3.20).

Corollary 3.2. Conditions for the secular stability of non-degenerate ($\delta^2 V_{c,\omega}(\mathbf{M}_0) \neq 0$) trivial invariant sets of stationary motions and relative equilibria always coincide.

Remark 3.7. The degree of instability of non-degenerate non-trivial invariant set $\mathbf{M}_0(\mathbf{c})$ cannot exceed that of the corresponding non-trivial invariant set $\mathbf{M}_0(\omega)$, but it can be smaller.

Remark 3.8. If $\mathbf{v} = (\dot{\mathbf{r}}, \dot{\mathbf{s}})$, where $\mathbf{r}$ and $\mathbf{s}$ are essential and cyclic generalized coordinates, the results presenred here coincide with the results presented in the Introduction (see section 1.4).

Chapter 4

A top on a horizontal plane with friction [30].

4.1 Equations of motion and effective potential of a top.

Let us consider a non homogeneous dynamically symmetric sphere (a top) on a horizontal plane with a sliding friction.

Let m be the mass of the sphere, A_1 and A_3 its equatorial and axial central moments of inertia, r its radius, c the distance from the sphere center of mass to its geometrical center and g the acceleration due to gravity. Denote the velocity of the sphere center of mass and its angular velocity about the center of mass by $\mathbf{v}$ and $\boldsymbol{\omega}$ respectively, and the unit vectors along the upward vertical and the axis of symmetry by $\boldsymbol{\gamma}$ and $\mathbf{e}$ respectively (see fig. 1). The equations of motion of the top may be written as follows

$$m\dot{\mathbf{v}} + \boldsymbol{\omega} \times m\mathbf{v} = -mg\gamma + \mathbf{R} \tag{4.1}$$

$$\Theta\dot{\omega} + \boldsymbol{\omega} \times \Theta\omega = \boldsymbol{\rho} \times \mathbf{R} \tag{4.2}$$

$$\dot{\gamma} + \boldsymbol{\omega} \times \gamma = 0 \tag{4.3}$$

$$(\gamma, (\mathbf{v} + \boldsymbol{\omega} \times \boldsymbol{\rho})) = 0 \tag{4.4}$$

Here $\Theta = \mathrm{diag}(A_1, A_1, A_3)$ is the central inertia tensor of the top, $\boldsymbol{\rho} = -r\boldsymbol{\gamma} + c\mathbf{e}$ is the radius - vector of the contact point of the top with the horizontal plane relative to its center of mass, $\mathbf{R}$ is the reaction of the supporting plane. Assume that $\mathbf{R} = N\boldsymbol{\gamma} + \mathbf{F}$ with $(\mathbf{F}, \boldsymbol{\gamma}) = 0$; $(\mathbf{F}, (\mathbf{v} + \boldsymbol{\omega} \times \boldsymbol{\rho})) \leq 0$ so N is a value of the normal reaction and $\mathbf{F}$ is the sliding friction.

Equations (4.1) and (4.2) express the laws of changing of momentum and moment of momentum of the top respectively, equation (4.3) expresses that the unit vector γ is the constant vector in the absolute space and equation (4.4) expresses that, for all motions, the top is in contact with the supporting horizontal plane.

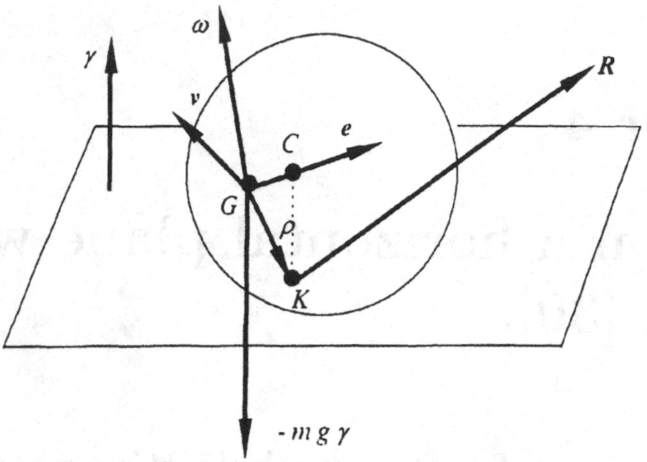

Fig. 1

System (4.1)–(4.4) is closed with respect to unknown functions $\mathbf{v}(t)$, $\boldsymbol{\omega}(t)$, $\boldsymbol{\gamma}(t)$ and $N(t)$ (for every given expression $\mathbf{F} = \mathbf{F}(N, \mathbf{v}, \boldsymbol{\omega}, \boldsymbol{\gamma})$ depending on a model of friction).

The total mechanical energy of the top

$$H(\mathbf{v}; \boldsymbol{\omega}; \boldsymbol{\gamma}) = \frac{1}{2}(m\mathbf{v}, \mathbf{v}) + \frac{1}{2}(\boldsymbol{\Theta\omega}, \boldsymbol{\omega}) - mg(\boldsymbol{\rho}, \boldsymbol{\gamma}) \leq h \tag{4.5}$$

is an ungrowing function because

$$\frac{dH}{dt} = (\mathbf{F}, (\mathbf{v} + \boldsymbol{\omega} \times \boldsymbol{\rho})) \leq 0$$

Equations (4.1)–(4.4) admit (for arbitrary $\mathbf{R}$) two first integrals: Jellet's

$$K(\boldsymbol{\omega}, \boldsymbol{\gamma}) = (\boldsymbol{\Theta\omega}, \boldsymbol{\rho}) = rk = \text{const} \quad (k \in \mathbf{R}) \tag{4.6}$$

and geometrical integral

$$\Gamma(\boldsymbol{\gamma}) = (\boldsymbol{\gamma}, \boldsymbol{\gamma}) = 1. \tag{4.7}$$

The latter can be considered as the configuration space $\mathbf{S}^2 \subset \mathbf{R}^3$ of essential coordinates $\boldsymbol{\gamma}$.

Thus a top on a horizontal plane with friction represents a dissipative mechanical system with symmetry (compare (4.5) and (4.6) with (3.1) and (3.2) respectively).

According to chapter 3 we can calculate the effective potential of the top

$$V_k(\boldsymbol{\gamma}) = \min_{\mathbf{v}, \boldsymbol{\omega}} H(\mathbf{v}, \boldsymbol{\omega}, \boldsymbol{\gamma})|_{K=k} = H(\mathbf{v}_k(\boldsymbol{\gamma}); \boldsymbol{\omega}_k(\boldsymbol{\gamma}); \boldsymbol{\gamma}) =$$

$$= -mg(\rho,\gamma) + \frac{1}{2}\cdot\frac{k^2r^2}{(\Theta\rho,\rho)} \quad \left(\mathrm{v}_k(\gamma)=0;\ \omega_k(\gamma)=\frac{kr}{(\Theta\rho,\rho)}\rho\right).$$

Thus the effective potential $V_k(\gamma)$ has the following form

$$V_k(\gamma) = mgr(1+f(\gamma));\quad f(\gamma) = -\varepsilon\gamma_3 + \frac{1}{2}\cdot\frac{\kappa^2}{\delta(\gamma_1^2+\gamma_2^2)+(\varepsilon-\gamma_3)^2}$$

$$\varepsilon = \frac{c}{r}\in(0,1);\quad \delta=\frac{A_1}{A_3}\in(\frac{1}{2};+\infty);\quad \kappa^2=(A_3mgr)^{-1}k^2\in[0;+\infty).$$

Here and further $\gamma_1,\gamma_2,\gamma_3$ and $\omega_1,\omega_2,\omega_3$ are components of vectors γ and ω in the principal central axes of the top inertia.

4.2 Invariant sets of stationary motions of the top and their stability.

According to theorem 3.1 invariant sets of stationary motions of the top correspond to solutions of the system

$$\frac{\partial W_k}{\partial\gamma}=0,\qquad \left(W_k = V_k + \frac{1}{2}\nu(\gamma^2-1)\right).$$

The latter system has the following form

$$\left[\nu-\kappa^2\delta\sigma^{-2}(\gamma)\right]\gamma_1=0,\qquad \left[\nu-\kappa^2\delta\sigma^{-2}(\gamma)\right]\gamma_2=0,$$

$$\nu\gamma_3+\kappa^2(\varepsilon-\gamma_3)\sigma^{-2}(\gamma)-\varepsilon=0,\quad \gamma^2=1,\quad \left(\sigma(\gamma)=\delta\left(\gamma_1^2+\gamma_2^2\right)+(\varepsilon-\gamma_3)^2\right).\ (4.8)$$

Obviously, system (4.8) has always two trivial solutions (for arbitrary value of parameter $\kappa(k)$:

$$\gamma_1=\gamma_2=0,\quad \gamma_3=\pm1\quad \left(\nu=\pm\varepsilon+\kappa^2(1\mp\varepsilon)^{-3}\right)\tag{4.9}$$

This system can have also no more than two families of nontrivial solutions

$$\gamma_1^2+\gamma_2^2=\sin^2\theta,\gamma_3=\cos\theta\qquad \left(\nu=k^2\delta\left[\delta\sin^2\theta+(\varepsilon-\cos\theta)^2\right]^{-2}\right)\tag{4.10}$$

where $\theta=\theta(\kappa^2)$ is a solution of the equation

$$\kappa^2\left[\varepsilon+(\delta-1)\cos\theta\right]=\varepsilon\left[\delta(1-\cos^2\theta)+(\varepsilon-\cos\theta)^2\right]^2\tag{4.11}$$

One can show that equation (4.11) respectively $\cos\theta$ has at most two real solution satisfying inequality $|\cos\theta|<1$ (a number of solutions depends on parameters ε and δ of the top and, of course, parameter $\kappa(k)$.

Solutions (4.9) and (4.10) correspond to the poles $\mathbf{P}_\pm$ and parallels $\mathbf{P}_\theta$ of the unit sphere $\mathbf{S}^2$ respectively. Thus the system has two invariant points $\mathbf{P}_\pm$ (zero - dimensional invariant sets) and no more than two invariant circles $\mathbf{P}_\theta$ (one - dimensional invariant sets) in the configuration space $\mathbf{S}^2$.

The points $\mathbf{P}_\pm$ correspond to permanent rotations of the top about its vertically situated axis of symmetry:

$$Q_\pm = \{\gamma_1 = \gamma_2 = 0, \quad \gamma_3 = \pm 1; \quad \omega_1 = \omega_2 = 0; \quad \omega_3 = \omega(1 \mp \varepsilon); \quad \mathbf{v} = 0\} \tag{4.12}$$

where $\omega = \omega(k^2)$:

$$\omega = A_3^{-1}(1 \pm \varepsilon)^{-1} k \tag{4.13}$$

The case $\gamma_3 = +1$ corresponds to the lowest position of the center of mass while the $\gamma_3 = -1$ corresponds to its highest position.

The circle $\mathbf{P}_\theta$ corresponds to regular precessions of the top:

$$Q_\theta = \{\gamma_1^2 + \gamma_2^2 = \sin^2\theta, \gamma_3 = \cos\theta; \omega_1 = \omega\gamma_1, \omega_2 = \omega\gamma_2, \omega_3 = \omega(\gamma_3 - \varepsilon); \mathbf{v} = 0\} \tag{4.14}$$

where $\omega = \omega(k^2)$:

$$k^2 = \{A_1\left[(1-\delta) - \varepsilon^2\right]\omega^4 + A_3^{-1}(mgc)^2\}(1-\delta)^{-1}\omega^{-3} \tag{4.15}$$

One can show that equation (4.15) respectively ω has at most two real solutions. Thus there are no more than two families of regular precessions of the top when it rotates about its axis of symmetry with angular velocity - $\omega\varepsilon$, the latter rotates about the vertical with angular velocity ω while the angle between these axes is equal θ.

The effective potential takes a local strict minimum (maximum) at the poles $\mathbf{P}_\pm$ of the unit sphere $\mathbf{S}^2$ if

$$[(1 \mp \varepsilon) - \delta](1 \pm \varepsilon)^{-2} k^2 \pm A_3 mgc > 0 \quad (< 0) \tag{4.16}$$

and on the circles $\mathbf{P}_\theta$ if $\dfrac{d\theta}{d(\kappa^2)} > 0 \quad (< 0)$, i.e, if

$$(A_1 - A_3)\frac{d\omega}{d(k^2)} > 0 \quad (< 0) \tag{4.17}$$

respectively (in the latter inequalities $\theta = \theta(k^2)$ is a solution of equation (4.11) and $\omega = \omega(k^2)$ is a solution of equation (4.15)). Using relations (4.13) and (4.15) inequalities (4.16) and (4.17) can be represented in the forms

$$[A_3(1 \mp \varepsilon) - A_1]\omega^2 \pm mgc > 0 \quad (< 0) \tag{4.18}$$

$$A_1\left[A_3(1-\varepsilon^2) - A_1\right]\omega^4 - 3(mgc)^2 < 0 \quad (> 0) \tag{4.19}$$

respectively.

One can show that the total mechanical energy preserves its initial value only on the set $\mathbf{Q} = \mathbf{Q}_+ \bigcup \mathbf{Q}_- \bigcup \mathbf{Q}_\theta$ if the sliding friction vanishes when the sliding velocity vanishes (i.e. if $\mathbf{F} = 0$ for $\mathbf{v} + \boldsymbol{\omega} \times \boldsymbol{\rho} = \mathbf{0}$). Thus the permanent rotations (4.12) and the regular precessions (4.14) are stable (unstable) on condition (4.16) and (4.17) or (4.18) and (4.19) correspondingly. Moreover the stable stationary motions are partially asymptotic stable. In particular, stable permanent rotations are asymptotically stable with respect to $\mathbf{v}, \boldsymbol{\gamma}, \omega_1$ and ω_2 (see theorems 3.2,3.4,3.5).

4.3 Bifurcations.

Invariant sets (4.12) and (4.14) can be represented on the plane $(k; h)$ (k is a constant of Jellet's integral and h is an initial value of the total mechanical energy) as lines $\Sigma_\pm$ and Σ_θ respectively. Lines $\Sigma_\pm$ are the two parabolas

$$h = \frac{k^2}{2A_3(1 - \mp\varepsilon^2)} \mp mgc$$

and lines Σ_θ are some curves with the following parametric equations (p is a parameter)

$$h = \frac{mgc}{2(1-\delta)p}\left[\delta(1-\varepsilon^2-\delta)p^2 - \varepsilon p + 3\right],$$

$$k^2 = \frac{A_3 mgc}{(1-\delta)^2 p^3}\left[\delta(1-\varepsilon^2-\delta)p^2 + 1\right]^2.$$

The existence of sets $\mathbf{P}_\theta$ and Σ_θ their form and the character of stationary values of the effective potential on the set $\mathbf{P}_\pm \bigcup \mathbf{P}_\theta$ depend on parameters ε and δ (and, of course, on value k of Jellet's integral). Consider the plane $(\delta; \varepsilon)$ of the parameters and separate it into six domains (see fig.2). These domains correspond to the following cases:

(1): $\delta > 1 + \varepsilon$ $\qquad (p > p_1)$

(2): $1 + \varepsilon > \delta > \varphi_+(\varepsilon)$ $\qquad (p \in (p_1; p_2))$

(3): $\varphi_+(\varepsilon) > \delta > \begin{cases} 1-\varepsilon, & 1 > \varepsilon > \varepsilon_0 \\ \varphi_+(\varepsilon) & 0 < \varepsilon < \varepsilon_0 \end{cases}$ $\qquad (p \in (p_2; p_1))$

(4): $\delta < \begin{cases} 1-\varepsilon, & 1 > \varepsilon > \varepsilon_0 \\ \varphi_+(\varepsilon) & 0 < \varepsilon < \varepsilon_0 \end{cases}$ $\qquad (p > p_2)$

(5): $\varphi_+(-\varepsilon) > \delta > 1 - \varepsilon, \quad 0 < \varepsilon < \varepsilon_0$ $\qquad (p \in (p_2; p_1))$

(6): $1 - \varepsilon > \delta > \varphi_-(\varepsilon), \quad 0 < \varepsilon < \varepsilon_0$ $\qquad (p > p_2)$

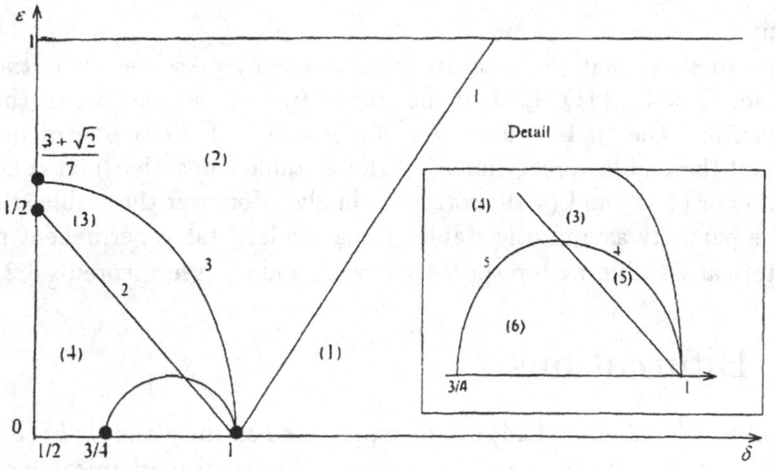

Fig. 2

$$\varphi_{\pm}(\varepsilon) = \frac{1-\varepsilon}{8}(7+\varepsilon \pm \sqrt{1+14\varepsilon+\varepsilon^2})$$

$$\varepsilon_0 = 7-\sqrt{48}, \quad p_1 = \frac{1}{\delta+\varepsilon-1}, \quad p_2 = \frac{1}{1+\varepsilon-\delta}$$

Corresponding to these domains Poincare - diagrams are given in the space $(k;\gamma_3)$ (see fig. 3) and Smale – diagram are given in the space $(k;h)$ (see fig. 4). The thick lines correspond to stable stationary motions, the thick points correspond to points of bifurcation. Latin members on fig. 4 indicate the corresponding topological types of domains of possible motions on the unit sphere $\mathbf{S}^2$.

Remark 4.1. If $1+\varepsilon > \delta > 1-\varepsilon$, then fast rotations of the top with the lowest position of the center of mass are unstable while such rotations of the top with the highest position of the center of mass are stable (see fig. 3 (2; 3; 5)).

Remark 4.2. If for given parameters ε, δ and k there exists only one stable invariant set (see, for example, fig. 3 (2) and fig. 4 (2)) then almost all motions of the top evaluate to the stable invariant set corresponding to these parameters. If there exist two stable invariant sets (see, for example, fig. 3 (3) and fig.4 (3)) then one cannot predict the limiting motion of the top using only values h and k.

4.4 Invariant sets of relative equilibria of the top and their stability.

Assume that the restriction

$$\frac{(\boldsymbol{\omega},\boldsymbol{\rho})r}{(\boldsymbol{\rho},\boldsymbol{\rho})} = \omega = \text{const} \quad (\omega \in \mathbf{R}) \tag{4.20}$$

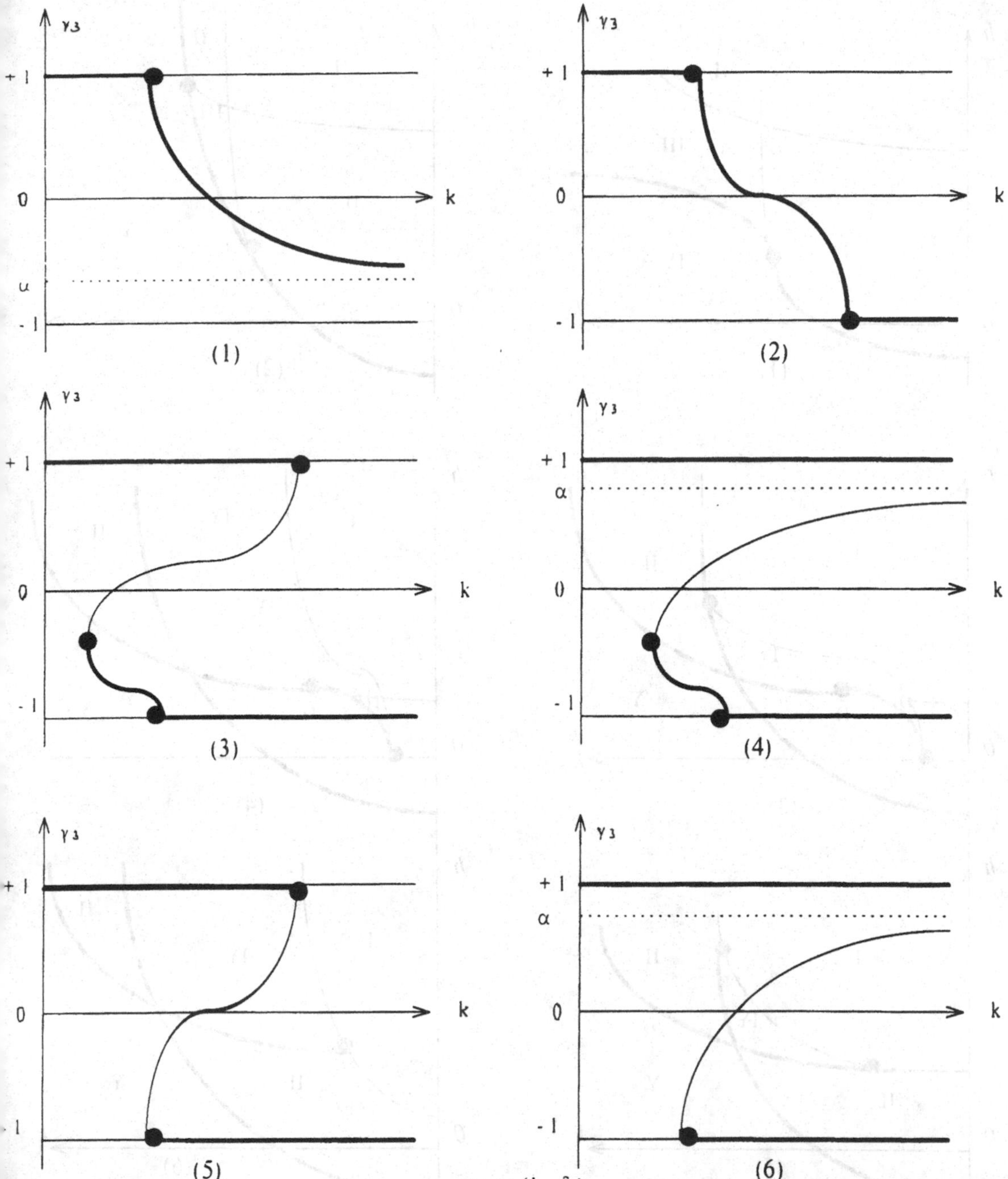

$\alpha = \varepsilon\ (1-\delta)$

Fig. 3.

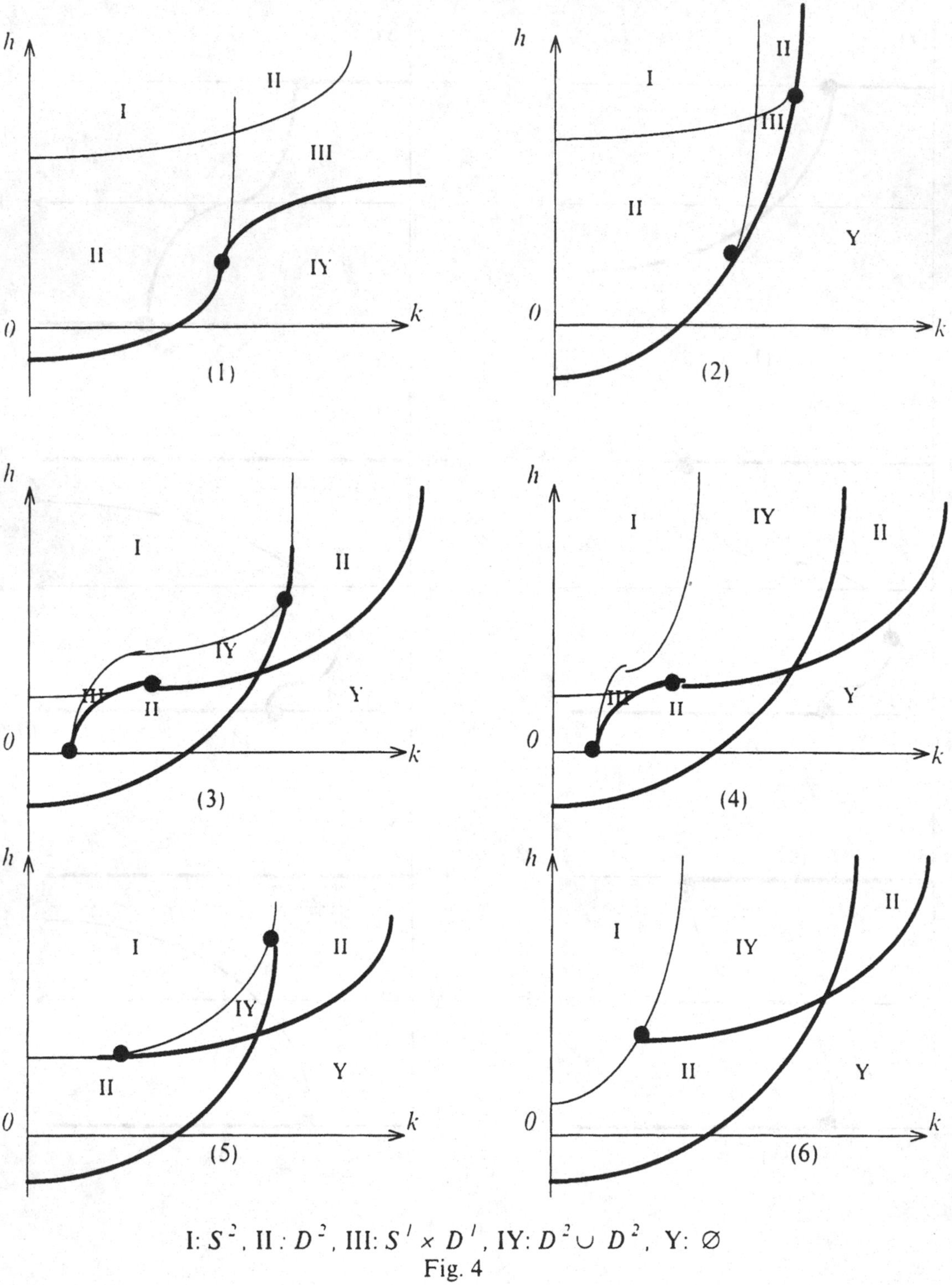

I: S^2, II : D^2, III: $S^1 \times D^1$, IY: $D^2 \cup D^2$, Y: $\varnothing$

Fig. 4

is fulfilled on all motions of the top. Then the changed potential of the restricted system is defined by

$$V_\omega(\gamma) = -mg(\rho, \gamma) - \frac{1}{2}(\Theta\gamma, \gamma)\omega^2 = mgr(1 + \varphi(\gamma));$$

$$\varphi(\gamma) = -\varepsilon\gamma_3 - \frac{1}{2}\left[\delta(\gamma_1^2 + \gamma_2^2) + (\varepsilon - \gamma_3)^2\right]\lambda^2 \quad (\lambda^2 = A_3(mgr)^{-1}\omega^2).$$

According to theorem 3.6 invariant sets of relative equilibria correspond to solutions of the system

$$\frac{\partial W_\omega}{\partial \gamma} = 0; \quad \gamma^2 = 1 \quad \left(W_\omega = V_\omega + \frac{1}{2}\nu(\gamma^2 - 1)\right).$$

The latter system has the following form

$$\left(\nu - \delta\lambda^2\right)\gamma_1 = 0, \quad \left(\nu - \delta\lambda^2\right)\gamma_2 = 0,$$

$$(\nu - \lambda^2)\gamma_3 + \varepsilon(\lambda^2 - 1) = 0, \quad \gamma^2 = 1. \tag{4.21}$$

Obviously, system (4.21) has always two trivial solutions (for arbitrary value of parameter $\lambda(\omega)$) :

$$\gamma_1 = \gamma_2 = 0, \quad \gamma_3 = \pm 1, \quad \left(\nu = \pm\varepsilon + (1 \mp \varepsilon)\lambda^2\right) \tag{4.22}$$

and can have one family of nontrivial solutions

$$\gamma_1^2 + \gamma_2^2 = \sin^2\theta, \quad \gamma_3 = \cos\theta \quad \left(\nu = \delta\lambda^2\right) \tag{4.23}$$

where $\theta = \theta(\lambda^2)$ is the solution of the equation

$$\lambda^2\left[\varepsilon - (1 - \delta)\cos\theta\right] = \varepsilon. \tag{4.24}$$

By the results presented previously (see theorem 1.11), the set of critical points (4.9) and (4.10) of the effective potential V_k coincides with the set of critical points (4.22) and (4.23) of the changed potential V_ω if constants k and ω are related as

$$k = (\Theta\gamma_\omega^0, \gamma_\omega^0)\omega; \quad \omega = (\Theta\gamma_k^0, \gamma_k^0)^{-1}k \tag{4.25}$$

where $\gamma_k^0\,(\gamma_\omega^0)$ are defined by relations (4.9),(4.10) ((4.22),(4.23)). In explicit form relations (4.25) are given by (4.13) and (4.15).

Note that equation (4.24) is uniquely solvable for $\cos\theta$, while equation (4.11) may have at most two solutions; in the configuration space $\mathbf{S}^2$, however, these relations (taking (4.15) into account) define the same set.

The changed potential V_ω takes a local strict minimum (maximum) at poles $\mathbf{P}_\pm$ if

$$[A_3(1 \mp \varepsilon) - A_1]\,\omega^2 \pm mgc > 0 \quad (< 0) \tag{4.26}$$

and on the circle $\mathbf{P}_\theta$ if

$$A_3 - A_1 < 0 \quad (> 0) \tag{4.27}$$

Evidently, condition (4.26) coincides with condition (4.18) as expected (see theorem 3.13), but the stability condition (4.27) ($A_3 < A_1$) is more restrictive than the corresponding condition (4.19); the latter is always correct if (4.27) holds, but can be correct even if $A_3 > A_1$ (compare with theorem 3.12).

Thus, the stability conditions of permanent rotations (trivial invariant sets) of a free top on a plane with friction and the corresponding relative equilibria of the top restricted by relation (4.20) are the same. Regular precessions (nontrivial invariant sets) of the free top are always stable if the inertia ellipsoid is prolate along the axis of symmetry, but they may be stable even if the ellipsoid is oblate (see, for example, fig.3 (2) and fig.2). At the same time regular precessions of the restricted top are stable (unstable) if the inertia ellipsoid of the top is prolate (oblate).

Chapter 5

Bifurcation and stability of the steady motions and relative equilibria of a rigid body in a central gravitation field [31]-[33].

5.1 Effective potential and equations of steady motions.

We will consider the problem of the translational - rotational motion of a rigid body with a triaxial ellipsoid of inertia in a central gravitational field. We will model the body as a collection of point masses $\frac{m_s}{2}$ situated at the opposite ends of three mutually perpendicular diameters d_s, $s = (1,2,3)$ of massless sphere of radius a. Without loss of generality, we assume that $m_1 > m_2 > m_3$.

Let $O\xi\eta\zeta$ be a fixed system of coordinates with origin at the attracting center, and let $Cx_1x_2x_3$ be a system of coordinates attached to the body with origin at its mass center and axes along the diameters d_1, d_2, d_3. The position of the mass center of the body relative to the fixed system of coordinates will be defined by spherical coordinates r, θ, ψ where $r > a$ is the length of the radius vector **OC**, θ is the angle between the vector **OC** and the plane $O\zeta\xi$ and ψ is the angle between the axis $O\zeta$ and the projection of the vector **OC** onto the plane $O\zeta\xi$. The orientation of the orbit of the body's mass center and the orientation of the body will be defined by the projections β_s and γ_s of the unit vectors $\boldsymbol{\beta}$ and $\boldsymbol{\gamma}$ in the direction of the $O\eta$ axis and the radius vector **OC**, respectively, onto the principal central inertia axes Cx_s, $s = (1,2,3)$ of the body. Obviously, $\sin\theta = (\boldsymbol{\beta}, \boldsymbol{\gamma}) = \sum_s \beta_s \gamma_s$.

The kinetic energy T and the potential energy V of the body are

$$2T = \left[m\left(\dot{r}^2 + r^2\dot{\psi}^2\cos^2\theta + r^2\dot{\theta}^2\right) + J_1\omega_1^2 + J_2\omega_2^2 + J_3\omega_3^2\right]$$

$$2V = -fM\sum_{s=1}^{3} m_s\left[F_s(a)+F_s(-a)\right], \quad F_s(a) = \left(r^2+a^2+2ar\gamma_s\right)^{-\frac{1}{2}}$$

where $m = m_1+m_2+m_3$ is the mass of the body, $J_i = (m_j+m_k)\,a^2$ is the moment of inertia of the body about the axis Cx_i $(i \neq j \neq k;\ i,j,k \in S_3,$ $S_3 = \{1,2,3\})$, ω_s is the projection of the absolute angular velocity $\boldsymbol{\omega}$ of the body onto the axis Cx_s, $s = (1,2,3)$, f is the gravitational constant, and M is the mass of the attracting center.

The system admits of two first integrals: $H = T + V = \text{const}$ (energy) and $K = \dfrac{\partial T}{\partial \dot{\psi}}$ (area). Setting $\boldsymbol{\omega} = \dot{\psi}\boldsymbol{\beta} + \boldsymbol{\Omega}$ where $\boldsymbol{\Omega}$ is the angular velocity of the body relative to a system of coordinates rotating uniformly about the $O\eta$ axis, one find the effective potential of the body (we recall that $\sin\theta = \sum_s \beta_s\gamma_s$)

$$W_k = \min_{\dot{r},\dot{\psi},\dot{\theta},\Omega} H|_{K=k} = V + \frac{k^2}{2J}, \quad J = mr^2\left(1-\left(\sum_{s=1}^{3}\gamma_s\beta_s\right)^2\right) + \sum_{s=1}^{3} J_s\beta_s^2$$

By Routh's theorem, the critical points $(r_0, \boldsymbol{\gamma}_0, \boldsymbol{\beta}_0)$ of the effective potential $W_k(r,\boldsymbol{\beta},\boldsymbol{\gamma})$ on the manifold $\boldsymbol{\gamma}^2 = 1$, $\boldsymbol{\beta}^2 = 1$ are the steady motions of the body

$$r = r_0, \quad \boldsymbol{\gamma} = \boldsymbol{\gamma}_0, \quad \boldsymbol{\beta} = \boldsymbol{\beta}_0, \quad \dot{\psi} = \omega_0$$

$$\left(\theta = \theta_0 = \arcsin\left(\boldsymbol{\gamma}_0, \boldsymbol{\beta}_0\right), \omega_0 = \frac{k}{J_0}, \boldsymbol{\Omega} = \mathbf{0}\right) \tag{5.1}$$

Under conditions (5.1), the mass center of the body uniformly describes a circle of radius $r_0\cos\theta_0$ in a plane parallel to the plane $O\zeta\xi$ at a distance $r_0|\sin\theta_0|$ from it; the orientation of the body remains constant during the motion.

To determine the critical points of W_k on the manifold $\{\boldsymbol{\beta}^2 = 1, \boldsymbol{\gamma}^2 = 1\}$, we consider the function

$$W = (fM)^{-1}W_k + \frac{p\left(\boldsymbol{\gamma}^2-1\right)}{2} + \frac{q\left(\boldsymbol{\beta}^2-1\right)}{2}$$

(p and q are undetermined Lagrange multipliers) and write down the conditions for this function to be stationary $\left(\kappa^2 = \dfrac{k^2}{fM}\right)$

$$\frac{\partial W}{\partial r} = \frac{1}{2}\sum_{s=1}^{3} m_s\left(\left(r+a\gamma_s\right)F_s^3(a) + \left(r-a\gamma_s\right)F_s^3(-a)\right) - \\ - \frac{\kappa^2 mr}{J^2}\left(1-\left(\sum_{s=1}^{3}\gamma_s\beta_s\right)^2\right) = 0 \tag{5.2}$$

$$\frac{\partial W}{\partial \gamma_s} = \frac{1}{2} m_s \left(F_s^3(a) - F_s^3(-a) \right) ra + p\gamma_s + \frac{\kappa^2 mr}{J^2} \beta_s \sum_{\sigma=1}^{3} \gamma_\sigma \beta_\sigma = 0 \qquad (5.3)$$

$$\frac{\partial W}{\partial \beta_s} = \frac{\kappa^2}{J^2} \left(mr^2 \gamma_s \sum_{\sigma=1}^{3} \gamma_\sigma \beta_\sigma - J_s \beta_s \right) + q\beta_s = 0 \;\; (s = 1, 2, 3) \qquad (5.4)$$

5.2 Trivial steady motions and their stability.

Solutions of system (5.2) – (5.4) of the form

$$\gamma_i = \pm 1, \;\; \beta_j = \pm 1, \;\; (i \neq j \neq k), \;\; \gamma_j = \gamma_k = \beta_i = \beta_k = 0$$
$$(i, j, k \in S_3), \;\; r = r_{ij}^{\pm}(\kappa^2) \qquad (5.5)$$

correspond to trivial orientations of the body: the principal axis Cx_i coincides with radius - vector of the mass center, the principal axis Cx_j coincides with the normal of the plane of the orbit and the principal axis Cx_k coincides with the tangent of the orbit. The plane of the orbit of the body's mass center passes through the attracting center ($\sin\theta = 0$). To investigate the stability of the steady motions corresponding to orientations (5.5) we calculate the second variation of W over the manifold $\delta\gamma_i = 0, \; \delta\beta_j = 0 : \; 2\delta^2 W = \Sigma_1 + \Sigma_2; \; \Sigma_1 = C_1 (\delta r)^2 + C_2 (\delta\gamma_k)^2 + C_3 (\delta\beta_k)^2$; $\Sigma_2 = C_{44} (\delta\beta_i)^2 + C_{45} (\delta\beta_i)(\delta\gamma_j) + C_{55} (\delta\gamma_j)^2$. Thus sufficient conditions of stability of trivial steady motions have the following form

$$C_1 > 0, \;\; C_2 > 0, \;\; C_3 > 0, \;\; C_4 = C_{44}C_{55} - C_{45}^2 > 0;$$

$$C_1 = \frac{mr}{(mr^2 + J_j)^2} \frac{dK_{ij}}{dr}, \;\; C_2 = \left[m_i \frac{(3r^2 + a^2)}{(r^2 - a^2)^3} - \frac{3m_k r}{(r^2 + a^2)^{\frac{5}{2}}} \right] ra^2,$$

$$C_3 = \frac{\kappa^2 (J_j - J_k)}{(mr^2 + J_j)^2}, \;\; C_{44} = \frac{\kappa^2 (J_j - J_i + mr^2)}{(mr^2 + J_j)^2}, \;\; C_{45} = \frac{\kappa^2 mr^2}{(mr^2 + J_j)^2},$$

$$C_{55} = \frac{\kappa^2 mr^2}{(mr^2 + J_j)^2} + m_i ra^2 \frac{(3r^2 + a^2)}{(r^2 - a^2)^3} - \frac{3m_j r^2 a^2}{(r^2 + a^2)^{\frac{5}{2}}},$$

$$\kappa^2 = K_{ij}(r), \; K_{ij} = \frac{(mr^2 + J_j)^2}{mr} \left[m_i \frac{(r^2 + a^2)}{(r^2 - a^2)^2} - \frac{(m_j + m_k) r}{(r^2 + a^2)^{\frac{3}{2}}} \right],$$

$$r_{ij}^{+}\left(\kappa^2\right) > r_{ij}^0 > r_{ij}^{-}\left(\kappa^2\right) > a, \; r_{ij}^0 : K'_{ij}\left(r_{ij}^0\right) = 0;$$

$$\kappa^2 > \left(\kappa_{ij}^0\right)^2, \; \left(\kappa_{ij}^0\right)^2 = K_{ij}\left(r_{ij}^0\right).$$

One can show that the degree of instability of the trivial steady motions may vary not only at the branch points $r = r_{ij}^0 \; \left(C_1\left(r_{ij}^0\right) = 0, \, i, j \in S_3\right)$ of the solutions (5.5) as functions of r, but also at the points

$$r = r_{ik}^* \left(C_2\left(r_{ik}^*\right) = 0, i > k\right), \; r = \bar{r}_{ij}^* \left(C_4\left(\bar{r}_{ij}^*\right) = 0, i > j\right).$$

This means that system (5.2)–(5.4), besides the solutions (5.5), has solutions corresponding to non-trivial orientations.

5.3 Nontrivial steady motions of the first level.

According to bifurcation theory, the solutions of system (5.2)-(5.4) bifurcate at the points r^*_{ik} $(i > k)$ and $\bar{r}^*_{ij}$ $(i > j)$ as functions of γ_i and γ_k and as functions of γ_i, γ_j, β_i, β_j, respectively. Thus system (5.2)-(5.4) has solutions corresponding to orientations of the form

$$\gamma_i = \cos\varphi,\ \gamma_k = \sin\varphi,\ \gamma_j = \beta_i = \beta_k = 0,\ \beta_j = \pm 1\ (j \neq i > k \neq j) \tag{5.6}$$

$$\gamma_i = \cos\varphi,\ \gamma_j = -\sin\varphi,\ \beta_i = \sin(\varphi+\theta),\ \beta_j = \cos(\varphi+\theta),\ \gamma_k = \beta_k = 0 \quad (k \neq i > j \neq k) \tag{5.7}$$

Under conditions (5.6), the plane of the orbit of the body's mass center passes through the attracting center ($\sin\theta = 0$), the axis Cx_i points along the normal to the plane of the orbit, and the axes Cx_j and Cx_k are rotated through the same angle φ relative to the radius vector of the mass center and the tangent to the orbit. The angle $\varphi = \varphi_{ik}(\kappa^2)$ and the radius of the orbit $r = r_{ik}(\kappa^2)$ are determined from the system

$$m_k m_i^{-1} = \Phi_{ik}(r,\varphi),\ \kappa^2 = \frac{(mr^2 + J_j)^2}{2mr}\Psi_{ikj}(r,\varphi)$$

$$\Phi_{ik} = \frac{F_i^3(-a) - F_i^3(a)}{F_k^3(-a) - F_k^3(a)} \cdot \frac{\gamma_k}{\gamma_i},\quad \Psi_{ikj} = m_i G_i + m_k G_k + m_j \frac{2r}{(r^2+a^2)^{\frac{3}{2}}}$$

$$G_s = F_s^3(a)(r + a\gamma_s) + F_s^3(-a)(r - a\gamma_s)\ (s = i,k);\ \gamma_i = \cos\varphi, \gamma_k = \sin\varphi \tag{5.8}$$

which is obtained from (5.2)-(5.4) by substituting conditions (5.6) into the system.

One can show that

$$\varphi = \pi n \pm \varphi_{ik}(\kappa^2),\ (n = 0,1),\ r = r_{ik}(\kappa^2),\ a < r_{ik}(\kappa^2) < r^*_{ik}$$

$$0 < \varphi_{ik}(\kappa^2) < \varphi^*_{ik} < \frac{\pi}{4}$$

where φ^*_{ik} is the unique root in the interval $0 < \varphi < \frac{\pi}{2}$ of the equation $m_k m_i^{-1} = \Phi_{ik}(a,\varphi)$, and r^*_{ik} is the unique root (for $r > a$) of the equation $C_2(r) = 0$.

Under condition (5.7), the distance from the plane of the orbit of the body's mass center is $r|\sin\theta| \neq 0$, the radius of the orbit is $r\cos\theta$, the axis Cx_k is directed along the tangent to the orbit, the axis Cx_i makes an angle φ with the radius vector of the mass center, and the axis Cx_j makes an angle $\varphi + \theta$ with the normal to the plane of the orbit. The angle θ may be expressed explicity in terms of φ and the distance r from the mass center to the attracting center

$$\theta = \theta_{ij} = \frac{1}{2}\operatorname{arctg}\frac{\nu_{ij}(r)\sin 2\varphi}{1 - \nu_{ij}(r)\cos 2\varphi};\quad \nu_{ij} = \frac{m_i - m_j}{m}\cdot\frac{a^2}{r^2} \tag{5.9}$$

The angle φ and distance r are determined by solving a system analogous to system (5.8), but the function Φ_{ij} and Ψ_{ijk} are more complicated.

Proceeding as before, it can be shown that

$$\varphi = \pi n \pm \bar{\varphi}_{ij}\left(\kappa^2\right),\ (n = 0, 1),\ r = \bar{r}_{ij}\left(\kappa^2\right),\ a < \bar{r}_{ij}\left(\kappa^2\right) < \bar{r}^*_{ij} \tag{5.10}$$

$$0 < \bar{\varphi}_{ik}\left(\kappa^2\right) < \bar{\varphi}^*_{ik} < \varphi^*_{ij} < \frac{\pi}{4}$$

where $\bar{\varphi}^*_{ik}$ is the unique root in the range $0 < \varphi < \frac{\pi}{2}$ of the equation $m_j m_i^{-1} = \Phi_{ij}(a, \varphi)$, and r^*_{ij} is the unique root (for $r > a$) of the equation $C_4(r) = 0$.

The relative position of the points r^*_{ik}, $\bar{r}^*_{ij}$, and $r^0_{ik(j)}$ has a considerable effect on the degree of instability of the steady motions corresponding to orientations (5.5),(5.6) and (5.7), and depends on the relationships between the masses m_1, m_2, and m_3. Henceforth, for simplicity, we shall confine ourselves to the case of similar masses

$$m_2 = m_1(1 - u),\ m_3 = m_2(1 - v);\ 0 < u \ll 1,\ 0 < v \ll 1 \tag{5.11}$$

Under conditions (5.11), the ellipsoid of inertia is almost a sphere - the characteristic shape of many celestial bodies in nature.

Under these conditions one has $r^0_{ij(k)} \ll \bar{r}^*_{ij}\ (r^*_{ik})$ and, in addition, one can then show that the steady motions corresponding to orientations (5.6) and (5.7) exist when $\kappa^2 \in \left(\left(\kappa^2_{ik}\right)_*, \left(\kappa^2_{ik}\right)^*\right)$ and $\kappa^2 \in \left(\left(\bar{\kappa}^2_{ij}\right)_*, \left(\bar{\kappa}^2_{ij}\right)^*\right)$, respectively, where $\left(\kappa^2_{ik}\right)_*$ and $\left(\kappa^2_{ik}\right)^*$ are defined by the second equation of system (5.8) with $r = a$, $\varphi = \varphi^*_{ik}$ or $r = r^*_{ik}$, $\varphi = 0$. The definition of $\left(\bar{\kappa}^2_{ik}\right)_*$ and $\left(\bar{\kappa}^2_{ik}\right)^*$ are analogous.

5.4 Stability of nontrivial steady motions of the first level.

To investigate the stability of the steady motions corresponding to orientation (5.6), we calculate the second variation of W over the linear manifold

$$\delta\beta_j = 0,\quad \delta\gamma_i \cos\varphi + \delta\gamma_k \sin\varphi = 0:$$

$$\delta^2 W = \Sigma_1 + \Sigma_2;$$

$$2\Sigma_1 = C_{11}(\delta r)^2 + 2C_{12}(\delta r)(\delta\gamma_k) + C_{22}(\delta\gamma_k)^2,$$

$$2\Sigma_2 = C_{33}(\delta\gamma_j)^2 + 2C_{34}(\delta\gamma_j)(\delta\beta_i) + 2C_{35}(\delta\gamma_j)(\delta\beta_k) +$$
$$+ C_{44}(\delta\beta_i)^2 + 2C_{45}(\delta\beta_i)(\delta\beta_k) + C_{55}(\delta\beta_k)^2;$$

$$C_{11} = \frac{mr}{(mr^2 + J_j)^2}\frac{\partial}{\partial r}\left[-\frac{1}{2}\frac{(mr^2 + J_j)^2}{mr}\Psi_{ikj}(r, \varphi)\right],$$

$$C_{12} = -\frac{3}{2}m_k\frac{ra}{\delta^{(3)}_i}\left[\delta^{(3)}_i\left(r\sigma^{(5)}_k + a\gamma_k\delta^{(5)}_k\right) - \delta^{(3)}_k\left(r\sigma^{(5)}_i + a\gamma_i\delta^{(5)}_i\right)\right],$$

$$C_{22}=-\frac{1}{2}m_k\frac{ra}{\delta_i^{(3)}\gamma_i}\left[\frac{\delta_i^{(3)}\delta_k^{(3)}}{\gamma_i\gamma_k}+3ra\sigma_k^{(5)}\delta_i^{(3)}\gamma_i+\sigma_k^{(5)}\delta_i^{(3)}\gamma_k\right],$$

$$C_{33}=\frac{1}{2}r^2\left[m_i\sigma_i^{(3)}+m_k\sigma_k^{(3)}+\frac{2m_j\left(r^2-2a^2\right)}{\left(r^2+a^2\right)^{\frac{5}{2}}}\right],$$

$$C_{44}=\frac{\kappa^2}{J^2}\left[mr^2\cos^2\varphi+(J_j-J_i)\right],\quad C_{55}=\frac{\kappa^2}{J^2}\left[mr^2\sin^2\varphi+(J_j-J_k)\right],$$

$$C_{34}=\pm\frac{\kappa^2}{J^2}mr^2\cos\varphi,\quad C_{35}=\pm\frac{\kappa^2}{J^2}mr^2\sin\varphi,\quad C_{45}=\pm\frac{\kappa^2}{J^2}mr^2\cos\varphi\sin\varphi;$$

$$\delta_s^{(n)}=F_s^n(a)-F_s^n(-a),\quad \sigma_s^{(n)}=F_s^n(a)+F_s^n(-a),\quad (s=1,2,3;\ n=3,5);$$

where $\gamma_i=\cos\varphi$, $\gamma_k=\sin\varphi$ and r and φ satisfy the first relations of system (5.8) and depend on κ^2 (see the second equation of that system); $i,\ j,\ k\in S_3,\ i>k$.

Since $C_{22}<0$ all steady motions corresponding to orientation (5.6) are unstable in the secular sense. The degree of instability of these motions when $r<r_{ik}^*$, close to r_{ik}^* is identical with the degree of instability of the corresponding trivial steady motions when $r>r_{ik}^*$ $(i>k)$, and does not vary along the entire branch (5.8) $(a<r_{ik}(\kappa^2)<r_{ik}^*)$ if the determinant Δ of the quadratic form $\delta^2W=\Sigma_1+\Sigma_2$ does not change sign for all $r\in(a;r_{ik}^*)$, $\varphi\in(0;\varphi_{ik}^*)$. Since $\Delta=\Delta_1\Delta_2$, where $\Delta_{1,2}$ is the determinant of the quadratic form $\Sigma_{1,2}$, and moreover $\Delta_1\neq 0$ for all steady motions corresponding to orientations (5.6), it follows that Δ vanishes if and only if Δ_2 vanishes. The determinant Δ_2 does not vanish for orientations (5.6) when $i=2$, $j=3$, $k=1$; it vanishes when $i=3$, $j=1$, $k=2$, and may vanish when $i=3$, $j=2$, $k=1$ at some point $(r_{3k}^{**};\varphi_{3k}^{**})$, $a<r_{3k}^{**}<r_{3k}^*$, $0<\varphi_{3k}^{**}<\varphi_{3k}^*$ $(k=1,2)$. For $k=1$, $i=2$, $j=3$, this point exists for arbitrary values of the masses $m_1>m_2>m_3$, but for $k=2$, $i=3$, $j=1$ it exists only when $2m_2>m_1+m_3$ (i.e only when $v>u$; see (5.11)). Under those conditions the degree of instability of the steady motions for orientation (5.6) $(k=1,2)$ is one less for $a<r<r_{3k}^{**}$ than for $r_{3k}^{**}<r<r_{3k}^*$.

To investigate the stability of the steady motions corresponding to orientations (5.7), one must evaluate the second variation of W over linear manifold

$$\delta\gamma_i\cos\varphi-\delta\gamma_j\sin\varphi=0,\quad \delta\beta_j\cos(\theta+\varphi)-\delta\beta_i\sin(\theta+\varphi)=0:$$

$$\delta^2W=\Sigma_1+\Sigma_2;$$

$$2\Sigma_1=C_{11}(\delta r)^2+2C_{12}(\delta r)(\delta\gamma_j)+2C_{13}(\delta r)(\delta\beta_i)+$$
$$+C_{22}(\delta\gamma_j)^2+2C_{23}(\delta\gamma_j)(\delta\beta_i)+C_{33}(\delta\beta_i)^2,$$
$$\Sigma_2=C_{44}(\delta\gamma_k)^2+2C_{45}(\delta\gamma_k)(\delta\beta_k)+C_{55}(\delta\beta_k)^2;$$

(the coefficients C_{pq} are very cumbersome in form and are therefore not given here).

As before, it can be shown that all steady motions corresponding to orientations (5.7) are unstable in the secular sense, since $\Delta_1<0$. When $i=2$, $j=1$, $k=3$, the degree of instability of these motions does not vary $(\Delta_1 \neq 0, \Delta_2 \neq 0)$ along the entire branch (5.10) and is identical with the degree of instability of the trivial steady motions for $r > r^*_{21}$. When $i = 3,\ j = 2,\ k = 1$, the degree of instability varies, while when $i = 3,\ j = 2,\ k = 1$, it may vary at some point $\left(\bar{r}^{**}_{3j}; \bar{\varphi}^{**}_{3j}\right)$, $a < \bar{r}^{**}_{3j} < \bar{r}^{*}_{3j},\ 0 < \bar{\varphi}^{**}_{3j} < \bar{\varphi}^{*}_{3j},\ \left(\Delta_2\left(\bar{r}^{**}_{3j}, \bar{\varphi}^{**}_{3j}\right) = 0,\ j = 1,2\right)$. When $j = 2$ this point exists for arbitrary masses $m_1 > m_2 > m_3$, but when $j = 1$ it exists only when $2m_2 > m_1 + m_3$ (i.e. only when $v > u$; as in the previous case $\Delta_1 \neq 0$ for all orientations (5.7)). Under these conditions the degree of instability of steady motions for orientations (5.7) $(j = 1,2)$ when $r > \bar{r}^*_{3j}$, while when $a < r < \bar{r}^{**}_{3j}$ $(j = 1,2)$ it is one less then the latter.

5.5 Nontrivial steady motion of the second level

At the points $\left(r^{**}_{32}; \varphi^{**}_{32}\right)$, and $\left(\bar{r}^{**}_{3j}; \bar{\varphi}^{**}_{3j}\right)$, there will always be steady motions bifurcating from the motions corresponding to orientation (5.6) $(i = 3,\ j = 1,\ k = 2)$ and (5.7) $(i = 3,\ j = 2,\ k = 1)$; these are motions corresponding to orientations of the general form

$$\boldsymbol{\gamma} = \boldsymbol{\gamma}^0,\ \boldsymbol{\beta} = \boldsymbol{\beta}^0\ \left(\left(\boldsymbol{\gamma}^0, \boldsymbol{\beta}^0\right) \neq 0,\ \gamma^0_s \neq 0,\ \beta^0_s \neq 0, \forall s = 1,2,3\right) \qquad (5.12)$$

If in addition the condition $2m_2 > m_1 + m_3$ is satisfied, the analogous assertion is true then $(i = 3,\ j = 2,\ k = 1)$, $(i = 3,\ j = 1,\ k = 2)$ also, for orientations (5.6) and (5.7), respectively. It can be shown that these steady motions exist only when $a < r < \bar{r}^{**}_{3k},\ a < r < \bar{r}^{**}_{3j}$ and only when $\kappa^2 < \left(\kappa^2_{3k}\right)^{**},\ \kappa^2 < \left(\kappa^2_{3j}\right)^{**}$, respectively, where

$$\left(\kappa^2_{3k}\right)^{**} = \left[\frac{\left(mr^2 + J_j\right)^2}{2mr}\Psi_{3kj}(r, \varphi)\right]_{r = r^{**}_{3k}, \varphi = \varphi^{**}_{3k}} \qquad (k, j = 1,2; k \neq j)$$

$$\left(\kappa^2_{3j}\right)^{**} = \left[\frac{\left(mr^2 + J_j\right)^2}{2mr}\bar{\Psi}_{3jk}(r, \varphi)\right]_{r = \bar{r}^{**}_{3k}, \varphi = \bar{\varphi}^{**}_{3k}}$$

Steady motions of general form are determined from system (5.2)-(5.4). They are characterized by the fact that the plane of the orbit of the body's mass center does not pass through the attracting center (as in case (5.7)), and moreover none of the principal central axes of inertia of the body coincides with any of the axes of the orbital system of coordinates (unlike cases (5.5),(5.6) and (5.7)).

5.6 Bifurcations.

Equations (5.2)-(5.4) define a single-parameter family of steady motions of the body (the curve $L = \{r = r(\kappa^2), \boldsymbol{\gamma} = \boldsymbol{\gamma}(\kappa^2), \boldsymbol{\beta} = \boldsymbol{\beta}(\kappa^2)\}$ in the space $(r, \boldsymbol{\gamma}, \boldsymbol{\beta}, \kappa^2)$). Sections of this space by the hyperplanes (5.5) are shown in Figs 5-10. The solid curves correspond to those branches of L lying in the hyperplanes and corresponding to trivial steady motion. The dashes and dash-dot curves correspond to the projections of those branches of L that leave the hyperplanes and correspond to steady motions (5.6) and (5.7). The dotted curves correspond to the projections of those branches of L that always leave the aforenamed non-trivial branches and correspond to orientations (5.12). Figure 9 corresponds to the case when $3v > u > v$ (see (5.11)); when $3v < u$ one must interchange the dotted and dash-dot curves in Fig.9. When $v > u$ one must add to Figs 8 and 9 curves leaving the non-trivial branched and corresponding to additional steady motions of type (5.12). The digits 0, 1, 2 and 3 indicate the degree of instability of the steady motions of the body corresponding to the relevant orientations (5.5),(5.6),(5.7) or (5.12). The degree of instability of the latter is indicated in accordance with the general considerations of bifurcation theory.

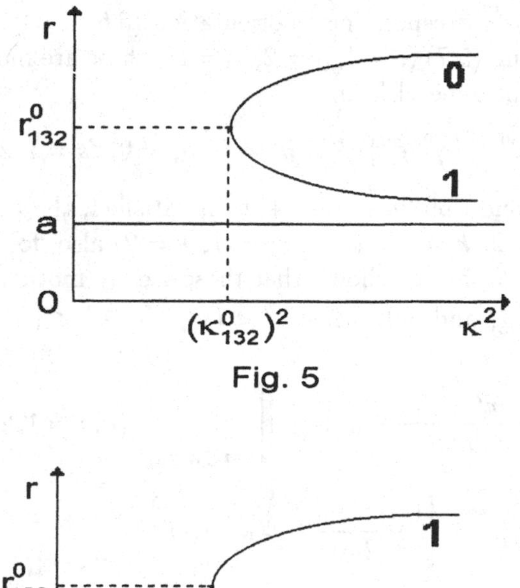

Fig. 5

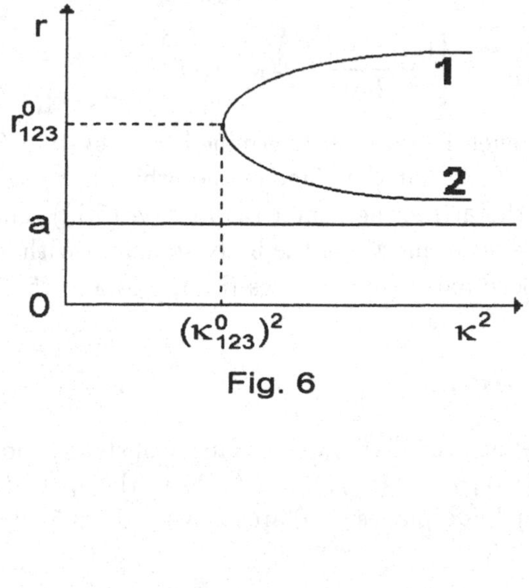

Fig. 6

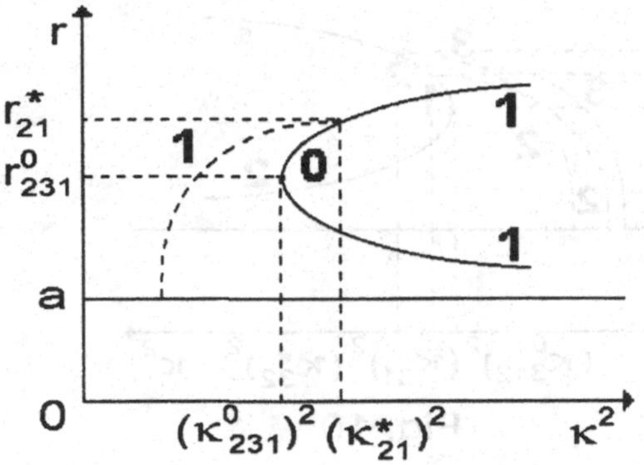

Fig. 7

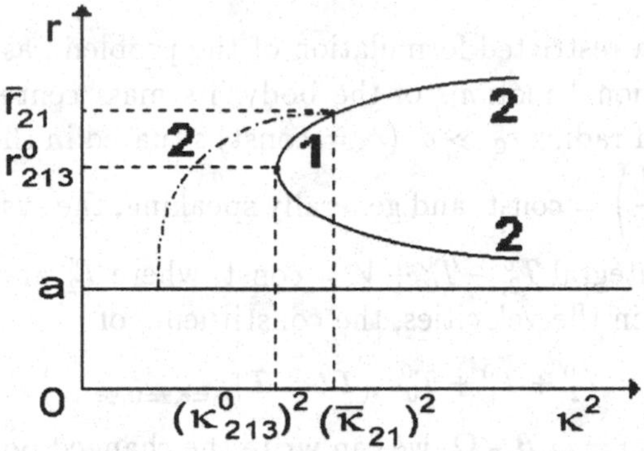

Fig. 8

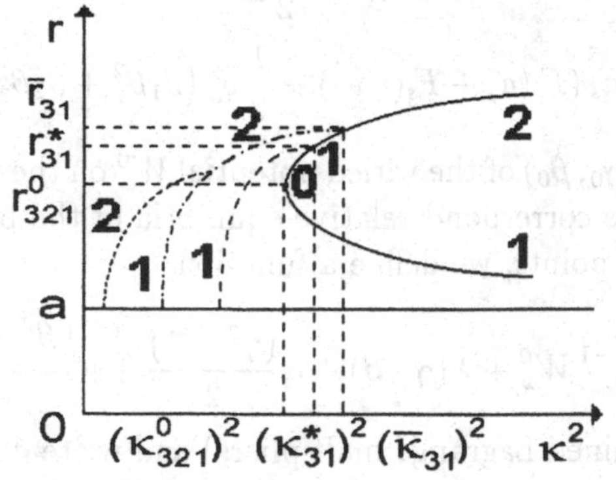

Fig. 9

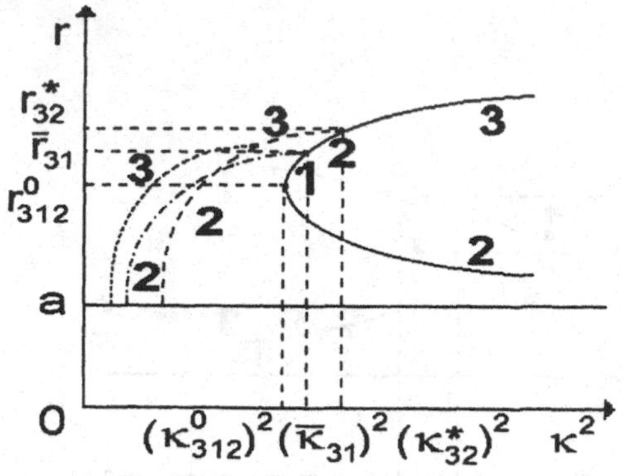

Fig. 10

5.7 Equation of relative equilibria.

We will now consider a restricted formulation of the problem, assuming that, independently of the rotational motions of the body, its mass center moves uniformly along a circular orbit of radius $r_0 \gg a$ $(r_0 = \text{const})$ situated in the $O\xi\zeta$ plane. Then $\theta = 0,\ \dot{\psi} \equiv \omega_0 = \left(\dfrac{fM}{r_0^3}\right) = \text{const}$, and generally speaking, the system admits of only a generalized energy integral $T_2^0 - T_0^0 + V = \text{const}$, where T_2^0 and T_0^0 second-degree and zero-degree forms in the velocities, the constituents of

$$T^0 = T_2^0 + T_1^0 + T_0^0 \quad \left(T^0 = T|_{r_0=r,\theta=0,\dot{\psi}=\omega_0}\right).$$

As in section 1 setting $\boldsymbol{\omega} = \omega_0\boldsymbol{\beta} + \boldsymbol{\Omega}$, we can write the changed potential of the body as

$$W_\omega^0 = V - \frac{J\omega_0^2}{2},$$

$$W_\omega^0 = -\frac{fM}{2}\sum_{s=1}^{3} m_s\left(F_s(a) + F_s(-a)\right) - \frac{1}{2}\omega_0^2\left(J_1\beta_1^2 + J_2\beta_2^2 + J_3\beta_3^2\right).$$

To the critical points $(\boldsymbol{\gamma}_0, \boldsymbol{\beta}_0)$ of the varied potential W_ω^0 on the manifold $\{\boldsymbol{\gamma}^2 = 1;\ \boldsymbol{\beta}^2 = 1;\ (\boldsymbol{\gamma}\cdot\boldsymbol{\beta}) = 0\}$ there correspond relative equilibria of the body in its circular orbit. To search for these points, we define a function

$$W^0 = (fM)^{-1}W_\omega^0 + \lambda(\boldsymbol{\gamma}\cdot\boldsymbol{\beta}) + \sigma\frac{(\boldsymbol{\gamma}^2 - 1)}{2} + \nu\frac{\left(\boldsymbol{\beta}^2 - 1\right)}{2}$$

(λ, σ and ν are undetermined Lagrange multipliers) and write down the conditions for it to be stationary (assuming, without loss of generality, that $fM = 1,\ m = 1,\ r_0 = 1$; then $\omega_0 = 1$ and $a \ll 1$)

$$\frac{\partial W^0}{\partial\gamma_s} = \frac{(am_s)\left(F_s^3(a) - F_s^3(-a)\right)}{2} + \lambda\beta_s + \sigma\gamma_s = 0$$

(5.13)

$$\frac{\partial W^0}{\partial \beta_s} = \beta_s(\nu - J_s) + \lambda\gamma_s = 0 \quad (s = 1,2,3)$$

5.8 Trivial relative equilibria and their stability.

System (5.13) admits solutions

$$\gamma_i = \pm 1,\ \beta_j = \pm 1,\ \gamma_j = \gamma_k = \beta_i = \beta_k = 0\,(i \neq j \neq k) \tag{5.14}$$

which correspond to trivial relative equilibria of the body (in which case $\lambda = 0$) analogous to the steady motion (5.5)

To investigate the stability of these relative equilibria, we will calculate the second variation of W^0 over the linear manifold

$$\delta\gamma_i = \delta\beta_j = 0;\ \delta\gamma_j = -\delta\beta_i.$$

We have

$$2\delta^2 W = C_1^0 (\delta\gamma_j)^2 + C_2^0 (\delta\gamma_k)^2 + C_3^0 (\delta\beta_k)^2;$$

$$C_1^0 = a^2 (m_i - m_j + bm_i - cm_j),\ C_1^0 = a^2 (m_i b - m_k c),$$

$$C_3^0 = a^2 (m_k - m_j);\ b = \frac{(3 + a^2)}{(1 - a^2)^3},\ c = \frac{3}{(1 + a^2)^{\frac{5}{2}}}.$$

Obviously, the conditions for stability of the relative equilibria (5.14) are $C_1^0 > 0$, $C_2^0 > 0$, $C_3^0 > 0$.

If $m_i > m_j$, then $C_1^0 > 0$; but if $m_i < m_j$, then $C_1^0 > 0$ ($C_1^0 < 0$) for $\mu_2 \equiv \frac{m_j}{m_i} < \mu_{ji}$ ($\mu_2 > \mu_{ji}$). If $m_i > m_k$, then $C_2^0 > 0$; but if $m_i < m_k$, then $C_2^0 > 0$ ($C_2^0 < 0$) for $\mu_1 \equiv \frac{m_k}{m_i} < \mu_{ki}$ ($\mu_1 > \mu_{ki}$). Finally, $C_3^0 > 0$ ($C_3^0 < 0$) for $m_k > m_j$ ($m_k < m_j$). We have used the following notation

$$\mu_{ji} = \mu_{ji}(a) = \frac{(4 - 2a^2 + 3a^4 + a^6)}{(c+1)(1 - a^2)^3} = 1 + \frac{35}{8}a^2 + o\left(a^2\right),$$

$$\mu_{ki} = \mu_{ki}(a) = \frac{b}{c} = 1 + \frac{35}{6}a^2 + o\left(a^2\right).$$

Thus, the relative equilibria (5.14) are

(**a**) always unstable if $i = 1,\ j = 2,\ k = 3$ ($m_i > m_j > m_k$; $\mu_1 < \mu_2 < 1$) (the degree of instability $\chi = 1$);

(**b**) always stable if $i = 1,\ j = 3,\ k = 2$ ($m_i > m_k > m_j$; $\mu_2 < \mu_1 < 1$) (then $\lambda = 0$);

(c) when $i = 2,\ j = 1,\ k = 3$ $(m_j > m_i > m_k;\ \ \mu_2 > 1 > \mu_1)$, they are unstable $(\chi = 1)$ if $\mu_2 < \mu_{ji}$, and unstable in the secular sense $(\chi = 2)$ if $\mu_2 > \mu_{ji}$;

(d) when $i = 2,\ j = 3,\ k = 1$ $(m_k > m_i > m_j;\ \ \mu_1 > 1 > \mu_2)$, they are stable $(\chi = 0)$ if $\mu_1 < \mu_{ki}$, and unstable $(\chi = 1)$ if $\mu_1 > \mu_{ki}$;

(e) when $i = 3,\ j = 1,\ k = 2$ $(m_j > m_k > m_i;\ \ \mu_2 > \mu_1 > 1)$, they are unstable $(\chi = 1)$ if $\mu_2 < \mu_{ji}$, or $(\chi = 3)$ if $\mu_1 > \mu_{ki}$ and unstable in the secular sense $(\chi = 2)$ if $\mu_1 < \mu_{ki}$ and $\mu_2 > \mu_{ji}$;

(f) when $i = 3,\ j = 2,\ k = 1$ $(m_k > m_j > m_i;\ \ \mu_1 > \mu_2 > 1)$, they are stable $(\chi = 0)$ if $\mu_1 < \mu_{ki}$ and $\mu_2 > \mu_{ji}$, unstable $(\chi = 1)$ if $\mu_1 > \mu_{ki}$ and $\mu_2 < \mu_{ji}$, or $\mu_1 < \mu_{ki}$ and $\mu_2 > \mu_{ji}$ and unstable in the secular sense $(\chi = 2)$ if $\mu_1 > \mu_{ki}$ and $\mu_2 > \mu_{ji}$.

Figure 11 shows the plane of the parameters μ_1 and μ_2 divided by solid lines into six domains corresponding to cases (a)-(f), respectively, with digits 0-3 indicated the degree of instability of the corresponding relative equilibria (5.14). The dashes lines indicate the bifurcation lines $\mu_1 = \mu_{ki}$ and $\mu_2 = \mu_{ji}$ across which the degree of instability of the trivial orientations changed (in cases (c)-(f)).

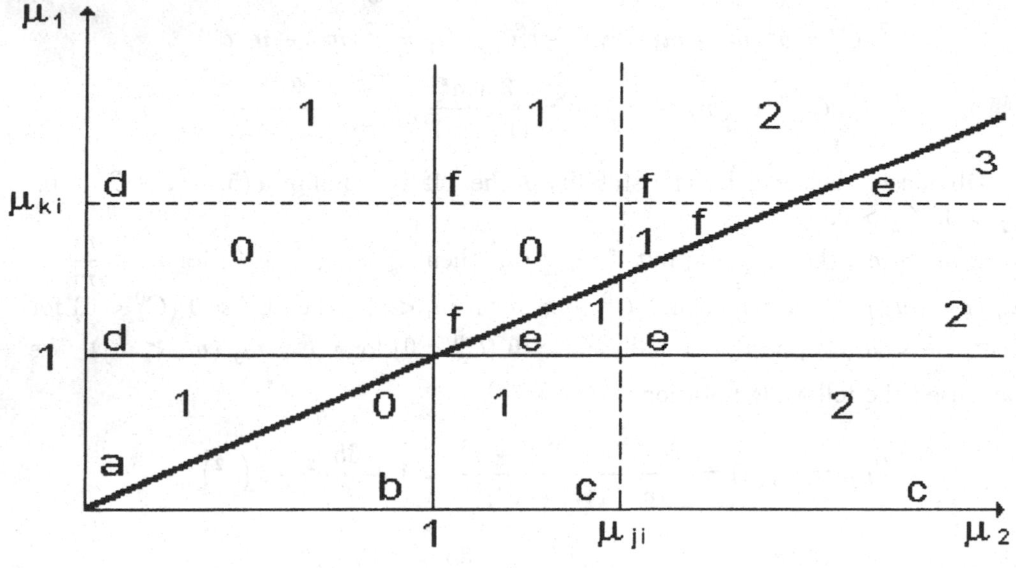

Fig. 11

5.9 Nontrivial relative equilibria of the first level and their stability.

The degree of instability of the trivial relative equilibria (5.14) corresponding to cases (d)-(f) changes when $\mu_1 = \mu_{ki}$. When that happens the coefficient C_2^0 vanishes and the following solutions bifurcate from the aforementioned trivial solutions of system (5.13) $(i > k)$

$$\gamma_i = \cos\varphi_0,\ \gamma_k = \sin\varphi_0,\ \gamma_j = \beta_i = \beta_k = 0,\ \beta_j = \pm 1. \tag{5.15}$$

The degree of instability of the trivial relative equilibria corresponding to cases (c), (e) and (f) changes when $\mu_2 = \mu_{ji}$. When that happens the coefficient C_1^0 vanishes and the following solutions bifurcate from the aforementioned trivial solutions of system (5.13)

$$\gamma_i = \cos\psi_0,\ \gamma_j = -\sin\psi_0,\ \beta_i = \sin\psi_0,\ \beta_j = \cos\psi_0,\ \gamma_k = \beta_k = 0. \tag{5.16}$$

Note that for the solutions (5.15) $\lambda = 0$, and for the solutions (5.16) $\lambda = a^2 \sin\psi \cos\psi\,(m_j - m_i) \neq 0$. The relative equilibria (5.15) are analogous to the steady motions (5.6) for which $\theta = 0$, but the relative equilibria (5.16) are essentially distinct from the corresponding steady motions (5.7), for which $\theta \neq 0$. Moreover, since $\lambda \neq 0$ for the solutions (5.16), a neccessary condition for their existence is the application of forces that keep the mass center of the body in the plane containing the attracting center.

The angles φ_0 and ψ_0 are found from the equations

$$\mu_1 = \operatorname{tg}\varphi_0 \cdot \frac{F_i^3(-a) - F_i^3(a)}{F_k^3(-a) - F_k^3(a)}, \tag{5.17}$$

$$\mu_2 = \operatorname{tg}\psi_0 \cdot \frac{F_i^3(-a) - F_i^3(a) + 2a\cos\psi_0}{F_j^3(-a) - F_j^3(a) + 2a\sin\psi_0} \tag{5.18}$$

(compare (5.17) with the first equation of system (5.8)). The properties of the solutions of equations (5.17) and (5.18) are analogous to those of the solutions (5.6) and (5.7).

Solutions (5.15) and (5.16) exist only when $\mu_1 < \mu_{ki}$ and $\mu_2 < \mu_{ji}$ respectively (the ellipsoid of inertia of the body is almost an ellipsoid of revolution); henceforth, therefore, we shall assume, without loss of generality, that

$$m_k = m_i\left(1 + ha^2\right) + o(a^2),\ m_j = m_i\left(1 + ga^2\right) + o(a^2). \tag{5.19}$$

To investigate the stability of the relative equilibria (5.15), we will calculate the second variation of W^0 over the linear manifold

$$\delta\beta_j = 0,\ \delta\gamma_j = -\operatorname{tg}\varphi_0\delta\gamma_k,\ \delta\gamma_j = -\cos\varphi_0\delta\beta_i - \sin\varphi_0\delta\beta_k.$$

We have

$$2\delta^2 W^0 = C_{11}^0\left(\delta\gamma_k\right)^2 + C_{22}^0\left(\delta\beta_i\right)^2 + 2C_{23}^0\left(\delta\beta_i\right)\left(\delta\beta_k\right) + C_{33}^0\left(\delta\beta_k\right)^2;$$

$$C_{11}^0 = -\frac{3a^2\left[\sin^2\varphi_0 m_i\left[F_i^5(a)+F_i^5(-a)\right]-\cos^2\varphi_0 m_k\left[F_k^5(a)+F_k^5(-a)\right]\right]}{2\cos^2\varphi_0}+$$
$$+\frac{m_i a\left[F_i^3(-a)-F_i^3(a)\right]}{2\cos^3\varphi_0},$$

$$C_{22}^0 = a^2(m_i-m_j)+a\cos^2\varphi_0\delta_{ij},\ C_{23}=a\sin\varphi_0\cos\varphi_0\delta_{ij},$$
$$C_{33}^0 = a^2(m_k-m_j)+a\sin^2\varphi_0\delta_{ij};$$
$$\delta_{ij} = \frac{m_i\left[F_i^3(-a)-F_i^3(a)\right]}{2\cos\varphi_0}-3cm_j.$$

Taking the first relation in (5.19) into account, we conclude that $C_{11}^0 < 0$ for all $i > k$, i.e. all the relative equilibria (5.15) are unstable in the secular sense. A detailed analysis of the other coefficients of the quadratic form $\delta^2 W^0$ shows that the relative equilibrium is

(d) always unstable if $i = 2,\ j = 3,\ k = 1$ (then $\chi = 1$);

(e) when $i = 3,\ j = 1,\ k = 2$, it is unstable ($\chi = 3$) if $g > g_+$, and unstable in the secular sense ($\chi = 2$) if $g < g_+$;

(f) when $i = 3,\ j = 2,\ k = 1$, it is unstable ($\chi = 1$) if $g > g_-$, and unstable in the secular sense ($\chi = 2$) if $g < g_-$.

We have used the notation

$$g_\pm = g_\pm(\varphi_0) = \frac{35}{48}\left[6\cos^4\varphi_0+\cos 2\varphi_0 \pm\left(\frac{9}{4}\sin^4 2\varphi_0+\cos^4 2\varphi_0\right)^{\frac{1}{2}}\right].$$

Similarly, one can investigate the stability of the relative equilibria (5.16) by analyzing the second variation of W^0 over the linear manifold

$$\delta\gamma_j = -\delta\beta_i,\ \delta\gamma_j = \delta\beta_j = -\operatorname{tg}\psi_0\delta\beta_k.$$

It turns out that all the relative equilibria (5.16) ($i > j$) are unstable in the secular sense, and the equilibrium (5.16) is:

(c) unstable in the secular sense when $i = 2,\ j = 1,\ k = 3$ (then $\chi = 2$);

(e) when $i = 3,\ j = 1,\ k = 2$, it is unstable ($\chi = 1$) if $h > h_-$, and unstable in the secular sense ($\chi = 2$) if $h < h_-$;

(f) when $i = 3,\ j = 2,\ k = 1$, it is unstable ($\chi = 1$) if $h < h_+$, and unstable in the secular sense ($\chi = 2$) if $h > h_+$.

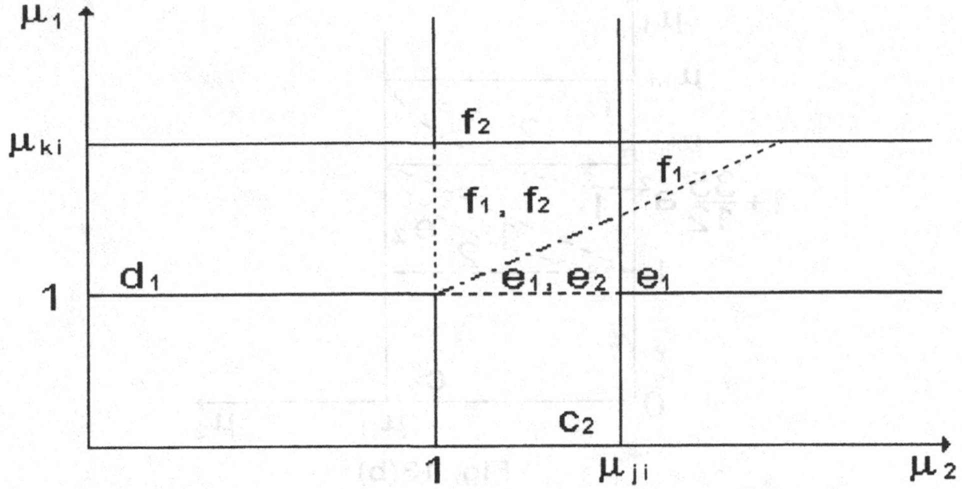

Fig. 12

We have used the notation

$$h_{\pm} = h_{\pm}(\psi_0) = \frac{35}{48}\left[8\cos^4\psi_0 - \cos 2\psi_0 \pm \left(4\sin^4 2\psi_0 + \cos^4 2\psi_0\right)^{\frac{1}{2}}\right].$$

The solid lines in Fig.12 define the domain in the μ_1, μ_2 plane in which the solutions (5.15) and (5.16) exist. The solutions (5.15) exist in the strip $\mu_1 \in (1;\ \mu_{ki})$, $\mu_2 > 0$, while the solutions (5.16) exist in the strip $\mu_2 \in (1;\ \mu_{ji})$, $\mu_1 > 0$. The first strip is divided by the dots into three domains, corresponding to cases (5.15) (d)-(f) (marked d_1-f_1), while the second strip is divided by dashes into the domains corresponding to cases (5.16) (c), (e) and (f) (marked c_2, e_2, f_2).

These strips are shown separately in Fig.13, with the degrees of instability of the respective relative equilibria (5.15) and (5.16) indicated. The dash-dot curves correspond to projections of the bifurcation curves across which the degree of instability of the equilibrium orientations (5.15) and (5.16) changes (in case (e) and (f)).

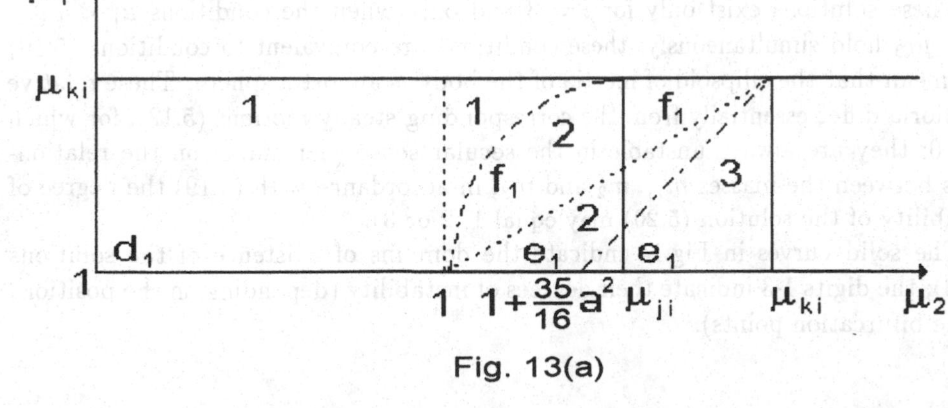

Fig. 13(a)

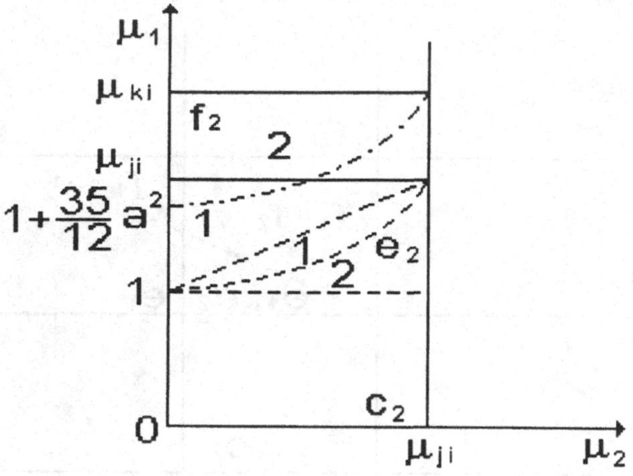

Fig. 13(b)

5.10 Nontrivial relative equilibria of the second level.

In cases (e) and (f) $(i = 3,\ j, k = 1, 2)$ there is a change in the degree of instability of the relative equilibria (5.15) (when $g = g_{\pm}$) and (5.16) (when $h = h_{\pm}$). In these cases solutions of the general form

$$\gamma_i = \cos\varphi_0\cos\psi_0;\quad \gamma_j = \sin\varphi_0\cos\psi_0\sin\theta_0 - \sin\psi_0\cos\theta_0;$$
$$\gamma_k = \sin\varphi_0\cos\psi_0\cos\theta_0 + \sin\psi_0\sin\theta_0;$$
$$\beta_i = \cos\varphi_0\sin\psi_0;\quad \beta_j = \sin\varphi_0\sin\psi_0\sin\theta_0 + \cos\psi_0\cos\theta_0; \tag{5.20}$$
$$\beta_k = \sin\varphi_0\sin\psi_0\cos\theta_0 - \cos\psi_0\sin\theta_0;$$
$$\left(\lambda = (J_i - J_j)\sin\psi_0\left(\cos\psi_0 + \frac{\sin\varphi_0\sin\psi_0\sin\theta_0}{\cos\theta_0}\right) \neq 0\right)$$

bifurcate.

These solutions exist only for $i = 3$ and only when the conditions $\mu_1 < \mu_{k3}$, $\mu_2 < \mu_{k3}$ hold simultaneously; these conditions are equivalent to conditions (5.19) and mean that the ellipsoid of inertia of the body is almost a sphere. These relative equilibria differ essentially from the corresponding steady motions (5.12), for which $\theta \neq 0$; they are always unstable in the secular sense (depending on the relationships between the masses m_1, m_2 and m_3, in accordance with (5.19) the degree of instability of the solution (5.20) may equal 1, 2 or 3).

The solid curves in Fig.14 indicate the domains of existence of the solutions (5.20); the digits 1-3 indicate their degrees of instability (depending on the positions of the bifurcation points).

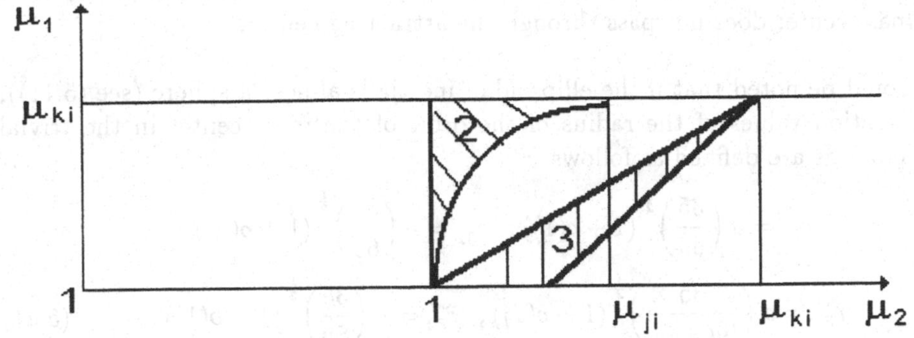

Fig. 14(a)

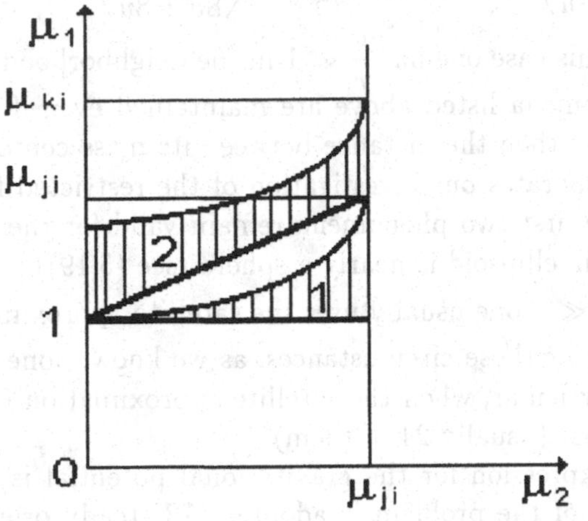

Fig. 14(b)

5.11 Conclusion.

Hence, the investigation of our model problem has revealed the following phenomenon, due to the use of an exact expression for the potential of the gravitation forces:

1. the existence of secular stability of the steady motions and relative equilibria of the body corresponding to trivial orientations, in cases in which not only the major axis of the body's ellipsoid of inertia ($i = 1, j = 3, k = 2$) but also the median ($i = 2, j = 3, k = 1$) and minor ($i = 3, j = 2, k = 1$) axes point along the radius vector of its mass center;

2. the existence of non-trivial steady motions and relative equilibria for which

at least two of the principal central axes of inertia are no axes of the orbital system of coordinates;

3. the existence of steady motions of the body for which the plane of the orbit of the mass center does not pass through the attracting center.

It should be noted that if the ellipsoid of inertia is almost a sphere (see (5.11)), the bifurcation values of the radius of the orbit of the mass center in the trivial steady motions are defined as follows

$$\begin{gathered} r_{21}^{*} = a\left(\frac{35}{6u}\right)^{\frac{1}{2}}(1+o(1)), \quad r_{32}^{*} = a\left(\frac{35}{6v}\right)^{\frac{1}{2}}(1+o(1)), \\ r_{31}^{*} = a\left(\frac{35}{6u+6v}\right)^{\frac{1}{2}}(1+o(1)), \quad \bar{r}_{32}^{*} = a\left(\frac{35}{8v}\right)^{\frac{1}{2}}(1+o(1)), \\ \bar{r}_{21}^{*} = a\left(\frac{35}{8u}\right)^{\frac{1}{2}}(1+o(1)), \quad \bar{r}_{31}^{*} = a\left(\frac{35}{8v+8u}\right)^{\frac{1}{2}}(1+o(1)). \end{gathered} \tag{5.21}$$

Consequently, in this case one has $\frac{a}{r} \ll 1$ in the neighborhood of the bifurcation points, and the phenomena listed above are maintained even if the dimensions of the body are much less then the distance between its mass center and the attracting center. This corroborates our investigation of the restricted formulation of the problem, in which the first two phenomena remain valid for the relative equilibria of a body whose inertia ellipsoid is nearly a sphere (see (5.19)).

Note that when $\frac{a}{r} \ll 1$ one usually uses the satellite approximation of the gravitational potential; under those circumstances, as we know, none of the above phenomena occurs. In particular, when the satellite approximation is used only trivial steady orientations exist (usually 24 of them).

When the exact expression for the gravitational potential is used and the unrestricted formulation of the problem is adopted, 72 steady orientations exist (at any rate, in our model problem): 24 trivial orientations (5.5), 24 "plane" ($\theta = 0$) non-trivial orientations (5.6), and 24 "three-dimentional" ($\theta \neq 0$) non-trivial orientations (5.7). In addition, if the inertia ellipsoid is almost a sphere (see (5.11)) then at least 32 additional steady motions "of general form" (5.12) exist (for arbitrary masses m_1, m_2 and m_3 that are sufficiently close together in value).

Similar conclusions hold for the restricted formulation of the problem. The only difference is that then all the relative equilibria are characterized by the condition $\theta \equiv 0$; as already pointed out, for the relative equilibria (5.14) and (5.15) corresponding to the "two-dimensional" ($\theta = 0$) steady motions (5.5) and (5.6), the reaction of the constrait $\theta = 0$ is zero, while for the relative equilibria (5.16) and (5.20), corresponding to the "three-dimentional" $\theta \neq 0$ steady motions (5.7) and (5.12), the reaction does not vanish (see the expression for the undetermined multiplier λ).

In addition, when $r \gg a$ non-trivial steady motions and relative equilibria exist only for bodies whose ellipsoid of inertia is nearly an ellipsoid of revolution (in particular, a sphere).

We also note that steady motions of the body for which the plane of the orbit of the mass center passes through the attracting center exist (for $m_1 \neq m_2 \neq m_3$) only when one of the principal central axes of inertia is orthogonal to the orbital plane.

Bibliography

1. Routh, E.J.: A treatise on the Stability of a Given State of Motion, MacMillan and Co., London 1877.

2. Routh, E.J.: The Advanced Part of a Treatise on the Dynamics of a System of Rigid Bodies, MacMillan and Co., London 1884.

3. Poincare, H.: Sur l'equilibre d'une masse fluide animee d'un mouvement de rotation, Acta Math., 7 (1885), 259-380.

4. Lyapunov, A.M.: On Constant Screw Motions of a Rigid Body in a Liquid, Izd. Khar'kov Mat. Obshch, Khar'kov 1888 (in Russian).

5. Lyapunov, A.M.: The General Problem of the Stability of Motion, Izd. Khar'kov Mat. Obshch, Khar'kov 1888 (in Russian).

6. Volterra, V.: Sur la theorie des variations des latitudes, Acta Math., 22 (1899), 257-273.

7. Levi-Civita, T.: Sur la recherche des solutions particulieres des systemes differentiels et sur les mouvements stationnaires, Prace Math. Fis., 17 (1906), 1-140

8. Salvadori, L.: Un osservazione su di un criteria di stabilita del Routh, Rend. Accad. Sci. Fis. Math. Napoli (IV), 20 (1953), 267-272.

9. Pozharitskii, G.K.: On the construction of the Lyapunov function from integrals of equations of perturbed motion, Prikl. Mat. Mekh., 22 (1958), 145-154 (in Russian)

10. Rumyantsev, V.V.: On the stability of permanent rotations of mechanical systems, Izv. AN SSSR, OTN, Mekh. Mash., 6 (1962), 113-121 (in Russian).

11. Rumyantsev, V.V.: On the stability of steady motions, Prikl. Mat. Mekh., 30 (1966), 922-923 (in Russian).

12. Salvadori, L.: Sull'estensione ai sistemi dissipativi del criterio di stabilita del Routh, Ric.mat, 15 (1966), 162-167.

13. Salvadori, L.: Un osservazione su di un criteria di stabilita del movimento, Mathematiche, 24 (1969), 218-239.

14. Karapetyan, A.V.: The Routh Theorem and its Extensoins, in: Colloq. Math. Soc. Janos Bolyai, 53: Qualitative Theory of Differential Equations, North Holland, Amsterdam and New York 1990, 271-290.

15. Karapetyan, A.V. and Rumyantsev, V.V.: Stability of Conservative and Dissipative Systems, in: Appl. Mech. Soviet Reviews, 1: Stability and Anal. Mech., Hemisphere, New York 1990, 3-144.

16. Smale, S.: Topology and mechanics, Invent. Math., 10 (1970), 305-311; 11 (1970), 45-64.

17. Abraham, R. and Marsden J.: Foundations of Mechanics, Benjamin, New York and Amsterdam 1978.

18. Karapetyan A.V.: Invariant Sets of Mechanical Systems with Symmetry, in: Vychisl. Mat and Inform., VTs RAN, Moscow 1996, 74-86 (in Russian).

19. Karapetyan A.V.: First integrals, invariant sets and bifurcations in dissipative systems, Regular and Chaotic Dynamics, 2 (1997), 75-80.

20. Tomson, W. and Tait, P.: Treatise on Natural Phylosophy, Cambridge Univ. Press, Cambridge (1879).

21. Chetaev, N.G.: The Stability of Motion, Pergamon Press, London (1961)

22. Krasovskii, N.N.: Problems of the Theory of Stability of Motion, Stanford Univ. Press, Stanford (1963).

23. Pascal, M.: Sur la recherche des mouvements stationnaires dans les systemes ayant des variables cyclic, Celest. Mech., 12 (1975), 337-358.

24. Stepanov, S.Ja.: On the Relationship of the Stability Conditions for Three Different Regimes of Cyclic Motions in a System, in: Probl. Anal. Mekh., Teor. Ust. i Upr., KAI, Kazan' (2) 1976, 303-308 (in Russian).

25. Hagedorn, P.: On the stability of steady motions in free and restricted dynamical systems, Trans. ASME, J. Appl. Mech., 46 (1979), 427-432.

26. Karapetyan, A.V. and Stepanov, S.Ja.: On the relationship of stability conditions of steady motions of a free system and positions of relative equilibrium of a restricted system, Sb. nauchn.-method. statei po teor. mekh., 20 (1990), 31-37 (in Russian)

27. Rubanovskii, V.N.: On the bifurcation and stability of steady motions. Teor. i Pricl.Mekh., 5 (1974), 67-79 (in Russian).

28. Kozlov, V.V.: On the degree of instability, Prikl. Math. Mekh., 57 (1993), 14-19 (in Russian).

29. Karapetyan, A.V. and Stepanov, S.Ja.: Steady motions and relative equilibria of mechanical systems with symmetry, Prikl. Mat. Mekh., 60 (1996), 736-743 (in Russian).

30. Karapetyan, A.V.: Qualitative investigation of the dynamics of a top on a plane with friction, Prikl. Mat. Mekh., 55 (1991), 698-701 (in Russian).

31. Abrarova, Ye.V. and Karapetyan, A.V.: Steady motions of a rigid body in a central gravitational field. Prikl. Mat. Mekh., 58 (1994), 5, 68-73. (in Russian).

32. Abrarova, Ye.V.: Stability of steady motions of a rigid body in a central field. Prikl. Mat. Mekh., 59 (1995), 6, 84-91. (in Russian).

33. Abrarova, Ye.V. and Karapetyan, A.V.: Bifurcation and stability of the steady motion and relative equilibria of a rigid body in a central gravitational field. Prikl. Mat. Mekh., 60 (1996), 3, 375-387. (in Russian).

PART V

ANALYTICAL METHODS IN DYNAMICAL SIMULATION OF FLEXIBLE MULTIBODY SYSTEMS

M. Pascal
Pierre et Marie Curie University, Paris, France

ABSTRACT

This part is devoted to the motion of hinged connected flexible bodies with special applications to satellites and robots. Some open problems arising in the dynamical formulation of these systems are discussed. Then the use of analytical methods for the vibrations analysis of flexible multibody systems is presented. Two special applications of this method in the field of astronautic are done. At last, the problem of control of flexible robots is investigated.

CHAPTER I

DYNAMICAL SIMULATION OF FLEXIBLE MULTIBODY SYSTEMS

1.1 INTRODUCTION

The aim of this work is to reviewe some open questions related to the flexible multibody systems simulation. First, by a detailed description of the kinematical motion in joint between two articulated flexible bodies, we show that the so-called : « floating reference frame » [1] attached to each deformable component can be chosen in a rather natural way.
Assuming then that the flexible bodies perform only small deformations, we show that the inclusion of non-linear terms in the strain energy occuring from the so-called geometric stiffness effects [2] is available only for slender bodies like beams or plates when they are subjected to axial / or in plane forces.
At last, we discuss the opportunity to include higher order terms in the kinetic energy of the system in case of large rates and large accelerations of the flexible vibrations.

1.2 PROBLEM FORMULATION

Let the multibody sytem consists of $n+1$ rigid or flexible bodies $(S_i)(i=0,..,n)$ articulated by n hinges $\ell a(a=1,.,n)$, assuming for the sake of simplicity tree topology. One hinge is supposed to connect only two bodies. The topology of the multibody system can be described by means of graph theory [3]. Every body (S_j) has an in board adjacent body $\left(S_{c(j)}\right)$ in the direct path going from (S_j) to (S_0). The hinge between $\left(S_{c(j)}\right)$ and (S_j) is denoted by (ℓ_j). The motion of the whole system is defined with respect to an inertial reference frame (R_0). Every component (S_j) of the system undergoes large displacements and small deformations with respect to a reference configuration $\left(S_j^0\right)$: in most cases, this reference configuration is a natural configuration without strains and stresses.
The kinematical description of the multibody system can be a global description or a relative description. In the first one, each component is described with respect to the inertial reference frame. In the relative description, the position of one body is defined with respect to his adjacent inboard body $\left(S_{c(j)}\right)$. This relative description leads to a minimal set of generalized coordinates.
In flexible multibody dynamics, the description of the relative motion of two flexible bodies linked by a hinge is not easy. In order to describe the relative motion, between two adjacent flexible bodies (S_j) and $\left(S_{c(j)}\right)$ connected by the hinge $\left(\ell_j\right)$, we assume

that one part (γ_j) of the boundary of (S_j) and one part $(\gamma_{c(j)})$ of the boundary of $(S_{c(j)})$ are undeformable. The relative position of (S_j) with respect to $(S_{c(j)})$ is well defined by the relative position of the rigid boundary (γ_j) with respect to the rigid boundary $(\gamma_{c(j)})$.

1.3 FLOATING REFERENCE FRAME

A second step in the formulation of the dynamics of flexible multibody system is the description of the deformations. Let us consider a deformable body (S_i) and his fixed reference configuration (S_i^0) at $t=0$. With respect to the inertial reference frame (R_0) (Fig. 1), let $\overrightarrow{r_i} = \overrightarrow{O_0M_i}$ and $\overrightarrow{a_i} = \overrightarrow{O_0M_{io}}$ the position of one material point M_i of (S_i) and the corresponding position of this material point in the reference configuration. The assumption of large displacement $\left(\overrightarrow{V_i} = \overrightarrow{r_i} - \overrightarrow{a_i}\right)$ and small strains leads to the standard method in which an intermediate configuration $(\tilde{S}_i)$ of the body is introduced : $(\tilde{S}_i)$ is deduced from the reference configuration (S_i^0) by means of a rigid transformation involving a translation $\overrightarrow{c_i}(t)$ and a rotation associated with the orthogonal tensor $P_i(t)$. For each material point M_i, we define the displacements field $\overrightarrow{u_i} = \overrightarrow{\tilde{M}_iM_i}$ where $\tilde{M}_i$ is rigidly connected to $(\tilde{S}_i)$. The inertial position of M_i is given by :

$$\begin{cases} \overrightarrow{r_i} = \overrightarrow{c_i} + \overrightarrow{x_i} + \overrightarrow{u_i} \\ \overrightarrow{c_i} = \overrightarrow{OO_i} \ , \quad \overrightarrow{x_i} = \overrightarrow{O_i\tilde{M}_i} \ , \quad \overrightarrow{u_i} = \overrightarrow{u_i}\left(\overrightarrow{x_i}, t\right) \end{cases} \tag{1.1}$$

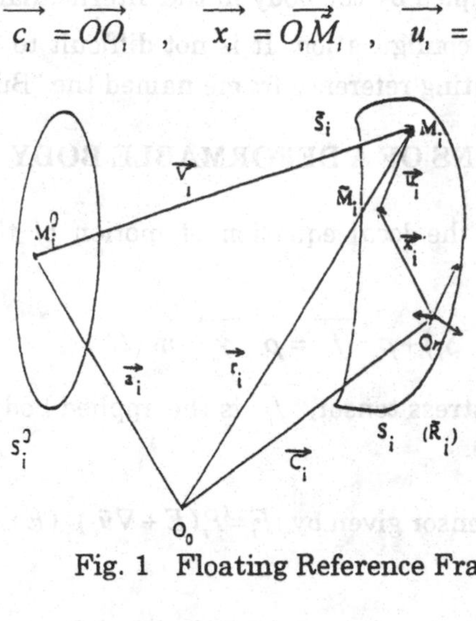

Fig. 1 Floating Reference Frame

O_i is the origin of a floating reference frame (R_i) rigidly connected to $(\tilde{S}_i)$; the intermediate configuration $(\tilde{S}_i)$ is chosen in such a way that the following estimations hold :

$$\begin{cases} \dfrac{\left\|\overrightarrow{u_i}\right\|}{L_i} << 1 & (L_i \text{ characteristic dimension of } (S_i)) \\ \left\|\overrightarrow{\nabla}\ \overrightarrow{u_i}\right\| << 1 & (\overrightarrow{\nabla} \text{ is the gradient operator}) \end{cases} \tag{1.2}$$

For an unrestrained flexible body, no kinematical boundaries conditions are given, so the displacement field is defined only through an arbitrary infinitesimal rigid displacement. Two kinds of methods are used to obtain an unique resolution of the displacements field $\overrightarrow{u_i}$ or equivalently to choose the floating reference frame.

- If on the body (S_i), there is a rigid part, we can choose the floating reference frame (R_i) connected to this rigid part. For example, if we assume that for each flexible component (S_i) linked to his inboard body $(S_{c(i)})$, one part (γ_i) of the boundary is undeformable, we can choose the floating reference frame rigidly connected to (γ_i). It results for the displacements field $\overrightarrow{u_i}$ the following kinematical boundaries conditions : $\overrightarrow{u_i} = \overrightarrow{0}$ on (γ_i)

- Another method assumes conditions for mean displacement of the body. These global conditions are the following

$$\int_{\Omega_i} \rho_i \overrightarrow{u_i}\, dx_i = 0 \ ; \ \int_{\Omega_i} \rho_i \overrightarrow{x_i} \wedge \overrightarrow{u_i}\, dx_i = 0 \tag{1.3}$$

where (Ω_i) is the domain occupied by the body in the intermediate configuration $(\tilde{S}_i)$, ρ_i is the mass density of this configuration. It is not difficult to show that these two conditions define a special floating reference frame named the "Buckens Frame" [4].

1.4 DYNAMICAL EQUATIONS OF A DEFORMABLE BODY

In the next step, we obtain the local equation of motion of the body (S_i) in the following form

$$\overrightarrow{\nabla}.(F_i\ S_i) + \rho_i \overrightarrow{f_i} = \rho_i \overrightarrow{\gamma_i} \ \text{ in } (\Omega_i) \tag{1.4}$$

where S_i is the second Piola stress tensor, $\overrightarrow{f_i}$ is the applied body force vector, $\overrightarrow{\gamma_i}$ is the acceleration of the particle.

At last F_i is the deformation tensor given by : $F_i = {}^tP_i(E + \nabla \vec{u}_i)$ (E unitarian tensor).

If we assume linear stress-strain relations with respect to a natural reference configuration $\left(S_i^0\right)$, the constitutive laws give :

$$\begin{cases} S_i = 2\mu_i \gamma_i + \lambda_i \, tr(\gamma_i)\, E \\ \lambda_i , \mu_i \quad \text{Lamé elastic constant} \end{cases} \tag{1.5}$$

Assuming small strains, we have : $F_i \; S_i \cong {}^t P_i \; S_i$ and $\gamma_i \cong 1/2(\nabla \vec{u}_i + {}^t\nabla \vec{u}_i)$.
In standard methods of simulation of flexible multibody systems, the displacements field $\overrightarrow{u_i}$ is approximated by

$$\overrightarrow{u_i} = \sum_{K=1}^{N_i} \overrightarrow{\varphi_{Ki}}\left(\overrightarrow{x_i}\right) q_{Ki}(t) \tag{1.6}$$

where $\overrightarrow{\varphi_{Ki}}\left(\overrightarrow{x_i}\right)$ are some shape functions (coming for finite element analysis or modal synthesis methods) and $q_{Ki}(t)$ are generalized coordinates. The shape functions used satisfy some boundaries conditions from which we can deduce the floating reference frame chosen.
The motions equations are obtained by a variational method, for example by means of Lagrange Equations :

$$\left\{ \frac{d}{dt}\left(\frac{\partial T_i}{\partial \dot{p}_i}\right) - \frac{\partial T_i}{\partial p_i} = Q_i \right.$$

$T_i = \frac{1}{2}\, {}^t\dot{p}_i M_i \dot{p}_i$ is the kinetics energy of the body, p_i is the column matrix of the generalized coordinates : $p_i = \begin{bmatrix} \alpha_i \\ q_i \end{bmatrix}$ and Q_i is the column matrix of the generalized applied forces. The coordinates p_i involve n_i rigid coordinates α_i and n_i elastic coordinates q_i. If a consistent representation of the displacements field is used to compute the mass matrix M_i, linear and quadratic terms with respect to the elastic coordinates occur in this matrix. The corresponding Lagrange equations are :

$$\begin{cases} M_i \ddot{p}_i + K_i p_i + \tilde{H}_i(p_i, \dot{p}_i) = \tilde{F}_i \\ M_i = \begin{bmatrix} M_{RR}^i & M_{RE}^i \\ {}^t M_{RE}^i & M_{RE}^i \end{bmatrix} \quad \tilde{F}_i = \begin{bmatrix} F_R^i \\ F_E^i \end{bmatrix} \\ K_i = \begin{bmatrix} O & O \\ O & K_E^i \end{bmatrix} \quad \tilde{H}_i = \begin{bmatrix} H_R^i \\ H_E^i \end{bmatrix} \end{cases}$$

where the subscripts R and E denote rigid and elastic respectively. K_E^i is a constant stiffness matrix, $\tilde{F}_i$ are applied forces and $\tilde{H}_i$ is a vector of non linear inertial (Coriolis and Centripetal) forces : $\tilde{H}_i = \dot{M}_i\, p_i - \dfrac{\partial T_i}{\partial p_i}$.

About the configuration dependant mass matrix M_i, it can be shown the following properties :

$$\begin{cases} M_{RR}^i = M_0^i + M_1^i + M_2^i \\ M_{RE}^i = \tilde{M}_0^i + \tilde{M}_1^i \end{cases}$$

Here the submatrices $M_0^i, \tilde{M}_0^i$ and M_{EE}^i are independant of the elastic variables q_i, M_1^i and $\tilde{M}_1^i$ are linear functions of these parameters and M_2^i is a quadratic function of these parameters. It results that for the elastic coordinates, the corresponding equations :

$$M_{RE}^i\, \ddot{\alpha}_i + M_{EE}^i\, \ddot{q}_i + K_E^i\, q_i + H_E^i = F_E^i \tag{1.7}$$

are linear with respect to these variables while for the rigid coordinates, the corresponding equations

$$M_{RR}^i\, \ddot{\alpha}_i + M_{RE}^i\, \ddot{q}_i + H_R^i = F_R^i \tag{1.8}$$

contain quadratic terms with respect to the elastic variables and their first and second derivatives with respect to time. Several studies [5], [6] deal with the possibility of dropping these terms. In limiting cases where the rigid motion is very slow with respect to the elastic vibrations, it is possible to introduce two characteristic time scales : the characteristic period T_1 of the free elastic vibrations and the characteristic time T_2 of the rigid motion.

By assuming the following estimates $K_E^i = \varepsilon^{-2} K_0^i$, $T_1 = \vartheta(\varepsilon^{-1})$, $T_2 = \vartheta(1)$, where K_0^i is a matrix with bounded elements and ε is a small dimensionless parameter, it is possible to obtain an approximate solution for the elastic vibrations in terms of two times scales :

$$q_i = \varepsilon^2 q_i^0(t) + \varepsilon^2 y_i(\tau) \;\; ; \;\; \tau = t\varepsilon^{-1} \quad \text{is the fast time} \tag{1.9}$$

$y_i(\tau)$ is the free elastic oscillations defined by the homogenous system associated with the equation (1.7) : $\varepsilon^2 M_{EE}^i\, \ddot{q}_i + K_0^i q_i = 0$. The related characteristic equation is of the form $\Delta(\wedge) \equiv \det\left(\wedge^2 M_{EE}^i + K_0^i\right) = 0$; $\wedge = \varepsilon\lambda$ and because the matrices M_{EE}^i and K_0^i are positive definite, the solution for the eigenvalues are $\wedge_j = \pm i \wedge_j^{'} \left(j = 1,..,n_i^{'}\right)$ $\left(\wedge_j^{'}\ \text{real number}\right)$.

We deduce for $y_i(\tau)$ the following solution : $y_i = \sum_j a_{ij} \cos\left(\wedge_j^{'}\, \tau\right) + b_{ij} \sin\left(\wedge_j^{'}\, \tau\right)$.

The other part of the solution is a quasi-static solution defined by $q_i^0(t) = \left(K_0^i\right)^{-1}\left(M_0^i - F_{EO}^i\right)$, where M_0^i and F_{EO}^i are computed from the rigid model, namely for $\alpha_i = \alpha_i^0(t)$ solution of equation (1.8) in which q_i is set to zero. In the obtained solution (1.9) for the elastic vibrations, the part depending on the fast time τ has a second derivative with respect to t of order zero with respect to ε. It results that the estimate $\left\| \frac{\overrightarrow{u_i}}{L_i \omega_i^2} \right\| << 1$ is no longer fullfilled if the characteristic velocity ω_i is chosen as the characteristic velocity of the rigid motion.

1.5 CENTRIFUGAL STIFFENING

When the reference configuration (S_i^0) is a prestressed state, instead of formula (1.5), the constitutive law for the body (S_i) has the following form : $S_i = \sigma_{Oi} + 2\mu_i \chi_i + \lambda_i \, tr(\chi_i) \, E$, where σ_{Oi} is the Cauchy stress tensor of the reference configuration. Assuming small strains with respect to the reference configuration, the following approximations can be made :

$$F_i \, S_i \cong {}^t P_i \sigma_{Oi} + {}^t P_i \Psi_i \left(\overrightarrow{u_i} \right) + {}^t P_i \nabla \overrightarrow{u_i} \sigma_{Oi} \tag{1.10}$$

where $\Psi_i(\overrightarrow{u_i}) = 2\mu_i \chi_i + \lambda_i \, tr \chi_i \, E$.

In formula (1.10), the first term comes from the initial stresses and is assumed to be finite, the second term is the usual term linear with respect to the strains and depending on the elastic coefficients and the third term linear with respect to the strains but depending on the initial stresses is the so-called centrifugal stiffening term.

An equivalent form of the local equations of motion can be obtained by means of variational methods, for example by Hamilton's principle. These methods need first the computation of the kinetic energy of the body and secondly the computation of the potential energy. In the case already investigated of a prestressed state, the elastic potential energy is given by the formula :

$$\begin{cases} \upsilon^i = \upsilon_E^i + \upsilon_{GE}^i \\ \upsilon_E^i = \dfrac{1}{2} \displaystyle\int_{\Omega_i} \sum_{\alpha,\beta=1}^{3} \psi^i_{\alpha\beta} \frac{\partial u_{i\alpha}}{\partial x_{i\beta}} \, d\Omega_i \\ \upsilon_{GE}^i = \dfrac{1}{2} \displaystyle\int_{\Omega_i} \sum_{\alpha,\beta,K=1}^{3} \sigma^i_{O\alpha\beta} \frac{\partial u_{iK}}{\partial x_{i\alpha}} \frac{\partial u_{iK}}{\partial x_{i\beta}} \, d\Omega_i \end{cases} \tag{1.11}$$

υ_E^i is the usual potential energy of the body, $\psi^i_{\alpha\beta}$ are the components of the stress tensor Ψ_i, $u_{i\alpha}(\alpha = 1,2,3)$ and $x_\beta(\beta = 1,2,3)$ being the coordinates in the floating reference frame of $\overrightarrow{u_i}$ and $\overrightarrow{x_i}$. υ^j_{GE} is the so-called geometric stiffness in which the components $\sigma^j_{O\alpha\beta}(\alpha,\beta=1,2,3)$ of the initial stress tensor σ_{Oi} occur.

Many papers have been published during the last twenties years about the problem of incoporating geometric stiffening in the dynamics of flexible multibody systems [5], [6]. Among the papers dealing with this geometric (or centrifugal) stiffening effects, most are concerned by bodies modeled as flexible beams. Very few are concerned with three dimensional elastic bodies. On the other hand, most of the papers discuss how it is possible to take into account these effects but very few authors discuss in what cases these centrifugal stiffening effects are significant.

If the assumed model is a three-dimensional elastic body (S_i), the centrifugal stiffening effects must be incorporated in the dynamical model only when the reference configuration (S_i^O) is a prestressed state. The first step of the modelisation is the determination of the initial Cauchy stress tensor σ_{Oi}. For example in the motion of a deformable vehicle like a train is modeled as small motions around a nominal motion. The initial stresses are produced by this nominal motion and it is assumed that the applied forces during this nominal motion are not time dependant. Another attempt to introduce the centrifugal stiffening effects was made in [7] : the authors assume that in the intermediate configuration $(\tilde{S}_i)$ of the body (the so-called floating reference configuration) there exist an initial stress produced by the inertial and external forces associated with the rigid motion of the body (motion of the floating reference frame). But the methodology used by the authors are not correct because theses forces are all included in the motion equations of the body and they produce the internal stresses and the corresponding strains in it. The only correct formulation will be first to define the initial stresses produced by the inertial and external forces exerted on the reference configuration ; then the determination of the motion of the body with respect to this reference configuration will be made but we had to take into account only the increments of the centrifugal and external forces between the actual configuration and the reference configuration.

If on the other hand, the assumed model of the body (S_i) is a flexible beam, the main phenomenom is that the axial rigidity of the beam is much more greater than its bending rigidity. It results that in many cases, the axial displacement is much more smaller than the transverse displacements. All the methods to take into account the centrifugal stiffening are approximate methods in which the axial displacement is neglected or assumed to be of second order with respect to the transverse displacements. The centrifugal stiffening effects are added to take into account the initial stresses produced by the axial forces [4].

1.6 CONCLUSIONS

In this paper, some open questions arising in dynamical simulation of flexible multibody systems are discussed, namely the choice of the so-called floating reference frame associated with each flexible component or equivalently the choice of boundaries conditions imposed to the displacements field of one flexible unrestrained

body in order to have a unique resolution of it. We discuss also the opportunity to include the centrifugal or geometric stiffness effects in the dynamical equations. The conclusion is that in almost all cases, these effects have to be taken into account only for slender bodies like beams (or plates) subjected to axial (or in plane) forces.
The last question discussed is the occurence of non linear terms with respect to the elastic parameters and their first and second derivatives with respect to time, in the global motion of the flexible body. When an approximate expression of the displacements field is used, the dynamical equations are obtained from variational methods leading to the computation of the kinetic energy of the system. The non linear terms with respect to elastic parameters in this case come from the expression of the mass matrix in which linear and quadratic terms with respect to the elastic parameters occur. It seems that only a non dimensional analysis, with some estimate about the order of magnitude of the different characteristic frequencies occuring in the system (free vibrations frequency of the components, frequency of the rigid body rotations) can give some satisfactory justifications.

Chapter II

VIBRATIONS ANALYSIS OF FLEXIBLE MULTIBODY SYSTEMS

2.1. - INTRODUCTION

The aim of this work is the computation of natural frequencies and the associated vibrations modes of a flexible multibody system. In structural dynamics, the finite element method is very often used but for flexible structures made up from standard elements such as beams, strings and membranes, the distributed element method has been successful, leading in several cases to less computer time. This paper will discuss new research perspectives for the use of analytic representation of structural dynamics.

2.2. - PROBLEM FORMULATION

Let the multibody systems consist of $(n + 1)$ bodies (S_i) $(i = 0, 1, ..., n)$ interconnected by n hinges ℓ_a $(a = 1, ..., n)$. The only external forces and torques are exerted on the first body which is assumed to be rigid. These external actions are represented by a force $\vec{F}_0(t)$ in the centre of mass G_0 of (S_0) and a torque $\vec{M}_0(t)$. The multibody system undergoes small vibrations around an equilibrium position in which the flexible parts are undeformed.

2.2.1. - Kinematics of motion of two contiguous bodies relative to one another

Let us assumed that each body (S_i) has at most two hinges attached to him (Fig. 2) and each hinge (ℓ_a) is supposed to connect only two bodies named $S_{i+(a)}$ and $S_{i-(a)}$. We assume that on one part $\gamma_{i+(a)}$ of the boundary of $S_{i+(a)}$, the elastic displacement is a rigid displacement ; the same assumption is made for one part $\gamma_{i-(a)}$ of the boundary of $S_{i-(a)}$. The relative motion of the flexible body $S_{i-(a)}$ with respect to $S_{i+(a)}$ is described by the relative motion of the rigid interface $\gamma_{i-(a)}$ with respect to the rigid interface $\gamma_{i+(a)}$: the relative position of $S_{i+(a)}$ with respect to $S_{i-(a)}$ depends on N_a degrees of freedom, with $1 \leq N_a \leq 6$ according to the kind of articulation between $S_{i+(a)}$ and $S_{i-(a)}$.

2.2.2. - Equations of motion for a flexible appendage

Let us first consider the motion of one flexible body (S_i), of mass m_i and centre of mass G_i, articulated to two other flexible bodies by means of rigid links (γ_i) and (γ_i') (Fig. 3). Let $(\mathcal{T}_i)$ and $(\mathcal{T}_i')$ with origins O_i and O_i' two triads rigidly connected to the rigid interfaces (γ_i) and (γ_i'). The flexible body (S_i) undergoes small motions around an equilibrium configuration in which this body is undeformed. Let $(\mathcal{T}_i^0)$ be the position of the triad $(\mathcal{T}_i)$ in the equilibrium configuration. With repsect to $(\mathcal{T}_i^0)$, the position of $(\mathcal{T}_i)$ is given by a translation vector $\vec{r}_i$ and a rotation vector $\overrightarrow{\alpha_i}$, $\overrightarrow{\delta_i}$ is the rotation vector of $(\mathcal{T}_i')$ with respect to $(\mathcal{T}_i)$. Let us denotes by $\overrightarrow{u_i}(\bar{P},t) = \overrightarrow{\bar{P}P}$ the elastic displacement of every material point P of (S_i), $\bar{P}$ being the undeformed position of P, with $\vec{x} = \overrightarrow{O_i\bar{P}}$. It is assumed that $\overrightarrow{r_i}$, $\overrightarrow{\alpha_i}$, $\overrightarrow{\delta_i}$ are small vectors, of the same order of smallness as the displacement field $\overrightarrow{u_i}$.

The linearized equations of motion of the flexible body (S_i) are given by [8] :

$$\left.\begin{aligned} L_i[u_i] + \rho_i(\ddot{r}_i - \tilde{x}\ddot{\alpha}_i + \ddot{u}_i) &= 0 \quad \text{in } (D_i) \\ u_i &= 0 \quad \text{on } (\gamma_i) \\ u_i &= u_{0_i'} - \left(\tilde{x} - \tilde{x}_{0_i'}\right)\delta_i \quad \text{on } (\gamma_i') \\ \sigma_i(u_i)\nu_i &= 0 \quad \text{on } (\Gamma_{F_i}) = \partial D_i - (\gamma_i) \cup (\gamma_i') \end{aligned}\right\} \qquad (2.1)$$

Here (D_i) the domain occupied by the flexible body in its undeformed configuration, with boundary ∂D_i, $\overrightarrow{\nu_i}$ is the unitarian vector of the outward normal to ∂D_i, ρ_i is the mass density of (S_i), L_i is a linear self-adjoint differential operator with respect to the components of vector $\vec{x}$ in the $(\mathcal{T}_i)$ frame. $\overrightarrow{u_{0_i'}}$ and $\overrightarrow{x_{0_i'}}$ are the elastic displacement of $0_i'$ and the vector locating the undeformed position of this point in the $(\mathcal{T}_i)$ frame. σ_i is the stress tensor.

For every vector $\vec{X}$, X denotes the matrix column of its components (X_1, X_2, X_3) in the $(\mathcal{T}_i)$ frame and $\tilde{X}_i$ the skew symmetric matrix

$$\tilde{X} = \begin{pmatrix} 0 & -X_1 & X_2 \\ X_3 & 0 & -X_2 \\ -X_2 & X_1 & 0 \end{pmatrix}$$

The components in the $(\mathcal{T}_i)$ frame of the resultant force $\overrightarrow{F_i}$ and resultant torque $\overrightarrow{M_i}$ exerted on (γ_i) are :

$$F_i = \int_{\gamma_i} \sigma_i(u_i)\nu_i \, dS \qquad M_i = \int_{\gamma_i} \tilde{x}_i \, \sigma_i(u_i)\nu_i \, dS \qquad (2.2)$$

Similar expressions are obtained for the force and the torque exerted on (γ_i') :

$$F_i' = \int_{\gamma_i'} \sigma_i(u_i)\nu_i \, dS \qquad M_i' = \int_{\gamma_i'} \left(\tilde{x} - \tilde{x}_{0_i'}\right) \sigma_i(u_i)\nu_i \, dS \quad (2.2)'$$

The global motion of (S_i) is deduced from (2.1), (2.2) and (2.2)' :

$$\left.\begin{aligned} m_i\left(\ddot{r}_i - \tilde{l}_i\ddot{\alpha}_i\right) + \int_{D_i} \rho_i\ddot{u}_i dx &= F_i + F_i' \\ m_i\tilde{l}_i\ddot{r}_i + I_i\ddot{\alpha}_i + \int_{D_i} \rho_i\tilde{x}\ddot{u}_i dx &= M_i + M_i' + \tilde{x}_{0_i'}F_i' \end{aligned}\right\} \quad (2.3)$$

Here $\overrightarrow{\ell_i}$ locates the undeformed position of G_i in (T_i) and I_i is the inertia tensor in 0_i of the body (S_i) in its undeformed configuration. By use of Fourier's transformation, we obtain from equations (2.1), (2.2), (2.2)', the following equations :

$$\left.\begin{aligned} L_i\left[\bar{v}_i\right] - \rho_i\omega^2\bar{v}_i &= 0 && \text{in } (D_i) \\ \bar{v}_i &= \bar{r}_i - \tilde{x}\,\bar{\alpha}_i && \text{on } (\gamma_i) \\ \bar{v}_i &= \bar{r}_i' - \left(\tilde{x} - x_{0_i'}\right)\bar{\alpha}_i' && \text{on } (\gamma_i') \\ \sigma_i\left(\bar{v}_i\right)\nu_i &= 0 && \text{on } (\Gamma_{F_i}) \end{aligned}\right\} \quad (2.4)$$

$$\bar{F}_i = \int_{\gamma_i} \sigma_i\left(\bar{v}_i\right)\nu_i dS \qquad \bar{M}_i = \int_{\gamma_i} \tilde{x}\sigma_i\left(\bar{v}_i\right)\nu_i dS \quad (2.5)$$

$$\bar{F}_i' = \int_{\gamma_i'} \sigma_i\left(\bar{v}_i\right)\nu_i \, dS \qquad \bar{M}_i' = \int_{\gamma_i} \left(\tilde{x} - \tilde{x}_{0_i'}\right)\sigma_i\left(\bar{v}_i\right)\nu_i \, dS \quad (2.5)'$$

Here $\bar{X}(\omega)$ is the Fourier's transform of $\overrightarrow{X_{(t)}}$,

$$\left.\begin{aligned} \bar{v}_i &= \bar{u}_i + \bar{r}_i - \tilde{x}\,\bar{\alpha}_i \\ \bar{r}_i' &= \bar{u}_{0_i'} + \bar{r}_i - \tilde{x}_{0_i'}\bar{\alpha}_i \\ \bar{\alpha}_i' &= \bar{\alpha}_i + \bar{\delta}_i \end{aligned}\right\}$$

2.2.3. - Equations of motion for the main rigid body

The linearized equations of motion of (S_0) are

$$\left.\begin{aligned} m_0\ddot{R} &= F_0 + F_0' \\ I_0\ddot{\theta} &= M_0 + \tilde{x}_{0_i}F_0' + M_0' \end{aligned}\right\} \quad (2.6)$$

Here m_0 and I_0 are the mass and the inertia tensor in G_0 of (S_0), R and θ are the column matrices of the $(\mathcal{T}_0)$ components of the vectors $\vec{R}$ and $\vec{\theta}$ giving the inertial position of a triad $(\mathcal{T}_0)$ rigidly connected to (S_0) with respect to an inertial reference frame. This inertial frame is chosen to have R and θ small.
(F_0, M_0) and (F_0', M_0') are the $(\mathcal{T}_0)$ components of $\left(\overrightarrow{F_0}, \overrightarrow{M_0}\right)$, $\left(\overrightarrow{F_0'}, \overrightarrow{M_0'}\right)$ where to two last vectors are the resultant force and the resultant torque in O_1 introduced by the link ℓ_1.

In frequency domain, we obtain :

$$\left.\begin{array}{rcl} -m_0\omega^2\bar{R} & = & \bar{F}_0 + \bar{F}_0' \\ -I_0\omega^2\bar{\theta} & = & \bar{M}_0 + \tilde{x}_{0_1}\bar{F}_0' + \bar{M}_0' \end{array}\right\} \qquad (2.6)'$$

2.3. - IMPEDANCE AND INERTANCE MATRICES

The problem formulated by equations (2.4), (2.5) and (2.5)' being a linear problem, we can define a linear transformation depending on the frequency ω giving the resultant forces and torques $\bar{Q}_i = {}^t\left(\bar{F}_i, \bar{M}_i, \bar{F}_i', \bar{M}_i'\right)$ exerted on the boundaries (γ_i) and (γ_i') in terms of the displacements $\bar{q}_i = {}^t\left(\bar{r}_i, \bar{\alpha}_i, \bar{r}_i', \bar{\alpha}_i'\right)$ of these boundaries :

$$\bar{Q} = Z_i(\omega)\bar{q}_i.$$

$Z_i(\omega)$ is a symmetrical matrix of dimensions 12 named the impedance matrix of (S_i) [9]. The inverse transformation is given by $\bar{q}_i = \mathcal{H}_i(\omega)\bar{Q}_i$.
Here $\mathcal{H}_i(\omega) = [Z_i(\omega)]^{-1}$ is the inertance matrix of (S_i).
For the main rigid body (S_0), the impedance matrix $Z_0(\omega)$ is of dimensions 6 :

$$\left.\begin{array}{rcl} \bar{Q}_0 & = & Z_0(\omega)\bar{q}_0 \\ \bar{Q}_0 & = & {}^t\left(\bar{F}_0 + \bar{F}_0', \bar{M}_0 + \bar{M}_0' + \tilde{x}_{0_1}\bar{F}_0'\right) \\ \bar{q}_0 & = & {}^t\left(\bar{R}, \bar{\theta}\right) \\ Z_0(\omega) & = & -\omega^2\begin{pmatrix} m_0 E & 0 \\ 0 & I_0 \end{pmatrix} \end{array}\right\}$$

E is the identity matrix of order 3.

2.4. - COMPONENT MODES

The component modes of a flexible body (S_i) are the modes of vibrations of the body when it vibrates independently with respect to the other parts of the whole system. Two sets of component modes can be defined :

2.4.1. - Constrained modes

Constrained modes are vibrations modes of (S_i) with the rigid links (γ_i) and (γ_i') fixed ($\bar{r}_i = \bar{\alpha}_i = \bar{r}_i' = \bar{\alpha}_i' = 0$). They are defined by :

$$\left.\begin{array}{rcll} v_i & = & V_{n_i}(x)\ cos\,(\Omega_{n_i} t) & n = 1, 2, ... \\ L_i\,[V_{n_i}] - \rho_i \Omega^2_{n_i} V_{n_i} & = & 0 & \text{in } (D_i) \\ V_{n_i} & = & 0 & \text{on } (\gamma_i) \\ V_{n_i} & = & 0 & \text{on } (\gamma_i') \\ \sigma_i\,(V_{n_i})\,\nu_i & = & 0 & \text{on } (\Gamma_{F_i}) \end{array}\right\}$$

We obtain an infinite set of constrained modes, with orthogonality property :

$$\int_{D_i} \rho_i\ {}^tV_{p_i} V_{n_i}\ dx\ =\ 0\ \text{ if }\ \Omega^2_{n_i}\ -\ \Omega^2_{p_i}\ \neq 0$$

The constrained modes can be normalized by setting.

$$\int_{D_i} \rho_i\ {}^tV_{n_i} V_{n_i}\ dx\ =\ 1 \qquad n = 1, 2, ...$$

2.4.2. - Global modes

Global modes are vibrations modes of (S_i) with the rigid links (γ_i) and (γ_i') free ($\bar{F}_i = \bar{M}_i = \bar{F}_i' = \bar{M}_i' = 0$). They are defined by :

$$\left.\begin{array}{rcll} v_i & = & v_{n_i}(x)\ cos\,(\omega_{n_i} t) & n = 1, 2, ... \\ L_i\,[v_{n_i}] - \rho\omega^2_{n_i} v_{n_i} & = & 0 & \text{in } (D_i) \\ \sigma_i\,(v_{n_i})\,\nu_i & = & 0 & \text{on } (\Gamma_{F_i}) \\ v_{n_i} & = & r_{n_i} - \tilde{x}\ \alpha_{n_i} & \text{on } (\gamma_i) \\ v_{n_i} & = & r'_{n_i} - \left(\tilde{x} - \tilde{x}_{0_i'}\right) \alpha'_{n_i} & \text{on } (\gamma_i') \end{array}\right\}$$

$$\left.\begin{array}{ll} \int_{\gamma_i} \sigma\,(v_{n_i})\,\nu_i\ dS = 0 & \int_{\gamma_i'} \sigma\,(v_{n_i})\,\nu_i\ dS = 0 \\ \int_{\gamma_i} \tilde{\sigma}\,(v_{n_i})\,\nu_i\ dS = 0 & \int_{\gamma_i'} \tilde{x}\sigma\,(v_{n_i})\,\nu_i\ dS = 0 \end{array}\right\}$$

Again, we obtain an infinite set of orthogonal vibrations modes.

2.5. - DERIVATION OF THE IMPEDANCE/INERTANCE MATRICES IN TERMS OF COMPONENT MODES

In some particular cases like beams in axial, torsional and bending vibrations, extensible strings or membranes under uniform tension [10], it is possible to obtain an analytical representation of the matrices $Z_i(\omega)$ and $\mathcal{H}_i(\omega)$. In more general cases, for which a closed form solution for these matrices is not available, we obtain a spectral expension of $Z_i(\omega)$ and $\mathcal{H}_i(\omega)$ in terms of components modes.

2.5.1. - Derivation of the impedance matrix in terms of constrained modes

A spectral expansion of the impedance matrix $Z_i(\omega)$ of each flexible appendage (S_i) is obtained in the form :

$$\left.\begin{aligned} Z_i(\omega) &= Z_i(0) - \omega^2 \mathcal{M}_{0_i} - \omega^4 \sum_1^{\infty} \frac{K_{n_i}}{\Omega_{n_i}^2 - \omega^2} \\ \mathcal{M}_{0i} &= \int_{D_i} \rho_i \, {}^t[\varphi_i] \, [\varphi_i] \, dx \\ K_{n_i} &= \Gamma_{n_i} \, {}^t\Gamma_{n_i} \\ \Gamma_{n_i} &= \int_{D_i} \rho_i \, {}^t[\varphi_i] \, V_{n_i} \, dx \end{aligned}\right\} \qquad (2.7)$$

Here $[\varphi_i]$ is the matrix of static deformation modes of (S_i) : $[\varphi_i]$ is a rectangular matrix of dimensions (3×12) :

$$[\varphi_i] = [\{\varphi_{i1}\}, \{\varphi_{i2}\}, \{\varphi_{i3}\}, \{\varphi_{i4}\}]$$

$\{\varphi_{ip}\}$ $(p = 1, 2, 3, 4)$ are defined by :

$$\left.\begin{aligned} L_i[\{\varphi_{ip}\}] &= 0 && \text{in } \; D_i) \\ \sigma_i[\{\varphi_{ip}\}]\,\nu_i &= 0 && \text{on } \; (\Gamma_{Fi}) \\ \{\varphi_{ip}\} &= E\,\delta_{p1} - \tilde{x}\,\delta_{p2} && \text{on } \; (\gamma_i) \\ \{\varphi_{ip}\} &= E\,\delta_{p3} - \left(\tilde{x} - \tilde{x}_{O_i'}\right)\delta_{p4} && \text{on } \; (\gamma_i') \end{aligned}\right\}$$

δ_{pK} $(p, K = 1, ..., 4)$ are the Kronecker's symbols.

A convergence property of the matrices K_{ni} (modal gains matrix of the nth constrained mode) holds :

$$\sum_1^{\infty} K_{n_i} = \mathcal{M}_{0_i}$$

This convergence property provides an upper bound of the residual modal gains when only a finite number of terms is used in the spectral expansion (2.7).

2.5.2. - Derivation of the inertance matrix in terms of global modes

A spectral expansion of the inertance matrix $\mathcal{H}_i(\omega)$ of each flexible appendage (S_i) is obtained in the form.

$$\left.\begin{array}{rcll} \mathcal{H}_i(\omega) & = & \mathcal{H}_{i0} - \dfrac{{}^tN_i\,\mathcal{M}_i^{-1}N_i}{\omega^2} + \displaystyle\sum_{n=1}^{\infty} \dfrac{k_{n_i}\omega_{n_i}^2\omega^2}{\omega_{n_i}^2 - \omega^2} & \\ N_i & = & \begin{pmatrix} E & 0 & E & 0 \\ 0 & E & \tilde{x}_{0_i'} & E \end{pmatrix} & \text{rectangular matrix } (6 \times 12) \\ \mathcal{M}_i & = & \begin{pmatrix} m_iE & -m_i\tilde{l}_i \\ m_i\tilde{l}_i & I_i \end{pmatrix} & \begin{array}{c}\text{mass matrix in} \\ 0_i \text{ of body } (S_i) \text{ undeformed}\end{array} \\ k_{n_i} & = & \gamma_{ni}\,{}^t\gamma_{ni} & \gamma_{ni} = \displaystyle\int_{D_i} \rho_i\,{}^t[P_i]\;v_{ni}\,dx \end{array}\right\} \quad (2.8)$$

Here $[P_i] = [\{P_{1i}\}, \{P_{2i}\}, \{P_{3i}\}, \{P_{4i}\}]$ is the matrix of attachment modes defined as static deformation modes of (S_i) subjected to unitarian forces or torques on the boundaries (γ_i) or (γ_i') and corresponding force elements equilibrating these actions.

A convergence property of the matrix k_{ni} (modal gains matrix of the nth global mode) holds :

$$\left.\begin{array}{rcl} \sum_1^{\infty} k_{n_i} & = & \mathcal{M}_{0i}^* \\ \mathcal{M}_{0i}^* & = & \displaystyle\int_{D_i} \rho_i\,{}^t[P_i][P_i]dx \end{array}\right\}$$

2.6. - REDUCED IMPEDANCE MATRIX OF THE WHOLE SYSTEM

For each body (S_i) $(i = 0, 1, ..., n)$ we obtain an impedance matrix $Z_i(\omega)$ giving the forces and torques exerted on the boundaries in terms of the displacements of these boundaries :

$$\overline{Q_i} = Z_i(\omega)\,\overline{q_i} \qquad i = (0, 1, ..., n)$$

It results for the whole system the following relation :

$$\left.\begin{array}{ll} \bar{Q} = Z(\omega)\bar{q} & \\ \bar{Q} = {}^t\left(\overline{Q_0}, \overline{Q_1}, ..., \overline{Q_n}\right) & Z_{(\omega)} = \begin{pmatrix} Z_0 & & 0 \\ & \ddots & \\ 0 & & Z_n \end{pmatrix} \\ \bar{q} = {}^t\left(\overline{q_0}, \overline{q_1}, ... \overline{q_n}\right) & \end{array}\right\}$$

$Z(\omega)$ is a symmetrical matrix of order $12n + 6$. The set of displacements q is not a set of independant displacements. Each link ℓ_i introduces kinetical linear constraints between the components of q.

We can write the general form of these constraints as : $q = T\,\hat{q}$. Here $\hat{q}$ is the column matrix of a set of independant displacements of the whole system and T is a constant rectangular matrix of maximum rank. It results for the whole system the relation

$$\left.\begin{aligned} \bar{\hat{Q}} &= \hat{Z}(\omega)\,\bar{\hat{q}} \\ \hat{Z}(\omega) &= {}^tT\,Z\,T \qquad\qquad \bar{\hat{Q}} = {}^tT\bar{Q} \end{aligned}\right\} \tag{2.9}$$

$\hat{Z}(\omega)$ is the reduced impedance matrix of the whole system ; $\bar{\hat{Q}}$ has the following form : $\bar{\hat{Q}} = \mathcal{F} + K(\omega)\bar{\hat{q}}$.

Here $\mathcal{F}$ is the column matrix of the external actions for the whole system (exerted only on the main rigid body (S_0)) and $K(\omega)$ is a stiffness matrix associated with the servomotors located in the links.

The final form of the equation (2.9) is

$$\left.\begin{aligned} H(\omega)\bar{\hat{q}} &= \bar{\mathcal{F}} \\ H(\omega) &= Z(\omega) - K(\omega) \end{aligned}\right\}$$

A modal analysis of the coupled system can be made from (2.9) ; the equation $det\ H(\omega) = 0$ gives the frequencies of vibrations of the whole system.

2.7. - CONCLUSIONS

An attempt to extend the theory of impedance matrix first introduced in [9] for particular structures to more general flexible multibody systems has been made. The more significant result is that it is possible to obtain an expansion of this impedance matrix by means of an infinite set of modes ; the theory gives also an upper bound of the residual modal gains when only a finite number of modes is used.

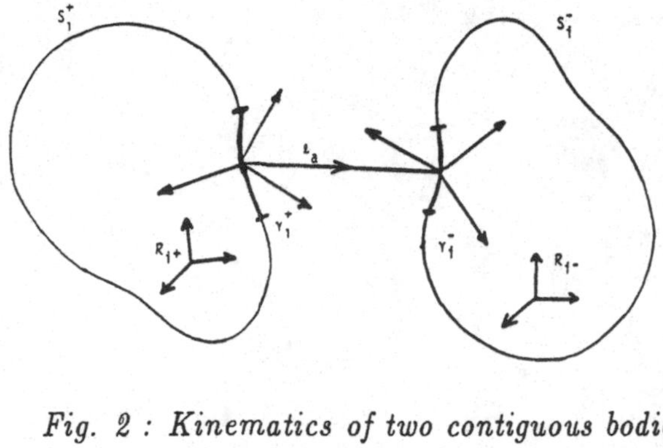

Fig. 2 : Kinematics of two contiguous bodies

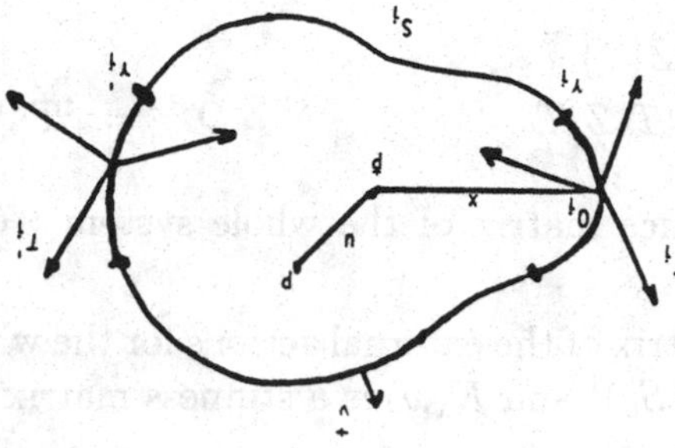

Fig. 3 : Dynamics of one flexible body

CHAPTER III

DYNAMICAL SIMULATION OF A FLEXIBLE MANIPULATOR ARM

3.1 INTRODUCTION

The subject of this work is the dynamics of a space manipulator designed to be set on a servicing satellite (S) and to grasp another satellite (P). A continuum approach (modal impedance) [9] is used to describe distributed flexibility in the manipulator. We investigate a simplified model of this manipulator in order to perform analysis behavior for control purpose.

3.2 SYSTEM MATHEMATICAL DESCRIPTION

The manipulator arm is composed of three flexible beams (B_i) (i=1,2,3) connected by pinned joints (Fig. 4). The satellite (S) and the payload (P) are assumed to be rigid bodies and the system to have planar motion. Every limb (B_i) is modelled as a flexible beam O_i, O'_i, where O_i and O'_i are joints (i=1,2,3). Let us denote by $O_o \equiv O_1$ the joint between body (S) and beam (B_1), $O'_1 \equiv O_2$ the joint between (B_1) and (B_2), $O'_2 \equiv O_3$ the joint between (B_2) and (B_3) and at last $O'_3 \equiv O_p$ the joint between (B_3) and the payload (P). A concentrated mass μ_i is introduced in joint O_{i+1} (i=1,2) in order to represent the servomotor and the joint itself. In each joint O_i (i=1,2,3) and O_p, a localized stiffness k_{i-1} (i=1,2,3) and k_p is introduced. The only external forces are assumed to act only on satellite (S) and payload (P). The system is assumed to undergo small vibrations around an equilibrium position in which the flexible limbs (B_i) are undeformed. Let us denote by $R_o(T;X_1X_2X_3)$ an inertial reference frame with X_3 perpendicular to the plane of motion, and $\mathcal{R}_o$: (O_o ,$x_oy_oz_o$) a reference frame rigidly connected to (S). With respect to the inertial frame, the satellite (S) is defined by two translational parameters $\vec{R}_o$ (R_{o1},R_{o2},0) locating the position of joint O_o and by the rotation matrix

$$V_o = \begin{pmatrix} \cos\theta_o & -\sin\theta_o & 0 \\ \sin\theta_o & \cos\theta_o & 0 \\ 0 & 0 & 1 \end{pmatrix}$$

giving the orientation of triad $\mathfrak{C}_0$.

We assume that in the equilibrium position, the frame $\mathfrak{C}_0$ is coincident with the inertial frame. It results that R_{o1}, R_{o2} and θ_o are small quantities.

To every flexible limb (B_i) (i=1,2,3), we associate a triad $\mathfrak{C}_i$: (O_i, $x_i y_i z_i$) rigidly connected to the undeformed configuration of B_i, with $O_i\,\overrightarrow{x_i}$ axis tangent in point O_i to the deformed configuration of B_i (Fig. 5). $\mathfrak{C}_i^o$ (O_i^o, $x_i^o y_i^o z_i^o$) is the equilibrium position of triad $\mathfrak{C}_i$, V_i the rotation matrix giving the inertial orientation of $\mathfrak{C}_i$, V_i^o the rotation matrix giving the inertial orientation of $\mathfrak{C}_i^o$.

$$V_i = \begin{pmatrix} \cos\theta_i & -\sin\theta_i & 0 \\ \sin\theta_i & \cos\theta_i & 0 \\ 0 & 0 & 1 \end{pmatrix} \qquad V_i^o = \begin{pmatrix} \cos\theta_i^o & -\sin\theta_i^o & 0 \\ \sin\theta_i^o & \cos\theta_i^o & 0 \\ 0 & 0 & 1 \end{pmatrix}$$

With respect to frame $\mathfrak{C}_i^o$, the position of frame $\mathfrak{C}_i$ is given by $\overrightarrow{r_i} = \overrightarrow{O_i^o O_i}$ and by the rotation matrix

$$A_i = \begin{pmatrix} \cos\alpha_i & -\sin\alpha_i & 0 \\ \sin\alpha_i & \cos\alpha_i & 0 \\ 0 & 0 & 1 \end{pmatrix} \qquad \alpha_i = \theta_i - \theta_i^o$$

It results that $\overrightarrow{r_i}$ and α_i are small quantities.

Let us denote by $\overrightarrow{u_i}(s,t) = u_i(s,t)\,\overrightarrow{y_i}$ ($0 \leqslant s \leqslant \ell_i$) the elastic displacement of beam B_i (ℓ_i is the lenght of B_i) and (r_{i1}, r_{i2}, 0) the $\mathfrak{C}_i$ components of vector $\overrightarrow{r_i}$.

At last, the inertial position of the rigid payload (P) is given by $\overrightarrow{R_p} = \overrightarrow{TO_p}$ and by the rotation matrix V_p between R_o and triad $\mathfrak{r}_p$: $(O_p ; x_p y_p z_p)$ rigidly connected to S_p.

$$V_p = \begin{pmatrix} \cos\theta_p & -\sin\theta_p & 0 \\ \sin\theta_p & \cos\theta_p & 0 \\ 0 & 0 & 1 \end{pmatrix}$$

$\mathfrak{r}_p^o (O_p^o ; x_p^o y_p^o z_p^o)$ being the equilibrium position of triad $\mathfrak{r}_p$, V_p^o the rotation matrix between R_o and this triad is given by :

$$V_p^o = \begin{pmatrix} \cos\theta_p^o & -\sin\theta_p^o & 0 \\ \sin\theta_p^o & \cos\theta_p^o & 0 \\ 0 & 0 & 1 \end{pmatrix}$$

It results that $\alpha_p = \theta_p - \theta_p^o$ and the $\mathfrak{r}_p$ components (r_{p1}, r_{p2}) of vector $\overrightarrow{r_p} = \overrightarrow{O_p^o O_p}$ are small quantities.

3.3 IMPEDANCE MATRIX OF EVERY INDIVIDUAL BODY IN THE SYSTEM

For every flexible beam (B_i), it is possible to obtain in frequency domain the exact impedance matrix $Z_i(\omega)$ giving the relationship between forces/torques and displacements in the boundaries ([8], [11]) :

$$Z_i(\omega)\, q_i = Q_i \quad (i=1,2,3) \tag{3.1}$$

$$q_i = \begin{pmatrix} r_{i1} \\ r_{i2} \\ \alpha_i \\ r'_{i2} \\ \alpha'_i \end{pmatrix} \qquad Q_i = \begin{pmatrix} F_{i1} + F'_{i1} \\ F_{i2} \\ M_i \\ F'_{i2} \\ M'_i \end{pmatrix}$$

Here $Z_i(\omega)$ is a 5 by 5 symmetric matrix depending on the frequency ω. $\overrightarrow{F_i}(F_{i1}, F_{i2}, 0)$, $\overrightarrow{M_i} = M_i \overrightarrow{z_i}$, $\overrightarrow{F'_i}(F'_{i1}, F'_{i2}, 0)$, $\overrightarrow{M'_i} = M'_i \overrightarrow{z_i}$ are forces and torques exerted by the joints in O_i and O'_i (all components are given in triad $\mathfrak{C}_i$) ; at last r'_{i2} and α'_i are displacement and rotation coming from flexibility of the end O'_i of beam (B_i).

Similarly, we obtain for the satellite (S) and the payload (P) the following relationsphips :

$$Z_o(\omega)q_o = Q_o \quad q_o = \begin{pmatrix} R_{o1} \\ R_{o2} \\ \theta_o \end{pmatrix} \quad Q_o = \begin{pmatrix} F_{o1} + F'_{o1} \\ F_{o2} + F'_{o2} \\ M_o + M'_o \end{pmatrix} \tag{3.2}$$

$$Z_p(\omega)q_p = Q_p \quad q_p = \begin{pmatrix} r_{p1} \\ r_{p2} \\ \alpha_p \end{pmatrix} \quad Q_p = \begin{pmatrix} F_{p1} + F'_{p1} \\ F_{p2} + F'_{p2} \\ M_p + M'_p \end{pmatrix} \tag{3.3}$$

Here, $\overrightarrow{F_o}(F_{o1}, F_{o2}, 0)$, $\overrightarrow{M_o} = M_o\overrightarrow{z_o}$, are external forces and torques in point O_o exerted on (S), $\overrightarrow{F'_o}(F'_{o1}, F'_{o2}, 0)$, $\overrightarrow{M'_o} = M'_o\overrightarrow{z_o}$ are forces and torques exterted on S by joint O_o (all components are given in $\mathfrak{C}_o$ frame).
Similarly, $\overrightarrow{F'_p}(F'_{p1}, F'_{p2}, 0)$, $\overrightarrow{M'_p} = M'_p\overrightarrow{z_p}$ are external forces and torques in point O_p exerted on payload (P), $\overrightarrow{F_p}(F_{1p}, F_{p2}, 0)$, $\overrightarrow{M_p} = M_p\overrightarrow{z_p}$ are forces and torques exerted on P by joints O_p (all components in $\mathfrak{C}_p$ frame).

We obtain a redundant matrix for the whole system in the form

$$\left\{ Z(\omega)q = Q \quad q = \begin{pmatrix} q_o \\ q_1 \\ q_2 \\ q_3 \\ q_p \end{pmatrix} \quad Q = \begin{pmatrix} Q_o \\ Q_1 \\ Q_2 \\ Q_3 \\ Q_p \end{pmatrix} \right.$$

$$Z = \begin{pmatrix} Z_o & & & & \\ & Z_1 & & 0 & \\ & & Z_2 & & \\ & 0 & & Z_3 & \\ & & & & Z_p \end{pmatrix}$$ is a symmetrical matrix of order 21.

3.4 REDUCE IMPEDANCE MATRIX OF THE WHOLE SYSTEM

The set of displacements q are not independent : these displacements are related by the kinematical constraints introduced by the links :

$$\begin{cases} R_o = V_1^o\, r_1 \;\; ; \;\; r'_1 = {}^tV_1^o\, V_2^o\, r_2 \\ r'_2 = {}^tV_2^o\, V_3^o\, r_3 \;\; ; \;\; r'_3 = {}^tV_3^o\, V_p^o\, r_p \end{cases} \tag{3.4}$$

Here $R_o = {}^t(R_{o1}, R_{o2}, 0)$, $r_i = {}^t(r_{i1}, r_{i2}, 0)$ (i=1,2,3), $r_p = {}^t(r_{p1}, r_{p2}, 0)$.
$r'_i = {}^t(r_{i1}, r'_{i2}, 0)$

We can write the general form of the kinematical constraints as

$$\begin{cases} q = T\,\hat{q} \\ \hat{q} = {}^t(R_o, \theta_o, \alpha_1, r'_{12}, \alpha'_1, \alpha_2, \alpha'_2, \alpha_3, \alpha'_3, r_p, \alpha_p) \end{cases} \tag{3.5}$$

$\hat{q}$ is the column matrix of a set of independent displacements and T is a constant rectangular matrix of dimensions (21 × 13).

On the other hand, in each joint $(0_i, 0_p)$ the torques exerted on the arms are given by

$$\begin{cases} M'_o = -M_1 = k_o(\alpha_1 - \theta_o) \\ M'_i = -M_{i+1} = k_i(\alpha_{i+1} - \alpha'_i) \quad (i=1,2) \\ M'_3 = -M_p = k_p(\alpha_p - \alpha'_3) \end{cases} \tag{3.6}$$

In joints O_1 and O_p, the action-reaction forces are connected by the following relations :

$$\begin{cases} F'_o + V_1^o\, F_1 = 0 \\ F'_3 + {}^tV_3^o\, V_p^o\, F_p = 0 \end{cases}$$

In joints O_2 and O_3, the motion of concentrated masses μ_1 and μ_2 in frequency domain gives :

$$\begin{cases} \mu_i\, \omega^2\, r_i = F'_i + {}^tV_i^o\, V_{i+1}^o\, F_{i+1} \\ (i = 1,2) \end{cases}$$

It results that the whole system satisfies in frequency domain the equation

$$\hat{Z}(\omega)\hat{q} = \hat{Q} \tag{3.7}$$

Here $\hat{Z} = {}^tTZT$ is the 13 by 13 non redundant impedance matrix of the whole system. $\hat{Q} = {}^tTQ$ has the following form : $\hat{Q} = \hat{\mathfrak{F}} + K\hat{q}$.
$\hat{\mathfrak{F}} = {}^t(F_{o1}, F_{o2}, M_o, 0 \ldots 0, F'_{p1}, F'_{p2}, M'_p)$ is the 13 by 1 column matrix of external forces exerted on satellite and payload and $K(\omega)$ is the 13 by 13 stiffness matrix associated with motion of concentrated masses in joints and servomotors.

Relation (3.7) gives

$$\hat{A}(\omega)\hat{q} = \hat{\mathfrak{F}} \; ; \; \hat{A} = \hat{Z} - K \tag{3.8}$$

The general form of equation (3.8) is

$$\begin{bmatrix} H_{bb} & H_{ba} \\ H_{ab} & H_{aa} \end{bmatrix} \begin{bmatrix} q_b \\ q_a \end{bmatrix} = \begin{bmatrix} \mathfrak{F}_b \\ 0 \end{bmatrix} \tag{3.9}$$

Here $q_b = {}^t(R_{o1}, R_{o2}, \theta_o, r_{p1}, r_{p2}, \alpha_p)$ is the 6 by 1 column matrix of the boundary variables of the system and $q_a = {}^t(\alpha_1, r'_{12}, \alpha'_1, \alpha_2, \alpha'_2, \alpha_3, \alpha'_3)$ is the 7 by 1 column matrix of the inner variables.

By a Guyan-type reduction, the equation (3.9) is transformed in

$$H_R \, q_b = \mathcal{F}_b \qquad (3.10)$$

where $H_R = H_{bb} - H_{ba} H_{aa}^{-1} H_{ab}$ is the reduced impedance matrix of order 6 of the whole system. The global modes of the system are given by

$$\det H_R = 0 \quad (\text{if } \det H_{aa} \neq 0)$$

The boundary variables q_b are obtained by $q_b = H_R^{-1} \mathcal{F}_b$ and then, the inner variables are given by $q_a = - H_{aa}^{-1} H_{ab} q_b$.

We can compute the torques in joints by means of formulas (3.6).

3.5 CONCLUSIONS

The dynamical simulation of a flexible manipulator arm in planar motion has been performed by means of exact continuum approach. The global frequencies of the system are given by the roots of the determinant of the impedance matrix of the system. In [12] some numerical results are given using datas coming from the late Hera project for the Hermès/MTFF composite dynamics.

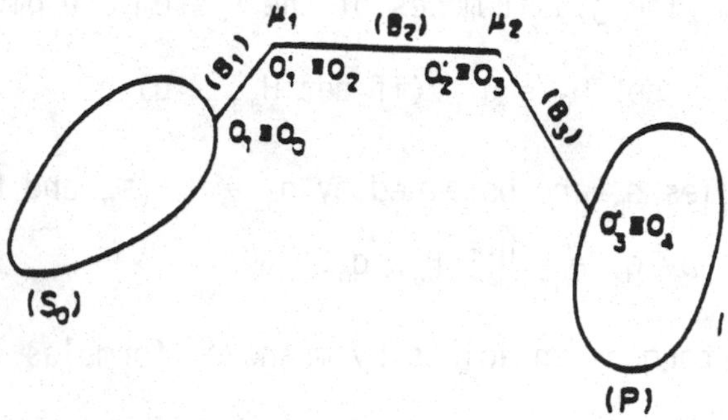

Fig. 4 Flexible manipulator arm

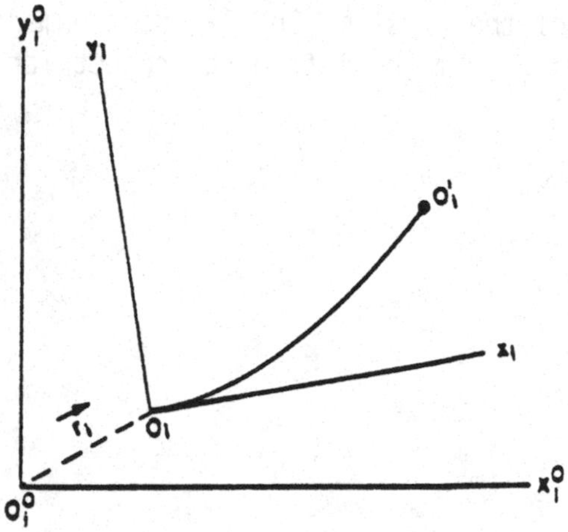

Fig. 5 Local reference frame for flexible limb B_1.

CHAPTER IV

MODAL ANALYSIS AND DYNAMICAL RESPONSE OF A FLEXIBLE SPACE STATION

4.1 INTRODUCTION

In modal analysis, a very interesting concept, first introduced by Hughes [9] is the concept of impedance matrix. For an hybrid system modelled as a main rigid body (B_0) rigidly connected to several flexible appendages and undergoing small motions around an equilibrium position, the impedance matrix gives in frequency domain the external forces and torques exerted on the system in terms of the displacements of the main body. This impedance matrix is obtained by a spectral expansion in terms of an infinite set of "cantilever" modes.

In this paper, this method is used to obtain the modal analysis of a complex spacecraft assuming that the orbital motion of the satellite can be neglected. The more general case where the spacecraft is gravity gradient stabilized and revolves around the Earth in a circular orbit with an constant angular velocity has been investigated in [13].

4.2 PROBLEM FORMULATION

Let us consider a system (Σ) modelled as a main rigid body (B_0) with p flexible appendages (B_K) rigidly connected to (B_0) along a rigid part (γ_K) of the boundary of (B_K). The system undergoes small motions around an equilibrium position with the flexible appendages undeformed. With respect to an inertial reference frame ($\mathfrak{R}$) $\equiv \left(T;\vec{X}_1,\vec{X},_2\vec{X}_3\right)$ the position of the main body (B_0) is defined by the three components (R_1,R_2,R_3) of the vector $\vec{R}=\overrightarrow{TO}$ (O mass center of (B_0)) and by the rotation matrix V_0 giving the inertial orientation of a triad ($\mathfrak{R}_0$)$\equiv\left(O;\vec{x}_0,\vec{y}_0,\vec{z}_0\right)$ rigidly connected to (B_0) . Assuming that in the equilibrium position, ($\mathfrak{R}_0$) is coincident with the inertial frame, it results that (R_1,R_2,R_3) are small quantities and the rotation matrix V_0 can be represented (to first order) by : $V_0 = E - \tilde{\theta}$.

E is the 3 by 3 unit matrix and $\tilde{\theta}$ is the skew symmetric matrix associated with the vector $\vec{\theta}(\theta_1, \theta_2, \theta_3)$. $\theta_i (i=1,2,3)$ ($(\mathcal{R}_0)$ components of $\vec{\theta}$) are small quantities.

For every flexible body (B_K), a local reference frame $(\mathcal{R}_K) \equiv (O_K ; \vec{x}_k, \vec{y}_K, \vec{z}_K)$ rigidly connected to (γ_K) is chosen. The inertial location of an arbitrary material point P of (B_K) is given by :

$$\begin{cases} z_K = R + {}^tV_K [d_K + x + u_K] \\ R = {}^t(R_1, R_2, R_3) \end{cases} \qquad k = 1, \ldots, p$$

where z_K is the column matrix of the inertial components of vector $\overrightarrow{TP}$, d_K is the column matrix of the $(\mathcal{R}_K)$ components of vector $\overrightarrow{OO_K}$, x and $u_K(x,t)$ are the column matrices of the $(\mathcal{R}_K)$ components of vector $\overrightarrow{O_K \bar{P}}$ ($\bar{P}$ undeformed position of P) and of the elastic displacement field in point P. At last, V_K is the rotation matrix giving the inertial orientation of $(\mathcal{R}_K)$: $V_K = V_0 A_K = (E - \tilde{\theta})\, A_K$.

Here A_K is the constant rotation matrix giving the orientation of $(\mathcal{R}_K)$ with respect to $(\mathcal{R}_0)$.

Assuming that no distributed forces act on the flexible appendages (B_K), we can write the linearized equations of motion of the system in the following form :

Global motion

$$M_\Sigma \ddot{\Theta} + \sum_{K=1}^{p} \int_{B_K} \rho_K S_K \ddot{U}_K \, dx = \mathcal{F}(t) \qquad (4.1)$$

Local motion

$$L_K[u_k] + \rho_K[\ddot{u}_K + {}^tS_K \, \ddot{\Theta}] = 0 \quad \text{in } (D_K) \qquad \Theta = {}^t[R, \theta] \qquad (4.2)$$

Boundary conditions

$$\begin{cases} u_K = 0 & on\ (\gamma_K) \\ \sigma_K(u_K)\nu_K = 0 & on\ \{(\partial D_K)/(\gamma_K)\} \end{cases} \tag{4.3}$$

Here M_Σ is the 6 by 6 mass matrix in point 0 of the whole system with the flexible appendages undeformed ; $S_K = {}^t[A_K - (\tilde{a}_K + \tilde{x})A_K]$ is the 6 by 3 matrix coupling the rigid body motion Θ and the displacement field u_K. $\mathcal{F} = {}^t[F,M]$ where F and M, are the column matrices of the inertial components of the resultant external force $\vec{F}$ and resultant external torque $\vec{M}$ in point 0 exerted on the main body (B_0) ; L_K is a linear differential operator ; (D_K) is the volume occupied by (B_K) undeformed ; σ_K is the stress tensor and ν_K is the unitarian vector of the outward normal on the boundary (∂D_K).

4.3 REPRESENTATION OF THE SYSTEM MOTION IN TERMS OF THE CANTILEVER MODES

The "cantilever" modes are the vibrations modes of each flexible appendage (B_K) when the main body (B_0) is fixed $(\Theta = 0)$.
We obtain an infinite set of cantilever modes defined by

$$u_K = U_{n_K} \cos(\Omega_{n_K} t) \qquad n_K = 1,2,\ldots$$

$$\begin{cases} L_K\ [U_{n_K}] - \rho_K\, \Omega^2_{n_K}\, U_{n_K} = 0 & in\ (D_K) \\ U_{n_K} = 0 & on\ (\gamma_K) \\ \sigma_K(U_{n_K})\nu_K = 0 & on\ \{(\partial D_K)/(\gamma_K)\} \end{cases}$$

with the orthogonality property

$$\int_{D_K} \rho_K\, {}^tU_{n_K}\, U_{m_K}\, dx = 0 \qquad if \quad \Omega^2_{n_K} - \Omega^2_{m_K} \neq 0$$

For each flexible body (B_K), the displacement field u_K is obtained by an expansion in terms of the cantilever modes : $u_K = \sum_{n_K=1}^{\infty} U_{n_K} q_{n_K}(t)$.

The generalized coordinates $q_{n_K}(t)$ are obtained by a set of decoupled equations :

$$\begin{cases} \ddot{q}_{n_K} + \Omega^2_{n_K} q_{n_K} = - {}^t\Gamma_{n_K} \ddot{\Theta} & n_K = 1, 2, \ldots \\ \Gamma_{n_K} = \int_{B_K} S_K U_{n_K} \rho_K \, dx & \textit{(participating factor of the } n_K\textit{th cantilever mode)} \end{cases}$$

Assuming harmonic external actions in the form $\mathcal{F}(t) = \bar{\mathcal{F}} e^{i\omega t}$, the response Θ of the main body (B_0) is given by

$$\begin{cases} \Theta = \bar{\Theta} e^{i\omega t} \\ - \omega^2 Z(\omega^2) \bar{\Theta} = \bar{\mathcal{F}} \end{cases} \tag{4.4}$$

$Z(\omega^2)$ is the reduced impedance matrix of the whole system. $Z(\omega^2)$ is obtained by a spectral expansion in terms of the cantilever modes :

$$\begin{cases} Z(\omega^2) = Z(0) + \omega^2 \sum_{K=1}^{p} \sum_{n_K=1}^{\infty} \dfrac{G_{n_K}}{\Omega^2_{n_K} - \omega^2} \\ G_{n_K} = \Gamma_{n_K} {}^t\Gamma_{n_K} \qquad Z(0) = M_\Sigma \end{cases} \tag{4.5}$$

G_{n_K} is the gain matrix of the n_K Th cantilever mode. The following convergence property holds :

$$\sum_{K=1}^{p} \sum_{n_K=1}^{\infty} G_{n_K} = M_A \tag{4.6}$$

where M_A is the 6 by 6 mass matrix in point 0 of the set of all flexible appendages undeformed.

The global frequencies ω_n and the corresponding global modes Θ_n of the system are obtained from the formula (4.4) when there are no external actions ($\bar{\mathcal{F}} = 0$) :

$$\begin{cases} \det Z(\omega_n^2) = 0 & n=1,\ldots,\infty \\ Z(\omega_n^2)\Theta_n = 0 & \end{cases}$$

4.4 APPLICATION TO A SPACE STATION

Let us consider a space station (Σ) modelled as a main rigid body (B_0) connected to two identical solar pannels B_1 and B_2 and one antenna (B_3) (Fig. 6). Each solar pannel (B_i) consists of a rectangular membrane (BL_i) held in tension by a central flexible mast (SP_i) and two rigid tip pieces (TP_i) and (γ_i) (i=1,2). The solar pannels can rotate around the central mast. The antenna (B_3) is modelled as a flexible mast rigidly connected to the main body at one end 0_3 and loaded at the other end $0_3'$ by a mass μ. For each solar pannel (B_i), the local reference frame ($0_i:\vec{x}_i,\vec{y}_i,\vec{z}_i$) chosen is rigidly connected to the tip piece (γ_i) (i=1,2), where $\vec{y}_i$ directed along the undeformed position of (SP_i) is parallel to $\vec{y}_0$ axis. We assume that the orbital motion of the space station can be neglected, and that the rotation of the solar pannels being very slow, we can assume that the pannel (B_i) is rigidly connected to the main body by means of the tip piece (γ_i).

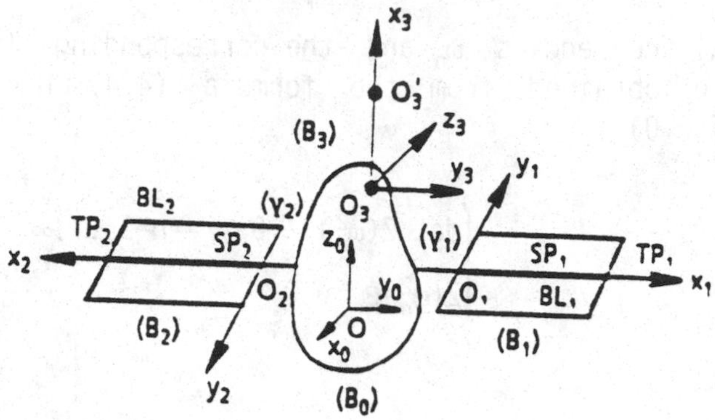

Fig. 6 Flexible Space Station

4.4.1 "Cantilever" modes

For each flexible appendage (B_1, B_2, B_3), the cantilever modes are obtained by an exact continuum method.

Solar pannels :
The length ℓ of (B_1) is assumed to be much greater than the width e. It is then possible to assume that the central mast (SP_1) and the blanket (BL_1) are inextensible : it results that the axial elastic displacement of the mast (SP_1) and the axial deformation of (BL_1) are of second order with respect to the other deformations.
Three set of cantilever modes are obtained for each pannel according to the motion of the tip piece (TP_1).

Bending out of plane :
The motion of the tip piece (TP_1) is a (small) translation parallel to $\vec{z_1}$ axis.

Lateral bending :
The motion of (TP_1) is a (small) translation parallel to $\vec{x_1}$ axis.
Twisting modes :
The motion of (TP_1) is a (small) rotation δ around the $\vec{y_1}$ axis.

Antenna :
Assuming that the undeformed antenna is directed along the $\vec{z_3}$ axis and that the antenna mast is symmetrical, we obtain two identical bending

modes $p_{nz}(s)$ $(0 \leqslant s \leqslant \tilde{\ell})$ in the $\overrightarrow{x_3}$ and $\overrightarrow{y_3}$ directions.

The corresponding frequencies, mode shapes, participating factors and modal gains are computed in analytical form.

4.4.2 Modal analysis of the space station

In concrete cases, only a few number of cantilever frequencies (namely low frequency modes) are computed. The impedance matrix Z of the whole system is obtained by the spectral expansion (4.5) where only a finite number of terms is retained. The convergence property (4.6) provides an upper bound of the residual modal gains.
The global frequencies ω_n are obtained as the roots of the determinant of the impedance matrix $Z(\omega)$.

4.5 NUMERICAL RESULTS [14] :

Several numerical simulations are performed according to different mass distributions of the system.
Some particular cases are first investigated : if the antenna is omitted and if the system is symmetric with respect to the plane $(0;\overrightarrow{x_0},\overrightarrow{y_0})$ the out of plane bending modes of the pannels are coupled only with roll motions (rotation of the main body with respect to $\overrightarrow{x_0}$ axis) and the twisting modes are coupled only with pitch motion (rotation of the main body with respect to $\overrightarrow{y_0}$ axis).
In more general cases, no decoupling in the motion of the system occurs. Several numerical investigations are performed according to the value of the angle φ giving the orientation of the solar pannels with respect to the main body (B_0).

CHAPTER V

DYNAMICAL CONTROL LAW FOR A FLEXIBLE ROBOT

5.1 INTRODUCTION

Weight minimization and large accelerations in many robotics applications lead to an increasing tendancy for elastic vibrations. Light structures in robotics allow a better ratio payload/weight of the robots and energetic reduction for the motors in the joints. In space applications, the very long and slender geometric dimensions induce important vibrations. Manipulators arms require a reasonable accuracy in the end-point positioning of the arm. This accuracy is deteriorated by structural deformations. For improving the performance of these flexible robots, the elastic deflections must be taken into account in the control law, in order to damp their motion. Classical control laws defined for rigid robots can be extended for flexible robots but they will not have the same ideal performance as the number of actuators is less than the number of variables to control.
Several different methods have been proposed in the last few years in order to solve this problem : Inverse dynamics and computed torque applied to the position control multilinks elastic robots [15], L Q R method based on a linearized behavior model and a state feed back [16], singular perturbation methods where the elastic and the rigid motions are supposed to be time separable [17], non linear decoupling methods based on the dynamical model of the robots [18], [19]. In this work, the last method is used because the non linearity in the behavior of the robots seems to be a predominant effect.

5.2 MODELISATION OF THE ROBOT [20]

Let us consider a multibody open chain robot composed of NB bodies $S_1,..S_{NB}$ articulated by $(NB-1)$ joints $L_1,..L_{NB-1}$.

5.2.1 Dynamical model of one flexible component

For each flexible components (S_i) of the robot, the motion is resolved into a rigid motion of a reference configuration (Ω_r) and a small elastic displacement $\vec{\varsigma}$ (Fig. 7).

The inertial position $\vec{u}$ of every material point M is given by :

$$\vec{u} = \overrightarrow{O_{iner}O_0} + \overrightarrow{O_0 O} + \overrightarrow{OM'} + \overrightarrow{M'M} = \vec{\alpha} + \vec{u}_{ref} + \vec{x} + \vec{\varsigma} \qquad (5.1)$$

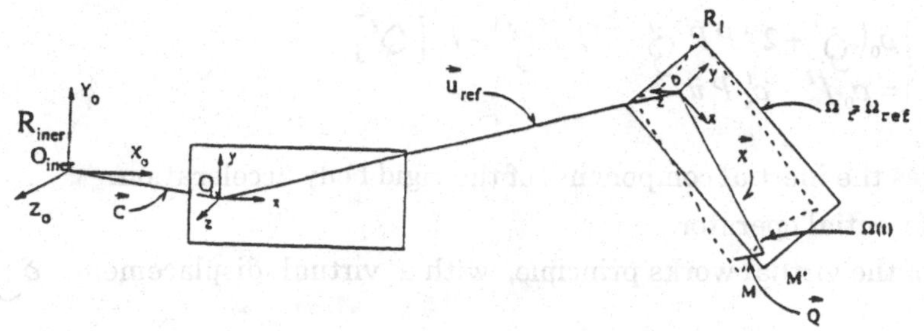

Fig. 7 Decomposition of the displacement

where $\bar{\mathfrak{g}}=\bar{\mathfrak{g}}(\bar{x},t)$ is the elastic displacement field related to an undeformed reference configuration (Ω_r) and $\bar{x}=\overrightarrow{OM}$ gives the position of the point M which is assumed to be rigidly connected to the reference configuration (Ω_r). $\bar{u}_{ref}$ gives the rigid body translation motion, $\bar{\alpha}$ is a constant vector. The choice of the reference configuration will be given later. Let us introduce the orthogonal matrix P giving the position of the local frame (R_ℓ) rigidly connected to the reference configuration (Ω_r) with respect to the inertial frame (R_{iner}). (${}^tP=P^{-1}$). P defines the rigid rotationnal motion of the body. The column matrix of the inertial components of vector $\bar{u}$ is given by :

$$u=\alpha+u_{ref}+P(x^{\ell}+\mathfrak{g}^{\ell}) \tag{5.2}$$

where x^{ℓ} and $\mathfrak{g}^{\ell}$ are the column matrices of the local components of vectors $\bar{x}$ and $\bar{\mathfrak{g}}$. The local motion equations of the flexible component are :

$$\overrightarrow{div}\,(\mathbb{F}\,\mathbb{S})+\rho_0\,\bar{f}=\rho_0\,\ddot{\bar{u}}\ \text{ in } (\Omega_0) \tag{5.3}$$

where (Ω_0) is the initial configuration of the body, ρ_0 is the initial mass density, $\mathbb{S}$ is the second Piola-Krichoff stress tensor, $\mathbb{F}$ is the strain gradient defined by $d\bar{u}=\mathbb{F}d\bar{x}$.

$$\mathbb{F}=P(\mathbb{E}+\nabla.\mathfrak{g}^{\ell})\quad (\mathbb{E}\ \text{unitarian tensor})$$

∇ is the gradient operator with respect to the local components of vector $\bar{x}$, $\bar{f}$ is the body applied force.

Assuming small strains, and linear constitutive law, the motion equation projected in the local reference frame (R_ℓ) has the following expression :

$$\begin{cases} \rho_0\left(\ddot{u}^\ell + 2\,{}^tP\dot{P}\,\dot{u}^\ell + {}^tP\ddot{P}u^\ell\right) + k_L[u^\ell] \\ = \rho_0 f^\ell - \rho_0^\ell\, P\,\ddot{u}_r \end{cases} \tag{5.4}$$

$\ddot{u}_r = \ddot{u}_{ref} + \ddot{P}x^\ell$ gives the inertial components of the rigid body acceleration.
k_L is a linear differential operator.
By application of the virtual works principle, with a virtual displacement δu^ℓ, we obtain :

$$\begin{cases} \int_{\Omega r} \rho_0\; {}^t\delta u^\ell\left(\ddot{u}^\ell + 2\,{}^tP\dot{P}\,\dot{u}^\ell + {}^tP\ddot{P}\,u^\ell\right)dv \\ + \int_{\Omega r} {}^t\delta u^\ell\, k_L[u^\ell]dv = \int_{\Omega r} {}^t\delta u^\ell\left(\rho_0 f^\ell - \rho_0\, {}^tP\,\ddot{u}_r\right)dv \end{cases} \tag{5.5}$$

A spacial discretisation of the reference domain (Ω_r) is performed by means of finite elements method : $u^\ell = N(x^\ell)q$; where q is the column matrix of the local components of the displacements of the nodes and $N(x^\ell)$ is the interpolating matrix.
Assuming a virtual displacement of the body in the form $\delta u^\ell = N\delta q$, the virtual works principle gives the following dynamical equations :

$$M\ddot{q} + C\dot{q} + Kq = F_e - F_R \tag{5.6}$$

where $M = \int_{\Omega r}\rho_0\; {}^tNN\,dv$ is the (constant) mass matrix ; $C = \int_{\Omega r}2\rho_0\; {}^tN\,{}^tP\dot{P}N\,dv$ is the gyroscopic matrix ; $K = K_s + K_d, K_s = \int_{\Omega r}{}^tNk_L[N]\,dv$ is the structural stiffness matrix ; $K_d = \int_{\Omega r}\rho_0\; {}^tN\,{}^tP\ddot{P}N\,dv$ is the dynamical stiffness matrix ; $F_e = \int_{\Omega r}\rho_0\; {}^tNf^\ell\,dv$ is related to the applied forces ; $F_R = \int_{\Omega r}\rho_0\; {}^tN\,{}^tP\,\ddot{u}_r\,dv$ is related to the inertial forces produced by the motion of the reference configuration (Ω_r).

The matrices C, K_d and F_R are time variant.

5.2.2 Incremental motion of the flexible component

The motion of the flexible component is defined by an incremental method : in each time interval $[t_n, t_{n+1}]$, the reference configuration (Ω_r) is the rigid configuration reached at the time t_n and the motion equations are linearized with respect to the increment of displacement q_n of this time interval (Fig. 8).

The increment q_n is splitted into a rigid increment q_r^n and a flexible increment q_d^n :

$$q_n = q_r^n + q_d^n$$

Let us consider the matrix X giving the modes shapes of the body assuming that the reference configuration (Ω_r) is fixed. X is a constant matrix defined by :

$$\begin{cases} \left(K_s - M\omega_j^2\right) X_j = 0 \quad (j = 1,..,N) \\ X = \left(X_1 \ldots X_N\right) \end{cases}$$

where N is the total degrees of freedom of the nodes, X_j is the mode shape associated with the pulsation ω_j. The modal matrix X is splitted into the matrix X_r of rigid body mode shapes and the matrix X_d of deformation mode shapes.
Assuming that the increment q_n is small, this increment is expressed as a linear combination of the rigid modes and the deformation modes :

$$q^n = \sum_{j=1}^{6} Y_{rj}^n X_{rj} + \sum_{j=1}^{nd} Y_{dj}^n X_{dj} = X_r Y_r^n + X_d Y_d^n = XY^n \tag{5.7}$$

$Y = \begin{bmatrix} Y_r \\ Y_d \end{bmatrix}$ gives the modal amplitude of the modes, nd is the number of deformation modes chosen : in most cases, only few deformation modes are used, namely low frequencies modes. The new generalized variables of the dynamical model are defined by the equations

$$\mathfrak{K}_n Y^n + \mathfrak{C}_n \dot{Y}^n + m \ddot{Y}^n = \mathfrak{I}_e^n - \mathfrak{I}_R^{n-1} \tag{5.8}$$

where $\mathfrak{K}_n = {}^tX K_n X$ is the generalized stiffness matrix ; $\mathfrak{C}_n = {}^tX C_n X$ is the generalized damping matrix ; $m={}^tXMX$ is the (constant) generalized mass matrix ; $\mathfrak{I}_e^n = {}^tX F_e^n$ and $\mathfrak{I}_R^{n-1} = {}^tX F_R^{n-1}$ are related to the applied forces and to the residual inertia forces respectively.

5.2.3 Dynamical model of the whole system

The dynamical model of the robot is deduced from Eq. (5.6) :

$$\begin{cases} \overline{M}\,\ddot{\overline{q}} + \overline{C}\,\dot{\overline{q}} + \overline{K}\,\overline{q} = \overline{F}_e - \overline{F}_R + \overline{\Gamma} + {}^t\overline{P}\left[{}^tJ\lambda\right] \; (i) \\ g(\overline{U}) = 0 \qquad (ii) \end{cases} \tag{5.9}$$

$\overline{q}$ is the column matrix of the local components of the displacements increments of the whole system, $\overline{\Gamma}$ is the column matrix of the local components of the forces or torques produced by the actuators, $\overline{F}_e$ and $\overline{F}_R$ are the column matrices of the applied forces and residual inertia forces for the whole system, $\overline{M}, \overline{C}, \overline{K}$ and $\overline{P}$ are block-diagonal matrices defined from M_j, C_j, K_i and $P_j (j = 1,..NB)$.
At last λ is the column matrix of the Lagrange multipliers associated with the constraint equations (ii) in which $\overline{U}$ is the column matrix of the inertial components of the nodes of the whole system and J is the Jacobian matrix of the function g.

The constraint equation (5.9) (*ii*) is linearized on the time interval $\left[t_n, t_{n+1}\right]$:

$$g\left(\overline{U}^{n+1}\right) = g\left(\overline{U}^{n}\right) + \left(\frac{\partial g}{\partial \overline{U}}\right)_n \Delta\, \overline{U}^{n+1} + \varepsilon\left(\left\|\Delta\, \overline{U}\right\|^2\right) = 0$$

$$\Rightarrow J_n\, \Delta\, \overline{U}^{n+1} = J_n\left(\overline{U}^{n+1} - \overline{U}^{n}\right) = 0$$

From the incremental motion of the system (Fig. 8), we deduce $\overline{U}^{n+1} = \overline{U}_r^n + \overline{P}_n\, \overline{q}^{n+1}$; $\overline{U}^n = \overline{U}_r^n + \overline{P}_{n-1}\, q_d^n$. $\overline{U}_r$ is the column matrix of the inertial components of the reference motion of the nodes for the whole system, $\overline{q}_d$ is the column matrix of the local components of the flexible increments of the nodes. By projection of the constraint equations (*ii*) on the modal basis of the whole system, we obtain $\pi_n\, \overline{Y}^{n+1} = \pi_n\, \tilde{Y}_d^n$; $\pi_n = J_n\, \overline{P}_n\, \overline{X}$. $\tilde{Y}_d$ is the column matrix of the modal amplitudes of the deformations modes.

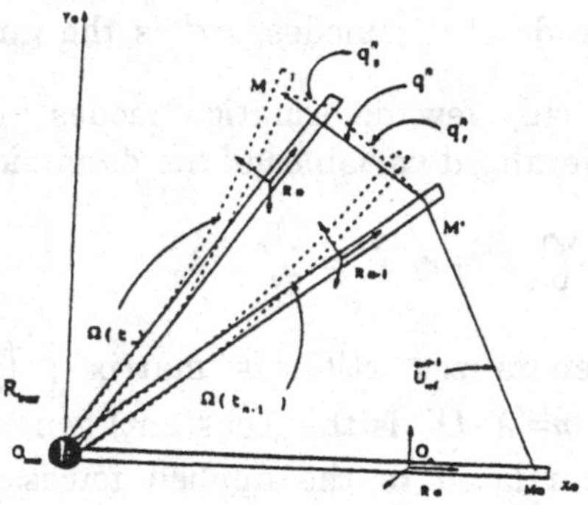

Fig. 8 Description of the Incremental Motion

On the other hand, the generalization of the equation (5.8) to the whole system gives the following equation :

$$\overline{K}_n\, \overline{Y}^n + \overline{C}_n\, \dot{\overline{Y}}^n + \overline{m}\, \ddot{\overline{Y}}^n = \overline{\mathfrak{I}}_e^{\,n} - \mathfrak{I}_R^{\,n-1} + {}^t\overline{X}\, \overline{\Gamma}^n + {}^t\pi_n\, \lambda^n \tag{5.10}$$

5.2.4 Numerical integration

The numerical code used is the PLEXUS Software [21] in which a Newmark semi-explicit scheme is chosen for the integration of the motion equations :

$$\begin{cases} \overline{Y}^{n+1} = \overline{Y}^n + \Delta t\, \dot{\overline{Y}}^n + \dfrac{\Delta t^2}{2}\, \ddot{\overline{Y}}^n \\ \dot{\overline{Y}}^n = \dot{\overline{Y}}^{n-1} + \dfrac{\Delta t}{2}\left(\ddot{\overline{Y}}^{n-1} + \ddot{\overline{Y}}^n\right) \end{cases} \tag{5.11}$$

From these formulas and from the constraint equations, we obtain :

$$\pi_n \ddot{\bar{Y}}^n = \frac{1}{\Delta t^2} \pi_n \left[\tilde{Y}_d^n - \bar{Y}^n - \Delta t \dot{\bar{Y}}^{n-1} - \frac{\Delta t^2}{2} \ddot{\bar{Y}}^{n-1} \right] \tag{5.12}$$

The acceleration $\ddot{\bar{Y}}^n$ is computed from the dynamical equations (5.10) in which $\dot{\bar{Y}}^n$ is estimated by its value in the middle of the step :

$$\begin{cases} \ddot{\bar{Y}}^n = \overline{W}_n + \overline{m}^{-1}\, {}^t\pi_n\, \lambda^n + \overline{m}^{-1}\, {}^t\overline{X}\, \overline{\Gamma}^n \\ \overline{W}_n = \overline{m}^{-1} \left(\overline{\mathfrak{I}}_e^{\,n} - \overline{\mathfrak{I}}_R^{\,n-1} - \overline{K}_n\, \overline{Y}^n - \mathcal{C}_n\, \dot{\bar{Y}}^{n-1/2} \right) \\ \dot{\bar{Y}}^{n-1/2} = \dot{\bar{Y}}^{n-1} + \frac{\Delta t}{2}\, \ddot{\bar{Y}}^{n-1} \end{cases} \tag{5.13}$$

From formulas (5.11) and (5.12), we deduce the Lagrange multipliers :

$$\begin{cases} \lambda^n = -\left[H_1^n\right]^{-1} \pi_n \overline{W}_n + \frac{\left[H_1^n\right]^{-1}}{\Delta t^2} \pi_n \left[\tilde{Y}_d^n - \overline{Y}^n - \Delta t\, \dot{\bar{Y}}^{n-1} - \frac{\Delta t^2}{2} \ddot{\bar{Y}}^{n-1} \right] \\ -\left[H_1^n\right]^{-1} \pi_n\, \overline{m}^{-1}\, {}^t\overline{X}\, \overline{\Gamma}^n \end{cases} \tag{5.14}$$

where $H_1^n = \pi_n\, \overline{m}^{-1}\, {}^t\pi_n$ is a square matrix of dimensions $(n_\ell \times n_\ell)$ (n_ℓ is the number of scalar constraints) which is supposed to be non-singular.
From formulas (5.13) and (5.14), we deduce the incremental acceleration at time t_n :

$$\begin{cases} \ddot{\bar{Y}}^n = \overline{Z}^n + \Phi^n\, {}^t\overline{X}\, \overline{\Gamma}^n \\ \overline{Z}^n = \left(E - H_2^n\right) \overline{W}_n + \frac{H_2^n}{\Delta t^2} \left(\tilde{Y}_d^n - \overline{Y}^n - \Delta t\, \dot{\bar{Y}}^{n-1} - \frac{\Delta t^2}{2} \ddot{\bar{Y}}^{n-1} \right) \\ \Phi^n = \left(E - H_2^n\right) \overline{m}^{-1} \end{cases} \tag{5.15}$$

where $H_2^n = \overline{m}^{-1}\, {}^t\pi_n \left[H_1^n\right]^{-1} \pi_n$ is a square matrix of dimensions $(\tilde{N} \times \tilde{N})$ ($\tilde{N}$ total number of degrees of freedom of the whole system) ; E is the $(\tilde{N} \times \tilde{N})$ unitarian matrix.

5.3 DYNAMICAL CONTROL LAWS IN THE JOINT SPACE [22]

The aim is to choose the control forces or torques produced by the motors at the joints in order to obtain a desired trajectory defined in the joint space and to damp the elastic vibrations. The main difficulty to solve this problem is that the actuators act only on the rigid body degrees of freedom.

To design the control laws, we use the dynamical model (5.15) in which a split of the rigid variables and the deformation variables are done in order to obtain the rigid body accelerations :

$$\begin{cases} \ddot{\bar{Y}}_r^n = \bar{Z}_r^n + \left(\Phi_{rr}^n \, {}^t\bar{X}_r + \Phi_{rd}^n \, {}^t\bar{X}_d \right) \bar{\Gamma}^n & (i) \\ \ddot{\bar{Y}}_d^n = \bar{Z}_d^n + \left(\Phi_{dr}^n \, {}^t\bar{X}_r + \Phi_{dd}^n \, {}^t\bar{X}_d \right) \bar{\Gamma}^n & (ii) \end{cases} \tag{5.16}$$

We define the following control vector at each time t_n :

$$\bar{V}^n = \ddot{\tilde{\bar{U}}}_c^n + \bar{G}_v^r \left(\dot{\tilde{\bar{U}}}_c^n - \dot{\tilde{\bar{U}}}_r^n \right) + \bar{G}_p^r \left({}^t\bar{U}_c^n - {}^t\bar{U}_r^n \right) - \bar{G}_v^d \, \dot{\tilde{\bar{U}}}_d^n - \bar{G}_p^d \, {}^t\bar{U}_d^n \tag{5.17}$$

In this formula, the notations $\dot{\tilde{U}}$ and $\ddot{\tilde{U}}$ give the relative velocities and the relative accelerations (in local reference frames) of the components of matrix ${}^t\bar{U}$.
The desired trajectory is defined by ${}^t\bar{U}_c = {}^t\bar{U}_{rc}$, ${}^t\bar{U}_{dc} = 0$.
At last, $\bar{G}_p^r$, $\bar{G}_v^r$, $\bar{G}_p^d$ and $\bar{G}_v^d$ are gains matrices. These matrices are chosen in order to minimize the rigid error $\varepsilon_r = {}^t\bar{U}_c - {}^t\bar{U}_r$ and the flexible error $\varepsilon_d = -{}^t\bar{U}_d$ and their first and second derivatives with respect to time. If the dynamical model (5.15) is used, we assume the following relation :

$$\bar{V}^n = \ddot{\tilde{\bar{U}}}_r^n = \ddot{\tilde{\bar{U}}}_r^{n-1} + \bar{X}_r \ddot{\bar{Y}}^n = \ddot{\tilde{\bar{U}}}_r^{n-1} + \bar{X}_r \bar{Z}_r^n + \left(\bar{X}_r \, \phi_{rr}^n \, {}^t\bar{X}_r + \bar{X}_r \, \phi_{rd}^n \, {}^t\bar{X}_d \right) \bar{\Gamma}^n$$

from which the control torque is obtained :

$$\begin{cases} \bar{\Gamma}^n = \bar{A}^n \left(\bar{V}^n - \ddot{\tilde{\bar{U}}}_r^{n-1} - \bar{X}_r \bar{Z}_r^n \right) \\ \bar{A}^n = \left(\bar{X}_r \, \phi_{rr}^n \, {}^t\bar{X}_r + \bar{X}_r \, \bar{\phi}_{rd}^n \, {}^t\bar{X}_d \right)^{-1} \end{cases}$$

Hence the obtained torque depends on the control vector and on the dynamical model and includes a feed back of the rigid and elastic variables.
For practical implementation of the obtained control command, only some special nodes will occur in the expression (5.17) of the control vector, namely the nodes where the actuators lie and some nodes where the deformations have to be damped. It results that in the feed back gains matrices $\bar{G}_v^r$, $\bar{G}_p^r$, $\bar{G}_v^d$, $\bar{G}_p^d$, only the terms related to these nodes are not equal to zero.
By substitution in the dynamical model, the errors on the rigid and flexible variables are given by :

$$\begin{cases} \ddot{\varepsilon}_r^n + \bar{G}_v^r \, \dot{\varepsilon}_r^n + \bar{G}_p^r \, \varepsilon_r^n = -\bar{G}_v^d \, \dot{\varepsilon}_d^n - \bar{G}_p^d \, \varepsilon_d^n \\ \ddot{\varepsilon}_d^n = L_2^n - L_1^n \, \ddot{\tilde{\bar{U}}}_c^n - L_1^n \left(\bar{G}_v^r \, \dot{\varepsilon}_r^n + \bar{G}_p^r \, \varepsilon_r^n + \bar{G}_v^d \, \dot{\varepsilon}_d^n + \bar{G}_p^d \, \varepsilon_d^n \right) \end{cases} \tag{5.18}$$

$$L_1^n = \left(\bar{X}_d \, \Phi_{dr}^n \, {}^tX_r + \bar{X}_d \, \Phi_{dd}^n \, {}^tX_d \right) \bar{A}_n ; \; L_2^n = L_1^n \left(\ddot{\tilde{\bar{U}}}_r^{n-1} + \bar{X}_r \, Z_r^n \right) - \bar{X}_d \, Z_d^n$$

In terms of the error on the state vector $\Delta X = {}^t({}^t\varepsilon_r, {}^t\varepsilon_d, {}^t\dot{\varepsilon}_r, {}^t\dot{\varepsilon}_d)$, the formula (5.18) takes the form :

$$\begin{cases} \Delta \dot{X}_n = \mathcal{A}(\Delta X_n) + \mathcal{B}(\Delta X_n)\mathcal{U}_n \\ \mathcal{U}_n = -G\,\Delta X_n \end{cases} \tag{5.19}$$

$$\mathcal{A} = \begin{pmatrix} O \\ O \\ O \\ L_2^n - L_1^n \ddot{\tilde{U}}_c^n \end{pmatrix} \qquad \mathcal{B} = \begin{pmatrix} O & O & I & O \\ O & O & O & I \\ I & I & I & I \\ L_1^n & L_1^n & L_1^n & L_1^n \end{pmatrix} \qquad G = \begin{pmatrix} G_p^r & O & O & O \\ O & \overline{G}_p^d & O & O \\ O & O & \overline{G}_v^r & O \\ O & O & O & \overline{G}_v^d \end{pmatrix}$$

The obtained equation is a non-linear equation.
The gains matrix G, which have to be chosen in order to minimize the rigid error ε_r and the magnitude of the elastic variables, can be obtained by linearization of equation (5.19) in the vicinity of a point of the desired trajectory and by using standard technics of computation of the gains matrix for linear systems.

5.4 CONCLUSIONS

In this work, a dynamical model for flexible robots is used to design a non-linear control law in order to follow a desired trajectory in the joints space and to damp the structural vibrations of the links. Several test examples [22] show the efficiency of this non linear control law.

REFERENCES

1. De Veubeke, B.F.: The Dynamics of Flexible Bodies, J. of Eng. Sci, 1976, Vol. 14, pp. 895-913.
2. Wallrapp, O. and Schertassek, R.: Representation of Geometric Stiffening in Multibody System Simulation, Int. J. for Numerical Methods in Engineering, 1991, Vol. 32, pp. 1833-1850.
3. Wittenburg, J.: Dynamics of Systems of Rigid Bodies, B.G. Teubner, Stuttgart 1977.
4. Pascal, M. : Non Linear Effects in Transient Dynamic Analysis of Flexible Multibody Systems. Proceed. of the 15th ASME Conf. on Mech. Vib. and Noise (1995), Boston (USA).
5. Wielanga, T.J.: Simplifications in the Simulation of Mechanisms Containing Flexible Members, 1984, P.H.D. Thesis, College of Engineering, The University of Michigan, Ann Arbor (USA).
6. Padilla, C.E., and Von Flotow, A.H.: Non linear Strain - Displacement Relations and Flexible Multibody Dynamics, J. Guidance, Control and Dynamics, 1992, Vol. 15, N° 1, pp. 128-136.
7. Banerjee, A.K. and Lemak, J.M.: Multi-Flexible Body Dynamics Capturing Motion-Induced Stiffness, ASME J. Applied Mechanics, 1991, Vol. 58, pp. 766-775.
8. Pascal, M.: Dynamics Analysis of a system of Hinge-connected Flexible Bodies, Celestial Mechanics, 41, (1988), 253-274.
9. Hughes, P.C.: Dynamics of Flexible Space Vehicles with Active Attitude Control, Celestial Mechanics, 9 (1974), 21-39.
10. Kolousek, V.: Dynamics in engineering structures. London, Butterworths, 197 3.
11. Poelaert, D.: Dynamics and Control of large flexible spacecraft, Proc. 4th Symposium, Blacksburg, Va (1983).
12. Pascal, M.: Dynamical Analysis of a Flexible Manipulator Arm, Acta Astr., 21, 3, (1990), 161-169.
13. Pascal, M.: Modal Analysis of a Rotating Flexible Space Station by a Continuous Approach, Z. Flugwiss. Welraumforsch 18 (1994), 1-6.
14. Pascal, M., Sylla, M.: Modèle dynamique par approche continue d'une grande structure spatiale, La Recherche Aérospatiale, 2, (1993), 67-77.
15. Bayo, E., Papadopoulos, P., Stubbe, J. and Serna, M. A.: Inverse Dynamics and Kinematics of Multilink Elastic Elastic Robots : An Iterative Frequency Domain Approach, Int. J. of Robotics Research, Vol. 8, No. 6, 1989, pp. 49-62.
16. Chaloub, N. G. and Ulsoy, A. G.: Control of a Flexible Arm : Experimental and Theoretical Results, J. Dyn. Syst. Meas. Control ASME, Vol. 109, 1987, pp. 299-310.
17. Siciliano, B. and Book, W. J.: A Singular Perturbation Approach to Control of Lightweight Flexible Manipulators, Int. J. of Robotics Research, Vol. 7, No. 4, 1988, pp. 79-90.

18. Chedmail, P., Aoustin, Y. and Chevallereau, C.: Modelling and Control of Flexible Robots, Int. J. For Num. Meth. in Eng., Vol. 32, 1991, pp. 1595-1619.
19. Modi, V. J., Karray, F. and Chan, J. K.: On the Control of a Class of Flexible Manipulators Using Feed back Linearization Approach, 42nd Congress of the Int. Astr. Fed., October 5 - 11, 1991, Montreal, Canada.
20. Azouz, N.: Modélisation des structures souples polyarticulées. Application à la simulation des robots, Thèse de l'Université Pierre et Marie Curie (1994).
21. Barraco, A., Cuny, B., Hoffmann, A., Jamet, P., Combescure, A., Lepareux, M. and Bung, H.: Plexus-Software for the Numerical Analysis of the Dynamical Behavior of Rigid and Flexible Mechanisms in Multibody Systems Handbook, Berlin - Springer - Verlag, 1990.
22. Yachou, B.: Contrôle dynamique des structures polyarticulées déformables, Thèse de l'Université Pierre et Marie Curie (1995).

PART VI

MODERN ANALYTICAL METHODS APPLIED TO MECHANICAL ENGINEERING SYSTEMS

P. Hagedorn and W. Seemann
Darmstadt University of Technology, Darmstadt, Germany

Abstract

In the following lectures modern analytical methods are applied to several industrial systems. The first application is the use of singular perturbation analysis in the eigenvalue problem of a vibrating string with a small bending stiffness. This models the behavior of overhead transmission lines. A more complicated system is the eigenvalue problem of a cylinder vibrating in an cylindrical duct, which is filled with a viscous fluid.

The next lecture shows the use of analytical methods in the modelling of ultrasonic motors. Of special interest is the coupling between the electric and the mechanical field in the piezoelectric patches. Here HAMILTON's principle for electromechanical systems is of great importance. It allows to find approximate solutions fulfilling additional constraint equations.

In overhead transmission lines also wind excited vibrations are important as they may lead to fatigue. Models for the corresponding mechanism are given as well as the analysis of special vibration absorbers designed by modern methods of vibration theory.

The last lecture deals with three problems of nonholonomic systems and of stability and instability theorems. The first problem shows that the augmented Lagrangian is not stationary for nonholonomic systems. The second gives a simple exercise in LIAPUNOV stability. The last deals with the LAGRANGE-DIRICHLET theorem and its inverses.

Chapter 1

Singular Perturbation Analysis

1.1 Introduction

Though perturbation analysis has been developed a long time ago, the method has achieved a growing interest recently due to the availability of computer algebra. The advantage of perturbation analysis over purely numerical methods such as for example Finite Element Methods (FEM) is that perturbation analysis gives directly (approximate) analytical formulae showing explicitly the dependence of the system's response on the material and geometrical parameters without intensive numerical studies. Normally, perturbation analysis is used as *regular perturbation* where a variable x is expressed as a power series in a small parameter ε. The small parameter ε is contained in the problem formulation and is therefore known in advance. After introducing the asymptotic series in the equations describing the problem, the equations may be ordered with respect to different powers of ε. Equating all factors of the different powers to zero yields equations which may be solved recursively. Examples may be found in the books of van Dyke [9] and Kevorkian/Cole [22] or for vibration analysis of nonlinear systems in the books of Hagedorn [13], Nayfeh [27] and Nayfeh/Mook [23].

Let us consider the DUFFING-equation

$$\ddot{x} + \omega_0^2 x + \varepsilon x^3 = 0 \tag{1.1}$$

of which we know that for small ε we obtain free vibrations with a slightly different frequency compared to the frequency for $\varepsilon = 0$. Introducing the asymptotic expansion

$$x(t) = x_0(t) + \varepsilon x_1(t) + \varepsilon^2 x_2(t) + \ldots \tag{1.2}$$

yields an equation of the form

$$(\ddot{x}_0 + \omega_0^2 x_0) + \varepsilon(\ddot{x}_1 + \omega_0^2 x_1 + x_0^3) + \varepsilon^2(\ddot{x}_2 + \omega_0^2 x_2 + 3x_0^2 x_1) + \ldots = 0. \tag{1.3}$$

Perturbation theory implies that equation (1.3) is fulfilled if all factors of the different powers of ε are zero. This leads to the equations

$$\ddot{x}_0 + \omega_0^2 x_0 = 0, \tag{1.4}$$

$$\ddot{x}_1 + \omega_0^2 x_1 = -x_0^3, \tag{1.5}$$

$$\ddot{x}_2 + \omega_0^2 x_2 = -3x_0^2 x_1, \tag{1.6}$$

$$\vdots$$

which may be solved recursively. Such equations can nowadays be derived easily via computer algebra. This is helpful especially for more complicated differential equations. The solutions of equations (1.4) to (1.6) are

$$x_0 = C_0 \sin(\omega_0 t + \gamma_0), \tag{1.7}$$

$$x_1 = \frac{3t}{8\omega_0} C_0^3 \cos(\omega_0 t + \gamma_0) - \frac{1}{32\omega_0^2} C_0^3 \sin(3\omega_0 t + 3\gamma_0) + C_1 \sin(\omega_0 t + \gamma_1), \tag{1.8}$$

$$\vdots$$

We observe that the finite sum

$$x(t) = x_0(t) + \varepsilon x_1(t) + \varepsilon^2 x_2(t) + \ldots + \varepsilon^m x_m(t) \tag{1.9}$$

is not periodic due to the secular terms contained in (1.8) which are increasing linearly with time. This leads to a modified approach in such a way that not only the dependent variable x is expressed by an asymptotic expansion but also the squared circular frequency ω^2:

$$x(t) = x_0(t) + \varepsilon x_1(t) + \varepsilon^2 x_2(t) + \ldots \tag{1.10}$$

$$\omega^2 = \omega_0^2 + \varepsilon e_1 + \varepsilon^2 e_2 + \ldots \tag{1.11}$$

Substituting (1.10) and (1.11) into (1.1), ordering with respect to the powers of ε and setting the coefficients to zero leads to a system of differential equations which again can be solved recursively. The parameters $e_1, e_2, \ldots$ are unknown at the moment and have to be determined later by the requirement of vanishing secular terms. The system of differential equations for the case of the DUFFING-equation is

$$\ddot{x}_0 + \omega^2 x_0 = 0, \tag{1.12}$$

$$\ddot{x}_1 + \omega^2 x_1 = -x_0^3 + e_1 x_0, \tag{1.13}$$

$$\ddot{x}_2 + \omega^2 x_2 = -3x_0^2 x_1 + e_2 x_0 + e_1 x_1, \tag{1.14}$$

$$\vdots$$

with the solutions

$$x_0 = C_0 \sin(\omega t + \gamma_0), \tag{1.15}$$

$$x_1 = C_1 \sin(\omega t + \gamma_1) - \frac{1}{32\omega^2} C_0^3 \sin(3\omega t + 3\gamma_0), \tag{1.16}$$

$$x_2 = C_2 \sin(\omega t + \gamma_2) + \frac{1}{1024\omega^4} C_0^5 [\sin(5\omega t + 5\gamma_0) - 3\sin(3\omega t + 3\gamma_0)]. \tag{1.17}$$

Equation (1.17) was obtained by setting $C_1 = 0$ and

$$e_1 = \frac{3}{4} C_0^2, \tag{1.18}$$

$$e_2 = \frac{3}{128\omega^2} C_0^4. \tag{1.19}$$

Equations (1.18) and (1.19) permit the determination of the circular frequency ω up to terms of ε^2. As expected from physics, a positive value of ε increases the frequency of the vibration whereas a degressive spring ($\varepsilon < 0$) diminishes the frequency with increasing amplitude.

This approach has been known for a long time and should only serve as an introduction to perturbation analysis. More interesting is the case of *singular* perturbation theory which may be applied to problems in which the coefficient of the highest derivative is small or equal to zero for the reduced or unperturbed system.

1.2 Singular Perturbation Theory for Eigenvalue Problems

Singular perturbation theory is applied to problems in which the coefficient of the highest derivative is a small parameter ε. The reduced problem ($\varepsilon = 0$) then usually does not allow fulfillment of all boundary conditions. Therefore, boundary layers are introduced. In the boundary layers the coordinates are stretched such that in the stretched coordinates the differential equations permit fulfillment of all boundary and initial conditions. If we have e.g. a space-dependent problem $W(x)$ the stretching is done by introducing a new coordinate $\tilde{x}$ related to the original coordinate in a problem dependent way. In many cases the transformation

$$\tilde{x} = \frac{x}{\varepsilon} \tag{1.20}$$

or more generally $\tilde{x} = x/\alpha(\varepsilon)$ is used. This leads to differential equations in $\tilde{x}$ for which the coefficient of the highest derivative is no longer a small quantity so that regular perturbation theory may be applied for the transformed equations.

The *outer solution* $W^o(x)$, for points far away from the boundaries, can be obtained by a regular perturbation analysis. The *inner solution* $W^i(x)$, near the boundaries, is calculated with the aid of the above mentioned transformation. Therefore, we have a solution near the boundaries which differs from the solution in the field far away from the boundaries. Nevertheless, both solutions should coincide in an overlapping region, at least up to a certain order of magnitude. It is therefore convenient to use the symbols $o(x)$ and $O(x)$, two symbols for the order of magnitude:

$$O(x) : \Rightarrow \lim_{x \to 0} \frac{O(x)}{x} \to \text{finite value} \tag{1.21}$$

$$o(x) : \Rightarrow \lim_{x \to 0} \frac{o(x)}{x} = 0 \tag{1.22}$$

$O(x)$ stands for terms of the same order as x and $o(x)$ for terms small compared to x. We assume the overlapping region to be characterized by

$$x_\eta = \eta x \qquad \Rightarrow \qquad \tilde{x}_\eta = \frac{\eta}{\varepsilon} x \tag{1.23}$$

with

$$\varepsilon \ll \eta \ll 1. \tag{1.24}$$

The difference $d(x_\eta)$ between the two solutions (inner and outer) at an intermediate point is calculated and ordered with respect to the different orders of magnitude $O(\frac{\eta}{\varepsilon})$, $O(1)$, $O(\eta)$, $O(\varepsilon)$, $O(\eta^2)$, $O(\eta\varepsilon)$, $O(\varepsilon^2)$, ... Demanding that all terms up to a certain order of magnitude have to vanish, yields equations for the determination of those integration constants which could not be determined before. In the case of an eigenvalue problem this also leads to the characteristic equation, and therefore to the eigenvalues. This is called the *matching of the inner and outer solutions.* Calculating the *common part* $C_p(x)$ of both allows to establish the complete or composite solution

$$W(x) = W^i(x) + W^o(x - C_p(x)) \tag{1.25}$$

which is valid in the whole domain except for exponentially small terms.

As examples we first consider a string with small bending stiffness. This may be a good model for the cables of transmission lines. These cables may be treated as simple strings without bending stiffness to solve most of the problems which occur for these systems, for example the sag between two towers. Nevertheless, near the clamps or near spacers the bending stiffness is important, because the vibration of the strings lead to bending stresses near the clamps and therefore to fatigue. In a primary step we look at symmetric vibration modes by assuming boundary conditions for a clamped string at one end of the cable and symmetry conditions in the middle. This leads to one boundary layer with the corresponding inner solution in form of an asymptotic expansion and an outer solution far away from the boundary. The calculation of the symmetric modes is simplified, because only one matching between the boundary layer and the outer field has to be done.

In a second step all modes of a string fixed rigidly at both ends are calculated. These modes fulfill all the boundary conditions at both ends of the cable. Two boundary layers have to be introduced at either end of the string and matching has to be done at points in the two intermediate regions between the boundary layers and the free field.

The second example is a cylinder vibrating in a fluid filled duct. If the fluid is incompressible and inviscid, the result for the vibration frequency is very simple. If the fluid is viscous the general solution of the differential equations and the corresponding frequency equation can be written in terms of BESSEL functions with complex arguments. For a very narrow gap, however, a numerical evaluation of the characteristic equation is difficult and could not be obtained for parameter values of interest. Therefore, a singular perturbation analysis was done using computer algebra as far as possible.

1.3 Vibrations of a String with Small Bending Stiffness

In the following section, the eigenfrequencies of a string with small bending stiffness are determined by singular perturbation analysis. Starting point is a string which is clamped at its left and right ends, see figure 1.1. The string is supposed to have a length of $2\,\ell$, mass per

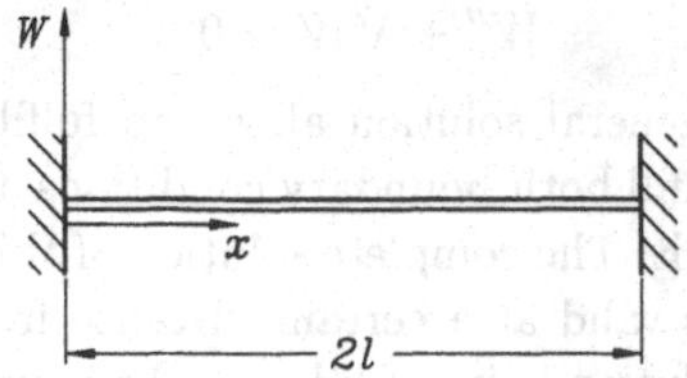

Figure 1.1: String with small bending stiffness, clamped at both ends

unit length ρA, a small bending stiffness EI, and a longitudinal tension T. The corresponding equation of motion is

$$\rho A\ddot{w}(\overline{x},t) + EIw^{IV}(\overline{x},t) - Tw''(\overline{x},t) = 0. \tag{1.26}$$

In addition we have the boundary conditions

$$\begin{aligned} w(0) &= w(2\ell) = 0, \\ w'(0) &= w'(2\ell) = 0. \end{aligned} \tag{1.27}$$

First, we introduce the nondimensional quantities $x = \frac{\bar{x}}{\ell}$, $\varepsilon = \sqrt{\frac{EI}{T}} \cdot \frac{1}{\ell} = \frac{\ell_{\text{char}}}{\ell}$ and an ansatz of the form

$$w(x,t) = W(x) \sin \omega t. \tag{1.28}$$

This yields an ordinary differential equation

$$\varepsilon^2 W^{IV} - W'' - \lambda^2 W = 0. \tag{1.29}$$

In this equation the nondimensional parameter $\lambda^2 = \omega^2 \frac{\rho A \ell^2}{T}$ was introduced. Dashes denote derivatives with respect to the nondimensional coordinate x.

1.3.1 Symmetric Modes of a String with Small Bending Stiffness

In this subsection we consider only symmetric modes of the vibrating string. Therefore, the differential equation is given by (1.29). The corresponding boundary and symmetry conditions yield for a symmetric problem

$$\begin{aligned} W(0) &= 0, \\ W'(0) &= 0, \\ W'(1) &= 0, \\ \varepsilon^2 W'''(1) &= 0. \end{aligned} \tag{1.30}$$

An ansatz of this form naturally allows only the determination of symmetric modes corresponding to modes of uneven order. In a first step we express the parameter λ as an asymptotic expansion of ε

$$\lambda = \lambda_0 + \varepsilon \lambda_1 + \varepsilon^2 \lambda_2 + \dots . \tag{1.31}$$

For λ^2 this yields

$$\lambda^2 = \lambda_0^2 + \varepsilon(2\lambda_0\lambda_1) + \varepsilon^2(2\lambda_0\lambda_2 + \lambda_1^2) + \dots . \tag{1.32}$$

The reduced problem ($\varepsilon = 0$)

$$W'' + \lambda^2 W = 0 \tag{1.33}$$

shows that the corresponding general solution allows to fulfill the boundary conditions at $x = 1$, but it is impossible to fulfill both boundary conditions at $x = 0$ identically. Instead at $x = 0$ a boundary layer will result. The complete solution of this problem is therefore divided into an outer solution, which is valid at a certain distance from the boundary or from the boundary layer, and an inner solution being valid near the boundary (in the boundary layer).

Outer Solution

For the outer solution the transverse displacement W is expressed as an asymptotic expansion in the form of a power series of ε

$$W(x) = h_0(x) + \varepsilon h_1(x) + \varepsilon^2 h_2(x) + \dots . \tag{1.34}$$

In this ansatz some a priori knowledge was used, as it would have also been possible to express the asymptotic expansion of W in the form of a power series of $\sqrt{\varepsilon}$. Later the computation however shows that an asymptotic expansion with respect to powers of ε suffices. The outer expansion is introduced into the differential equation (1.29) and ordered with respect to the powers of ε. Equating to zero every factor of the different powers of ε yields the following conditions

$$\varepsilon^0: \quad h_0'' + \lambda_0^2 h_0 = 0, \tag{1.35}$$
$$\varepsilon^1: \quad h_1'' + \lambda_0^2 h_1 = -2\lambda_0\lambda_1 h_0, \tag{1.36}$$
$$\varepsilon^2: \quad h_2'' + \lambda_0^2 h_2 = -(2\lambda_0\lambda_2 + \lambda_1^2)h_0 - 2\lambda_0\lambda_1 h_1 + h_0^{IV} \tag{1.37}$$
$$\vdots$$

The general solution for $h_0(x)$ and its corresponding derivative yield

$$h_0(x) = C\cos\lambda_0 x + S\sin\lambda_0 x,$$
$$h_0'(x) = -\lambda_0 C\sin\lambda_0 x + \lambda_0 S\cos\lambda_0 x. \tag{1.38}$$

For the zeroth order approximation $h_0(x)$ only one boundary condition at $x = 1$ has to be fulfilled $W'(1) = 0$. This yields

$$C\sin\lambda_0 - S\cos\lambda_0 = 0. \tag{1.39}$$

Formally one of the constants C or S remains undetermined. Nevertheless, we introduce a second condition in the form $W(1) = A$, meaning that we suppose the amplitude at $x = 1$ to be related to A. This yields a second equation of the form

$$C\cos\lambda_0 + S\sin\lambda_0 = A. \tag{1.40}$$

Therefore, the constants C and S can be determined in dependence of the amplitude A. Consequently the solution for $h_0(x)$ is

$$h_0(x) = A\cos\lambda_0\cos\lambda_0 x + A\sin\lambda_0\sin\lambda_0 x. \tag{1.41}$$

With the known solution $h_0(x)$ the differential equation for h_1 and the corresponding boundary conditions at $x = 1$ are

$$h_1'' + \lambda_0^2 h_1 = -2A\lambda_0\lambda_1(\cos\lambda_0\cos\lambda_0 x + \sin\lambda_0\sin\lambda_0 x),$$
$$h_1(1) = 0,$$
$$h_1'(1) = 0. \tag{1.42}$$

The general solution for h_1 is

$$h_1(x) = C_1\cos\lambda_0 x + S_1\sin\lambda_0 x + A\lambda_1(\cos\lambda_0\cos\lambda_0 x - \sin\lambda_0\sin\lambda_0 x)x \tag{1.43}$$

and can be written as

$$h_1(x) = A\lambda_1 x\sin[\lambda_0(1-x)] + C_1\cos\lambda_0 x + S_1\sin\lambda_0 x. \tag{1.44}$$

Fitting the general solution for h_1 to the boundary conditions mentioned above at $x = 1$ yields a system of linear equations for the integration constants C_1 and S_1.

$$C_1\cos\lambda_0 + S_1\sin\lambda_0 = 0,$$
$$-A\lambda_1\lambda_0 - C_1\lambda_0\sin\lambda_0 + S_1\lambda_0\cos\lambda_0 = 0. \tag{1.45}$$

The solution is

$$C_1 = -A\lambda_1 \sin \lambda_0, \quad S_1 = A\lambda_1 \cos \lambda_0. \tag{1.46}$$

As the result for h_1 we have

$$\begin{aligned} h_1(x) &= A\lambda_1 x \sin[\lambda_0(1-x)] - A\lambda_1 \sin[\lambda_0(1-x)], \\ h_1(x) &= -A\lambda_1(1-x)\sin[\lambda_0(1-x)]. \end{aligned} \tag{1.47}$$

Analogously the results for $h_0(x)$ and $h_1(x)$ are introduced into the differential equation for $h_2(x)$. This yields

$$\begin{aligned} h_2'' + \lambda_0^2 h_2 =& -(2\lambda_0\lambda_2 + \lambda_1^2)A\cos[\lambda_0(1-x)] \\ &+ 2\lambda_0\lambda_1^2 A(1-x)\sin[\lambda_0(1-x)] + \lambda_0^4 A\cos[\lambda_0(1-x)], \end{aligned} \tag{1.48}$$

or

$$h_2'' + \lambda_0^2 h_2 = (\lambda_0^4 - 2\lambda_0\lambda_2 - \lambda_1^2)A\cos[\lambda_0(1-x)] + 2\lambda_0\lambda_1^2 A(1-x)\sin[\lambda_0(1-x)]. \tag{1.49}$$

The inhomogeneous differential equation for $h_2(x)$ can be solved with an ansatz of the form

$$\begin{aligned} h_{2p} =& a(1-x)\cos[\lambda_0(1-x)] + b(1-x)\sin[\lambda_0(1-x)] \\ &+ c(1-x)^2\cos[\lambda_0(1-x)] \\ &+ d(1-x)^2\sin[\lambda_0(1-x)]. \end{aligned} \tag{1.50}$$

Taking into consideration the solution of the homogeneous differential equation the general solution is

$$\begin{aligned} h_2(x) =& C_2\cos\lambda_0 x + S_2\sin\lambda_0 x + A\frac{\lambda_0^3 - 2\lambda_2}{2}(1-x)\sin[\lambda_0(1-x)] \\ &- A\frac{\lambda_1^2}{2}(1-x)^2\cos[\lambda_0(1-x)] \end{aligned} \tag{1.51}$$

and the derivative of h_2 is given by

$$\begin{aligned} h_2'(x) =& -C_2\lambda_0\sin\lambda_0 x + S_2\lambda_0\cos\lambda_0 x \\ &+ A\frac{\lambda_0^3 - 2\lambda_2}{2}\{-\sin[\lambda_0(1-x)] + \lambda_0(1-x)\cos[\lambda_0(1-x)]\} \\ &- A\frac{\lambda_1^2}{2}\{-2(1-x)\cos[\lambda_0(1-x)] - (1-x)^2\sin[\lambda_0(1-x)]\}. \end{aligned} \tag{1.52}$$

The boundary conditions for h_2

$$\begin{aligned} h_2(1) &= 0, \\ h_2'(1) &= 0 \end{aligned} \tag{1.53}$$

lead to a system of homogeneous linear equations for the integration constants

$$\begin{aligned} C_2\cos\lambda_0 + S_2\sin\lambda_0 &= 0, \\ -C_2\lambda_0\sin\lambda_0 + S_2\lambda_0\cos\lambda_0 &= 0 \end{aligned} \tag{1.54}$$

with the homogeneous solution

$$\begin{aligned} C_2 &= 0, \\ S_2 &= 0. \end{aligned} \tag{1.55}$$

Inner Expansion

For the inner expansion we first have to transform the coordinates to the stretched coordinate $\tilde{x} = \frac{x}{\varepsilon}$. The derivative with respect to the coordinate x can then be expressed by

$$\frac{d}{dx} = \frac{1}{\varepsilon}\frac{d}{d\tilde{x}}. \tag{1.56}$$

Introducing the transformed coordinate into the differential equation results in

$$W_{\tilde{x}\tilde{x}\tilde{x}\tilde{x}} - W_{\tilde{x}\tilde{x}} - \varepsilon^2\lambda^2 W = 0. \tag{1.57}$$

For the solution in the boundary layer once again an asymptotic expansion of the form

$$W(\tilde{x}) = f_0(\tilde{x}) + \varepsilon f_1(\tilde{x}) + \varepsilon^2 f_2(\tilde{x}) + \dots \tag{1.58}$$

is assumed. Taking into consideration the asymptotic expansion for λ^2 as given in equation (1.32) yields an asymptotic expansion for the differential equation leading to the following system of differential equations

$$\begin{aligned}
\varepsilon^0 : &\quad f_0^{IV} - f_0'' = 0,\\
\varepsilon^1 : &\quad f_1^{IV} - f_1'' = 0,\\
\varepsilon^2 : &\quad f_2^{IV} - f_2'' = \lambda_0^2 f_0,\\
\varepsilon^3 : &\quad f_3^{IV} - f_3'' = \lambda_0^2 f_1 + 2\lambda_0\lambda_1 f_0,\\
&\quad \vdots
\end{aligned} \tag{1.59}$$

where a dash denotes a derivative with respect to the new coordinate $\tilde{x}$. The corresponding boundary conditions for f_i are

$$\begin{aligned}
f_0(0) = 0, &\quad f_0'(0) = 0,\\
f_1(0) = 0, &\quad f_1'(0) = 0,\\
f_2(0) = 0, &\quad f_2'(0) = 0,\\
&\quad \vdots
\end{aligned} \tag{1.60}$$

The general solution for $f_0(\tilde{x})$ is

$$\begin{aligned}
f_0(\tilde{x}) &= K_{10}e^{\tilde{x}} + K_{20}e^{-\tilde{x}} + K_{30}\tilde{x} + K_{40},\\
f_0'(\tilde{x}) &= K_{10}e^{\tilde{x}} - K_{20}e^{-\tilde{x}} + K_{30}.
\end{aligned} \tag{1.61}$$

Fitting the solution to the boundary conditions at $\tilde{x} = 0$ yields

$$\begin{aligned}
K_{10} + K_{20} + K_{40} &= 0,\\
K_{10} - K_{20} + K_{30} &= 0.
\end{aligned} \tag{1.62}$$

Thus two of the four integration constants can be arbitrarily chosen, e.g. the integration constants K_{20} and K_{10}. For K_{30} and K_{40}, respectively, this yields

$$\begin{aligned}
K_{30} &= K_{20} - K_{10},\\
K_{40} &= -K_{20} - K_{10}.
\end{aligned} \tag{1.63}$$

Taking into account that the term $e^{\tilde{x}}$ for $\varepsilon \to 0$ is ascending exponentially the integration constant K_{10} has to vanish and we obtain a solution for f_0

$$f_0(\tilde{x}) = K_0\left[e^{-\tilde{x}} + (\tilde{x} - 1)\right]. \tag{1.64}$$

Since both the differential equation and the boundary conditions for f_1 are the same as for f_0 this leads to the analogous result for $f_1(\tilde{x})$

$$f_1(\tilde{x}) = K_1 \left[e^{-\tilde{x}} + (\tilde{x} - 1) \right]. \tag{1.65}$$

Introducing the solution for f_0, equation (1.64), into the differential equation for f_2 yields

$$f_2^{IV} - f_2'' = \lambda_0^2 K_0 \left[e^{-\tilde{x}} + (\tilde{x} - 1) \right]. \tag{1.66}$$

The boundary conditions are $f_2(0) = 0$ and $f_2'(0) = 0$. The particular solution can be obtained by an ansatz in the form of

$$f_{2p}(\tilde{x}) = a\tilde{x}e^{-\tilde{x}} + b\tilde{x}^2 + c\tilde{x}^3, \tag{1.67}$$

which leads to

$$f_{2p}(\tilde{x}) = -\frac{\lambda_0^2 K_0}{2}\tilde{x}e^{-\tilde{x}} + \frac{\lambda_0^2 K_0}{2}\tilde{x}^2 - \frac{\lambda_0^2 K_0}{6}\tilde{x}^3, \tag{1.68}$$

so that the general solution is given by

$$f_2(\tilde{x}) = K_{12}e^{\tilde{x}} + K_{22}e^{-\tilde{x}} + K_{32}\tilde{x} + K_{42} + \frac{\lambda_0^2 K_0}{2}\left(-\tilde{x}e^{-\tilde{x}} + \tilde{x}^2 - \frac{1}{3}\tilde{x}^3 \right). \tag{1.69}$$

Fitting the solution to the boundary conditions again yields a system of equations with four unknowns

$$\begin{aligned} K_{12} + K_{22} + K_{42} &= 0, \\ K_{12} - K_{22} + K_{32} &= -\frac{\lambda_0^2}{2}K_0. \end{aligned} \tag{1.70}$$

For arbitrary K_{12} and K_{22} the integration constants K_{42} and K_{32} may be determined by

$$\begin{aligned} K_{42} &= -K_{12} - K_{22}, \\ K_{32} &= -\frac{\lambda_0^2 K_0}{2} + K_{22} - K_{12}. \end{aligned} \tag{1.71}$$

As $e^{\tilde{x}}$ is increasing exponentially for $\varepsilon \to 0$ the integration constant K_{12} has to vanish, leading to the conditions

$$\begin{aligned} K_{42} &= -K_{22} \\ K_{32} &= K_{22} - \frac{\lambda_0^2 K_0}{2}. \end{aligned} \tag{1.72}$$

Hence the solution for $f_2(\tilde{x})$ is determined except for an integration constant K_2 which has to be determined later:

$$f_2(\tilde{x}) = K_2 \left[e^{-\tilde{x}} + \tilde{x} - 1 \right] + \frac{\lambda_0^2 K_0}{2}\left[-\tilde{x}e^{-\tilde{x}} - \tilde{x} + \tilde{x}^2 - \frac{1}{3}\tilde{x}^3 \right]. \tag{1.73}$$

Having determined the inner solution and the outer solution up to the order of ε^2 both solutions have to be matched at an intermediate point.

Matching of the Inner and Outer Expansions

The matching of the inner and outer expansion at an intermediate point x_η is the main difficulty in eigenvalue problems in singular perturbation analysis. For this matching one assumes overlapping of the regions of validity of the inner and outer solution. This overlapping region is characterized by the coordinate x_η. For x_η we have

$$x = \eta x_\eta \quad \text{with} \quad \varepsilon \ll \eta \ll 1. \tag{1.74}$$

Therefore, the difference between both solutions at a fixed point x_η is expressed as an asymptotic expansion and the matching is done such that the terms of the asymptotic expansion are equated to zero up to a certain order. This difference is given by

$$d(x_\eta) = h_0(x_\eta) + \varepsilon h_1(x_\eta) + \varepsilon^2 h_2(x_\eta) - f_0(x_\eta) - \varepsilon f_1(x_\eta) - \varepsilon^2 f_2(x_\eta) \tag{1.75}$$

if all of the terms up to ε^2 in the outer and inner expansion are taken into consideration. As a result we obtain

$$\begin{aligned} d(x_\eta) =& A\cos\lambda_0\cos\lambda_0\eta x_\eta + A\sin\lambda_0\sin\lambda_0\eta x_\eta \\ &+ \varepsilon\left[-A\lambda_1(1-\eta x_\eta)\sin\{\lambda_0(1-\eta x_\eta)\}\right] \\ &+ \varepsilon^2\left[A\frac{\lambda_0^3-2\lambda_2}{2}(1-\eta x_\eta)\sin\{\lambda_0(1-\eta x_\eta)\}\right. \\ &\left.-A\frac{\lambda_1^2}{2}(1-\eta x_\eta)^2\{\lambda_0(1-\eta x_\eta)\}\right] \\ &- K_0\left[e^{-\frac{\eta}{\varepsilon}x_\eta} + \frac{\eta}{\varepsilon}x_\eta - 1\right] - \varepsilon K_1\left[e^{-\frac{\eta}{\varepsilon}x_\eta} + \frac{\eta}{\varepsilon}x_\eta - 1\right] \\ &- \varepsilon^2\left[K_2\left(e^{-\frac{\eta}{\varepsilon}x_\eta} + \frac{\eta}{\varepsilon}x_\eta - 1\right)\right. \\ &\left.+\frac{\lambda_0^2K_0}{2}\left(-\frac{\eta}{\varepsilon}e^{-\frac{\eta}{\varepsilon}x_\eta} - \frac{\eta}{\varepsilon}x_\eta + \frac{\eta^2}{\varepsilon^2}x_\eta^2 - \frac{1}{3}\frac{\eta^3}{\varepsilon^3}x_\eta^3\right)\right]. \end{aligned} \tag{1.76}$$

At this point it has to be underlined that all exponentially small terms vanish because

$$\lim_{\varepsilon\to 0} e^{-\frac{\eta}{\varepsilon}x_\eta} = 0, \tag{1.77}$$

due to

$$\frac{\eta}{\varepsilon} \to \infty. \tag{1.78}$$

In addition we have to take into account that the functions $\cos\lambda_0\eta x_\eta$ and $\sin\lambda_0\eta x_\eta$ for $\eta\to 0$ are not identically one and zero. Instead these functions have to be expressed in series form

$$\begin{aligned} \cos\lambda_0\eta x_\eta &= 1 - \frac{\lambda_0^2\eta^2x_\eta^2}{2} + \frac{\lambda_0^4\eta^4x_\eta^4}{24} - \dots, \\ \sin\lambda_0\eta x_\eta &= \lambda_0\eta x_\eta - \frac{\lambda_0^3\eta^3x_\eta^3}{6} + \dots. \end{aligned} \tag{1.79}$$

Furthermore, we have to take into consideration

$$\begin{aligned} \sin\{\lambda_0(1-\eta x_\eta)\} &= \sin\lambda_0\cos\lambda_0\eta x_\eta - \cos\lambda_0\sin\lambda_0\eta x_\eta, \\ \cos\{\lambda_0(1-\eta x_\eta)\} &= \cos\lambda_0\cos\lambda_0\eta x_\eta + \sin\lambda_0\sin\lambda_0\eta x_\eta. \end{aligned} \tag{1.80}$$

With these relations the difference $d(x_\eta)$ can be expressed as

$$\begin{aligned}
d(x_\eta) =& A\cos\lambda_0\left[1-\frac{\lambda_0^2\eta^2x_\eta^2}{2}+\ldots\right]+A\sin\lambda_0\left[\lambda_0\eta x_\eta-\frac{\lambda_0^3\eta^3x_\eta^3}{6}+\ldots\right]\\
&+\varepsilon\left[(-A\lambda_1\sin\lambda_0)\left(1-\frac{\lambda_0^2\eta^2x_\eta^2}{2}\ldots\right)+\left(A\lambda_1\cos\lambda_0)(\lambda_0\eta x_\eta-\frac{\lambda_0^3\eta^3x_\eta^3}{6}+\ldots\right)\right.\\
&+A\lambda_1\eta x_\eta\sin\lambda_0\left(1-\frac{\lambda_0^2\eta^2x_\eta^2)}{2}+\ldots\right)\\
&\left.+(-A\lambda_1\eta x_\eta\cos\lambda_0)\left(\lambda_0\eta x_\eta-\frac{\lambda_0^3\eta^3x_\eta^3}{6}+\ldots\right)\right]\\
&+\varepsilon^2\left[\left(A\frac{\lambda_0^3-2\lambda_2}{2}\sin\lambda_0\right)\left(1-\frac{\lambda_0^2\eta^2x_\eta^2}{2}+\ldots\right)\right.\\
&-A\frac{\lambda_0^3-2\lambda_2}{2}\cos\lambda_0\left(\lambda_0\eta x_\eta-\frac{\lambda_0^3\eta^3x_\eta^3}{6}+\ldots\right)\\
&-A\frac{\lambda_0^3-2\lambda_2}{2}\eta x_\eta\sin\lambda_0\left(1-\frac{\lambda_0^2\eta^2x_\eta^2}{2}+\ldots\right)\\
&+A\frac{\lambda_0^3-2\lambda_2}{2}\eta x_\eta\cos\lambda_0\left(\lambda_0\eta x_\eta-\frac{\lambda_0^3\eta^3x_\eta^3}{6}+\ldots\right)\\
&-A\frac{\lambda_1^2}{2}(1-2\eta x_\eta+\eta^2x_\eta^2)\cos\lambda_0\left(1-\frac{\lambda_0^2\eta^2x_\eta^2}{2}+\ldots\right)\\
&\left.-A\frac{\lambda_1^2}{2}(1-2\eta x_\eta+\eta^2x_\eta^2)\sin\lambda_0\left(\lambda_0\eta x_\eta-\frac{\lambda_0^3\eta^3x_\eta^3}{6}+\ldots\right)\right]\\
&-K_0\left[\frac{\eta}{\varepsilon}x_\eta-1\right]-\varepsilon K_1\left[\frac{\eta x_\eta}{\varepsilon}-1\right]\\
&-\varepsilon^2\left[K_2\left(\frac{\eta}{\varepsilon}x_\eta-1\right)+\frac{\lambda_0^2K_0}{2}\left(-\frac{\eta}{\varepsilon}x_\eta+\frac{\eta^2}{\varepsilon^2}x_\eta^2-\frac{1}{3}\frac{\eta^3}{\varepsilon^3}x_\eta^3\right)\right].
\end{aligned}\tag{1.81}$$

Ordering the difference $d(x_\eta)$ with respect to different orders of magnitude yields

$$O\left(\frac{\eta}{\varepsilon}\right): \quad -K_0 x_\eta, \tag{1.82a}$$

$$O(1): \quad A\cos\lambda_0, \tag{1.82b}$$

$$O(\eta): \quad A\sin\lambda_0 \cdot \lambda_0 x_\eta - K_1 x_\eta, \tag{1.82c}$$

$$O(\varepsilon): \quad -A\lambda_1 \sin\lambda_0 + K_1, \tag{1.82d}$$

$$O(\eta^2): \quad -A\cos\lambda_0 \cdot \frac{\lambda_0^2 x_\eta^2}{2}, \tag{1.82e}$$

$$O(\varepsilon\eta): \quad A\lambda_1\lambda_0 x_\eta \cos\lambda_0 + A\lambda_1 x_\eta \sin\lambda_0 - K_2 x_\eta, \tag{1.82f}$$

$$O(\varepsilon^2): \quad A\frac{\lambda_0^3 - 2\lambda_2}{2}\sin\lambda_0 - A\frac{\lambda_1^2}{2}\cos\lambda_0 + K_2. \tag{1.82g}$$

This ordering may be simplified by using computer algebra, e.g. MATHEMATICA. To obtain the equations (1.82a–1.82g) some of the results of calculations of higher orders of magnitude are utilized, e.g. the relation

$$K_0 = 0, \tag{1.83}$$

which follows from equation (1.82a). Equation (1.82b) is valid for arbitrary λ_0 only if $A = 0$, corresponding to the trivial solution. At this point it has to be noted that the problem at hand is an eigenvalue problem implying that the eigenvalue λ_0 has to be determined instead of the amplitude A. Thus equation (1.82b) serves for the determination of the eigenvalue of zeroth order λ_0. Consequently λ_0 has to be an uneven multiple of $\pi/2$.

$$\lambda_0 = \frac{2n+1}{2}\pi, \quad n = 0, 1, 2, \ldots . \tag{1.84}$$

Introducing this into equation (1.82c) yields

$$K_1 = A\lambda_0 \sin\lambda_0 = A\frac{2n+1}{2}\pi(-1)^n. \tag{1.85}$$

From equation (1.82d) we obtain

$$\lambda_1 = \frac{K_1}{A\sin\lambda_0} = \lambda_0 = \frac{2n+1}{2}\pi. \tag{1.86}$$

Equation (1.82e) is identically fulfilled if λ_0 corresponds to equation (1.84). From equation (1.82f) K_2 may be determined

$$K_2 = A\lambda_1(\lambda_0\cos\lambda_0 + \sin\lambda_0) = A\left(\frac{2n+1}{2}\pi\right)(-1)^n, \tag{1.87}$$

so that from equation (1.82g) a second order approximation for λ_2 may be computed

$$A\frac{\lambda_0^3 - 2\lambda_2}{2}(-1)^n + A\lambda_0(-1)^n = 0,$$

$$\lambda_2 = \lambda_0 + \frac{\lambda_0^3}{2} = \left[\frac{2n+1}{2}\pi\right]\left[1 + \frac{1}{2}\left[\frac{2n+1}{2}\pi\right]^2\right]. \tag{1.88}$$

Complete or Composite Solution

It is desirable to construct a solution valid not only in the boundary layer or in the outer field, but in both of them. Adding the inner and outer expansion and subtracting the common part we obtain such a solution.

$$W(x) = W^o(x) + W^i(x) - C_p(x). \tag{1.89}$$

In equation (1.89) C_p denotes the part common to both the outer and inner solution. This *common part* can be obtained if either the inner or the outer solution is expanded up to those orders of magnitude which are considered during the matching process. Exponentially small terms are neglected. In the present case these orders of magnitude are $O(\eta^2)$, $O(\varepsilon\eta)$, $O(\varepsilon^2)$. This yields

$$\begin{aligned} C_p = {} & A\left(1 - \frac{\lambda_0^2\eta^2 x_\eta^2}{2}\right)\cos\lambda_0 + A\lambda_0\eta x_\eta \sin\lambda_0 \\ & + \varepsilon\left[-A\lambda_1 \sin\lambda_0 + A\lambda_1\lambda_0\eta x_\eta \cos\lambda_0 + A\lambda_1\eta x_\eta \sin\lambda_0\right] \\ & + \varepsilon^2\left[A\frac{\lambda_0^3 - 2\lambda_2}{2}\sin\lambda_0 - A\frac{\lambda_1^2}{2}\cos\lambda_0\right]. \end{aligned} \tag{1.90}$$

Taking into account

$$\begin{aligned} \cos\lambda_0 &= 0, \\ \sin\lambda_0 &= (-1)^n, \\ \lambda_1 &= \lambda_0, \\ \lambda_2 &= \lambda_0 + \frac{\lambda_0^3}{2}, \\ \eta x_\eta &= x \end{aligned} \tag{1.91}$$

the common part can be simplified as

$$C_p = A\lambda_0 x(-1)^n + \varepsilon\left[-A\lambda_0(-1)^n + A\lambda_0 x(-1)^n\right] + \varepsilon^2\left[-A\lambda_0(-1)^n\right],$$

or

$$C_p = A\lambda_0\left[x + \varepsilon(x-1) - \varepsilon^2\right](-1)^n. \tag{1.92}$$

Of course the same result would be obtained if the solution in the boundary layer is expanded and the relations for K_1, K_2, λ_0, λ_1, λ_2 are introduced. Thus we obtain a solution valid in the whole range of definition, i.e. the eigenmode

$$\begin{aligned} W(x) = {} & A\left\{\sin\lambda_0 \sin\lambda_0 x - \varepsilon\lambda_0(1-x)\sin\left[\lambda_0(1-x)\right]\right. \\ & + \varepsilon^2(-\lambda_0)(1-x)\sin\left[\lambda_0(1-x)\right] - \varepsilon^2\frac{\lambda_0^2}{2}(1-x)^2\cos\left[\lambda_0(1-x)\right] \\ & + \varepsilon\lambda_0\sin\lambda_0\left[\mathrm{e}^{-\frac{x}{\varepsilon}} + \frac{x}{\varepsilon} - 1\right] + \varepsilon^2\lambda_0\sin\lambda_0\left[\mathrm{e}^{-\frac{x}{\varepsilon}} + \frac{x}{\varepsilon} - 1\right] \\ & \left. - \lambda_0\sin\lambda_0\left[x + \varepsilon(x-1) - \varepsilon^2\right]\right\}. \end{aligned} \tag{1.93}$$

More important than the mode however is the circular eigenfrequency ω which has to be determined in eigenvalue problems. If the parameters λ_0, λ_1, λ_2 are substituted into the asymptotic expansion for λ, the eigenvalue is determined as

$$\lambda = \lambda_0 + \varepsilon\lambda_0 + \varepsilon^2\lambda_0\left(1 + \frac{\lambda_0^2}{2}\right), \quad \lambda_0 = \frac{2n+1}{2}\pi, \quad n = 0, 1, 2, \ldots \tag{1.94}$$

For the frequency of the vibrating string we have the relation

$$\lambda^2 = \omega^2 \ell^2 \frac{\rho A}{T} = \omega^2 \frac{\ell^2}{c^2}. \tag{1.95}$$

In equation (1.95) the eigenfrequency ω and the eigenvalue λ are related so that with the square of the speed of wave propagation of a string without bending stiffness

$$c^2 = \frac{T}{\rho A} \tag{1.96}$$

we obtain

$$\lambda = \omega \frac{\ell}{c}. \tag{1.97}$$

Equation (1.97) may be solved for the eigenfrequency ω

$$\omega = \frac{c}{\ell} \lambda. \tag{1.98}$$

Taking into account that the small parameter ε corresponds to $\varepsilon = \frac{\ell_{char}}{\ell}$ the eigenfrequency for the vibrating string with small bending stiffness is

$$\omega = \frac{c}{\ell} \frac{2n+1}{2} \pi \left(1 + \frac{\ell_{char}}{\ell} + \ldots \right). \tag{1.99}$$

This relation clearly shows that the frequency of a string with small bending stiffness is slightly larger than the frequency of a string without bending stiffness. Nevertheless the increase in frequency is not only obtained as a numerical value, as it could have been done by using the Finite Element Method, but instead in an analytical form. The result shows that the eigenfrequency of a vibrating string with small bending stiffness is equal to the eigenfrequency of a vibrating string without bending stiffness but with a length reduced by the characteristic value ℓ_{char}. A comparison with the exact solution will be given later.

1.3.2 Computation with Two Boundary Layers

If we want to determine not only the symmetric modes but all modes of a string with small bending stiffness we start with the differential equation

$$\varepsilon^2 W^{IV} - W'' - \lambda^2 W = 0$$

and the corresponding boundary conditions

$$\begin{aligned} W(0) &= W(2) = 0, \\ W'(0) &= W'(2) = 0. \end{aligned} \tag{1.100}$$

Once again we use an asymptotic expansion for the eigenvalue λ in the form

$$\lambda = \lambda_0 + \varepsilon \lambda_1 + \varepsilon^2 \lambda_2 + \ldots . \tag{1.101}$$

Inner Solution in the Boundary Layer at x = 0

As shown in the previous section we introduce a new variable

$$\tilde{x} = \frac{x}{\varepsilon} \tag{1.102}$$

in the boundary layer at $x = 0$ and express the solution as an asymptotic expansion

$$W^l(\tilde{x}) = f_0(\tilde{x}) + \varepsilon f_1(\tilde{x}) + \varepsilon^2 f_2(\tilde{x}) + \dots \tag{1.103}$$

Of course this leads to the same differential equations and boundary conditions for the different approximations f_0, f_1, f_2 as in the previous section so that the solutions for $f_0(\tilde{x})$, $f_1(\tilde{x})$ and $f_2(\tilde{x})$ can be written as

$$\begin{aligned} f_0(\tilde{x}) &= K_0 \left[e^{-\tilde{x}} + (\tilde{x} - 1)\right], \\ f_1(\tilde{x}) &= K_1 \left[e^{-\tilde{x}} + (\tilde{x} - 1)\right], \\ f_2(\tilde{x}) &= K_2 \left[e^{-\tilde{x}} + (\tilde{x} - 1)\right] + \frac{\lambda_0^2 K_0}{2} \left[-\tilde{x}e^{-\tilde{x}} - \tilde{x} + \tilde{x}^2 - \frac{1}{3}\tilde{x}^3\right]. \end{aligned} \tag{1.104}$$

Outer solution

An asymptotic expansion of the outer solution

$$W^o(x) = h_0(x) + \varepsilon h_1(x) + \varepsilon^2 h_2(x) + \dots \tag{1.105}$$

also leads to the same differential eqautions as given for the outer solution in the previous section

$$\begin{aligned} h_0'' + \lambda_0^2 h_0 &= 0, \\ h_1'' + \lambda_0^2 h_1 &= -2\lambda_0\lambda_1 h_0, \\ h_2'' + \lambda_0^2 h_2 &= -(2\lambda_0\lambda_2 + \lambda_1^2)h_0 - 2\lambda_0\lambda_1 h_1 + h_0^{IV}. \end{aligned} \tag{1.106}$$

The difference is that now we do not have any boundary conditions for $h_0(x)$ so that we obtain the general solution

$$h_0(x) = C_0 \cos \lambda_0 x + S_0 \sin \lambda_0 x. \tag{1.107}$$

Introducing $h_0(x)$ into the differential equation for $h_1(x)$ leads to

$$h_1(x) = C_1 \cos \lambda_0 x + S_1 \sin \lambda_0 x + \lambda_1 S_0 x \cos \lambda_0 x - \lambda_1 C_0 x \sin \lambda_0 x, \tag{1.108}$$

so that we finally obtain the result for $h_2(x)$

$$\begin{aligned} h_2(x) =& C_2 \cos \lambda_0 x + S_2 \sin \lambda_0 x \\ &+ \left(-\frac{\lambda_1^2 S_0}{2\lambda_0} + \lambda_1 S_1 - \frac{\lambda_0^4 - 2\lambda_0\lambda_2 - \lambda_1^2}{2\lambda_0} S_0\right) x \cos \lambda_0 x \\ &+ \left(\frac{\lambda_1^2 C_0}{2\lambda_0} - \lambda_1 C_1 + \frac{\lambda_0^4 - 2\lambda_0\lambda_2 - \lambda_1^2}{2\lambda_0} C_0\right) x \sin \lambda_0 x \\ &- \frac{\lambda_1^2 S_0}{2} x^2 \sin \lambda_0 x - \frac{\lambda_1^2 C_0}{2} x^2 \cos \lambda_0 x. \end{aligned} \tag{1.109}$$

Inner Solution in the Boundary Layer at $x = 2$

The perturbation analysis shows that it is convenient to introduce a new coordinate in the boundary layer at $x = 2$, which is zero at the boundary. For the calculation of the inner solution in the boundary layer at $x = 2$ we introduce a new coordinate of the form

$$y = 2 - x, \tag{1.110}$$

or, solved for x,

$$x = 2 - y. \tag{1.111}$$

The transformation of the differential equation to the new coordinate y yields the original differential equation

$$\varepsilon^2 \hat{W}_{yyyy} - \hat{W}_{yy} - \lambda^2 \hat{W} = 0, \tag{1.112}$$

in which the indices denote the derivatives with respect to the corresponding coordinate. A transformation of the boundary conditions results in

$$\begin{aligned} \hat{W}(0) &= 0, \\ \hat{W}'(0) &= 0. \end{aligned} \tag{1.113}$$

In the differential equation (1.112) and in the boundary conditions (1.113) a new variable $\hat{W}$ was introduced to indicate that it is a function of the new coordinate y. In the next step we stretch the coordinate by introducing

$$\tilde{y} = \frac{y}{\varepsilon} \tag{1.114}$$

leading us to the same differential equation and the same boundary conditions as for the inner solution in the boundary layer at the left end. An asymptotic expansion of $\hat{W}(\tilde{y})$ in the form

$$\hat{W}(\tilde{y}) = g_0(\tilde{y}) + \varepsilon g_1(\tilde{y}) + \varepsilon^2 g_2(\tilde{y}) + \ldots \tag{1.115}$$

yields the same differential equations which we obtained for the inner solution at $x = 0$. Results for $g_0(\tilde{y})$, $g_1(\tilde{y})$ and $g_2(\tilde{y})$ can thus be given immediately

$$\begin{aligned} g_0(\tilde{y}) &= J_0 \left[e^{-\tilde{y}} + (\tilde{y} - 1) \right], \\ g_1(\tilde{y}) &= J_1 \left[e^{-\tilde{y}} + (\tilde{y} - 1) \right], \\ g_2(\tilde{y}) &= J_2 \left[e^{-\tilde{y}} + (\tilde{y} - 1) \right] + \frac{\lambda_0^2 J_0}{2} \left[-\tilde{y} e^{-\tilde{y}} + \tilde{y} + \tilde{y}^2 - \frac{1}{3} \tilde{y}^3 \right]. \end{aligned} \tag{1.116}$$

Matching of the Different Solutions

For the matching of the solutions between inner and outer expansion near the left boundary layer we once again introduce

$$x_\eta = \frac{x}{\eta} \tag{1.117}$$

with

$$\varepsilon \ll \eta \ll 1. \tag{1.118}$$

As in the previous section, the difference of both solutions at $x = \eta x_\eta$ is taken

$$\begin{aligned} d(\eta x_\eta) =& C_0 \cos \lambda_0 \eta x_\eta + S_0 \sin \lambda_0 \eta x_\eta \\ &+ \varepsilon \left[C_1 \cos \lambda_0 \eta x_\eta + S_1 \sin \lambda_0 \eta x_\eta + \lambda_1 S_0 \eta x_\eta \cos \lambda_0 \eta x_\eta - \lambda_1 C_0 \eta x_\eta \sin \lambda_0 \eta x_\eta\right] \\ &+ \varepsilon^2 \Bigg[C_2 \cos \lambda_0 \eta x_\eta + S_2 \sin \lambda_0 \eta x_\eta \\ &+ \left(-\frac{\lambda_1^2 S_0}{2\lambda_0} + \lambda_1 S_1 - \frac{\lambda_0^4 - 2\lambda_0\lambda_2 - \lambda_1^2}{2\lambda_0} S_0\right) \eta x_\eta \cos \lambda_0 \eta x_\eta \\ &+ \left(\frac{\lambda_1^2 C_0}{2\lambda_0} - \lambda_1 C_1 + \frac{\lambda_0^4 - 2\lambda_0\lambda_2 - \lambda_1^2}{2\lambda_0} C_0\right) \eta x_\eta \sin \lambda_0 \eta x_\eta \\ &- \frac{\lambda_1^2 S_0}{2} \eta^2 x_\eta^2 \sin \lambda_0 \eta x_\eta - \frac{\lambda_1^2 C_0}{2} \eta^2 x_\eta^2 \cos \lambda_0 \eta x_\eta\Bigg] \\ &- K_0 \left[e^{-\frac{\eta}{\varepsilon} x_\eta} + \frac{\eta}{\varepsilon} x_\eta - 1\right] - \varepsilon K_1 \left[e^{-\frac{\eta}{\varepsilon} x_\eta} + \frac{\eta}{\varepsilon} x_\eta - 1\right] \\ &- \varepsilon^2 \Bigg[K_2 \left(e^{-\frac{\eta}{\varepsilon} x_\eta} + \frac{\eta x_\eta}{\varepsilon} - 1\right) + \frac{\lambda_0^2 K_0}{2} \Bigg(-\frac{\eta}{\varepsilon} x_\eta e^{-\frac{\eta}{\varepsilon} x_\eta} \\ &- \frac{\eta}{\varepsilon} x_\eta + \frac{\eta^2}{\varepsilon^2} x_\eta^2 - \frac{1}{3} \frac{\eta^3}{\varepsilon^3} x_\eta^3\Bigg)\Bigg] . \end{aligned} \tag{1.119}$$

As only terms up to the order η^2 and ε^2 were taken into account, higher order terms are neglected in the following. Similarly, the exponentially small terms are neglected so that the difference $d(\eta x_\eta)$ yields

$$\begin{aligned} d(\eta x_\eta) =& C_0 \left[1 - \frac{\lambda_0^2 \eta^2 x_\eta^2}{2}\right] + S_0 \lambda_0 \eta x_\eta \\ &+ \varepsilon C_1 + \varepsilon S_1 \lambda_0 \eta x_\eta + \varepsilon \lambda_1 S_0 \eta x_\eta \\ &+ \varepsilon^2 C_2 - K_0 \frac{\eta}{\varepsilon} x_\eta + K_0 - K_1 \eta x_\eta + \varepsilon K_1 \\ &- \varepsilon K_2 \eta x_\eta + \varepsilon^2 K_2 + \varepsilon \eta \frac{\lambda_0^2 K_0 x_\eta}{2} - \eta^2 x_\eta^2 \frac{\lambda_0 K_0}{2}. \end{aligned} \tag{1.120}$$

Ordering of equation (1.120) with respect to the different orders of magnitude yields

$$-K_0 x_\eta = 0 \Rightarrow K_0 = 0 \tag{1.121}$$

for the order of $O\left(\frac{\eta}{\varepsilon}\right)$ so that we obtain for the other orders of magnitude

$$\begin{aligned} O(1): &\quad C_0 = 0, \\ O(\eta): &\quad S_0 \lambda_0 x_\eta - K_1 x_\eta = 0 \;\Rightarrow\; K_1 = \lambda_0 S_0, \\ O(\varepsilon): &\quad C_1 + K_1 = 0 \;\Rightarrow\; C_1 = -K_1 = -\lambda_0 S_0, \\ O(\eta^2): &\quad C_0 \frac{\lambda_0^2 x_\eta^2}{2} = 0, \\ O(\eta\varepsilon): &\quad S_1 \lambda_0 x_\eta + S_0 \lambda_1 x_\eta - K_2 x_\eta = 0 \;\Rightarrow\; K_2 = S_1 \lambda_0 + S_0 \lambda_1, \\ O(\varepsilon^2): &\quad C_2 + K_2 = 0 \;\Rightarrow\; C_2 = -K_2 = -S_1 \lambda_0 - S_0 \lambda_1. \end{aligned} \tag{1.122}$$

At this point it is convenient to write the outer solution explicitly, since later it also has to be transformed to the new coordinate y

$$\begin{aligned} W^o(x) =& S_0 \sin \lambda_0 x + \varepsilon \left[-\lambda_0 S_0 \cos \lambda_0 x + S_1 \sin \lambda_0 x + \lambda_1 S_0 x \cos \lambda_0 x\right] \\ &+ \varepsilon^2 \left[(-S_1\lambda_0 - S_0\lambda_1) \cos \lambda_0 x + S_2 \sin \lambda_0 x \right. \\ &+ \left(-\frac{\lambda_0^4 - 2\lambda_0\lambda_2}{2\lambda_0} S_0 + \lambda_1 S_1\right) x \cos \lambda_0 x - \frac{\lambda_1^2}{2} S_0 x^2 \sin \lambda_0 x \\ &\left. + \lambda_0\lambda_1 S_0 x \sin \lambda_0 x\right]. \end{aligned} \tag{1.123}$$

For the matching of the outer and inner solution at the right boundary the outer solution, $W^o(x)$ is transformed to the new boundary layer coordinate y

$$\begin{aligned} \widehat{W}^o(y) =& S_0 \sin\left[\lambda_0(2-y)\right] \\ &+ \varepsilon \left\{-\lambda_0 S_0 \cos\left[\lambda_0(2-y)\right] + S_1 \sin\left[\lambda_0(2-y)\right] + \lambda_1 S_0 (2-y) \cos\left[\lambda_0(2-y)\right]\right\} \\ &+ \varepsilon^2 \left\{(-S_1\lambda_0 - S_0\lambda_1) \cos\left[\lambda_0(2-y)\right] + S_2 \sin\left[\lambda_0(2-y)\right] \right. \\ &+ \left(-\frac{\lambda_0^4 - 2\lambda_0\lambda_2}{2\lambda_0} S_0 + \lambda_1 S_1\right) (2-y) \cos\left[\lambda_0(2-y)\right] \\ &\left. - \frac{\lambda_1^2}{2} S_0 (2-y)^2 \sin\left[\lambda_0(2-y)\right] + \lambda_0\lambda_1 S_0 (2-y) \sin\left[\lambda_0(2-y)\right]\right\}. \end{aligned} \tag{1.124}$$

As for the matching at the left end, we introduce a fixed coordinate y_η between the right boundary layer and the outer field, at which both solutions should coincide. Therefore we have

$$y_\eta = \frac{y}{\eta} \tag{1.125}$$

or, solved for y,

$$y = \eta y_\eta. \tag{1.126}$$

Further on the relations

$$\sin\left[\lambda_0(2-y)\right] = \sin 2\lambda_0 \cos \lambda_0 y - \cos 2\lambda_0 \sin \lambda_0 y, \tag{1.127}$$

$$\cos\left[\lambda_0(2-y)\right] = \cos 2\lambda_0 \cos \lambda_0 y + \sin 2\lambda_0 \sin \lambda_0 y \tag{1.128}$$

are needed. The difference $\hat{d}(\eta y_\eta)$ between the outer and the inner expansion at an intermediate point between the right boundary layer and the free field is

$$\begin{aligned}\hat{d}(\eta y_\eta) =& S_0 \sin 2\lambda_0 \left(1 - \frac{\lambda_0^2 \eta^2 y_\eta^2}{2}\right) - \lambda_0 \eta y_\eta \cdot S_0 \cos 2\lambda_0 \\ &+ \varepsilon \{-\lambda_0 S_0 [\cos 2\lambda_0 + \lambda_0 \eta y_\eta \cdot \sin 2\lambda_0] \\ &+ S_1 \sin 2\lambda_0 - S_1 \lambda_0 \eta y_\eta \cdot \cos 2\lambda_0 \\ &+ S_0 \lambda_1 (2 - \eta y_\eta)[\cos 2\lambda_0 + \lambda_0 \eta y_\eta \sin 2\lambda_0]\} \\ &+ \varepsilon^2 \{(-S_1 \lambda_0 - S_0 \lambda_1) \cos 2\lambda_0 + S_2 \sin 2\lambda_0 \\ &+ \left(-\frac{\lambda_0^4 - 2\lambda_0 \lambda_2}{2\lambda_0} S_0 + \lambda_1 s_1\right) 2 \cdot \cos 2\lambda_0 \\ &- \frac{\lambda_1^2}{2} S_0 \cdot 4 \cdot \sin 2\lambda_0 + \lambda_0 \lambda_1 S_0 \cdot 2 \cdot \sin 2\lambda_0 \Big\} \\ &- J_0 \left[\frac{\eta}{\varepsilon} y_\eta - 1\right] - \varepsilon J_1 \left[\frac{\eta}{\varepsilon} y_\eta - 1\right] - \varepsilon^2 \Big\{ J_2 \left[\frac{\eta}{\varepsilon} y_\eta - 1\right] \\ &- \frac{\lambda_0^2 J_0}{2} \left[-\frac{\eta}{\varepsilon} y_\eta + \frac{\eta^2}{\varepsilon^2} y_\eta^2 - \frac{1}{3} \frac{\eta^3}{\varepsilon^3} y_\eta^3\right]\Big\} .\end{aligned} \tag{1.129}$$

In equation (1.129) the exponentially small terms have already been neglected and terms of higher order than η^2, $\varepsilon\eta$ or ε^2 have been partially neglected. From the highest order term $O\left(\frac{\eta}{\varepsilon}\right)$ follows $J_0 = 0$, and thus for the other orders of magnitude

$$O(1): \quad S_0 \sin 2\lambda_0 = 0, \tag{1.130a}$$

$$O(\eta): \quad - S_0 \lambda_0 y_\eta \cos 2\lambda_0 - J_1 y_\eta = 0, \tag{1.130b}$$

$$O(\varepsilon): \quad - \lambda_0 S_0 \cos 2\lambda_0 + S_1 \sin 2\lambda_0 + 2 S_0 \lambda_1 \cos 2\lambda_0 + J_1 = 0, \tag{1.130c}$$

$$O(\eta^2): \quad - S_0 \frac{\lambda_0^2 y_\eta^2}{2} \sin 2\lambda_0 = 0, \tag{1.130d}$$

$$\begin{aligned}O(\varepsilon\eta): \quad & - S_0 \lambda_0^2 y_\eta \sin 2\lambda_0 - S_1 \lambda_0 y_\eta \cos 2\lambda_0 - S_0 \lambda_1 y_\eta \cos 2\lambda_0 \\ & + 2 S_0 \lambda_0 \lambda_1 y_\eta \sin 2\lambda_0 - J_2 y_\eta = 0,\end{aligned} \tag{1.130e}$$

$$\begin{aligned}O(\varepsilon^2): \quad & (-S_1 \lambda_0 - S_0 \lambda_1) \cos 2\lambda_0 + S_2 \sin 2\lambda_0 \\ & + 2\left(-\frac{\lambda_0^4 - 2\lambda_0\lambda_2}{2\lambda_0} S_0 + \lambda_1 S_1\right) \cos 2\lambda_0 \\ & - 2\lambda_1^2 S_0 \sin 2\lambda_0 + 2\lambda_0 \lambda_1 S_0 \sin 2\lambda_0 + J_2 = 0.\end{aligned} \tag{1.130f}$$

Equation (1.130a) implies that $S_0 \neq 0$ can only be chosen if

$$\sin 2\lambda_0 = 0 \tag{1.131}$$

or, expressed in terms of λ_0,

$$\lambda_0 = \frac{n\pi}{2}, \quad n = 1, 2, 3, \dots . \tag{1.132}$$

Introducing (1.132) into (1.130b) yields

$$J_1 = -S_0 \lambda_0 \cos 2\lambda_0 = -S_0 \lambda_0 (-1)^n . \tag{1.133}$$

In addition, the other equations require

$$\lambda_1 = \lambda_0, \tag{1.134}$$

$$J_2 = -(S_1\lambda_0 + \lambda_1 S_0)\cos 2\lambda_0, \tag{1.135}$$

$$\lambda_2 = \left(1 + \frac{\lambda_0^2}{2}\right)\lambda_0. \tag{1.136}$$

For the eigenvalue λ we therefore obtain the same result as in the previous section where we used the symmetry condition and took into consideration the boundary layer at $x = 0$ only. This result being valid for eigenvalues that are uneven multiples of $\pi/2$. Of course the integration constant S_1 remains arbitrary, as $S_0 + \varepsilon S_1$ may be expressed as a new constant $\overline{S}_0$ comprising all the sinusoidal parts of the outer expansion.

Complete Solution

With the solutions obtained for the two boundary layers and the free field, we construct a solution valid in the whole region between $x = 0$ and $x = 2$. The procedure is the same as in the previous section, where we only considered symmetric modes. Therefore we write

$$W(x) = W^l(x) + W^o(x) + W^r(x) - C_{p1}(x) - C_{p2}(x) \tag{1.137}$$

with W^l as the solution in the left boundary layer, W^o the outer solution and W^r the solution in the right boundary layer. C_{p1} and C_{p2} are the common parts of the outer expansion and the left boundary layer and of the outer expansion and the right boundary layer, respectively. In the common parts once again terms up to the order of η^2, $\varepsilon\eta$ and ε^2 are taken into consideration. The different parts are

$$\begin{aligned}
C_{p1}(x) &= S_0\lambda_0 x + \varepsilon\lambda_0[-S_0 + S_1 x + S_0 x] + \varepsilon^2\lambda_0[-S_1 - S_0],\\
C_{p2}(x) &= \{-S_0\lambda_0(2-x) + \varepsilon[-\lambda_0 S_0 - \lambda_0 S_1(2-x) + S_0\lambda_0 x]\\
&\quad +\varepsilon^2[\lambda_0 S_0 + \lambda_0 S_1]\}\cos 2\lambda_0,\\
W^l(x) &= \varepsilon\lambda_0 S_0\left[e^{-\frac{x}{\varepsilon}} + \frac{x}{\varepsilon} - 1\right] + \varepsilon^2(S_1\lambda_0 + S_0\lambda_0)\left[e^{-\frac{x}{\varepsilon}} + \frac{x}{\varepsilon} - 1\right],\\
W^o(x) &= S_0\sin\lambda_0 x + \varepsilon[-\lambda_0 S_0\cos\lambda_0 x + S_1\sin\lambda_0 x + \lambda_0 S_0 x\cos\lambda_0 x]\\
&\quad \varepsilon^2\Big[(-S_1\lambda_0 - S_0\lambda_0)\cos\lambda_0 x + S_2\sin\lambda_0 x\\
&\quad + (\lambda_0 S_1 + \lambda_0 S_0)x\cos\lambda_0 x\\
&\quad + \lambda_0^2 S_0 x\sin\lambda_0 x - \frac{\lambda_0^3 S_0}{2}x^2\sin\lambda_0 x\Big],\\
W^r(x) &= -\varepsilon S_0\lambda_0\cos 2\lambda_0\left[e^{-\frac{2-x}{\varepsilon}} + \frac{2-x}{\varepsilon} - 1\right]\\
&\quad - \varepsilon^2(S_1\lambda_0 + S_0\lambda_0)\cos 2\lambda_0\left[e^{-\frac{2-x}{\varepsilon}} + \frac{2-x}{\varepsilon} - 1\right].
\end{aligned} \tag{1.138}$$

They should be substituted into equation (1.137).

1.3.3 Comparison with the Exact Solution

The simple eigenvalue problem described by the differential equation

$$W'''' - \frac{1}{\varepsilon^2}W'' - \frac{\lambda^2}{\varepsilon^2}W = 0 \tag{1.139}$$

and the boundary conditions

$$\begin{aligned} W(0) &= W(2) = 0, \\ W'(0) &= W'(2) = 0 \end{aligned} \tag{1.140}$$

can of course be solved analytically using an ansatz

$$W = Ce^{\kappa x} \tag{1.141}$$

Introducing equation (1.141) into equation (1.139) leads to

$$\kappa^4 - \frac{\kappa^2}{\varepsilon^2} - \frac{\lambda^2}{\varepsilon^2} = 0, \tag{1.142}$$

and thus to

$$\kappa^2 = \frac{1}{2\varepsilon^2}(1 \pm \sqrt{1 + 4\varepsilon^2\lambda^2}). \tag{1.143}$$

With

$$\kappa_1^2 = \frac{1}{2\varepsilon^2}(1 + \sqrt{1 + 4\varepsilon^2\lambda^2}), \tag{1.144}$$

$$\kappa_2^2 = \frac{1}{2\varepsilon^2}(-1 + \sqrt{1 + 4\varepsilon^2\lambda^2}) \tag{1.145}$$

the general solution can be written as

$$W(x) = C_1 \sin \kappa_2 x + C_2 \cos \kappa_2 x + C_3 \sinh \kappa_1 x + C_4 \cosh \kappa_1 x. \tag{1.146}$$

The boundary conditions lead to the characteristic equation

$$2\frac{\lambda}{\varepsilon}[\cos 2\kappa_2 \cosh 2\kappa_1 - 1] - \frac{1}{\varepsilon^2} \sin 2\kappa_2 \sinh 2\kappa_1 = 0. \tag{1.147}$$

As κ_1 and κ_2 are implicit functions of the eigenvalue λ

$$\kappa_1 = f(\lambda), \tag{1.148}$$

$$\kappa_2 = g(\lambda) \tag{1.149}$$

possible eigenvalues can be obtained numerically, e.g. with MATLAB or MATHEMATICA. For the calculation of the mode shapes, e.g. C_1 can be set to unity to determine the remaining integration constants by the following relations

$$\begin{aligned} C_2 &= -\frac{\kappa_1 \sin 2\kappa_2 - \kappa_2 \sinh 2\kappa_1}{\kappa_1 \cos 2\kappa_2 - \kappa_1 \cosh 2\kappa_1} C_1, \\ C_3 &= -\frac{\kappa_2}{\kappa_1} C_1, \\ C_4 &= -C_2. \end{aligned} \tag{1.150}$$

Thus the analytical solution is known and can be plotted e.g. with MATHEMATICA or MATLAB, though these programs may face difficulties if the characteristic length ℓ_{char} is small compared to the length ℓ of the string.

1.3.4 Numerical Results

The following figures show the solutions for the outer field and for the boundary layers together with the common parts as well a comparison of the complete solution with the exact solution for the mode shapes. The comparisons are done both for the case in which only symmetric modes are calculated by taking into consideration only one boundary layer at the left end of the string as well as for the case in which the complete string was considered with two boundary layers (one at either end of the string). In Figure 1.2 the first mode is considered which was determined by analyzing the symmetric problem both for $\varepsilon = 0.2$ as well as $\varepsilon = 0.05$. In the upper figures the outer solution can be seen, the solution in the boundary layer is depicted in the mid figures together with the common part, and a comparison of the complete solution with the exact solution which was obtained numerically. For the parameters used, these two solutions are indistinguishable in the figures (lower figures). In Figure 1.3 the results are shown for the second symmetric mode which corresponds to mode three of the complete string. Figures 1.4, 1.5 and 1.6 show the results for the outer solution, the inner solutions in comparison with the common parts and a comparison of the complete solutions with the exact solutions for the first three modes of a vibrating string with a small bending stiffness. It can bee seen that even for moderate values of ε the exact solution and a solution by singular perturbation theory agree very well. At this point it has to be said that for very small values of ε the exact solution, which has to be calculated numerically, sometimes leads to error messages by MATLAB.

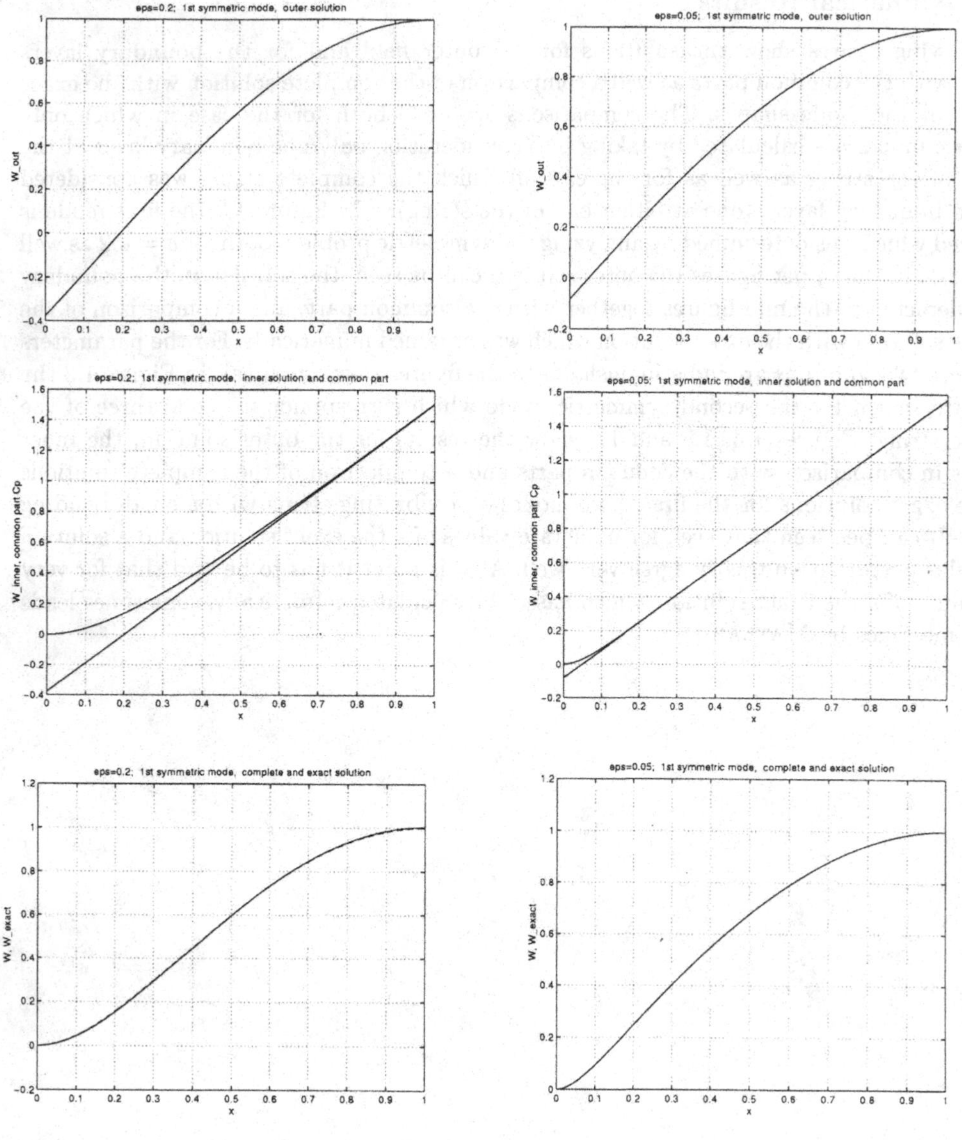

Figure 1.2: Outer solution, inner solution, and common part as well as a comparison of the complete solution and the exact solution obtained numerically for the first symmetric mode determined by considering the left boundary layer and symmetry conditions at $x = 1$. Left figures for $\varepsilon = 0.2$, right figures for $\varepsilon = 0.05$.

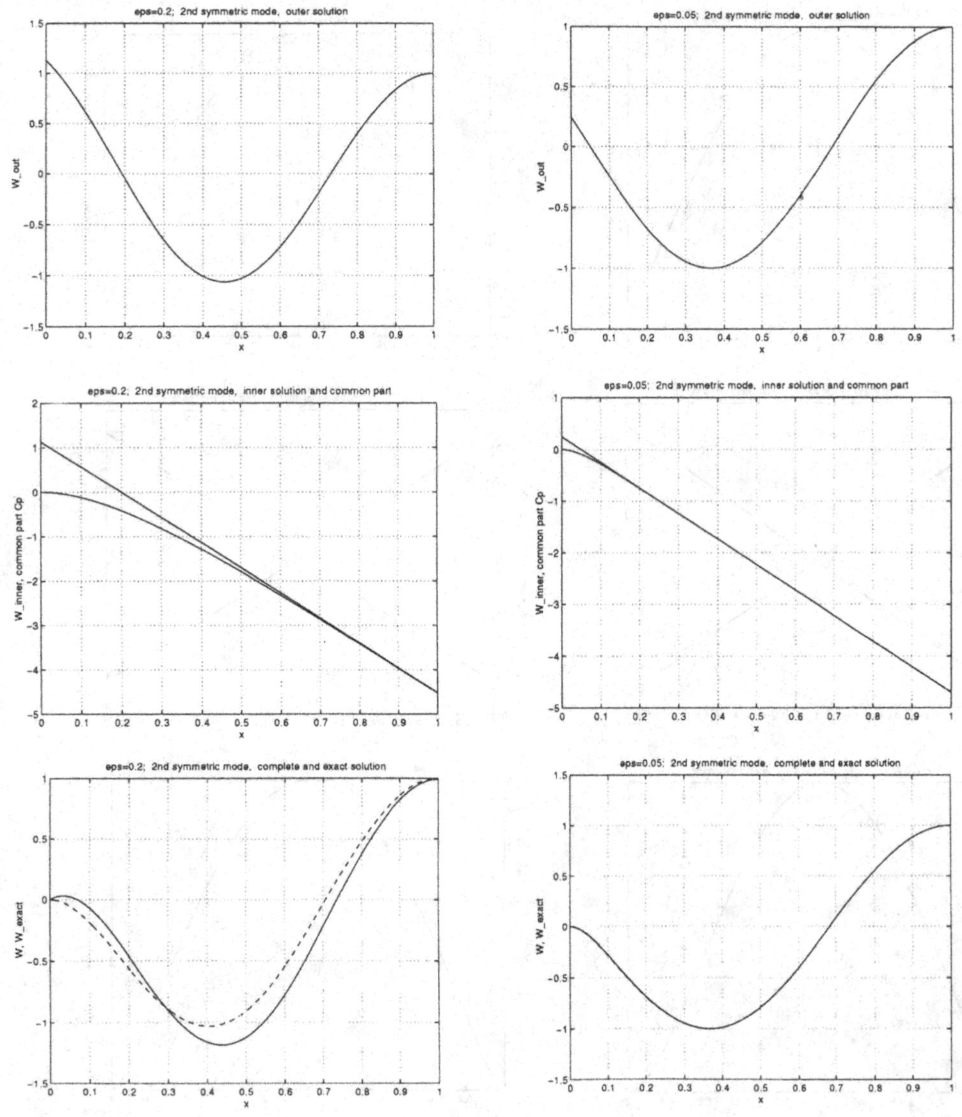

Figure 1.3: Outer solution, inner solution, and common part as well as a comparison of the complete solution and the exact solution obtained numerically for the second symmetric mode determined by considering the left boundary layer and symmetry conditions at $x = 1$. Left figures for $\varepsilon = 0.2$, right figures for $\varepsilon = 0.05$.

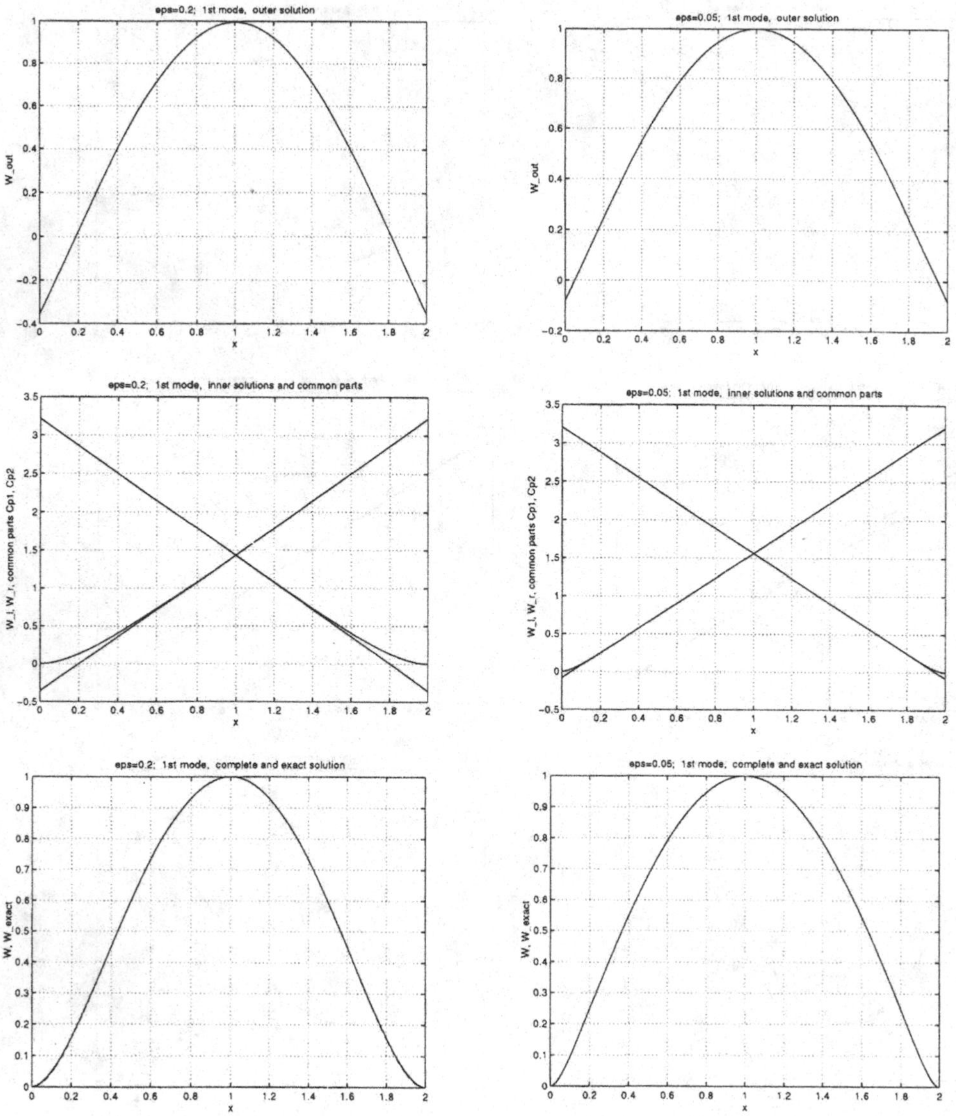

Figure 1.4: Outer solution, inner solutions, and common parts of the two boundary layers and a comparison of the complete and the exact numerical solution. First mode, $\varepsilon = 0.2$, $\varepsilon = 0.05$

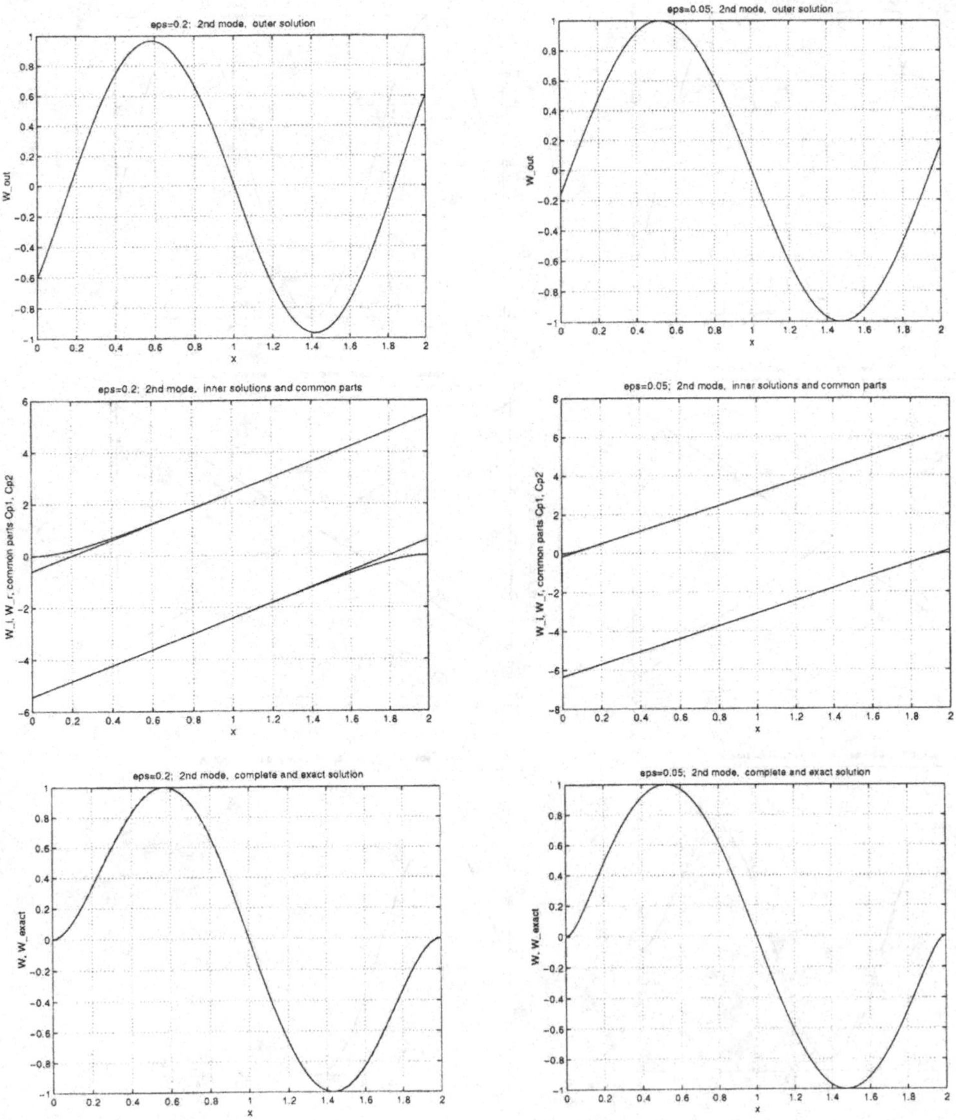

Figure 1.5: Outer solution, inner solutions, and common parts of the two boundary layers and a comparison of the complete and the exact numerical solution. Second mode, $\varepsilon = 0.2$, $\varepsilon = 0.05$

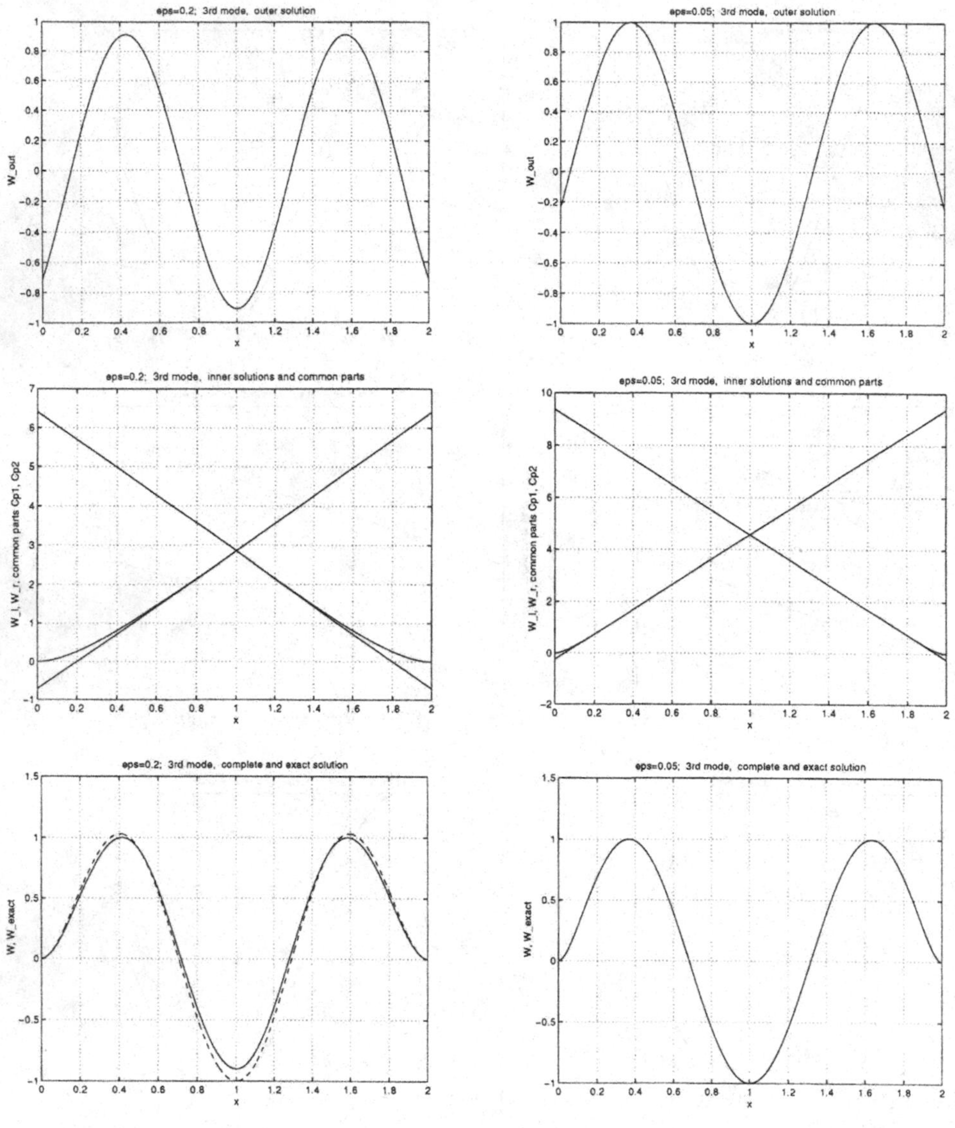

Figure 1.6: Outer solution, inner solutions, and common parts of the two boundary layers and a comparison of the complete and the exact numerical solution. Third mode, $\varepsilon = 0.2$, $\varepsilon = 0.05$

1.4 Vibrating Cylinder in a Cylindrical Fluid-filled Duct

1.4.1 Introduction

Another technical system which may be analyzed by using singular perturbation theory is a vibrating cylinder in a cylindrical duct filled with an incompressible fluid with low viscosity. In order to get a two-dimensional system a circular rigid cylinder of radius r_i of infinite length and a coaxial cylindrical duct of radius r_o is considered, see figure 1.7. The cylinder of mass

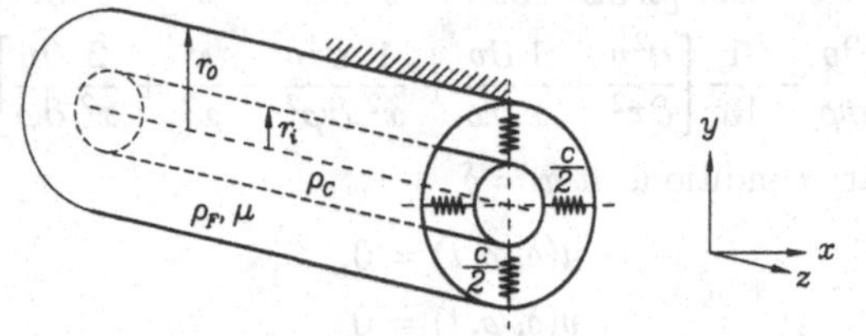

Figure 1.7: Cylinder vibrating in a duct filled with a fluid of low viscosity

density ρ_c is elastically supported by a foundation with linear stiffness c per unit length. The annular volume between cylinder and container is filled with an incompressible viscous fluid of mass density ρ_f and viscosity μ. Purely transversal oscillations of the cylinder are considered. Therefore the fluid velocity has no axial component. For very low viscosities, that are often found in practice, numerical difficulties arise because the coefficient of the highest derivative of the governing differential equation is proportional to μ. Also, the influence of the parameters on the eigenvalue cannot be seen directly if the system is analyzed by the Finite Element Method. Applying singular perturbation theory thus is convenient to analyze the dependence of the eigenvalues on the very low viscosity and the other system parameters.

1.4.2 Formulation of the Problem

In order to reduce the number of parameters, it is convenient to introduce nondimensional variables

$$x = \frac{r}{r_i}, \quad \delta = \frac{r_o}{r_i}, \tag{1.151}$$

$$q = \frac{\bar{q}}{r_i}, \quad u = \frac{\bar{u}}{r_i\lambda_0}, \quad v = \frac{\bar{v}}{r_i\lambda_0}, \tag{1.152}$$

$$t = \bar{t}\lambda_0, \quad p = \frac{\bar{p}}{\rho_F r_i^2 \lambda_0^2} \tag{1.153}$$

$$\mathrm{Re} = \frac{\rho_F \lambda_0 r_i^2}{\mu}, \quad \alpha = \frac{\rho_F}{\rho_C} \tag{1.154}$$

with

$$\lambda_0^2 = \frac{c}{\rho_C r_i^2}. \tag{1.155}$$

In equation (1.151) x denotes the nondimensional radial coordinate of a cylindrical r,φ,z coordinate system, t the nondimensional time. u and v are nondimensional velocities of the fluid in radial and circumferential direction and q is the nondimensional displacement of the cylinder, respectively. p is a pressure disturbance. Re is a REYNOLDS number which tends to infinity if the fluid is inviscid, and α represents the ratio of the mass densities.

The differential equations for the fluid are the continuity equation

$$\frac{\partial u}{\partial x} + \frac{1}{x}\frac{\partial v}{\partial \varphi} + \frac{u}{x} = 0, \tag{1.156}$$

and the linearized NAVIER-STOKES equations

$$\frac{\partial u}{\partial t} + \frac{\partial p}{\partial x} - \frac{1}{\mathrm{Re}}\left[\frac{1}{x}\frac{\partial}{\partial x}(x\frac{\partial u}{\partial x}) + \frac{1}{x^2}\frac{\partial^2 u}{\partial \varphi^2} - \frac{u}{x^2} - \frac{2}{x^2}\frac{\partial v}{\partial \varphi}\right] = 0, \tag{1.157}$$

$$\frac{\partial v}{\partial t} + \frac{1}{x}\frac{\partial p}{\partial \varphi} - \frac{1}{\mathrm{Re}}\left[\frac{\partial^2 v}{\partial x^2} + \frac{1}{x}\frac{\partial v}{\partial x} + \frac{1}{x^2}\frac{\partial^2 v}{\partial \varphi^2} - \frac{v}{x^2} + \frac{2}{x^2}\frac{\partial u}{\partial \varphi}\right] = 0. \tag{1.158}$$

In addition, the boundary conditions at $x = \delta$

$$u(\delta, \varphi, t) = 0, \tag{1.159}$$

$$v(\delta, \varphi, t) = 0 \tag{1.160}$$

and the transition conditions between the fluid and the cylinder

$$\frac{d^2 q}{dt^2} + q - \frac{\varepsilon}{\pi}\int_0^{2\pi}\left\{\left[\frac{2}{\mathrm{Re}}\frac{\partial u}{\partial x} - \frac{1}{Re}\left(\frac{\partial u}{\partial x} + \frac{1}{x}\frac{\partial v}{\partial \varphi} + \frac{u}{x}\right) - p\right]_{x=1}\sin\varphi + \frac{1}{\mathrm{Re}}\left[\frac{\partial v}{\partial x} - \frac{v}{x} + \frac{1}{x}\frac{\partial u}{\partial \varphi}\right]_{x=1}\cos\varphi\right\} d\varphi = 0, \tag{1.161a}$$

$$u(1, \varphi, t) = \frac{dq}{dt}\sin\varphi, \tag{1.161b}$$

$$v(1, \varphi, t) = \frac{dq}{dt}\cos\varphi \tag{1.161c}$$

have to be fullfilled.

1.4.3 Analytical Exact Solution

Introducing a stream function ψ defined by

$$u(x, \varphi, t) = -\frac{1}{x}\frac{\partial \psi}{\partial \varphi}, \quad v(x, \varphi, t) = \frac{\partial \psi}{\partial x} \tag{1.162}$$

gives a differential equation of the form

$$\Delta\Delta\psi - \mathrm{Re}\Delta\psi_t = 0. \tag{1.163}$$

It can be written as

$$\Delta(\Delta\psi - \mathrm{Re}\psi_t) = 0 \tag{1.164}$$

with the solution

$$\psi = [AJ_1(\kappa x) + BY_1(\kappa x) + Cx + \frac{D}{x}]e^{i\omega t}e^{in\varphi} \tag{1.165}$$

where J_1 and Y_1 denote the well known BESSEL functions. The parameter

$$\kappa = \sqrt{i\omega\mathrm{Re}} = (i+1)\sqrt{\frac{\omega\mathrm{Re}}{2}} \tag{1.166}$$

is a function of ω and Re. Due to the kinematical transition conditions (1.161b) and (1.161c), only the circumferential wavenumber $n = 1$ is used. Assuming a corresponding solution

$$q(t) = Ce^{i\omega t} \tag{1.167}$$

for the motion of the cylinder and applying all boundary conditions (1.159) and (1.161) leads to the governing characteristic equation in the form of a vanishing determinant. Obviously, for a low viscosity μ, i.e. high REYNOLDS numbers Re, the coefficient of the highest derivative of the equation of motion (1.164) is a small number. Therefore, the characteristic equation for the determination of the eigenvalues, which is obtained by fitting the solution (1.165) to the boundary conditions at $x = \delta$, and the transition conditions at $x = 1$ is difficult to evaluate numerically. Even for moderate viscosities the solution of the characteristic equation by MATHEMATICA failed. Therefore, it was necessary to get approximate solutions by means of singular perturbation theory.

1.4.4 Solution by Perturbation Theory

For the perturbation analysis it is convenient not to introduce a stream function but to work with the velocity components. Then together with $q(t)$ according to equation (1.167) we assume the quantities u, v and p to be of the form

$$u(x,\varphi,t) = U(x)e^{i\omega t}\sin\varphi, \tag{1.168}$$

$$v(x,\varphi,t) = V(x)e^{i\omega t}\cos\varphi, \tag{1.169}$$

$$p(x,\varphi,t) = P(x)e^{i\omega t}\sin\varphi, \tag{1.170}$$

which leads to

$$0 = i\omega(3x^2U' + x^3U'') + \frac{1}{\text{Re}}(-x^3U'''' - 6x^2U''' - 3xU'' + 3U'), \tag{1.171}$$

$$V(x) = U(x) + xU'(x), \tag{1.172}$$

$$P(x) = -xi\omega[U(x) + xU'(x)] + \frac{1}{\text{Re}}[4xU''(x) + x^2U'''(x)] \tag{1.173}$$

and the boundary conditions

$$U(1) = C_1 i\omega, \tag{1.174a}$$

$$V(1) = C_1 i\omega, \tag{1.174b}$$

$$(1-\omega^2)C_1 - \alpha\left[i\omega[U(1) + U'(1)] + \frac{3}{2Re}[3U'(1) - 3U''(1) - U'''(1)]\right] = 0. \tag{1.174c}$$

$$U(\delta) = 0, \tag{1.174d}$$

$$V(\delta) = 0, \tag{1.174e}$$

where a prime denotes a derivative with respect to x. Now, the small parameter ε has to be introduced. It is convenient to choose

$$\varepsilon^2 = \frac{1}{\text{Re}} \tag{1.175}$$

and the differential equation then reads

$$i\omega(3x^2\hat{U}' + x^3\hat{U}'') + \varepsilon^2(-x^3\hat{U}'''' - 6x^2\hat{U}''' - 3x\hat{U}'' + 3\hat{U}') = 0. \tag{1.176}$$

According to singular perturbation theory first the solution is computed separately in a region away from the boundaries and in the boundary layers near $x = 1$ and $x = \delta$. For both the outer and inner solutions the eigenvalue ω is assumed to be of the form

$$\omega = \omega_0 + \varepsilon\omega_1 + \varepsilon^2\omega_2 + \dots . \tag{1.177}$$

Outer Solution

The outer solution $\hat{U}(x)$ is expressed in the form

$$\hat{U} = \hat{u}_0(x) + \varepsilon\hat{u}_1(x) + \varepsilon^2\hat{u}_2(x) + \dots . \tag{1.178}$$

The differential equation (1.171) for the outer solution reads

$$i\omega(3x^2\hat{U}' + x^3\hat{U}'') + \varepsilon^2(-x^3\hat{U}'''' - 6x^2\hat{U}''' - 3x\hat{U}'' + 3\hat{U}') = 0, \tag{1.179}$$

which is exactly the original differential equation. After introducing ω and $\hat{U}$ in this differential equation and ordering equal powers of ε a system of differential equations

$$3\hat{u}_0' + x\hat{u}_0'' = 0, \tag{1.180}$$

$$3\hat{u}_1' + x\hat{u}_1'' = -\frac{\omega_1}{\omega_0}(3\hat{u}_0' + x\hat{u}_0''), \tag{1.181}$$

$$\begin{aligned} 3\hat{u}_2' + x\hat{u}_2'' = &-\frac{\omega_2}{\omega_0}(3\hat{u}_0' + x\hat{u}_0'') - \frac{\omega_1}{\omega_0}(3\hat{u}_1' + x\hat{u}_1'') \\ &- \frac{1}{i\omega_0 x^2}(-x^3\hat{u}_0'''' - 6x^2\hat{u}_0''' - 3x\hat{u}_0'' + 3\hat{u}_0'), \end{aligned} \tag{1.182}$$

$$\begin{aligned} 3\hat{u}_3' + x\hat{u}_3'' = &-\frac{\omega_3}{\omega_0}(3\hat{u}_0' + x\hat{u}_0'') - \frac{\omega_2}{\omega_0}(3\hat{u}_1' + x\hat{u}_1'') \\ &- \frac{\omega_1}{\omega_0}(3\hat{u}_2' + x\hat{u}_2'') - \frac{1}{i\omega_0 x^2}(-x^3\hat{u}_1'''' - 6x^2\hat{u}_1''' - 3x\hat{u}_1'' + 3\hat{u}_1'), \end{aligned} \tag{1.183}$$

$$\vdots$$

results. The solution of (1.180) is given by

$$\hat{u}_0 = C_{10} + \frac{C_{20}}{x^2}. \tag{1.184}$$

Introducing the solution for $\hat{u}_0$ into equation (1.181) yields

$$3x\hat{u}_1' + x^2\hat{u}_1'' = 0 \tag{1.185}$$

with the solution

$$\hat{u}_1 = C_{11} + \frac{C_{21}}{x^2}. \tag{1.186}$$

Careful inspection of the recursive system shows that all right-hand sides vanish due to solutions calculated before. Therefore, the solution for $\hat{u}_j$, $j > 1$ is also given by

$$\hat{u}_j = C_{1j} + \frac{C_{2j}}{x^2}. \tag{1.187}$$

The integration constants cannot be determined yet. This will be done later by matching the solutions between the different regions.

Inner Solution at $x = 1$

For the inner solution near $x = 1$ the radial coordinate x is transformed to $\tilde{x}$

$$\tilde{x} = \frac{1}{\varepsilon}(x - 1). \tag{1.188}$$

The solution $\tilde{U}$ in the boundary layer is expanded as

$$\tilde{U} = \tilde{u}_0(x) + \varepsilon\tilde{u}_1(x) + \varepsilon^2\tilde{u}_2(x) + \dots . \tag{1.189}$$

Introducing all expressions into the differential equation (1.171) and collecting equal powers of ε gives a system of differential equations

$$-\tilde{u}_0'''' + i\omega_0\tilde{u}_0'' = 0, \tag{1.190}$$

$$-\tilde{u}_1'''' + i\omega_0\tilde{u}_1'' = 3\tilde{x}\tilde{u}_0'''' + 6\tilde{u}_0''' - 3\tilde{x}i\omega_0\tilde{u}_0'' - i\omega_1\tilde{u}_0'' - 3i\omega_0\tilde{u}_0', \tag{1.191}$$

$$\begin{aligned} -\tilde{u}_2'''' + i\omega_0\tilde{u}_2'' = {} & 3\tilde{x}\tilde{u}_1'''' + 6\tilde{u}_1''' + 12\tilde{u}_0''' + 3\tilde{u}_0'' - i\omega_2\tilde{u}_0'' - i\omega_1\tilde{u}_1'' \\ & - 3\tilde{x}i\omega_1\tilde{u}_0'' - 3\tilde{x}i\omega_0\tilde{u}_1'' - 3i\omega_1\tilde{u}_0' - 3i\omega_0\tilde{u}_1' \\ & - 6\tilde{x}i\omega_0\tilde{u}_0' + 3\tilde{x}^2\tilde{u}_0'''' - 3\tilde{x}^2 i\omega_0\tilde{u}_0'', \end{aligned} \tag{1.192}$$

$$\vdots \tag{1.193}$$

where now the prime denotes a differentiation with respect to the coordinate $\tilde{x}$. In addition, the transition conditions (1.174) at $\tilde{x} = 0$ are

$$\tilde{U}(0) = C_1 i\omega, \tag{1.194}$$

$$\tilde{V}(0) = \tilde{U}(0) + \frac{1}{\varepsilon}\tilde{U}'(0) = C_1 i\omega, \tag{1.195}$$

$$\begin{aligned} 0 = {} & (1 - \omega^2)C_1 - \alpha\Big[i\omega[\tilde{U}(0) + \frac{1}{\varepsilon}\tilde{U}'(0)] \\ & + \varepsilon^2[3\frac{1}{\varepsilon}\tilde{U}'(0) - 3\frac{1}{\varepsilon^2}\tilde{U}''(0) - \frac{1}{\varepsilon^3}\tilde{U}'''(0)]\Big] \end{aligned} \tag{1.196}$$

or by collecting equal powers of ε

$$\begin{aligned} \tilde{u}_0(0) &= C_1 i\omega_0, \\ \tilde{u}_1(0) &= C_1 i\omega_1, \\ \tilde{u}_2(0) &= C_1 i\omega_2, \end{aligned} \tag{1.197}$$

$$\vdots$$

$$\tilde{u}_0'(0) = 0, \tag{1.198a}$$

$$\tilde{u}_0(0) + \tilde{u}_1'(0) = C_1 i\omega_0, \tag{1.198b}$$

$$\tilde{u}_1(0) + \tilde{u}_2'(0) = C_1 i\omega_1, \tag{1.198c}$$

$$\vdots$$

$$-\alpha[-\tilde{u}_0'''(0) + i\omega_0\tilde{u}_0'(0)] = 0, \tag{1.199a}$$

$$-\alpha[-\tilde{u}_1'''(0) + i\omega_0\tilde{u}_1'(0)] = -(\lambda_0^2 - \omega_0^2)C_1 + \alpha\left[i\omega_0\tilde{u}_0(0) + i\omega_1\tilde{u}_0'(0) - 3\tilde{u}_0''(0)\right], \tag{1.199b}$$

$$\begin{aligned}-\alpha[-\tilde{u}_2'''(0) + i\omega_0\tilde{u}_2'(0)] = 2\omega_0\omega_1 C_1 + \alpha &\left[i\omega_0\tilde{u}_1(0) + i\omega_1\tilde{u}_0(0) + i\omega_1\tilde{u}_1'(0)\right.\\ &\left.+ i\omega_2\tilde{u}_0'(0) + 3\tilde{u}_0'(0) - 3\tilde{u}_1''(0)\right],\end{aligned} \tag{1.199c}$$

$$\vdots$$

The general solution of equation (1.190) is given by

$$\tilde{u}_0(\tilde{x}) = K_{10} + K_{20}\tilde{x} + K_{30}\mathrm{e}^{-\kappa\tilde{x}} + K_{40}\mathrm{e}^{\kappa\tilde{x}} \tag{1.200}$$

with the parameter

$$\kappa = \sqrt{i\omega_0}. \tag{1.201}$$

Fitting the solution to the boundary conditions (1.198a) and (1.198b) yields

$$\tilde{u}_0(\tilde{x}) = C_1\kappa^2 - 2K_{40} + K_{40}\mathrm{e}^{-\kappa\tilde{x}} + K_{40}\mathrm{e}^{\kappa\tilde{x}}. \tag{1.202}$$

The differential equation (1.192) for $\tilde{u}_1$ is

$$-\tilde{u}_1'''' + \kappa^2\tilde{u}_1'' = \left(-3\kappa^3 - \frac{\omega_1}{\omega_0}\kappa^4\right)K_{40}\mathrm{e}^{-\kappa\tilde{x}} + \left(3\kappa^3 - \frac{\omega_1}{\omega_0}\kappa^4\right)K_{40}\mathrm{e}^{\kappa\tilde{x}} \tag{1.203}$$

with the solution satisfying the boundary conditions (1.198b) and (1.199b)

$$\begin{aligned}\tilde{u}_1(\tilde{x}) &= K_{11} + K_{21}\tilde{x} + K_{31}\mathrm{e}^{-\kappa\tilde{x}} + K_{41}\mathrm{e}^{\kappa\tilde{x}}\\ &+ \frac{1}{2}\left(-3 - \frac{\omega_1}{\omega_0}\kappa\right)\left(\frac{2}{\kappa} + \tilde{x}\right)K_{40}\mathrm{e}^{-\kappa\tilde{x}}\\ &- \frac{1}{2}\left(3 - \frac{\omega_1}{\omega_0}\kappa\right)\left(-\frac{2}{\kappa} + \tilde{x}\right)K_{40}\mathrm{e}^{\kappa\tilde{x}},\end{aligned} \tag{1.204}$$

$$K_{11} = \left(\kappa^2\frac{\omega_1}{\omega_0} - \frac{1-\omega_0^2}{\alpha\kappa^3} + \kappa\right)C_1 + (2\frac{\omega_1}{\omega_0} - \frac{3}{\kappa})K_{40} - 2K_{41}, \tag{1.205}$$

$$K_{21} = \frac{1-\omega_0^2}{\alpha\kappa^2}C_1 - \kappa^2 C_1, \tag{1.206}$$

$$K_{31} = \frac{1-\omega_0^2}{\alpha\kappa^3}C_1 - \kappa C_1 + \frac{3}{\kappa}K_{40} + K_{41}. \tag{1.207}$$

The solution for $\tilde{u}_2$ can be expressed in the form

$$\tilde{u}_2 = K_{12} + K_{22}\tilde{x} + K_{32}e^{-\kappa\tilde{x}} + K_{42}e^{\kappa\tilde{x}}$$
$$+\left(\frac{b}{2\kappa}+\frac{d}{4\kappa^2}\right)\left(\frac{2}{\kappa^3}+\frac{\tilde{x}}{\kappa^2}\right)e^{-\kappa\tilde{x}}+\left(-\frac{c}{2\kappa}+\frac{e}{4\kappa^2}\right)\left(-\frac{2}{\kappa^3}+\frac{\tilde{x}}{\kappa^2}\right)e^{\kappa\tilde{x}}$$
$$+\frac{d}{4\kappa}\left(\frac{6}{\kappa^4}+4\frac{\tilde{x}}{\kappa^3}+\frac{\tilde{x}^2}{\kappa^2}\right)e^{-\kappa\tilde{x}}-\frac{e}{4\kappa}\left(\frac{6}{\kappa^4}-4\frac{\tilde{x}}{\kappa^3}+\frac{\tilde{x}^2}{\kappa^2}\right)e^{\kappa\tilde{x}}$$
$$+\frac{a}{\kappa^2}\frac{\tilde{x}^2}{2}, \tag{1.208}$$

$$K_{12} = -2K_{42} - \frac{3}{2}C_1 + \frac{3}{2}\frac{C_1}{\alpha\kappa^4}(1-\omega_0^2) - \frac{9}{2\kappa^2}K_{40} - \frac{3}{\kappa}K_{41} + 2\frac{\omega_1}{\omega_0}K_{41}$$
$$+\frac{\omega_1}{\omega_0}\frac{C_1}{\alpha\kappa^3}\left(\frac{3}{2}1+\frac{1}{2}\omega_0^2\right)+\frac{\omega_1}{\omega_0}\frac{9}{2\kappa}K_{40}+\frac{1}{2}\frac{\omega_1}{\omega_0}\kappa C_1+\frac{\omega_2}{\omega_0}\kappa^2C_1-2\left(\frac{\omega_1}{\omega_0}\right)^2K_{40}+2\frac{\omega_2}{\omega_0}K_{40}, \tag{1.209}$$

$$K_{22} = -\frac{\omega_1}{\omega_0}\left(\frac{1+\omega_0^2}{\alpha\kappa^2}+\kappa^2\right)C_1, \tag{1.210}$$

$$K_{32} = K_{42} - \frac{3}{2}C_1 + \frac{3}{2}\frac{C_1}{\alpha\kappa^4}(1-\omega_0^2) + \frac{9}{2\kappa^2}K_{40} + \frac{3}{\kappa}K_{41} - \frac{3}{2\kappa}\frac{\omega_1}{\omega_0}K_{40}$$
$$-\frac{3}{2}\frac{\omega_1}{\omega_0}\kappa C_1 - \frac{\omega_1}{\omega_0}\frac{C_1}{2\alpha\kappa^3}\left(1+3\omega_0^2\right), \tag{1.211}$$

$$a = -\frac{3C_1}{\alpha}(1-\omega_0^2) + 3\kappa^4C_1, \tag{1.212}$$

$$b = -\frac{3}{\alpha}(1-\omega_0^2)C_1 - \frac{39}{2}\kappa^2K_{40} - 3\kappa^3K_{41} + 3\kappa^4C_1 - \frac{\omega_1}{\omega_0}\kappa^4K_{41}$$
$$-\frac{9}{2}\kappa^3\frac{\omega_1}{\omega_0}K_{40} - \frac{\omega_1}{\omega_0}\frac{\kappa}{\alpha}(1-\omega_0^2)C_1 + \frac{\omega_1}{\omega_0}\kappa^5C_1 - \frac{\omega_2}{\omega_0}\kappa^4K_{40}, \tag{1.213}$$

$$c = -\frac{21}{2}\kappa^2K_{40} + 3\kappa^3K_{41} - \frac{\omega_1}{\omega_0}\kappa^4K_{41} + \frac{3}{2}\frac{\omega_1}{\omega_0}\kappa^3K_{40} - \frac{\omega_2}{\omega_0}\kappa^4K_{40}, \tag{1.214}$$

$$d = \frac{15}{2}\kappa^3K_{40} + 3\frac{\omega_1}{\omega_0}\kappa^4K_{40} + \left(\frac{\omega_1}{\omega_0}\right)^2\frac{\kappa^5}{2}K_{40}, \tag{1.215}$$

$$e = -\frac{15}{2}\kappa^3K_{40} + 3\frac{\omega_1}{\omega_0}\kappa^4K_{40} - \left(\frac{\omega_1}{\omega_0}\right)^2\frac{\kappa^5}{2}K_{40}. \tag{1.216}$$

Now, the inner solution in the boundary layer at $x = 1$ is determined up to terms of the order of ε^2. In these expressions C_1, K_{40}, K_{41}, K_{42} are integration constants which have to be determined later by matching between the different regions.

Solution in the Boundary Layer at $x = \delta$

For the solution in the boundary layer at $x = \delta$ once more a transformation of the variable x is introduced

$$y = \delta - x, \tag{1.217}$$
$$\bar{y} = \frac{1}{\varepsilon}y = \frac{1}{\varepsilon}(\delta - x). \tag{1.218}$$

Using this transformation a derivative with respect to x can be expressed by a derivative with respect to $\tilde{y}$ by

$$\frac{d}{dx} = -\frac{1}{\varepsilon}\frac{d}{d\bar{y}}. \tag{1.219}$$

Assuming the solution of $\bar{U}$ in form of an asymptotic expansion

$$\bar{U} = \bar{u}_0(\bar{y}) + \varepsilon \bar{u}_1(\bar{y}) + \varepsilon^2 \bar{u}_2(\bar{y}) + \dots \tag{1.220}$$

and collecting equal powers of ε in the differential equations yields a system of linear differential equations

$$-\bar{u}_0'''' + i\omega_0 \bar{u}_0'' = 0, \tag{1.221}$$

$$-\bar{u}_1'''' + i\omega_0 \bar{u}_1'' = -\frac{6}{\delta}\bar{u}_0''' - i\omega_1 \bar{u}_0'' + 3\frac{i\omega_0}{\delta}\bar{u}_0' - \frac{3}{\delta}\bar{y}\bar{u}_0'''' + 3\frac{i\omega_0}{\delta}\bar{y}\bar{u}_0'', \tag{1.222}$$

$$\begin{aligned} -\bar{u}_2'''' + i\omega_0 \bar{u}_2'' = & -\frac{6}{\delta}\bar{u}_1''' - i\omega_1 \bar{u}_1'' + 3\frac{i\omega_0}{\delta}\bar{u}_1' - \frac{3}{\delta}\bar{y}\bar{u}_1'''' + 3\frac{i\omega_0}{\delta}\bar{y}\bar{u}_1'' \\ & + \frac{3}{\delta^2}\bar{u}_0'' - i\omega_2 \bar{u}_0'' + 3\frac{i\omega_1}{\delta}\bar{u}_0' + \frac{12}{\delta^2}\bar{y}\bar{u}_0''' + 3\frac{i\omega_1}{\delta}\bar{y}\bar{u}_0'' \\ & - 6\frac{i\omega_0}{\delta^2}\bar{y}\bar{u}_0' + \frac{3}{\delta^2}\bar{y}^2\bar{u}_0'''' - 3\frac{i\omega_0}{\delta^2}\bar{y}^2\bar{u}_0'', \end{aligned} \tag{1.223}$$

where a prime denotes a derivative with respect to $\bar{y}$. In addition, the boundary conditions (1.174) read

$$\bar{u}_0(0) = 0, \quad \bar{u}_1(0) = 0, \quad \bar{u}_2(0) = 0, \dots \tag{1.224}$$

$$\bar{u}_0'(0) = 0, \quad \bar{u}_1'(0) = 0, \quad \bar{u}_2'(0) = 0, \dots \tag{1.225}$$

The solution of equation (1.221) is given by

$$\bar{u}_0(\bar{y}) = (L_{30} + L_{40}) + \kappa(L_{30} - L_{40})\bar{y} + L_{30}\mathrm{e}^{-\kappa\bar{y}} + L_{40}\mathrm{e}^{\kappa\bar{y}}. \tag{1.226}$$

Taking into account (1.226) the differential equation (1.222) can be integrated to

$$\begin{aligned} \bar{u}_1 = & \; L_{11} + L_{21}\bar{y} + L_{31}\mathrm{e}^{-\kappa\bar{y}} + L_{41}\mathrm{e}^{\kappa\bar{y}} \\ & + \frac{3}{\delta}\kappa(L_{30} - L_{40})\frac{\bar{y}^2}{2} - \left(\frac{3}{\delta}\kappa^3 - \frac{\omega_1}{\omega_0}\kappa^4\right) L_{30}\left(\frac{\bar{y}}{\kappa^2} + \frac{1}{2\kappa^3}\right)\mathrm{e}^{-\kappa\bar{y}} \\ & + \left(-\frac{3}{\delta}\kappa^3 - \frac{\omega_1}{\omega_0}\kappa^4\right) L_{40}\left(-\frac{\bar{y}}{\kappa^2} + \frac{1}{2\kappa^3}\right)\mathrm{e}^{\kappa\bar{y}}, \end{aligned} \tag{1.227a}$$

$$L_{11} = -L_{31} - L_{41} + \frac{\omega_1}{\omega_0}(L_{30} + L_{40}) - \frac{3}{\delta\kappa}(L_{30} - L_{40}), \tag{1.227b}$$

$$L_{21} = \kappa(L_{31} - L_{41}) + \frac{3}{2\delta}(L_{30} + L_{40}) - \frac{\kappa}{2}\frac{\omega_1}{\omega_0}(L_{30} - L_{40}). \tag{1.227c}$$

Similarly, the differential equation for $\bar{u}_2$ can be obtained

$$\begin{aligned}\bar{u}_2 &= L_{12} + L_{22}\bar{y} + L_{32}\mathrm{e}^{-\kappa\bar{y}} + L_{42}\mathrm{e}^{\kappa\bar{y}} \\ &+ \left(\frac{c_2}{2\kappa} + \frac{e_2}{4\kappa^2}\right)\left(\frac{2}{\kappa^3} + \frac{\bar{y}}{\kappa^2}\right)\mathrm{e}^{-\kappa\bar{y}} + \left(-\frac{d_2}{2\kappa} + \frac{f_2}{4\kappa^2}\right)\left(-\frac{2}{\kappa^3} + \frac{\bar{y}}{\kappa^2}\right)\mathrm{e}^{\kappa\bar{y}} \\ &+ \frac{e_2}{4\kappa}\left(\frac{6}{\kappa^4} + 4\frac{\bar{y}}{\kappa^3} + \frac{\bar{y}^2}{\kappa^2}\right)\mathrm{e}^{-\kappa\bar{y}} - \frac{f_2}{4\kappa}\left(\frac{6}{\kappa^4} - 4\frac{\bar{y}}{\kappa^3} + \frac{\bar{y}^2}{\kappa^2}\right)\mathrm{e}^{\kappa\bar{y}} \\ &+ \frac{a_2}{\kappa^2}\frac{\bar{y}^2}{2} + \frac{b_2}{\kappa^2}\frac{\bar{y}^3}{6},\end{aligned} \tag{1.228a}$$

$$\begin{aligned}L_{12} &= -L_{32} - L_{42} + \left(\frac{\omega_2}{\omega_0} - \frac{9}{2\delta^2\kappa^2} - \left(\frac{\omega_1}{\omega_0}\right)^2\right)(L_{30} + L_{40}) + \frac{9}{2\delta\kappa}\frac{\omega_1}{\omega_0}(L_{30} - L_{40}) \\ &- \frac{3}{\delta\kappa}(L_{31} - L_{41}) + \frac{\omega_1}{\omega_0}(L_{31} + L_{41}),\end{aligned} \tag{1.228b}$$

$$\begin{aligned}L_{22} &= \kappa(L_{32} - L_{42}) + \left(\frac{3}{8\delta^2\kappa} + \frac{3}{8}\kappa\left(\frac{\omega_1}{\omega_0}\right)^2 - \frac{\kappa}{2}\frac{\omega_2}{\omega_0}\right)(L_{30} - L_{40}) \\ &+ \frac{3}{2\delta}(L_{31} + L_{41}) - \frac{3}{2\delta}\frac{\omega_1}{\omega_0}(L_{30} + L_{40}) - \frac{\kappa}{2}\frac{\omega_1}{\omega_0}(L_{31} - L_{41}),\end{aligned} \tag{1.228c}$$

$$a_2 = \frac{9}{2}\frac{\kappa^2}{\delta^2}(L_{30} + L_{40}) + 3\frac{\kappa^3}{\delta}(L_{31} - L_{41}) - \frac{3}{2}\frac{\kappa^3}{\delta}\frac{\omega_1}{\omega_0}(L_{30} - L_{40}), \tag{1.228d}$$

$$b_2 = 12\frac{\kappa^3}{\delta^2}(L_{30} - L_{40}), \tag{1.228e}$$

$$c_2 = -\frac{21}{2}\frac{\kappa^2}{\delta^2}L_{30} + 3\frac{\kappa^3}{\delta}L_{31} - \kappa^4\frac{\omega_1}{\omega_0}L_{31} + \frac{3}{2}\frac{\kappa^3}{\delta}\frac{\omega_1}{\omega_0}L_{30} - \kappa^4\frac{\omega_2}{\omega_0}L_{30}, \tag{1.228f}$$

$$d_2 = -\frac{21}{2}\frac{\kappa^2}{\delta^2}L_{40} - 3\frac{\kappa^3}{\delta}L_{41} - \kappa^4\frac{\omega_1}{\omega_0}L_{41} - \frac{3}{2}\frac{\kappa^3}{\delta}\frac{\omega_1}{\omega_0}L_{40} - \kappa^4\frac{\omega_2}{\omega_0}L_{40}, \tag{1.228g}$$

$$e_2 = \frac{15}{2}\frac{\kappa^3}{\delta^2}L_{30} - 3\frac{\kappa^4}{\delta}\frac{\omega_1}{\omega_0}L_{30} + \frac{\kappa^5}{2}\left(\frac{\omega_1}{\omega_0}\right)^2 L_{30}, \tag{1.228h}$$

$$f_2 = -\frac{15}{2}\frac{\kappa^3}{\delta^2}L_{40} - 3\frac{\kappa^4}{\delta}\frac{\omega_1}{\omega_0}L_{40} - \frac{\kappa^5}{2}\left(\frac{\omega_1}{\omega_0}\right)^2 L_{40}. \tag{1.228i}$$

Matching of the Functions near $x = 1$

Now, the functions in the different regions have to be matched. Between the boundary layer at $x = 1$ and the outer region the solutions are assumed to be equal at a point

$$x = 1 + \eta x_\eta \tag{1.229}$$

or, expressed in $\tilde{x}$, at

$$\tilde{x}_\eta = \frac{\eta}{\varepsilon}x_\eta \tag{1.230}$$

with

$$\varepsilon \ll \eta \ll 1. \tag{1.231}$$

As $\frac{\eta}{\varepsilon}$ tends to infinity also $\mathrm{e}^{\kappa\tilde{x}_\eta}$ tends to infinity and $\mathrm{e}^{-\kappa\tilde{x}_\eta}$ is exponentially small, if the real part of κ is greater than zero. Thus, a bounded solution is possible only if all integration constants K_{4i}, $(i = 0, 1, \dots)$ vanish. The solution in the boundary layer at x_η is

$$\begin{aligned}\tilde{U}(x_\eta) =& K_{10} + K_{20}\frac{\eta}{\varepsilon}x_\eta \\ &+ \varepsilon\left[\left(\kappa^2\frac{\omega_1}{\omega_0} - \frac{1-\omega_0^2}{\alpha\kappa^3} + \kappa\right)C_1 + \left(\frac{1-\omega_0^2}{\alpha\kappa^2} - \kappa^2\right)C_1\frac{\eta}{\varepsilon}x_\eta\right] \\ &+ \varepsilon^2\left[-\frac{3}{2}C_1 + \frac{3}{2}\frac{C_1}{\alpha\kappa^4}(1-\omega_0^2) + \frac{\omega_1}{\omega_0}\frac{C_1}{\alpha\kappa^3}\left(\frac{3}{2}1 + \frac{1}{2}\omega_0^2\right) + \frac{1}{2}\frac{\omega_1}{\omega_0}\kappa C_1 + \frac{\omega_2}{\omega_0}\kappa^2 C_1\right. \\ &\left. - \frac{\omega_1}{\omega_0}\left(\frac{1+\omega_0^2}{\alpha\kappa^2} + \kappa^2\right)C_1\frac{\eta}{\varepsilon}x_\eta + \left(-\frac{3C_1}{\alpha}(1-\omega_0^2) + 3\kappa^4 C_1\right)\frac{1}{2\kappa^2}\left(\frac{\eta}{\varepsilon}\right)^2 x_\eta^2\right] \\ &+ O(\eta^3, \eta^2\varepsilon, \eta\varepsilon^2, \varepsilon^3). \end{aligned} \tag{1.232}$$

The outer solution at x_η is

$$\hat{U}(x_\eta) = C_{10} + \frac{C_{20}}{(1+\eta x_\eta)^2} + \varepsilon\left(C_{11} + \frac{C_{21}}{(1+\eta x_\eta)^2}\right) + \varepsilon^2\left(C_{12} + \frac{C_{22}}{(1+\eta x_\eta)^2}\right) + \dots \tag{1.233}$$

Sorting both solutions at x_η for equal powers of ε, η and products of both gives

	outer solution	inner solution	
$\varepsilon^0:$	$C_{10} + C_{20}$	$C_1\kappa^2,$	(1.234a)
$\varepsilon^1:$	$C_{11} + C_{21}$	$\left(\frac{\omega_1}{\omega_0}\kappa^2 - \frac{1-\omega_0^2}{\alpha\kappa^3} + \kappa\right)C_1,$	(1.234b)
$\eta^1:$	$-2C_{20}x_\eta$	$\left(\frac{1-\omega_0^2}{\alpha\kappa^2} - \kappa^2\right)C_1 x_\eta,$	(1.234c)
$\varepsilon^2:$	$C_{12} + C_{22}$	$-\frac{3}{2}C_1 + \frac{3}{2}\frac{C_1}{\alpha\kappa^4}(1-\omega_0^2) + \frac{1}{2}\frac{\omega_1}{\omega_0}\kappa C_1 + \frac{\omega_1}{\omega_0}\frac{C_1}{\alpha\kappa^3}\left(\frac{3}{2}1 + \frac{1}{2}\omega_0^2\right) + \frac{\omega_2}{\omega_0}\kappa^2 C_1,$	(1.234d)
$\varepsilon\eta:$	$-2C_{21}x_\eta$	$-\frac{\omega_1}{\omega_0}\left(\frac{1+\omega_0^2}{\alpha\kappa^2} + \kappa^2\right)C_1 x_\eta,$	(1.234e)
$\eta^2:$	$3C_{20}x_\eta^2$	$\left(-\frac{3}{\alpha}(1-\omega_0^2) + 3\kappa^4\right)\frac{C_1}{2\kappa^2}x_\eta^2$	(1.234f)
		$\vdots$	

For a correct matching both columns have to be equal for every power of ε and η.

Matching of the Functions near $x = \delta$

In order to match the solutions between the outer solution and the boundary layer solution near $x = \delta$, it is assumed that both solutions are equal at a point $y = \xi y_\xi$ so that

$$\bar{y}_\xi = \frac{\xi}{\varepsilon}y_\xi, \tag{1.235}$$

$$x_\xi = \delta - \xi y_\xi \tag{1.236}$$

with

$$\varepsilon \ll \xi \ll 1. \tag{1.237}$$

Thus, $\frac{\xi}{\varepsilon}$ tends to infinity. For this reason the integration constants of the terms $e^{\kappa\bar{y}}$ must vanish

$$L_{4i} = 0, \qquad i = 0, 1, \dots \tag{1.238}$$

and at y_θ the terms $e^{-\kappa\bar{y}_\xi}$ are exponentially small. Therefore both solutions at y_ξ are

$$\begin{aligned}\bar{U}(x_\xi) =& L_{30} + \kappa L_{30}\frac{\xi}{\varepsilon}y_\xi \\ &+ \varepsilon\left[\left(-L_{31} + \frac{\omega_1}{\omega_0}L_{30} - \frac{3}{\delta\kappa}L_{30}\right) + \left(\kappa L_{31} + \frac{3}{2\delta}L_{30} - \frac{\kappa}{2}\frac{\omega_1}{\omega_0}L_{30}\right)\frac{\xi}{\varepsilon}y_\xi + \frac{3\kappa}{2\delta}L_{30}(\frac{\xi}{\varepsilon})^2 y_\xi^2\right] \\ &+ \varepsilon^2\left[\left(-L_{32} + \left\{\frac{\omega_2}{\omega_0} - \frac{9}{2\delta^2\kappa^2} - (\frac{\omega_1}{\omega_0})^2\right\}L_{30} + \frac{9}{2\delta\kappa}\frac{\omega_1}{\omega_0}L_{30} + \left\{\frac{\omega_1}{\omega_0} - \frac{3}{\delta\kappa}\right\}L_{31}\right)\right. \\ &+ \left(\kappa L_{32} + \left\{\frac{3}{8\delta^2\kappa} + \frac{3}{8}\kappa(\frac{\omega_1}{\omega_0})^2 - \frac{3}{2\delta}\frac{\omega_1}{\omega_0} - \frac{\kappa}{2}\frac{\omega_2}{\omega_0}\right\}L_{30} + \left\{\frac{3}{2\delta} - \frac{\kappa}{2}\frac{\omega_1}{\omega_0}\right\}L_{31}\right)\frac{\xi}{\varepsilon}y_\xi \\ &+ \left.\left(\frac{9}{2}\frac{\kappa^2}{\delta^2}L_{30} - \frac{3}{2}\frac{\kappa^3}{\delta}\frac{\omega_1}{\omega_0}L_{30} + \frac{3\kappa^3}{\delta}L_{31}\right)\frac{1}{2\kappa^2}(\frac{\xi}{\varepsilon})^2 y_\xi^2 + 12\frac{\kappa^3}{\delta^2}L_{30}\frac{1}{6\kappa^2}(\frac{\xi}{\varepsilon})^3 y_\xi^3\right] + \dots\end{aligned} \tag{1.239}$$

for the inner solution and

$$\hat{U}(\hat{x}_\xi) = C_{10} + \frac{C_{20}}{(\delta - \xi y_\xi)^2} + \varepsilon\left[C_{11} + \frac{C_{21}}{(\delta - \xi y_\xi)^2}\right] + \varepsilon^2\left[C_{12} + \frac{C_{22}}{(\delta - \xi y_\xi)^2}\right] + \dots \tag{1.240}$$

for the outer solution. Again equal powers of ε and ξ or products of both are collected in the two solutions. The largest terms are related to a term $\kappa L_{30}y_\xi$ in the order of $O\frac{\xi}{\varepsilon}$ in the inner solution:

$$O\frac{\xi}{\varepsilon}: \quad \kappa L_{30}y_\xi. \tag{1.241}$$

This term has no corresponding term in the outer solution, which means that the only possibility for the vanishing of this term is

$$L_{30} = 0, \tag{1.242}$$

which is taken into consideration in the following in order to simplify the analysis. Collecting equal powers of ε, η or $\varepsilon\eta$ yields

	outer solution	inner solution	
$\varepsilon^0:$	$C_{10} + \frac{C_{20}}{\delta^2}$	$0,$	(1.243)
$\varepsilon^1:$	$C_{11} + \frac{C_{21}}{\delta^2}$	$-L_{31},$	(1.244)
$\xi:$	$\frac{2C_{20}}{\delta^3}y_\xi$	$\kappa L_{31}y_\xi,$	(1.245)
$\varepsilon^2:$	$C_{12} + \frac{C_{22}}{\delta^2}$	$-L_{32} + \left(\frac{\omega_1}{\omega_0} - \frac{3}{\delta\kappa}\right)L_{31},$	(1.246)
$\varepsilon\xi:$	$\frac{2C_{21}}{\delta^3}y_\xi$	$\left(\kappa L_{32} + \frac{3}{2\delta}L_{31} - \frac{\kappa}{2}\frac{\omega_1}{\omega_0}L_{31}\right)y_\xi,$	(1.247)
$\xi^2:$	$\frac{3C_{20}}{\delta^4}y_\xi^2$	$\frac{3}{2}\frac{\kappa}{\delta}L_{31}y_\xi^2$	(1.248)

if matching is done up to terms of order ξ^2, $\varepsilon\xi$ and ε^2.

Evaluation of the Eigenvalue ω

Equations (1.234a), (1.243) and (1.234c) form a set of three homogeneous equations for C_{10}, C_{20} and C_1 with nontrivial solutions if the determinant vanishes

$$\begin{vmatrix} 1 & 1 & -\kappa^2 \\ 0 & -2x_\eta & -\left(\frac{1-\omega_0^2}{\alpha\kappa^2} - \kappa^2\right) x_\eta \\ 1 & \frac{1}{\delta^2} & 0 \end{vmatrix} = 0. \tag{1.249}$$

Taking into consideration that $\kappa^2 = i\omega_0$ gives

$$\omega_0^2 = \frac{\delta^2 - 1}{\delta^2 - 1 + \alpha(\delta^2 + 1)} \tag{1.250}$$

which is identical to the normalized frequency of a cylinder vibrating in an inviscid, incompressible fluid. Choosing C_1 as a free constant, the relations

$$C_{10} = -\frac{\kappa^2}{\delta^2 - 1} C_1, \quad C_{20} = \frac{\kappa^2 \delta^2}{\delta^2 - 1} C_1 \tag{1.251}$$

hold. Equations (1.234c) and (1.234f) are linearly dependent, as well as (1.245) and (1.248), so that the equations (1.234b), (1.234e), (1.244) and (1.245) form a set of four homogeneous linear equations for C_{11}, C_{21}, L_{31} and C_1 which have nontrivial solutions for special values of ω_1. The characteristic equation is

$$\begin{vmatrix} 1 & 1 & 0 & -\left(\frac{\omega_1}{\omega_0}\kappa^2 - \frac{1-\omega_0^2}{\alpha\kappa^3} + \kappa\right) \\ 0 & -2 & 0 & \frac{\omega_1}{\omega_0}\left(\frac{1+\omega_0^2}{\alpha\kappa^2} + \kappa^2\right) \\ 1 & \frac{1}{\delta^2} & 1 & 0 \\ 0 & 0 & -\kappa & \frac{2\kappa^2}{\delta(\delta^2-1)} \end{vmatrix} = 0 \tag{1.252}$$

with the root

$$\frac{\omega_1}{\omega_0} = \frac{-2}{\kappa} \frac{\delta^2(1-\omega_0^2) + \alpha\delta^2\omega_0^2 + \frac{2\alpha\delta\omega_0^2}{\delta^2-1}}{(\delta^2-1)(1+\omega_0^2) + \alpha\omega_0^2(\delta^2+1)}. \tag{1.253}$$

This formula can be simplified by introducing ω_0^2

$$\frac{\omega_1}{\omega_0} = -\frac{2}{\kappa} \cdot \frac{\alpha(\delta^4 - \delta)}{\alpha(\delta^4 - 1) + (\delta^2 - 1)^2}, \tag{1.254}$$

and finally the expression for ω is given by

$$\omega = \omega_0 \left[1 - \sqrt{\frac{1}{\mathrm{Re}}} \frac{2}{\sqrt{i\omega_0}} \frac{\alpha(\delta^4 - \delta)}{\alpha(\delta^4 - 1) + (\delta^2 - 1)^2} + O\left(\frac{1}{\mathrm{Re}}\right) \right]. \tag{1.255}$$

The root of $i\omega_0$ in the denominator can be expressed by

$$\frac{1}{\sqrt{i\omega_0}} = \frac{1}{\sqrt{2\omega_0}}(1 - i), \tag{1.256}$$

which means that the real and the imaginary part of ω_1 are of equal magnitude. It is obvious that the imaginary part of ω is positive, resulting in damped vibration of the cylinder.

1.4.5 Results

Numerical results for the natural frequency ω_0 of the cylinder vibrating in an inviscid fluid and the first correction ω_1 due to the viscosity are shown in Figure 1.9 and 1.10 as a function of the nondimensional outer radius δ. Plots for different values of α are given. For ω_1 only the imaginary part is given, since the absolute values of the real part and the imaginary part are equal due to (1.257). In addition, several limiting cases can be examined.

Unbounded Fluid

If e.g. the outer radius tends to infinity ($\delta \to \infty$) which corresponds to an unbounded fluid. The results are

$$\omega_0^2(\delta \to \infty) = \frac{1}{1+\alpha}, \tag{1.257a}$$

$$\left.\frac{\omega_1}{\omega_0}\right|_{\delta\to\infty} = -\frac{2}{\sqrt{i\omega_{0\infty}}} \cdot \frac{\alpha}{\alpha+1}. \tag{1.257b}$$

If besides α tends to infinity ($\alpha \to \infty$) being possible if the cylinder is hollow ω_1 and ω_0 tend to

$$\omega_0(\delta, \alpha \to \infty) \to 0, \tag{1.258a}$$

$$\left.\frac{\omega_1}{\omega_0}\right|_{\delta,\alpha\to\infty} \to \infty. \tag{1.258b}$$

Narrow Gap

For the special case of a narrow gap the limiting values $\delta \to 1$ can be expressed as functions of

$$\Delta = \delta - 1 \ll 1 \tag{1.259}$$

in the form of

$$\omega_0^2(\delta \to 1) = \frac{\Delta}{\alpha}, \tag{1.260a}$$

$$\left.\frac{\omega_1}{\omega_0}\right|_{\delta\to 1} = -\frac{3}{4}\left(\frac{2}{\sqrt{i\omega_0(\delta \to 1)}}\right) \tag{1.260b}$$

Using e.g. FEM instead of singular perturbation theory these results could only have been obtained via a larger number of numerical computations. Singular perturbation theory here directly gives the dependence of the eigenvalues on the system's parameters. The complex modes are not given here, they may be calculated via an analysis similar to that given for the vibrating string with small bending stiffness.

As a conclusion we see that the investigation of a cylinder in a cylindrical fluid filled duct leads to a numerically ill conditioned problem if the fluid is almost inviscid. Applying singular perturbation theory allows to derive explicit formulae for the real and imaginary parts, so that the influence of the different parameters can be directly seen. For a first correction term ω_1, it is necessary to calculate terms up to the second order approximation. Though the analysis involves complicated expressions due to the inhomogeneous differential equations, the results for the eigenvalue are very simple. Several limiting cases can be investigated. The

analysis is less tedious if computerized symbolic manipulation is used. Though the use of symbolic manipulation was not explicitly shown, it was used in obtaining the above results. A computation of the eigenfunctions is possible but was not done here.

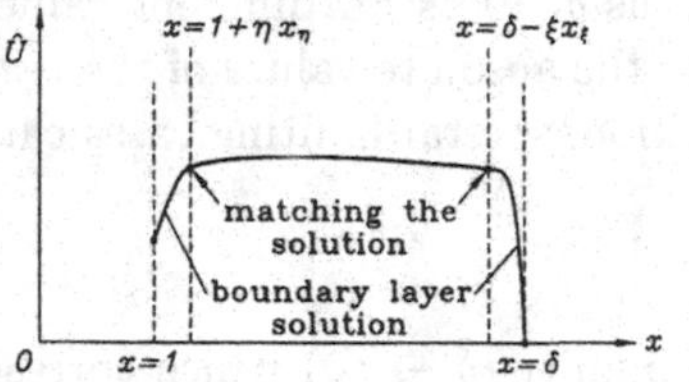

Figure 1.8: The boundary layers and the outer region

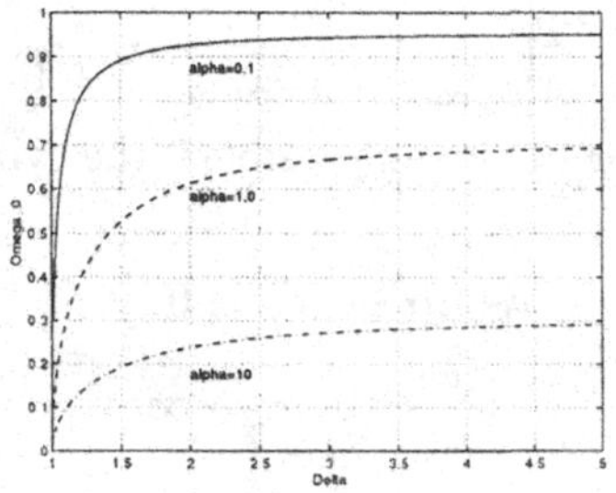

Figure 1.9: ω_0 as a function of δ

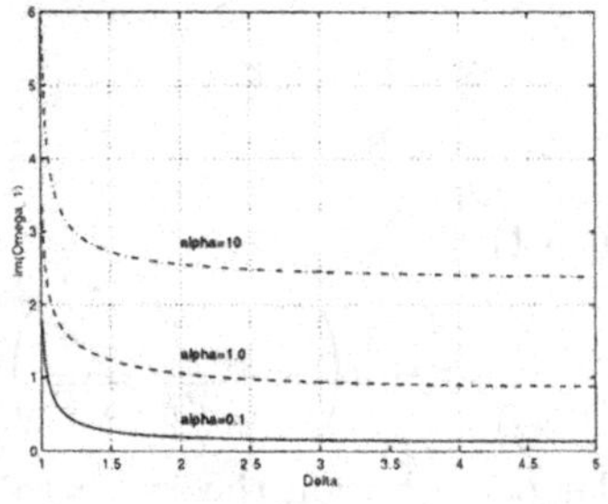

Figure 1.10: $\mathcal{Im}(\omega_1/\omega_0)$ as a function of δ

Chapter 2

Problems in the Mathematical Modelling of Ultrasonic Motors

2.1 Introduction

With the development of new piezoceramic materials a basis for the realization of efficient new types of motors was established. The working principle of these ultrasonic motors is totally different from that of electrodynamic motors. In electrodynamic motors the forces and moments are generated by magnetic fields whereas in ultrasonic motors piezoceramic elements are used for energy transformation and to obtain high frequency vibrations for a relative motion between the different parts of the system.

Different working principles are possible. Common to all types of ultrasonic motors is that they consist of two parts: A stator which is vibrating, but fixed in space in such a way that it does not undergo a rigid body motion. The second part is the moving part: rotor or slider. Normally, the stator vibrates simultaneously in two modes with a temporal phase shift, so that the points on the stator's surface move on elliptic orbits.

The advantages of these motors are that they have a speed of rotation in the range of 60 to 200 rpm and high torques without speed reducing gears. This is achieved with excitation frequencies of several ten Kilohertz. Even without an electrical excitation these motors have a high holding torque. If the vibrations are excited by piezoceramics, no magnetic field is present, which may be advantageous in several applications. The motors may be compact or miniaturized with a high ratio of torque to weight. They even may be used directly as joints in industrial applications such as light weight robots.

Nowadays the high frequencies of 40 to 200 kHz are generated by piezoelectric plates or elements. The use of piezoceramic materials for the excitation mechanism was eased by developments in power electronics. First industrial applications of ultrasonic motors were realized by japanese companies. They used traveling wave motors in autofocus lenses of cameras, in motors of headrests and in other applications. Nevertheless, the optimized use of piezoceramic elements as elements of high power transmission requires new detailed models for the excitation mechanism of the elastic structure.

2.2 Working Principles of Ultrasonic Motors

2.2.1 Traveling Wave Principle

The working principle of the traveling wave motor has been discussed in detail in recent articles ([16], [42], [36]). Here, it is briefly described for a better understanding of the basic requirements for the stator of a traveling wave motor. Consider a BERNOULLI-EULER beam desribed by the differential equation

$$EIw_{xxxx} + \rho A w_{tt} = 0. \tag{2.1}$$

A solution may be given in the form of a bending wave

$$w(x,t) = C\sin(\omega t - kx), \tag{2.2}$$

traveling in the direction of the positive x-axis. In (2.2), $w(x,t)$ is the transverse displacement of the centroid of the beam's cross-section, x and t are the spatial and time coordinates, C is the wave amplitude and ω, k are the frequency and the wave number, respectively. The wave number k and the wavelength λ are related through

$$k = \frac{2\pi}{\lambda}. \tag{2.3}$$

In EULER-BERNOULLI beam theory a cross-section of the beam remains perpendicular to the beam axis after deformation (see figure 2.1). A point at a distance y from the neutral axis therefore undergoes an axial displacement u

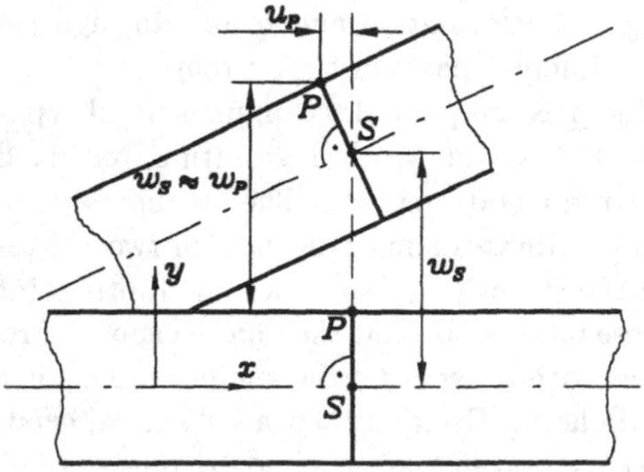

Figure 2.1: Displacements of a BERNOULLI-EULER beam

$$u(x,y,t) = -yw'(x,t) = Cky\cos(\omega t - kx). \tag{2.4}$$

For the above mentioned BERNOULLI-EULER beam the wavenumber k is

$$k = \left(\frac{\rho A \omega^2}{EI}\right)^{1/4} \tag{2.5}$$

with the mass per unit length ρA and the bending stiffness EI. If the transverse displacement (2.4) and the axial displacement (2.2) are divided by the corresponding amplitudes, squared and summed, it can be seen that the result is an equation for an ellipse

$$\left(\frac{w(x,t)}{C}\right)^2 + \left(\frac{u(x,y,t)}{Cky}\right)^2 = \text{const.} \tag{2.6}$$

Therefore any point not on the neutral axis performs an elliptic motion, and this also holds for points on the surface of the beam at $y = \pm h/2$. In reality, shear deformation may also be of some importance. Nevertheless the trajectory of a surface point is still elliptic, but with axes different from those computed by BERNOULLI-EULER beam theory as long as a linearized description is used.

The axial velocity of a surface point at $y = h/2$ can be easily obtained by differentiating the axial displacement

$$u_t(x, \frac{h}{2}, t) = -Ck\frac{h}{2}\omega \sin(\omega t - kx) = -Ck\frac{h}{2}\omega \cdot w(x,t). \tag{2.7}$$

It is obvious that the axial velocity achieves its maximum simultaneously with the transverse displacement. Therefore, a slider pressed to the beam contacts the beam at the points with maximum transverse displacement and is thus transported along the beam.

Normally, the traveling waves in a traveling wave motor are excited by piezoceramic elements bonded to the stator. Only standing waves are thus generated and two standing waves of equal amplitude and a temporal phase shift have to be superimposed to obtain the traveling wave

$$\begin{aligned} w(x,t) &= C \sin\omega t \sin kx + C\sin(\omega t + \varphi_0)\sin(kx + \xi_0) \\ &= \frac{C}{2}[\cos(kx - \omega t) - \cos(kx + \omega t) - \cos(kx + \omega t + \xi_0 + \varphi_0) + \cos(kx - \omega t + \xi_0 - \varphi_0)]. \end{aligned} \tag{2.8}$$

Equation (2.8) shows that for every combination of $\xi_0 = \pm\pi/2$ and $\varphi_0 = \pm\pi/2$ the superposition of two standing waves leads to a traveling wave. If the amplitudes are different or if either ξ_0 or φ_0 is not equal to $\pm\pi/2$, then (2.8) results in a combination of a traveling and a standing wave. Thus, the displacements in the transverse direction differ according to the standing waves, and therefore the trajectories of points on the surface also differ over one wavelength. This can be shown by plotting the trajectories of surface points for

$$w(x,t) = C_1 \sin kx \sin\omega t + C_2 \sin(kx + \xi_0)\sin(\omega t + \varphi_0) \tag{2.9}$$

for various ratios of C_2/C_1 or various phase shifts φ_0. Figure 2.2 shows these trajectories for several amplitude ratios and in figure 2.3 the influence of the phase can be seen. Each ellipse corresponds to a certain point on the surface. Though there are some small differences the two figures are very similar and it can be concluded that the excitation of a pure traveling wave with identical trajectories on the surface points is achieved if two standing harmonic waves of equal amplitudes and a spatial and temporal phase shift of $\pi/2$ are superimposed.

At this point it should be kept in mind that due to the reflections at the ends of a finite beam a pure traveling wave cannot be easily excited in a beam of finite length. Either the waves have to be damped at the ends of the beam, which results in high energy losses, or some sort of waveguide has to be constructed.

The simplest form of a waveguide can be achieved if instead of a beam a circular plate is used as the stator. With the help of plate theory it can easily be verified that for each eigenfrequency of a circular plate there always exist two modes which are phase shifted by 90° along the circumference of the plate. Of course, for a given number n of nodal diameters there always exist an infinite number of eigenfrequencies and eigenmodes. Nevertheless, for the ultrasonic traveling wave motor the number of nodal diameters is determined by the number of the piezoceramic elements which are bonded to the stator's surface along the

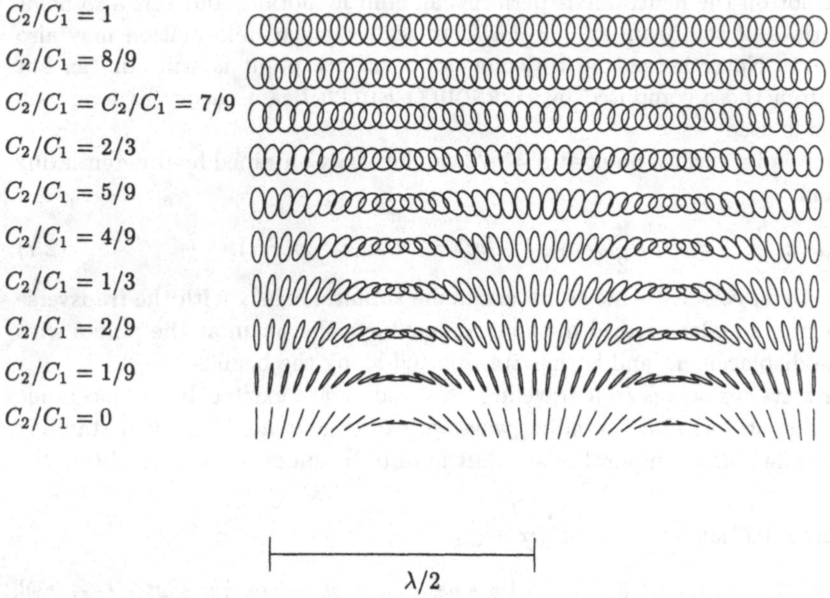

Figure 2.2: Trajectories of the points along the surface of the beam for different ratios of C_2/C_1

circumference. For a given number of nodal diameters only the lowest eigenfrequency is used. Detailed analytical models for the excitation mechanism of bending waves by piezoceramic elements will be given later.

For the case of a circular plate, of course, a bending wave traveling in circumferential direction can easily be obtained if the two standing waves with a phase shift of 90° along the circumference are excited with a temporal phase shift of 90°. As shown previously, the points with maximum deflection move with the maximum velocity in circumferential direction, so that a rotor, pressed against the stator, will be in contact with the stator only at those points. Therefore, due to frictional forces the rotor will be transported along the circumference. An excitation of the stator in resonance is necessary in order to obtain sufficiently high amplitudes. Knowledge of the eigenfrequencies of the stator is therefore important for the construction of a traveling wave motor. Of course, the determination of the eigenfrequencies and the corresponding mode shapes of a circular plate for given boundary conditions is known and will not be shown here in detail. An example of a stator for a rotational motor and the corresponding piezoceramic ring with elements of different polarizations are shown in Figure 2.4. An exploded view of a rotational traveling wave motor is given in Figure 2.5. Also traveling wave motors for a linear motion have been developed. Examples of stators are shown in Figure 2.6.

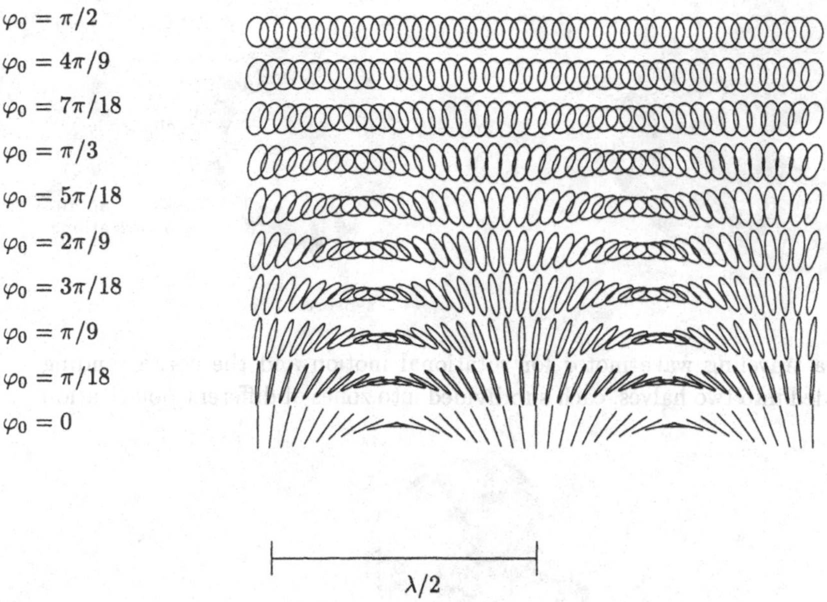

Figure 2.3: Trajectories of the points along the surface of the beam for different values of φ_0

2.2.2 Motors with Rod Resonators

The elliptic orbits of the stator's surface may not only be generated by traveling bending waves but also by a superposition of two vibration modes with a temporal phase shift. If the stator has the shape of a rod bending and longitudinal or torsional and longitudinal vibration may be used.

A simple system may consist of two stacks of piezoceramics which vibrate with a temporal phase shift, see Figure 2.7. Due to friction and the elliptic orbit of the connecting part a motion of the slider is generated. The advantage of such a micro-push motor is that the actuation distance of the slider is very large (up to several inches). With an additional control unit the position precision is very fine by using the microactuation possibilities of the piezoceramics ($\leq 0.01\mu$m).

Another motor with a rod resonator was realized using a two mode resonator simultaneously undergoing bending and longitudinal vibration. The stator shown in Figure 2.8 has diagonally connected electrodes so that both vibration modes may be excited in resonance.

2.3 Constitutive Equations for Piezoceramic Material

In a piezoceramic material the stress tensor σ and the strain tensor ε are coupled to the electric field E and the electric displacement D. If we consider large strains and large electric fields, of course the equations describing the coupling are nonlinear and hysteretic effects

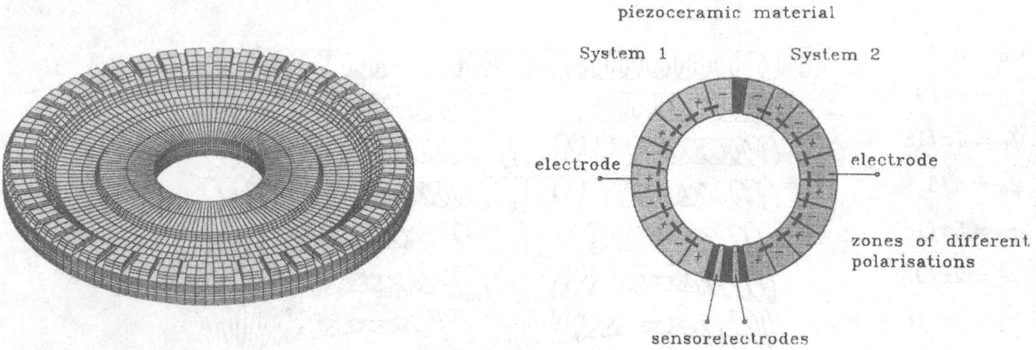

Figure 2.4: Stator of a traveling wave motor for rotational motion and the corresponding piezoceramic ring divided into two halves, each subdivided into zones of different polarization

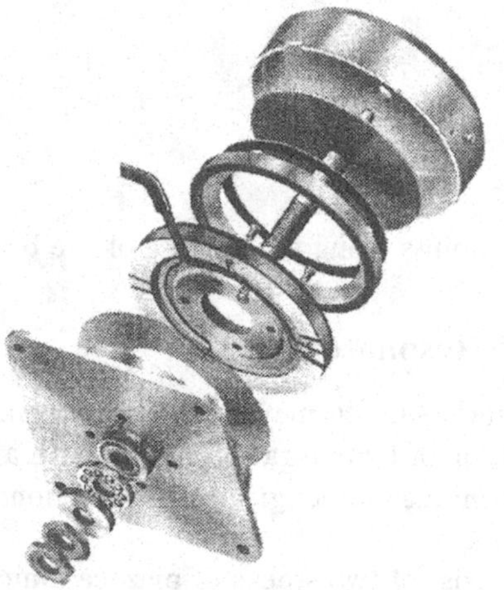

Figure 2.5: Exploded view of a traveling wave motor for rotational motion

may be important. Normally, these nonlinear relations can only be handled and incorporated into the analysis of the structure by using finite elements. In addition, it is very difficult to obtain the nonlinear parameters by experiments. Therefore, we consider here only the linear constitutive equations. First, we introduce the strain tensor ε or its components ε_{ij} as functions of the diplacements

$$\varepsilon_{ij} = \frac{1}{2}\left(u_{i,j} + u_{j,i}\right) \tag{2.10}$$

in which $()_{,i}$ is the derivative of $()$ with respect to the coordinate i. The constitutive equations may then be given in various forms, e.g.

$$\varepsilon_{ij} = s^E_{ijkl}\sigma_{kl} + d_{ijk}E_k, \tag{2.11}$$

$$D_i = d_{kli}\sigma_{kl} + \epsilon^{\sigma}_{ij}E_j \tag{2.12}$$

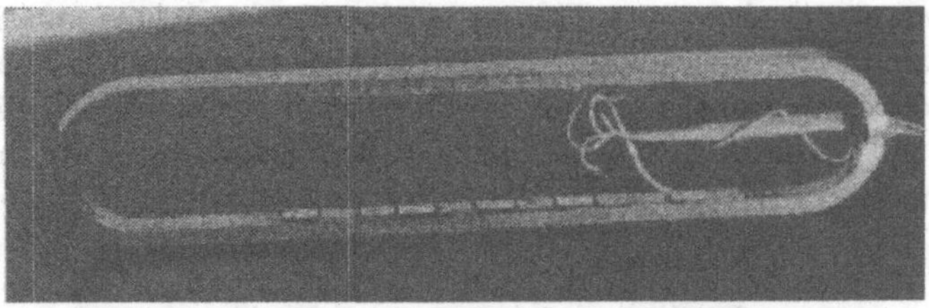

Figure 2.6: Stator of a traveling wave motor for linear motion

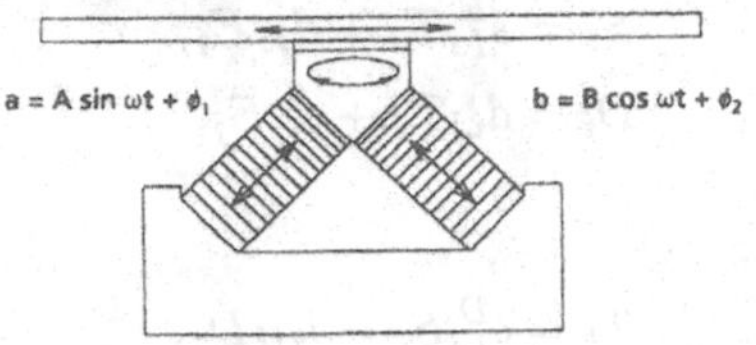

Figure 2.7: Rod resonator consisting of two piezoceramic stacks

or by

$$\sigma_{ij} = c^D_{ijkl}\varepsilon_{kl} - h_{ijk}D_k, \tag{2.13}$$
$$E_i = -h_{kli}\varepsilon_{kl} + \beta^{\varepsilon}_{ij}D_j \tag{2.14}$$

as well as

$$\sigma_{ij} = c^E_{ijkl}\varepsilon_{kl} - e_{ijk}E_k, \tag{2.15}$$
$$D_i = e_{kli}\varepsilon_{kl} + \epsilon^{\varepsilon}_{ij}E_j \tag{2.16}$$

or

$$\varepsilon_{ij} = s^D_{ijkl}\sigma_{kl} + g_{ijk}D_k, \tag{2.17}$$
$$E_i = -g_{kli}\sigma_{kl} + \beta^{\sigma}_{ij}D_j. \tag{2.18}$$

In these tensor equations EINTEIN's summation convention is used, which means that elements with equal indices have to be summed. c_{ijkl} is the stiffness tensor, s_{ijkl} the compliance tensor, ε_{ij} the permittivity tensor, β_{ij} is the dielectric impermeability tensor and d_{ijk}, h_{ijk}, e_{ijk} and g_{ijk} are the piezoelectric coupling tensors. It can be shown by thermodynamic considerations that the coupling tensors occuring in each set of two equations are mutually equal. Of course, each pair of equations corresponds to a system of linear equations for a given mechanical

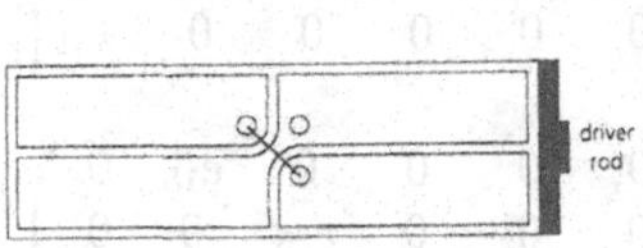

Figure 2.8: Stator of a motor for linear motion using bending and longitudinal vibration

quantity and a given electrical quantity. This means that one set of these equations is sufficient to describe the state in the piezoceramic material. As the stress tensor is symmetric and has only six independent components, and as the strain tensor is symmetric as well, very often the abbreviated form of the constitutive equations is used. Thus we introduce the stress matrix T and the strain matrix S according to

$$T = (\sigma_{11}, \sigma_{22}, \sigma_{33}, \sigma_{23}, \sigma_{13}, \sigma_{12})^T, \tag{2.19}$$

$$S = (\varepsilon_{11}, \varepsilon_{22}, \varepsilon_{33}, 2\varepsilon_{23}, 2\varepsilon_{13}, 2\varepsilon_{12})^T. \tag{2.20}$$

The tensor equations then take the form

$$S_I = s^E_{IJ} T_J + d_{jI} E_j, \tag{2.21}$$

$$D_i = d_{iJ} T_J + \epsilon^T_{ij} E_j, \tag{2.22}$$

or

$$T_I = c^D_{IJ} S_J - h_{jI} D_j, \tag{2.23}$$

$$E_i = -h_{iJ} S_J + \beta^S_{ij} D_j, \tag{2.24}$$

or

$$T_I = c^E_{IJ} S_J - e_{jI} E_j, \tag{2.25}$$

$$D_i = e_{iJ} S_J + \epsilon^S_{ij} E_j, \tag{2.26}$$

and

$$S_I = s^D_{IJ} T_J + g_{jI} D_j, \tag{2.27}$$

$$E_i = -g_{iJ} T_J + \beta^T_{ij} D_j \tag{2.28}$$

Once again, EINSTEIN's summation convention is used. Instead of the stiffness and compliance tensor we now have the stiffness and the compliance matrix which for an arbitrary anisotropic material have 21 indepent components each. Also the piezoceramic coupling tensors which have 27 components are described by a corresponding 6×3 matrix which of course has only few independent components.

As we restrict ourselves to piezoceramic material, in the following investigations instead of arbitrary piezoelectric material, the stiffness matrix and the piezoelectric coupling matrix are given explicitly as

$$(c^E_{IJ}) = \begin{pmatrix} c^E_{11} & c^E_{12} & c^E_{13} & 0 & 0 & 0 \\ c^E_{12} & c^E_{22} & c^E_{13} & 0 & 0 & 0 \\ c^E_{13} & c^E_{13} & c^E_{33} & 0 & 0 & 0 \\ 0 & 0 & 0 & c^E_{44} & 0 & 0 \\ 0 & 0 & 0 & 0 & c^E_{44} & 0 \\ 0 & 0 & 0 & 0 & 0 & \frac{1}{2}(c^E_{11} - c^E_{12}) \end{pmatrix}, \tag{2.29}$$

$$(e_{kJ}) = \begin{pmatrix} 0 & 0 & 0 & 0 & e_{15} & 0 \\ 0 & 0 & 0 & e_{15} & 0 & 0 \\ e_{31} & e_{31} & e_{33} & 0 & 0 & 0 \end{pmatrix}. \tag{2.30}$$

This is due to the fact that in these equations the indices i, j range from 1 to 3 and I, J from 1 to 6.

In many publications the electric field in the piezoceramic elements is assumed to be given. In reality however, not the electric field but the electric potential applied at the piezoceramic is prescribed. Therefore, an exact analysis requires that within the piezoceramic material the divergence of the electric displacement vanishes

$$\nabla D = 0. \tag{2.31}$$

If beams or plates are considered, to which piezoceramic patches are bonded, then often the constitutive equations given above are used to derive bending moments, normal forces and other integral components. For many applications however, it is advantageous not to use the synthetic form to derive the equations of motion but for instance HAMILTON's principle. If a piezoceramic material is involved, HAMILTON's principle for electromechanical systems has to be used

$$\delta \int_{t_0}^{t_1} (T - H)dt = 0 \tag{2.32}$$

with the kinetic energy T and the electric enthalpy H

$$H = \int_V \frac{1}{2} \left(T_I S_I - D_i E_i\right) dV = \int_V \left(\frac{1}{2} c_{IJ}^E S_I S_J - e_{iJ} E_i S_J - \frac{1}{2} \varepsilon_{ij}^S E_i E_j\right) dV \tag{2.33}$$

as described in [41], for the case of linear constitutive equations. In equation (2.33) again EINSTEIN's summation rule was used. If the static case is considered, the kinetic energy vanishes and HAMILTON's principle is reduced to

$$\delta \int_V \left(\frac{1}{2} c_{IJ}^E S_I S_J - e_{iJ} E_i S_J - \frac{1}{2} \varepsilon_{ij}^S E_i E_j\right) dV = 0. \tag{2.34}$$

HAMILTON's principle is particularly useful if not only bending moments, normal forces and shear forces but also higher order stress resultants and higher order approximations and the corresponding boundary conditions have to be derived.

2.4 Pin Force Models for Piezo-Actuated Beams

In this section we give a survey on the simple pin force models found in the literature. The name comes from the fact, that they lead to discrete shear forces at the ends of the piezoceramic elements. They are applied for bending actuation in beams as shown in Figure 2.9. The system under consideration is a beam with two piezoceramic elements bonded to the upper and lower surfaces. The beam of length ℓ may be subdivided into three sections, each having a length ℓ_i and with a local coordinate x_i. The thickness of the beam is $2h$, the thickness of the piezoceramic elements h_P. Ideal bonding of the piezoceramic elements to the beam is usually assumed for these models. Therefore, the displacements of the beam and of the ceramic at corresponding points of the bonding surface are equal. In reality, the bonding layer is of finite thickness, which reduces the stiffness by a certain amount. This effect will be neglected here.

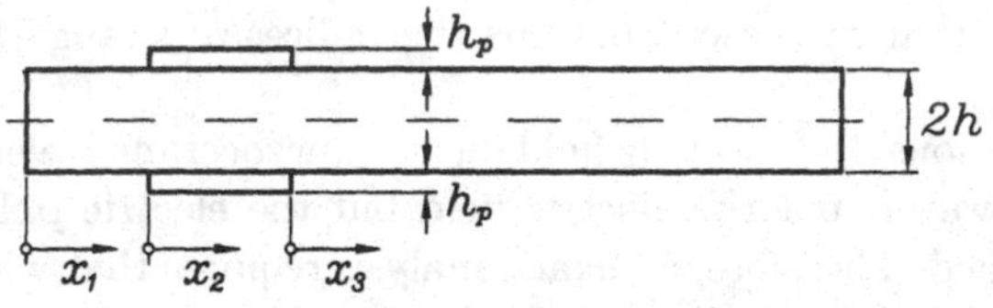

Figure 2.9: Geometry of the beam with piezoceramic elements

For an orientation of the polarization and the electric field in such a way as to give a bending deformation of the beam, mainly two models were investigated in the past. In the first model it is assumed that the thickness of the piezoceramic elements is small, so that the strain in axial direction does not depend on the z-coordinate. It is therefore denoted as *constant strain model*. For this model the strain and stress distributions are shown qualitatively as functions of the transverse coordinate in Figure 2.10. According to Figure 2.10 the strain in

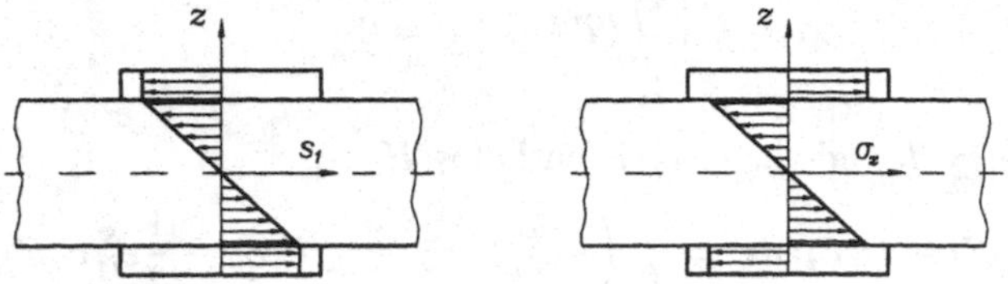

Figure 2.10: Strain and stress distribution in the beam and in the piezoceramics for the BERNOULLI-EULER model

the beam is

$$S_1 = -w''(x)z \tag{2.35}$$

and in the piezoceramic

$$S_1 = -w''(x)(\pm h). \tag{2.36}$$

All other strains vanish identically. In equation (2.36) the plus sign corresponds to the upper and the minus sign to the lower ceramic. The polarizations of the piezoceramic elements are in direction of the z-axis and the electric field has only an E_3 component, also in direction of the z-axis (upper ceramic) or in opposite direction (lower ceramic). Introducing the strain and the electric field into HAMILTON's principle (2.34) leads to

$$\begin{aligned}&\int_0^{\ell_2} \left(c_{11B}\frac{2}{3}bh^3w'' + 2bh_Pc_{11P}^Eh^2w'' - 2e_{31}E_3bh_Ph\right)\delta w''dx_2\\ &+\int_0^{\ell_1} c_{11B}\frac{2}{3}bh^3w''\delta w''dx_1 + \int_0^{\ell_3} c_{11B}\frac{2}{3}bh^3w''\delta w''dx_3 = 0.\end{aligned} \tag{2.37}$$

The integration by parts may be done, but as only the static case is investigated without any mechanical loads it can clearly be seen that the curvatures in zones 1 and 3 have to vanish. The indices B and P correspond to the beam and the piezoceramic, respectively. In zone 2 the calculus of variation yields

$$c_{11B}\frac{2}{3}bh^3w'' + 2bh_Pc_{11P}^Eh^2w'' - 2e_{31}E_3bh_Ph = 0, \tag{2.38}$$

or solving the equation with respect to the curvature

$$w'' = \frac{2e_{31}E_3bh_Ph}{c_{11B}\frac{2}{3}bh^3 + 2bh_Pc_{11P}^Eh^2}. \tag{2.39}$$

The second pin force model assumes a linearly varying strain not only in the beam but also in the piezoceramic elements. This model is refered to as the BERNOULLI-EULER model. Both the strain and the corresponding stress distributions are indicated in Figure 2.11. The

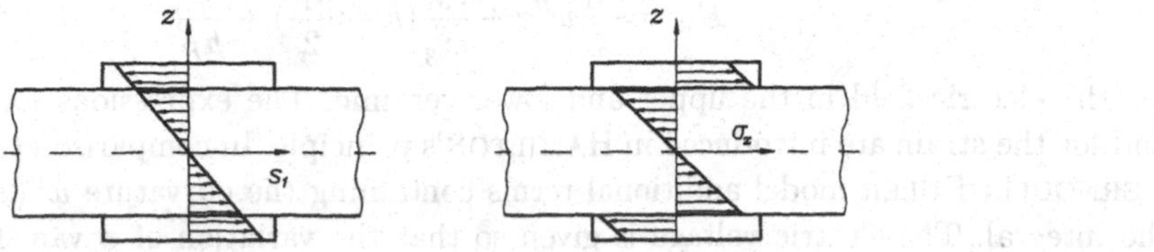

Figure 2.11: Strain and stress distribution in the beam and in the piezoceramics for the constant strain model

strain in the beam and in the piezoceramics is therefore

$$S_1 = -w''(x)z. \tag{2.40}$$

This strain and the given electric field are introduced into HAMILTON's principle. A variation according to (2.34) again leads to constant and therefore vanishing curvatures in zones 1 and 3 and according to

$$\int_0^{\ell_2}\Big(b\frac{2}{3}h^3c_{11B}w'' + 2b\left(h^2h_P + hh_P^2 + \frac{h_P^3}{3}\right)c_{11P}^Ew'' - 2e_{31}E_3b(hh_P + \frac{h_P^2}{2})\Big)\delta w''dx_2 = 0 \tag{2.41}$$

to an expression for the curvature of the form

$$w'' = \frac{2e_{31}E_3b(hh_P + \frac{h_P^2}{2})}{b\frac{2}{3}h^3c_{11B} + 2b\left(h^2h_P + hh_P^2 + \frac{h_P^3}{3}\right)c_{11P}^E}. \tag{2.42}$$

In the above models the electric field was assumed to be constant and given within the piezoceramic elements. For the BERNOULLI-EULER model, however, the strain in the ceramic is a function of z and so the electric field is not constant if a voltage is applied. Due to equation (2.31) the condition

$$D_{3,z} = \varepsilon_{33}^SE_{3,z} + e_{31}S_{1,z} = 0 \tag{2.43}$$

has to be satisfied. Introducing the strain (2.36) into this equation yields

$$E_3 = -\frac{e_{31}w''}{\varepsilon_{33}^S}z + C_1 \tag{2.44}$$

with an integration constant C_1 still to be determined. A second integration leads to the electric potential $\bar{\phi}$

$$\bar{\phi} = \frac{e_{31} w''}{\varepsilon_{33}^S} \frac{z^2}{2} - C_1 z + C_2. \tag{2.45}$$

The applied voltage corresponds to the difference

$$\phi = \bar{\phi}(h + h_P) - \bar{\phi}(h) \tag{2.46}$$

so that C_1 may be determined and we obtain

$$E_{3u} = \frac{e_{31}}{\varepsilon_{33}^S} w'' z - \frac{e_{31}}{\varepsilon_{33}^S}(h + \frac{h_P}{2}) - \frac{\phi}{h_P}, \tag{2.47}$$

$$E_{3l} = \frac{e_{31}}{\varepsilon_{33}^S} w'' z + \frac{e_{31}}{\varepsilon_{33}^S}(h + \frac{h_P}{2}) + \frac{\phi}{h_P} \tag{2.48}$$

for the electric field in the upper and lower ceramic. The expressions for the electric field and for the strain are introduced in HAMILTON's principle. In comparison to the conventional BERNOULLI-EULER model additional terms containing the curvature $w''(x)$ are obtained in the integral. The electric voltage is given so that the variation of ϕ vanishes. Therefore we obtain

$$\begin{aligned} &\int_0^{\ell_1} b\frac{2}{3}h^3 c_{11B} w'' \delta w'' dx_1 + \int_0^{\ell_3} b\frac{2}{3}h^3 c_{11B} w'' \delta w'' dx_3 \\ &+ \int_0^{\ell_2} [b\frac{2}{3}h^3 c_{11B} + 2b\left(h^2 h_P + h h_P^2 + \frac{h_P^3}{3}\right) c_{11P}^E \\ &-2\varepsilon_{33}^S b E_{3a}^2 \frac{3h^2 h_P + 3h h_P^2 + h_P^3}{3} - 2e_{31} b E_{3b}(2h h_P + h_P^2)] w'' \delta w'' \\ &-2\varepsilon_{33}^S b(E_{3b} w'' + E_{30}) h_P E_{3b} \delta w'' - e_{31} b E_{30}(2h h_P + h_P^2) \delta w'' dx_2 = 0. \end{aligned} \tag{2.49}$$

Once again only region 2 is of interest and the corresponding curvature is

$$w'' = \frac{-2e_{31}E_3(h h_P + \frac{h_P^2}{2})}{\frac{2}{3}h^3 c_{11B} + 2\left(h^2 h_P + h h_P^2 + \frac{h_P^3}{3}\right) c_{11P}^E + \frac{e_{31}^2}{6\varepsilon_{33}^S} h_P^3}. \tag{2.50}$$

It can be seen that the influence of the variable electric field leads to an additional term in the denominator of the curvature in comparison to the curvature determined by BERNOULLI-EULER theory and constant electric field, equation (2.42).

A comparison of the results for the different models is given in Figure 2.12. In this figure the curvature of a beam with given geometry is plotted as a function of the relative thickness h_P/h of the piezoceramics and the beam. It can be seen that the curvature for the constant strain model is increasing monotonically with increasing thickness of the piezoceramic elements. The difference between the BERNOULLI-EULER models with constant electric field and given electric potential is small, so that it is plotted separately in Figure 2.13. The relative difference between the results of these two models is in the range of 0.1 %, so that the effect of the nonconstant electric field may be neglected for a static excitation of the system.

The different pinforce models yield good results for the displacement. Nevertheless, they are not able to describe, for instance the stress distribution within the bonding layer.

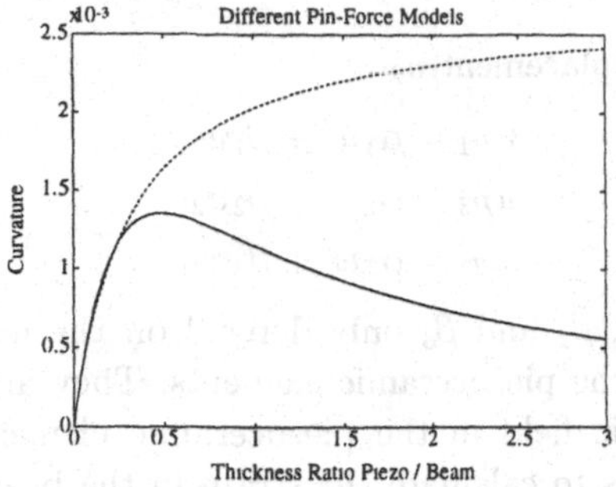

Figure 2.12: Comparison of the curvature for the static case for different pin-force models. (- - -) constant strain model, (—) BERNOULLI-EULER model with constant electric field

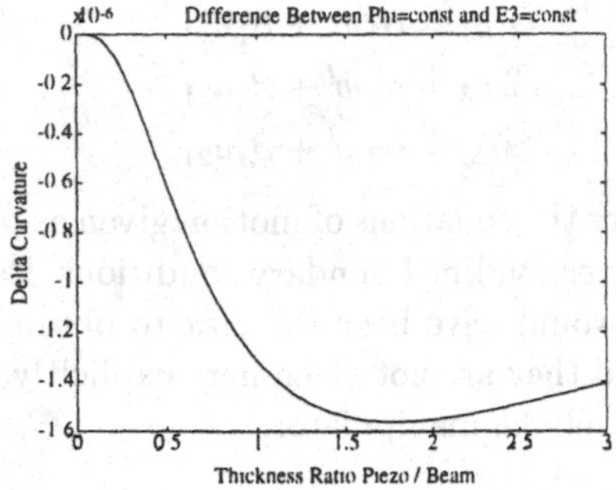

Figure 2.13: Difference of the curvatures for the BERNOULLI-EULER models with constant electric field and given electric potential ϕ

To get an approximation of the stresses in the bonding layer, a higher order beam theory has to be used. E.g. if we consider the zone of the beam to which the piezoceramic elements are bonded, then we may introduce displacement functions as follows

$$\begin{aligned} w_{2B}(x,z,t) &= w_2(x,t), \\ u_{2B}(x,z,t) &= u_{21}(x,t)z + u_{23}(x,t)z^3, \\ w_{2P}(x,z,t) &= w_2(x,t), \\ u_{2P}(x,z,t) &= u_{P1}(x,t)z + u_{P3}(x,t)z^3, \end{aligned} \tag{2.51}$$

which were chosen because a bending deformation is considered. This choice for the displacement functions automatically allows to fulfill several conditions. E.g. a vanishing shear stress is obtained at the outer surface of the piezoceramics as well as shear stress continuity at the bonding layer and continuity of the displacement in axial and transverse direction

$$T_{5P}(x, h + h_P, t) = 0, \tag{2.52}$$

$$T_{5B}(x,h,t) = T_{5P}(x,h,t), \tag{2.53}$$

$$u_{2B}(x,h,t) = u_{2P}(x,h,t). \tag{2.54}$$

Therefore, all displacement functions may be calculated as a function of the transverse displacement w and the axial displacement u_{21}

$$u_{P1} = \alpha_1 w' + \beta_1 u_{21}, \tag{2.55}$$

$$u_{P3} = \alpha_2 w' + \beta_2 u_{21}, \tag{2.56}$$

$$u_{23} = \alpha_3 w' + \beta_3 u_{21}. \tag{2.57}$$

The constants α_1, α_2, α_3, β_1, β_2 and β_3 only depend on the material and on geometrical parameters of the beam and the piezoceramic elements. They are not given here explicitly. If we assume that the electric field in the piezoceramic elements is given, then we may use the displacement functions to calculate the strain in the beam and in the piezoceramic elements and introduce these quantities into HAMILTON's principle. We may not only treat the static case, but we may introduce also the kinetic energy which may be calculated using the derivatives of the displacement functions with respect to the time:

$$\dot{u}_{P1} = \alpha_1 \dot{w}' + \beta_1 \dot{u}_{21}, \tag{2.58}$$

$$\dot{u}_{P3} = \alpha_2 \dot{w}' + \beta_2 \dot{u}_{21}, \tag{2.59}$$

$$\dot{u}_{23} = \alpha_3 \dot{w}' + \beta_3 \dot{u}_{21}. \tag{2.60}$$

HAMILTON's principle then gives the equations of motion given as partial differential equations for w and u_{21}, as well as the corresponding boundary conditions. Both the equations of motion and the boundary conditions would have been difficult to obtain by synthetic methods. As these equations are complicated they are not given here explicitly. Nowadays their derivation can be simplified by using a symbolic manipulator.

2.5 Piezoceramic Element Bonded to a Half-Space

The previous section, in which we described the pin force models, showed that it is difficult to obtain the stress distribution within the bonding layer. At the end of the previous section a higher order beam theory was introduced to get an approximation. On the other hand, if the piezoceramic elements are very thin then similarly to HERTZ' contact theory we may assume that we have a thin piezoceramic element bonded to an elastic half-space, if the dimension of the elastic structure is much larger than the length of the piezoceramic element. If the thickness of the piezoceramic element is very small compared to its length, we may assume that we have a uniform stress distribution at each cross-section of the element. For simplicity we restrict ourselves to a static excitation.

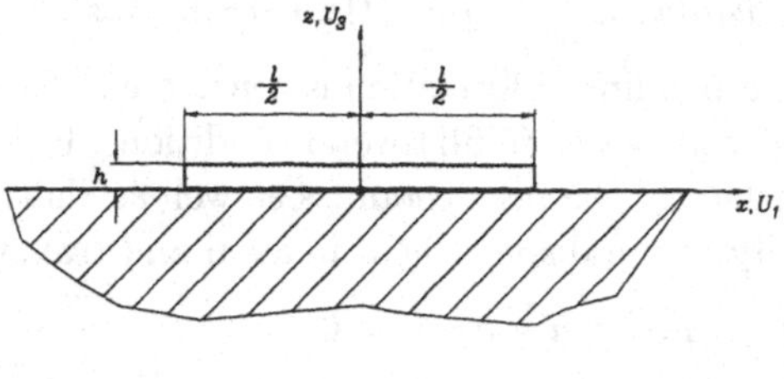

Figure 2.14: The piezoceramic element bonded to a half-space

2.5.1 Thin Ceramic Element on an Elastic Half-space

If we consider a very thin ceramic element on an elastic half-space, it may be assumed that the normal stress T_1 in the ceramic element is uniformly distributed over the cross-section. A free body diagram of a differential element of the ceramic and the corresponding shear stress on the half-space are shown in Fig 2.15. The stress T_3 in the bonding layer is neglected. The balance of the force in horizontal direction leads to

$$T_5 = h\frac{dT_1}{dx} \tag{2.61}$$

in which T_5 is the shear stress in the bonding layer and T_1 denotes the stress in the piezoceramic element. The strain S_1 at the surface of the half-space resulting from the shear stress

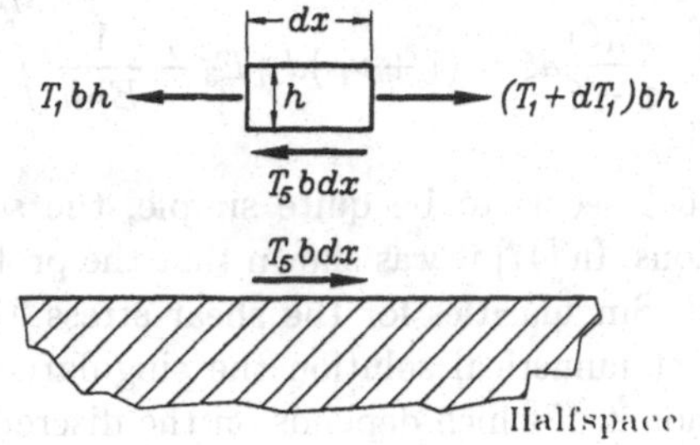

Figure 2.15: Differential element of the ceramic element and the corresponding shear stress on the half-space

along the bonding layer can be determined by integration

$$S_1(x) = -\frac{2(1-\nu^2)}{\pi E}\int_{-\ell/2}^{\ell/2}\frac{T_5(\xi)}{x-\xi}d\xi \tag{2.62}$$

as shown in [20] or any other textbook on elasticity. The constitutive equations for $T_3 = 0$ and $S_2 = 0$ yield

$$T_1 = \frac{1}{s_{11}(1-\nu_P^2)}[S_1 - (1+\nu_P)d_{31}E_3]. \tag{2.63}$$

The boundary conditions are $T_1(-\ell/2) = 0$ and $T_1(\ell/2) = 0$. Integration of (2.61) using (2.62) and (2.63) with

$$E_P = \frac{1}{s_{11}(1-\nu_P^2)}, \qquad E^* = \frac{E}{1-\nu^2} \tag{2.64}$$

leads to

$$\int_{-\ell/2}^{x} T_5(\xi)d\xi = -\frac{2E_P h}{\pi E^*}\int_{-\ell/2}^{\ell/2}\frac{T_5(\xi)}{x-\xi}d\xi - (1+\nu_P)d_{31}E_3E_P h. \tag{2.65}$$

As T_5 is an odd function

$$\int_{-\ell/2}^{\ell/2} T_5(\xi)d\xi = 0 \tag{2.66}$$

the integral may be split

$$\int_{-\ell/2}^{x} T_5(\xi)d\xi = -\int_{x}^{\ell/2} T_5(\xi)d\xi \tag{2.67}$$

so that we obtain

$$-\frac{2}{\pi E^*}\int_{-\ell/2}^{\ell/2} \frac{T_5(\xi)}{x-\xi}d\xi = (1+\nu_P)d_{31}E_3 - \frac{1}{E_P h}\int_{x}^{\ell/2} T_5(\xi)d\xi. \tag{2.68}$$

Though Equation (2.68) seems to be quite simple, the solution requires a detailed knowledge of integral equations. In [47] it was shown that the problem leads to a FREDHOLM equation of the second kind. Singularities for the shear stress T_5 occur at the ends of the ceramic element. For a direct numerical solution the singularities have to be replaced by discrete shear forces, the amount of which depends on the discretization near the ends.

2.5.2 Piezoceramic Element on a Rigid Substrate

The previous section showed a stress singularity at the ends of the piezoceramic element. Of course, this cannot be observed in reality due to yielding or debonding. If, in addition, it is assumed that the stiffness of the elastic half-space is much higher than the stiffness of the piezoceramic material, then a different model is obtained if it is assumed that we have a piezoceramic element ideally bonded to a rigid substrate. Therefore, displacements occur only within the piezoceramic element. For an approximate solution for a thin piezoceramic element we assume a displacement field of the form

$$u_1(x,z) = \bar{u}_1(x)z + \hat{u}_1(x)z^2, \tag{2.69}$$

$$u_3(x,z) = \bar{u}_3(x)z + \hat{u}_3(x)z^2 \tag{2.70}$$

with unknown functions $\bar{u}_1$, $\hat{u}_1$, $\bar{u}_3$ and $\hat{u}_3$. With these displacements the strain is calculated and introduced into HAMILTON's principle. Once again, we assume that the electric field in the piezoceramic element is given. This leads to four ordinary coupled differential equations and eight boundary conditions at $x = -\ell/2$ as well as eight boundary conditions at $x = \ell/2$. Therefore, this approach assures that the stresses at the ends of the ceramic element are minimized, but stresses may still occur on the upper surface of the piezoceramic element, which in reality should be stress-free. As the length of the piezoceramic element is much greater than its thickness ($\ell \gg h_p$) a better approximation will be obtained if the condition of vanishing stresses T_3 and T_5 on the upper surface is fulfilled in advance. This leads to two constraint equations for $\bar{u}_1$, $\hat{u}_1$, $\bar{u}_3$ and $\hat{u}_3$ in the form of two differential equations

$$C_{55}\left(\bar{u}_1 + 2\hat{u}_1 h_p + \bar{u}_3' h_p + \hat{u}_3' h_p^2\right) = 0, \tag{2.71}$$

$$C_{13}\left(\bar{u}_1' h_p + \hat{u}_1' h_p^2\right) + C_{33}\left(\bar{u}_3 + 2\hat{u}_3 h_p\right) - e_{33}E_3 = 0. \tag{2.72}$$

If we denote the first by g_1 and the second by g_2 we have

$$g_1 = 0, \tag{2.73}$$
$$g_2 = 0. \tag{2.74}$$

A direct elimination of two of the displacement functions is not possible. These constraint equations may however be easily incorporated into HAMILTON's principle by the use of LAGRANGE-multipliers Λ_1 and Λ_2

$$\delta \int_V (H + \Lambda_1 g_1 + \Lambda_2 g_2) dV = 0. \tag{2.75}$$

After variation and integration by parts this yields four coupled differential equations

$$L_1(\bar{u}_1, \hat{u}_1, \bar{u}_1, \hat{u}_1'', \bar{u}_3', \hat{u}_3', \Lambda_1, \Lambda_2') = 0, \tag{2.76}$$
$$L_2(\bar{u}_1, \hat{u}_1, \bar{u}_1'', \hat{u}_1'', \bar{u}_3', \hat{u}_3', \Lambda_1, \Lambda_2') = 0, \tag{2.77}$$
$$L_3(\bar{u}_1', \hat{u}_1', \bar{u}_3, \hat{u}_3, \bar{u}_3'', \hat{u}_3'', \Lambda_1', \Lambda_2, E_3) = 0, \tag{2.78}$$
$$L_4(\bar{u}_1', \hat{u}_1', \bar{u}_3, \hat{u}_3, \bar{u}_3'', \hat{u}_3'', \Lambda_1', \Lambda_2, E_3) = 0, \tag{2.79}$$

or, explicitly:

$$2C_{55}\bar{u}_1 + 2C_{55}h_p\hat{u}_1 - \frac{2}{3}C_{11}h_p^2\bar{u}_1'' - \frac{1}{2}C_{11}h_p^3\hat{u}_1'' + (C_{55} - C_{13})h_p\bar{u}_3'$$
$$+ \left(\frac{2}{3}C_{55} - \frac{4}{3}C_{13}\right) h_p^2\hat{u}_3' - 2C_{13}\Lambda_2' h_p + 2\Lambda_1 = 0, \tag{2.80}$$

$$2C_{55}\bar{u}_1 + \frac{8}{3}C_{55}h_p\hat{u}_1 - \frac{1}{2}C_{11}h_p^2\bar{u}_1'' - \frac{2}{5}C_{11}h_p^3\hat{u}_1'' + \left(\frac{4}{3}C_{55} - \frac{2}{3}C_{13}\right) h_p\bar{u}_3'$$
$$+(C_{55} - C_{13})h_p^2\hat{u}_3' - 2C_{13}\Lambda_2' h_p + 4\Lambda_1 = 0, \tag{2.81}$$

$$2C_{33}\bar{u}_3 + 2C_{33}h_p\hat{u}_3 - \frac{2}{3}C_{55}h_p^2\bar{u}_3'' - \frac{1}{2}C_{55}h_p^3\hat{u}_3'' + (C_{13} - C_{55})h_p\bar{u}_1'$$
$$+\frac{2}{3}(C_{13} - 2C_{55})h_p^2\hat{u}_1' + 2C_{33}\Lambda_2 - 2\Lambda_1' h_p = 2e_{33}E_3, \tag{2.82}$$

$$2C_{33}\bar{u}_3 + \frac{8}{3}C_{33}h_p\hat{u}_3 - \frac{1}{2}C_{55}h_p^2\bar{u}_3'' - \frac{2}{5}C_{55}h_p^3\hat{u}_3'' + \left(\frac{4}{3}C_{13} - \frac{2}{3}C_{55}\right) h_p\bar{u}_1'$$
$$+(C_{13} - C_{55})h_p^2\hat{u}_1' + 4C_{33}\Lambda_2 - 2\Lambda_1' h_p = 2e_{33E_3}. \tag{2.83}$$

These differential equations permit the elimination of Λ_1 and Λ_2

$$-2\Lambda_1 = \frac{1}{6}C_{11}\bar{u}_1''h_p^2 + \frac{1}{10}C_{11}\hat{u}_1''h_p^3 + \frac{2}{3}C_{55}\hat{u}_1 h_p + \frac{1}{3}(C_{13} + C_{55})\left(\bar{u}_3' h_p + \hat{u}_3' h_p^2\right), \tag{2.84}$$
$$-2C_{33}\Lambda_2 = \frac{1}{6}C_{55}\bar{u}_3''h_p^2 + \frac{1}{10}C_{55}\hat{u}_3''h_p^3 + \frac{2}{3}C_{33}\hat{u}_3 h_p + \frac{1}{3}(C_{13} + C_{55})\left(\bar{u}_1' h_p + \hat{u}_1' h_p^2\right) \tag{2.85}$$

so that the results are two equations of the form

$$k_{11}\bar{u}_1'' + k_{12}\bar{u}_1 + k_{13}\hat{u}_1'' + k_{14}\hat{u}_1 + k_{15}\bar{u}_3''' + k_{16}\bar{u}_3' + k_{17}\hat{u}_3''' + k_{18}\hat{u}_3' = 0, \tag{2.86}$$
$$k_{21}\bar{u}_1''' + k_{22}\bar{u}_1' + k_{23}\hat{u}_1''' + k_{24}\hat{u}_1' + k_{25}\bar{u}_3'' + k_{26}\bar{u}_3 + k_{27}\hat{u}_3'' + k_{28}\hat{u}_3 = 2e_{33}E_3 \tag{2.87}$$

and the two constraint equations

$$k_{31}\bar{u}_1 + k_{32}\hat{u}_1 + k_{33}\bar{u}_3' + k_{34}\hat{u}_3' = 0, \tag{2.88}$$
$$k_{41}\bar{u}_1' + k_{42}\hat{u}_1' + k_{43}\bar{u}_3 + k_{44}\hat{u}_3 = e_{33}E_3. \tag{2.89}$$

The coefficients k_{ij} are constant and depend on the stiffness coefficients C_{11}, C_{13}, C_{33}, C_{55} and on the thickness h_p

$$
\begin{aligned}
k_{11} &= \frac{C_{13}^2 + C_{13}C_{55}}{3C_{33}} h_p^2 - \frac{5}{6} C_{11} h_p^2, & k_{21} &= \frac{1}{6} C_{11} h_p^3, \\
k_{12} &= 2C_{55}, & k_{22} &= \frac{2}{3} C_{13} h_p - \frac{4}{3} C_{55} h_p, \\
k_{13} &= \frac{C_{13}^2 + C_{13}C_{55}}{3C_{33}} h_p^3 - \frac{3}{5} C_{11} h_p^3, & k_{23} &= \frac{1}{10} C_{11} h_p^4, \\
k_{14} &= \frac{4}{3} C_{55} h_p, & k_{24} &= \frac{1}{3} C_{13} h_p^2 - C_{55} h_p^2, \\
k_{15} &= \frac{C_{13}C_{55}}{6C_{33}} h_p^3, & k_{25} &= \frac{1}{3} C_{13} h_p^2 - \frac{1}{2} C_{55} h_p^2, \\
k_{16} &= \frac{2}{3} C_{55} h_p - \frac{4}{3} C_{13} h_p, & k_{26} &= 2C_{33}, \\
k_{17} &= \frac{C_{13}C_{55}}{10C_{33}} h_p^4, & k_{27} &= \frac{1}{3} C_{13} h_p^3 - \frac{8}{30} C_{55} h_p^3, \\
k_{18} &= \frac{1}{3} C_{55} h_p^2 - C_{13} h_p^2, & k_{28} &= \frac{4}{3} C_{33} h_p, \\
\\
k_{31} &= 1, & k_{41} &= C_{13} h_p, \\
k_{32} &= 2h_p, & k_{42} &= C_{13} h_p^2, \\
k_{33} &= h_p, & k_{43} &= C_{33}, \\
k_{34} &= h_p^2, & k_{44} &= 2C_{33} h_p.
\end{aligned}
\tag{2.90}
$$

In addition we now only have four boundary conditons at either end of the piezoceramic element

$$
\left.
\begin{aligned}
B_1(\bar{u}_1, \bar{u}_1'', \hat{u}_1, \hat{u}_1'', \bar{u}_3', \hat{u}_3') &= 0, \\
B_2(\bar{u}_1', \hat{u}_1', \bar{u}_3, \bar{u}_3'', \hat{u}_3, \hat{u}_3'', E_3) &= 0, \\
B_3(\bar{u}_1, \bar{u}_1'', \hat{u}_1, \hat{u}_1'', \bar{u}_3', \hat{u}_3') &= 0, \\
B_4(\bar{u}_1', \hat{u}_1', \bar{u}_3, \bar{u}_3'', \hat{u}_3, \hat{u}_3'', E_3) &= 0,
\end{aligned}
\right\}
\begin{array}{c} x = \ell/2 \\ \text{and} \\ x = -\ell/2 \end{array}
\tag{2.91}
$$

which take the form

$$\frac{2}{3} C_{55} h_p \bar{u}_3' + \frac{1}{2} C_{55} h_p^2 \hat{u}_3' + C_{55} \bar{u}_1 + \frac{4}{3} C_{55} h_p \hat{u}_1 + 2\Lambda_1 = 0, \tag{2.92}$$

$$\frac{1}{2} C_{55} h_p \bar{u}_3' + \frac{2}{5} C_{55} h_p^2 \hat{u}_3' + \frac{2}{3} C_{55} \bar{u}_1 + C_{55} h_p \hat{u}_1 + 2\Lambda_1 = 0, \tag{2.93}$$

$$\frac{1}{2} C_{11} h_p \bar{u}_1' + \frac{2}{5} C_{11} h_p^2 \hat{u}_1' + \frac{2}{3} C_{13} \bar{u}_3 + C_{13} h_p \hat{u}_3 - \frac{2}{3} e_{31} E_3 + 2C_13\Lambda_2 = 0, \tag{2.94}$$

$$\frac{2}{3} C_{11} h_p \bar{u}_1' + \frac{1}{2} C_{11} h_p^2 \hat{u}_1' + C_{13} \bar{u}_3 + \frac{4}{3} C_{13} h_p \hat{u}_3 - e_{31} E_3 + 2C_{13}\Lambda_2 = 0. \tag{2.95}$$

A solution of the coupled differential equations (2.86) to (2.89) is easily obtained by standard methods. A particular solution is

$$(\bar{u}_1, \hat{u}_1, \bar{u}_3, \hat{u}_3)_p^T = (0, 0, e_{33} E_3 / C_{33}, 0)^T. \tag{2.96}$$

For the solution of the homogeneous differential equations an ansatz

$$(\bar{u}_1, \hat{u}_1, \bar{u}_3, \hat{u}_3)_h^T = (\bar{U}_1, \hat{U}_1, \bar{U}_3, \hat{U}_3)^T e^{\lambda x} \tag{2.97}$$

with constant amplitudes $\bar{U}_1$, $\hat{U}_1$, $\bar{U}_3$, $\hat{U}_3$ leads to the general solution

$$(\bar{u}_1, \hat{u}_1, \bar{u}_3, \hat{u}_3)_h^T = \sum_{i=1}^{8} C_i \cdot (\bar{U}_{1i}, \hat{U}_{1i}, \bar{U}_{3i}, \hat{U}_{3i})^T e^{\lambda_i x} \tag{2.98}$$

with eight integration constants C_i. These integration constants may be determined by fitting the solution to the boundary conditions (2.91). Stress distributions in the bonding layer for T_3 and T_5 are shown Figure 2.16. Lines of equal shear stress T_5 in the piezoceramic element

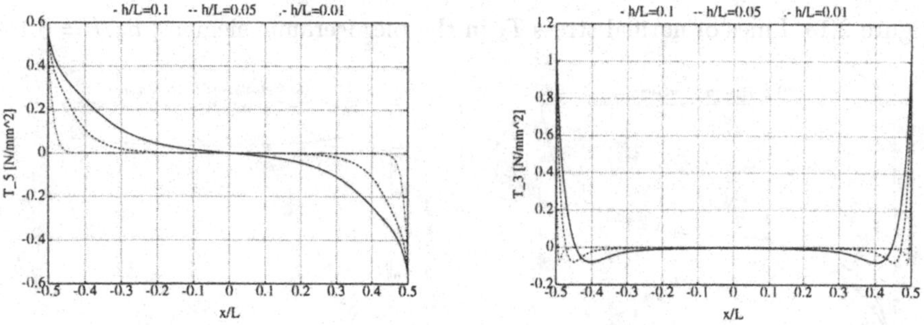

Figure 2.16: Shear stress T_5 and normal stress T_3 in the bonding layer; — $h_p/\ell = 0.1$, – – – $h_p/\ell = 0.05$, – · – $h_p/\ell = 0.01$

and of equal normal stress T_1 and T_3 are shown in Figures 2.17 and 2.18. A comparison of

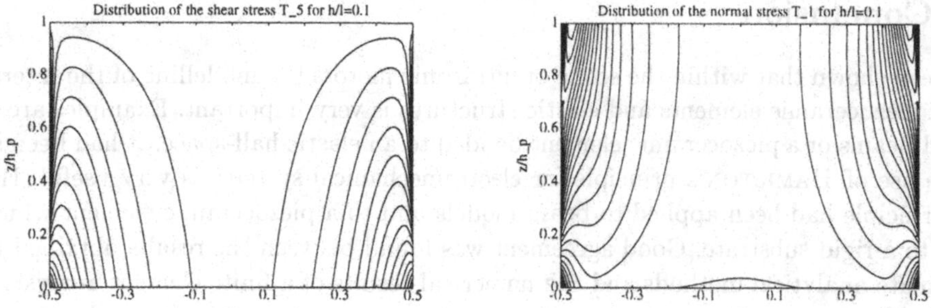

Figure 2.17: Lines of equal shear stress T_5 and equal normal stress T_1 in the piezoceramic element, $h_p/\ell = 0.1$

the stress normalized with respect to the maximum stress for the approximate solution with a solution by the FEM shows good agreement, see Figure 2.19

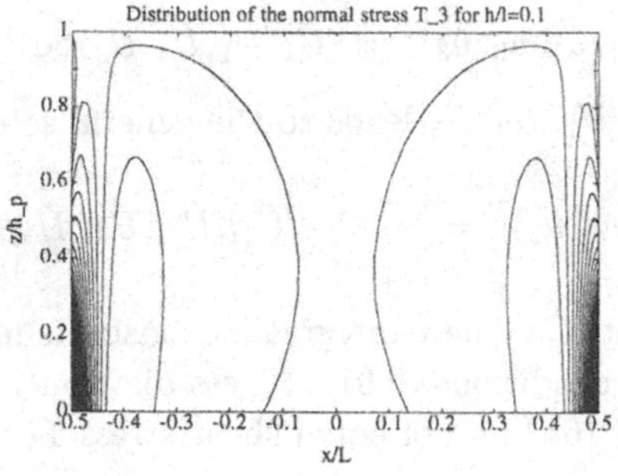

Figure 2.18: Lines of normal stress T_3 in the piezoceramic element, $h_p/\ell = 0.1$

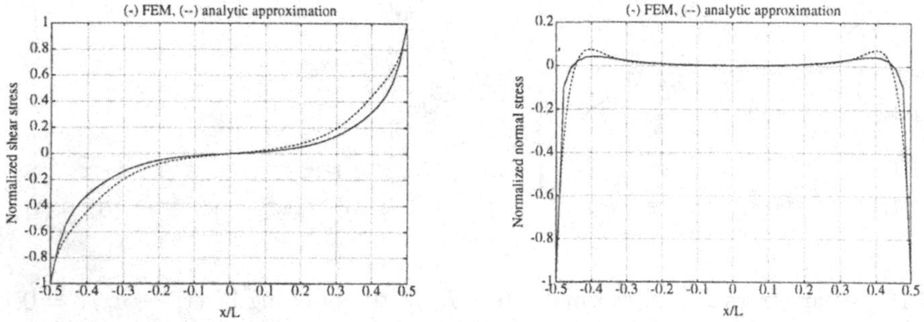

Figure 2.19: Shear stress T_5 and normal stress T_3 in the bonding layer

2.6 Conclusion

It has been shown that within the scope of ultrasonic motors the modelling of the interaction between piezoceramic elements and elastic structures is very important. Examples are piezo-actuated beams or a piezoceramic element bonded to an elastic half-space. It had been shown that the use of HAMILTON's principle for electromechanical systems is very useful. HAMILTON's principle had been applied to beam models and to a piezoceramic element which was bonded to a rigid substrate. Good agreement was found between the results obtained by the approximate analytical methods and the numerical results of a finite element analysis.

Chapter 3

Mathematical Modelling Problem in Vortex-Excited Oscillations of the Conductors of Transmission Lines

3.1 Single Conductor: Comparison of Different Mathematical Models

3.1.1 Introduction

In overhead transmission lines, different types of mechanical vibrations may occur. The most common type corresponds to wind-excited vibrations in the frequency range of 10Hz to 50Hz, caused by vortex shedding. Since these vibrations occur quite often, they may give rise to material fatigue, thus limiting the life time of the cables. As capital investment is very high — even in small countries such as West Germany surpassing for high voltage level transmission that directly connected with power generation — the problem of cable vibration and fatigue merit being given their proper attention. Low-frequency "galloping" vibrations (f <1Hz) are not addressed in this paper.

Since approximately 1930 the STOCKBRIDGE damper and similar devices have been used successfully to damp out cable vibrations. However, cases are known in which serious damage was caused to the cables at the points of attachment of the dampers, i.e. at their clamps. Obviously this was due to the fact that the dynamic characteristics of damper and cable had been improperly matched. Sufficient attention should be given to good modelling of this aspect of the problem. As we shall see, this can be done in different ways and at different levels. Only single cables will be modelled in this paper, the problem of bundled conductors, which are widely used, is more involved but can be treated in a similar manner.

3.1.2 Modelling of the Cable Dynamics

Figure 3.1 shows a typical transmisson line equipped with a damper which is usually mounted near the suspension clamp; there also may be more than one cable per damper in each span.

The span 1, i.e. the distance between suspension towers, is usually in the order of 300m to 1000m, while the cable's sag is small, in general a few percent of the span. On the other hand, the frequencies under consideration (10–50Hz) correspond to wave lengths of a few meters only, so that for the purpose of their study the sag can be disregarded, the cable being modelled as a straight flexible continuous system. From observations it is known that although the cable vibrations are not strictly planar, they occur predominantly in the vertical direction (see [8]), and this agrees well with what is known about vortex shedding (see section 3.1.3). For this reason, the dynamic behaviour of overhead transmission line cables can be conveniently studied in the plane.

Beam Under Tension

The mathematical model which is most often used for the cable is therefore the one of a beam with bending stiffness EI under a large normal force, i.e. a large tension T, whose transverse

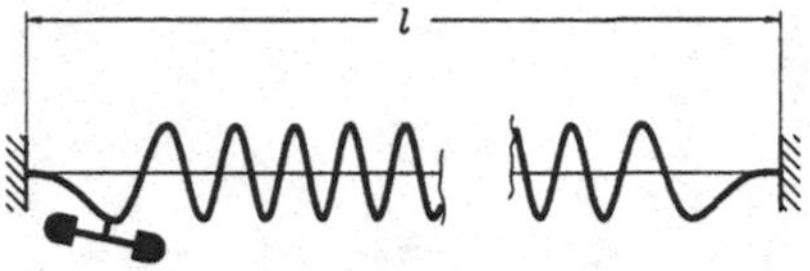

Figure 3.1: Overhead transmisson line with damper

vibrations are described by

$$EIw^{IV}(x,t) - Tw''(x,t) + m\ddot{w}(x,t) = q(x,t) + d(w,\dot{w},t). \tag{3.1}$$

In (3.1), $w(x,t)$ is the transverse displacement at the location x of the cable at time t (see Figure 3.2), m is the mass per unit length, $q(x,t)$ are the forces acting on the cable due to the wind action and vortex shedding and $d(w,\dot{w},t)$ stands for the cable's self damping. The

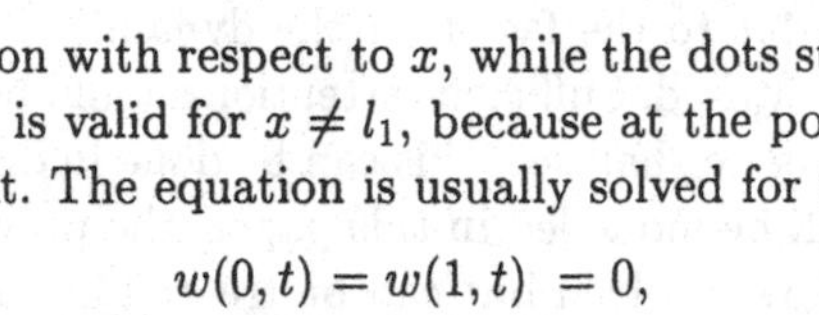

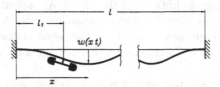

Figure 3.2: Cable with damper

primes indicate differentiation with respect to x, while the dots stand for differentiation with respect to t. Equation (3.1) is valid for $x \neq l_1$, because at the point $x = l_1$ the damper force has to be taken into account. The equation is usually solved for the boundary conditions

$$\begin{aligned} w(0,t) &= w(1,t) = 0, \\ w'(0,t) &= w'(1,t) = 0, \end{aligned} \tag{3.2}$$

which means that the suspension clamps are assumed as fixed during the vibrations. This is not necessarily so during actual vibrations, but it certainly is a case which may occur in reality (due to symmetric span for example) and it is therefore taken as a reference in the calculations.

Supposing $q(x,t)$ and $d(w,\dot{w},t)$ to be known as well as the damper force acting on the cable, one would have to solve the boundary value problem formed by (3.1) and (3.2). The bending stiffness in (3.1) is a parameter which can usually only be determined experimentally due to the complicated structure of the cables: they are formed by stranded wires which are neither completely free to slide with respect to the other wires nor do they form a rigid cross-section. The bending stiffness consequently lies somewhere in between the corresponding two extreme values. It is small but essential if one wishes to calculate bending strains, which are responsible for the fatigue damages. It has however only negligible influence on the eigenfrequencies and eigenmodes of the free cable vibrations (i.e. with $q(x,t) = 0$, $d(w,\dot{w},t) = 0$) which are almost exactly those of a string without bending stiffness (see [3]). In a string (without bending stiffness) fixed at both ends, the eigenfrequencies are all the integer multiples of the first eigenfrequency, which in the case of a transmission line is typically of the order of 0.1Hz. This means that the frequency range of 10Hz to 50Hz corresponds to the interval between the 100th and 500th eigenfrequencies of the cable. All these modes then do have to be modelled properly in the numerical solution of the nonlinear boundary value problems formed by (3.1) and (3.2), independently of the method of solution. This implies a large number of elements, if FEM techniques are used, or a high number of functions in the RITZ method for example. For bundled conductors, the numerical effort of course increases much more, almost becoming prohibitive, particularly in view of the fact that the wind forces are only modelled in a very rough manner, as we shall see in section 3.1.3.

There is therefore a strong interest in simplifying the model (without loosing essential information if possible). This goal is reached by using the fact that $e = \sqrt{EI/Tl^2} \ll 1$.

Flexible String With Zero Bending Stiffness and Boundary Layers

With the dimensionless factor e introduced above one can write (1) as

$$e^2 w^{IV}(x,t) + \frac{1}{Tl^2}\left[-Tw''(x,t) + m\ddot{w}(x,t) - q(x,t) - d(w,\dot{w},t)\right] = 0. \tag{3.3}$$

Recalling that e^2 is a very small number, one may try to solve (3.3) by using perturbation theory. This is in fact possible, it simplifies the problem enormously and gives excellent results. For $e = 0$ equation (3.3) describes the vibrations of a flexibel string (with zero bending stiffness). Of course, the order of the partial differential equation is then lower, so that not all the four boundary conditions (3.2) can be satisfied. Only the first two boundary conditions are fulfilled by the zero order solution to the problem.

In this zero order problem, no bending strains can of course be calculated since there is no bending stiffness. With the zero order solution known, higher order approximations can be however constructed by standard perturbational techniques. Since the order of the partial differential equation (3.3) changes with e^2 equal to or different from zero, this is a case of singular perturbations and large bending strains only exist at the points at which the zero order solution $w_0(x,t)$ has discontinuous slope, as is the case precisely at the points where concentrated forces act on the string, i.e. at the ends and at the point of attachment of the damper. At these places boundary layers occur, in which the bending strains decrease exponentially from a large value to almost zero. The locations of the bending boundary layers are shown in Figure 3.3. It is at these points that fatigue may occur and where the bending strains should be checked.

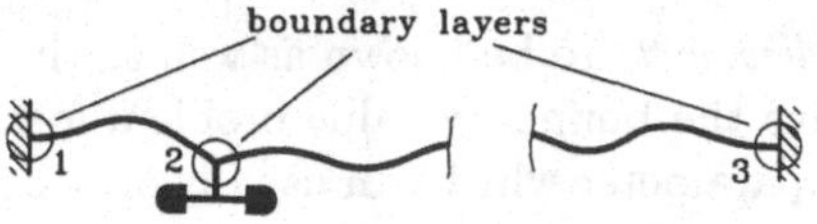

Figure 3.3: Bending boundary layers in a cable

In [14] it is shown, for example, that the bending strain at the right hand end of the cable, i.e. in the boundary layer 3 is given by

$$w''(l,t) = \frac{T}{EI} w_0'(\bar{l},t), \tag{3.4}$$

where w_0' is the slope at $x = \bar{l}$ calculated from (3.3) with $e^2 = 0$. Since $\sqrt{T/EI}$ is small compared to the length l and also to the wave lengths — it typically is of the order of a few centimeters while the wavelength is of the order of a few meters — the formulas of the perturbation analysis like (3.4) give excellent results. The bending strains in the free field, i.e. far away from the bending boundary layers may be as small as one fiftieth of the bending strains in the boundary layers.

A comparison of the bending strains calculated with the full equations with those computed with the perturbational approach for realistic cable parameters reveals that the difference is extremely small, far below one percent. Due to this fact, the formulas obtained by applying perturbational theory are now widely used, since the considerably simplify the mathematical treatment (see details in [14]).

Energy Balance in the Infinite or Semi-Infinite String

Form many observations of cable vibrations caused by vortex shedding under steady wind conditions it is known that one frequency clearly predominates at each wind speed. As we shall see in section 3.1.3, this frequency is proportinal to the speed of the wind blowing transversally to the cable. The wind may vary slowly in time over a wide range. For given wind conditions the cable can therefore be considered to oscillate with a single frequency. The eigenfrequencies of the cable on the other hand are very dense, being spaced for example at 0.1Hz intervals. Wind speeds which lead to resonant cable vibrations are therefore very close to speeds which correspond to anti-resonances. Also, the cable length between two suspension clamps is usually not well known and it is not even constant since it changes with temperature; the tension T in the cable is subject to even more severe changes with temperature.

If a cable oscillates with a frequency of 32.1Hz e.g., it is not possible for practical purposes to distinguish whether it vibrates resonantly in the 322nd, the 321st or the 320th mode. However, this is not very relevant, since the wave lengths corresponding to these neighboring modes are almost equal and so will be the maximum bending strains for given vibration amplitudes in the free field. Keeping in mind that a large part of the computational effort stems from the fact that a boundary value problem is being solved, one is therefore tempted to eliminate all the boundary conditions, at least during parts of the calculations.

In fact, harmonic or quasiharmonic monofrequent vibrations being assumed, the vibration amplitudes completely determine the power introduced by the wind forces into the cable. It turns out that the average energy level (averaged over one wavelength) is almost constant along most of the cable. If this average energy or amplitude level is known for the free field, the bending strains for example at $x = l$ can be easily calculated. It is more or less obvious that the boundary conditions at $x = 0$ do not affect the bending strains at $x = l$ for the high modes being considered. For the infinite string the relation between the wavelenght λ and circular frequency ω is

$$\lambda = \omega\sqrt{T/m} \tag{3.5}$$

and the only effect of the cable's length and boundary conditions is that the resonant frequencies are slightly shifted.

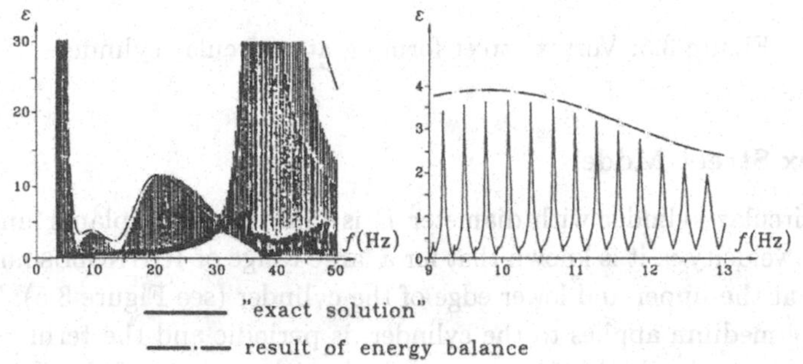

Figure 3.4: Bending strains: results of energy balance vs. boundary value problem

A simplified approach for the computation of the bending strains for a given wind speed is therefore obtained if one assumes that *any* frequency can be a resonant frequency for the cable, computing the vibration amplitudes A in the free field from an energy balance of the type

$$P_W(A) = P_D(A) + P_C(A). \tag{3.6}$$

In (3.6) $P_W(A)$ is the power introduced into the cable by the wind forces (for a given wind speed and frequency), $P_D(A)$ the power dissipated in the STOCKBRIDGE damper and $P_C(A)$ the power dissipated due to the cable's self-damping properties. Expressions for $P_W(A)$ and $P_D(A)$ will be given in sections 3.1.3 and 3.1.4. The power $P_C(A)$ is small compared to $P_D(A)$ and will be disregarded in this paper. Once the amplitude A has been determined from (3.6), the bending strains in the boundary layers can be computed by using simple perturbation analysis (see [14]).

An example of the results obtained with this simplified model is given in Figure 3.4. It is seen that the vibration amplitude found in this manner as a function of the frequency (or wind speed) is the envelope to the amplitude curve from the boundary value problem. The resonant peaks are close to this envelope. In comparing both results one should keep in mind that equation (3.6) can be easily solved by using a pocket calculator, while the solution to the boundary value problem is a much more formidable task further complicated by the fact that the wind forces cannot be precisely delineated.

3.1.3 Modelling of the Wind Forces

It is well known that mainly vortex shedding is involved in the mechanisms of excitation of the cable vibrations in the frequency range under consideration. Surprisingly enough, relatively little data is available on the vibrations of a string or cable even under the action of a laminar, steady transverse wind, particularly if the vibration amplitudes are as large as one cable diameter, which is the case in transmission line vibrations.

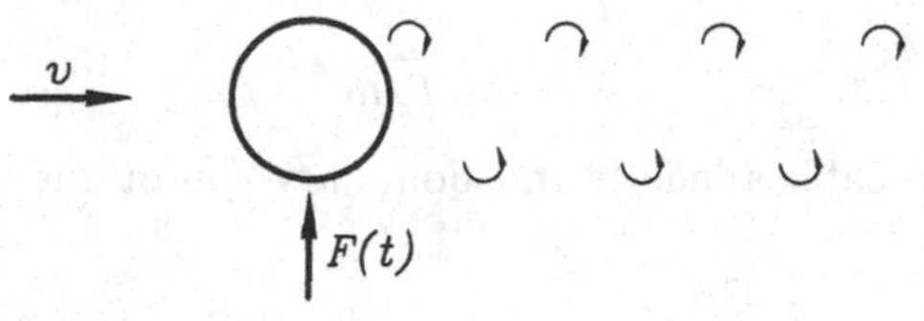

Figure 3.5: Vortex street forming at a circular cylinder

Kármán Vortex Street Model

If a stationary circular cylinder with diameter D is immersed in a planar uniform and stationary flow with velocity v, it is known that for a large range of REYNOLDS numbers vortices form alternately at the upper and lower edge of the cylinder (see Figure 3.5). The force $F(t)$ which the flowing medium applies to the cylinder, is periodic and the term with the fundamental frequency very clearly dominates over all the other terms of the FOURIER series of $F(t)$. This dominant part of $F(t)$ can be written as

$$F(t) = \frac{1}{2}\rho D L v^2 c_L \sin 2\pi f_S t \tag{3.7}$$

with ρ as the density of the flowing medium, D the cylinder's diameter, L its length, v the velocity of the unperturbed flow, c_L a coefficient which is of the order of 0.2–1.0 and f_S the STROUHAL frequency given by

$$f_S = c_S v / D, \tag{3.8}$$

($c_S = 0.19$). The expression (3.7) is sometimes used in the computation of the wind-excited vibrations of overhead transmission lines. One should however keep in mind that it is valid only for a stationary cylinder, or possibly also for a cylinder with small vibration amplitudes. In the cable vibrations, the amplitudes may become as large as D and (3.7) cannot be immediately applied.

The Oscillating Cylinder and the Oscillating Cable

Several researchers have measured the forces applied by the flow to a transversally oscillating cylinder (see [7], [40]). The results differ considerably and the most complete information seems to be contained in [40]. Also with a cylinder moved harmonically in transverse direction with amplitude A and frequency f, the fundamental harmaonic in general dominates the other terms in the FOURIER series of the force applied to the cylinder and $F(t)$ can be approximated as

$$F(t) = \frac{1}{2}\rho D L v^2 c_L \sin(2\pi f t + \varphi) \tag{3.9}$$

where now c_L and also the phase angle φ between the force and the cylinder displacement depend on the ratios $a = A/D$ and $r = f/f_S$. The power corresponding to this force is positive for $0.9 < r < 1.3$ and $0 < a < 1$ approximately, it becomes negative otherwise. At

$r = 1$ the "lock-in" phenomenon takes place and force amplitudes may become much larger than in the KÁRMÁN vortex street observed at the stationary cylinder.

With the aid of (3.9), the power introduced by the wind in a transversally oscillating cable can be estimated as

$$P_W = l f_S^3 D^4 \sum_{i=1}^{10} a_i \left(\frac{A}{D}\right)^i , \tag{3.10}$$

where a_i are coefficients which are determined by experiments. In deriving (3.10) it is taken into account that the amplitude is a function of x in the cable's standing wave oscillation and in the expression of the power of the wind forces acting on the cylinder. The cylinder's amplitude is substituted by $A \sin 2\pi x/\lambda$, followed by averaging over the wave length.

In his experiments, Staubli [40] observed that the force acting on the cylinder depends strongly on the flow being planar or not. If in the experiments, by some accidentally small asymmetry the vortices happened not to be shedded along a straight line parallel to the axis of the rigid cylinder, but in some other fashion, the magnitude of the force dropped considerably. In cable vibrations however, the flow cannot be planar and vortices will not form along straight lines. Points of the cable at a distance of one wavelength even will oscillate in phase opposition, and it is not at all clear how the vortices will develop. In [33] experimental data is presented which is obtained from measurements on a flexible free beam oscillating in its first mode and covering a broad range of wind speeds. The expression for the power which can be derived from this data does not differ too much from the cylinder experiments. In [34] a mathematical model was presented for the aerodynamic forces acting on a cable in which the force depends on the local displacement amplitude and phase. The model uses

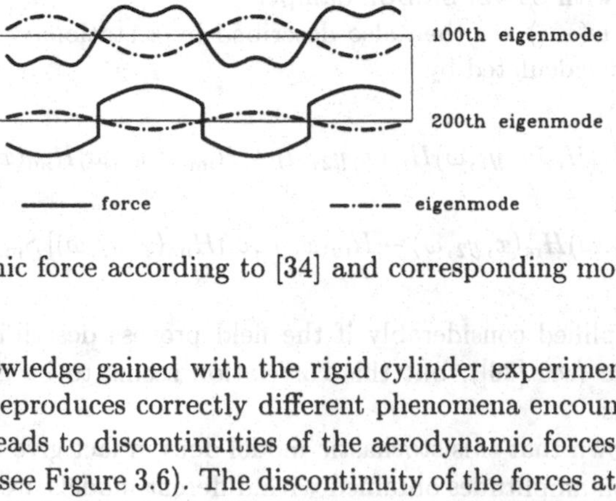

Figure 3.6: Aerodynamic force according to [34] and corresponding mode shape for different modes

only the empirical knowledge gained with the rigid cylinder experiments and tries to apply them to the cable. It reproduces correctly different phenomena encountered in transmission line vibrations but it leads to discontinuities of the aerodynamic forces in the high modes of vibration of the cable (see Figure 3.6). The discontinuity of the forces at the nodes of the high modes is of course unsatisfactory. It would imply a discontinuity in the vortices, which at one side of the node would be generated at the upper edge and simultaneously on the other side of the node at the lower edge of the cylinder. This is of course not realistic.

In a more realistic model for wind forces, these forces should not only be a function of the cable's motion at a given point but should reflect also the effect of motion in adjacent sections.

A Stochastic Model

In reality the wind velocity is not completely constant but has always a stochastic component, so that the modelling of wind forces by a stationary stochastic field process $q(x,t)$ seems reasonable. The effect that the wind forces at a given point of the cable are related to the forces at other points, and that this relation decreases with the distance, can be satisfied for example by assuming the crossrelation of $q(x,t)$ as

$$k(x_1, x_2, t) = f(x_1)f(x_2)k(t)\mathrm{e}^{-\frac{(x_1-x_2)^2}{\mu^2}} \tag{3.11}$$

with a suitable function $f(x)$, and where $k(t)$ is the autocorrelation function of some scalar stochastic process; the factor μ can be interpreted as correlation coefficient. The cross spectral density of the field process $q(x,t)$ is then

$$S(x_1, x_2, \omega) = f(x_1)f(x_2)s(\omega)\mathrm{e}^{-\frac{(x_1-x_2)^2}{\mu^2}}, \tag{3.12}$$

with $s(\omega)$ as the FOURIER-transform of $k(t)$.

In [25], the problem of transmission line vibrations was treated in this manner, the process $q(x,t)$ was assumed to be ergodic and gaussian and the power spectrum $s(\omega)$ was taken to be of the narrow band-type, which corresponds to the KÁRMÁN vortex street. In these calculations, GREEN's function $H(x,y,\omega)$ for the cable oscillations was used. It can be defined as the complex vibration amplitude of the cable at the section x, due to an exciting harmonic unit point force with frequency ω acting at the section y and can easily be calculated for the string with zero bending stiffness (some more computational work is involved in calculating it for the beam). It is real if no damping is taken into account and complex for the cable with STOCKBRIDGE damper.

The oscillations $w(x,t)$ are then also described by a stationary gaussian field process and its variance can be calculated by

$$\begin{aligned}\sigma_w^2(x) = \frac{1}{\pi} \int_{y_1=0}^{1} \int_{y_2=0}^{1} \int_{\Omega_1}^{\Omega_2} & [H_{\mathrm{re}}(x,y_1,\omega)H_{\mathrm{re}}(x,y_2,\omega) + H_{\mathrm{im}}(x,y_1,\omega)H_{\mathrm{im}}(x,y_2,\omega)]\, S_{\mathrm{re}(y_1,y_2,\omega)} \\ & + [H_{\mathrm{im}}(x,y_1,\omega)H_{\mathrm{re}}(x,y_2,\omega) - H_{\mathrm{re}}(x,y_1,\omega)H_{\mathrm{im}}(x,y_2,\omega)]\, S_{\mathrm{im}}(y_1,y_2,\omega)\, d\omega dy_1 dy_2 .\end{aligned} \tag{3.13}$$

The expression is simplified considerably if the field process describing the wind forces is not correlated in space (see [26]), but this correlation seems to be an essential feature in wind-excited vibrations.

In [25] it was shown that this stochastic model does in fact give realistic results and in Figure 3.7 the vibration amplitudes obtained with different models are compared. Of course, additional information is needed on the aerodynamic forces in order to use the stochastic model for a reliable prediction of the cable's dynamic behaviour. Once this information is available, the method should however become important for studying the vibrations leading to material fatigue and also for predicting damage accumulation levels. It finally should be observed that the approach taken in [25], [26] for studying the stochastic vibrations of a continuous system is new in that the system was not discretized and no modal analysis was used. Instead an integral representation was adopted.

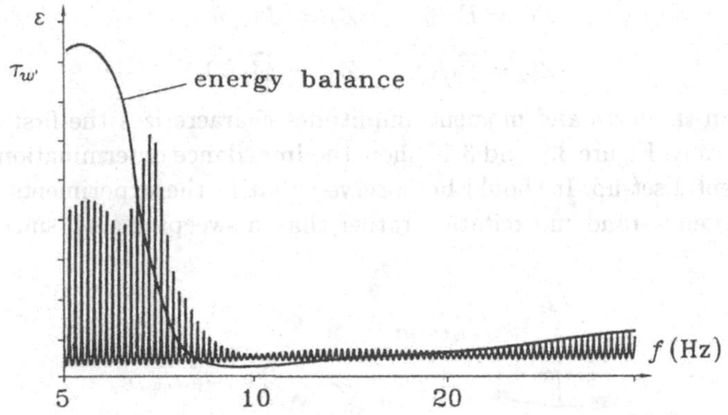

Figure 3.7: Bending strains: Results from stochastic model vs. energy balance

3.1.4 Modelling of the Stockbridge Damper

Dampers of the STOCKBRIDGE and of similar types are formed by two rigid masses fixed at the ends of a "messenger cable" which is clamped to the line (see Figure 3.8). The messenger cable consists of several steel wires and is built in such a way that a high amount of energy is dissipated in the deformation of these cables, i.e. when the end masses move. The dampers are usually placed near the suspension clamps, at a distance which should be less than one half wave length of the highest-frequency-mode to be damped, so that they are never placed at a vibration node of the undamped cable.

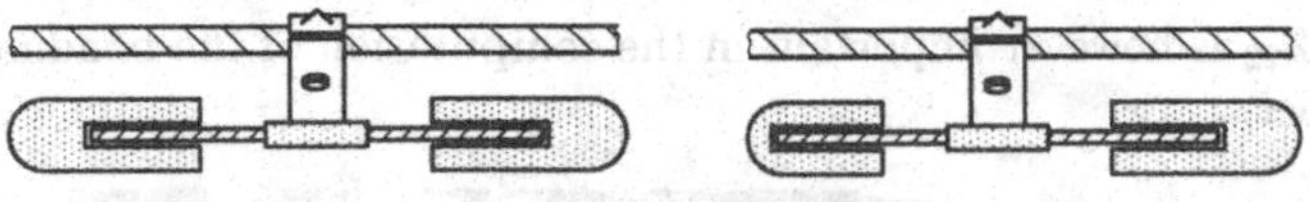

Figure 3.8: Typical dampers

Representation of the Damper by its Impedance Matrix

In an actual line, the damper clamp usually executes a transverse translational and a rotational motion when in operation. In both types of motion, mechanical energy is dissipated in the damper. With the assumption of linearity the dynamic behavior of the damper can be charcterized by the damper's driving point impedance matrix which can be measured in the laboratory. To this end, two tests are carried out [15]: in the first one, the damper clamp executes a translational harmonic oscillation with complex velocity amplitude $\widehat{\dot{y}}$, in the second one, a rotational oscillation with angular velocity amplitude $\widehat{\dot{Q}}$ and in both cases the moment and force amplitudes at the clamp $\widehat{M}$ and $\widehat{F}$ are measured. The elements Z_{ij} of the complex impedance matrix are then defined by

$$Z_{11} = \widehat{F}_1/\widehat{\dot{y}}, \qquad Z_{21} = \widehat{M}_1/\widehat{\dot{y}},$$
$$Z_{12} = \widehat{F}_2/\widehat{\dot{Q}}, \qquad Z_{22} = \widehat{M}_2/\widehat{\dot{Q}}, \tag{3.14}$$

where the index in the force and moment amplitudes characrerizes the first and second experiment, respectively. Figure 3.9 and 3.10 show the impedance determination of the damper and the experimental set-up. It should be observed that in the experiments it may be convenient to use a pseudo-random excitation rather than a sweeping sine, since this is usually much faster.

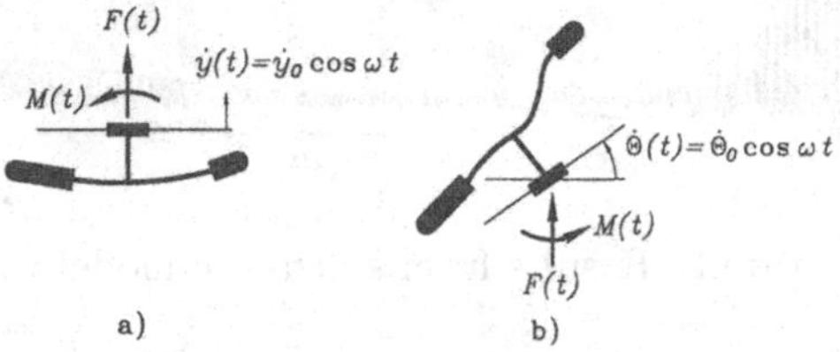

Figure 3.9: Impedance determination for the damper: a) vertical clamp motion, b) rotatory clamp motion

The complete information with regard to the action of the damper can then easily be deduced from the 2×2 impedance matrix $Z(\omega)$, as is shown in [15]. For a completely symmetric damper, the off-diagonal terms of the Z matrix vanish identically. In [15] it was also shown that the moment impedance Z_{22} does not seriously affect the overall energy balance. For the calculation of the vibration amplitudes in the free field and the bending strains at the suspension clamp, it is therefore sufficient to consider only the term Z_{11} of the impedance matrix. The term Z_{22} is however important in the computation of the bending strains at the clamp of the damper.

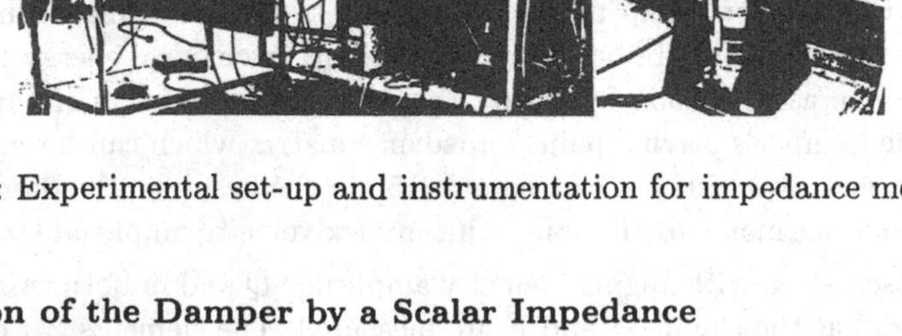

Figure 3.10: Experimental set-up and instrumentation for impedance measurements

Representation of the Damper by a Scalar Impedance

The approach usually taken in studying the effect of the dampers on the cable vibrations is to consider only the single scalar impedance $Z_{11}(\omega)$, which suffices to estimate the vibration amplitudes correctly. An additional advantage of this approach is that the cable's vibration analysis can be carried out with the string model (with zero bending stiffness), and the bending strains are computed a posteriori via singular perturbations. If the complete impedance matrix is used, point moments are applied to the cable, so that the beam has be to used to model the cable.

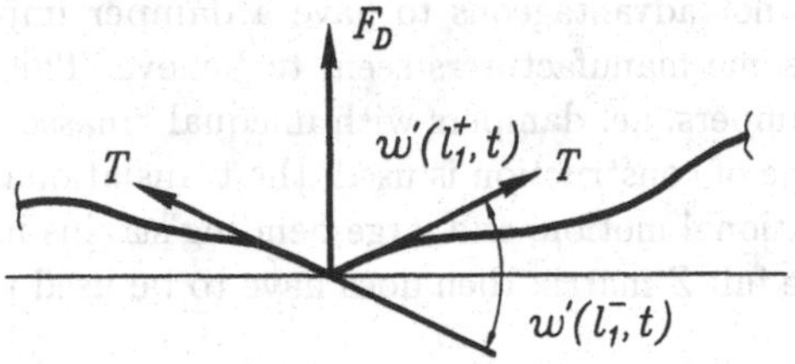

Figure 3.11: Force acting on the damper

The force acting on the damper located at $x = l_1$ can be written as (see Figure 3.11):

$$F_D(t) = T\left[w'(l_1^+, t) - w'(l_1^-, t)\right]. \tag{3.15}$$

If the cable oscillates harmonically with frequency ω, one has

$$w(x,t) = \text{Re}\left[W(x)e^{j\omega t}\right] \tag{3.16}$$

and

$$F_D(t) = T\left[W'(l_1^+, t) - W'(l_1^-, t)\right]\cos\omega t. \tag{3.17}$$

On the other hand, the damper force can be related to the velocity $\dot{w}(l_1, t)$ by means of the damper impedance:

$$F_D(t) = \text{Re}\left[Z_{11}W(l_1)j\omega e^{j\omega t}\right], \tag{3.18}$$

so that $F_D(t)$ can be eliminated from (3.17) and (3.18). In [14] this approach is used to calculate the power P_D absorbed by the damper under given conditions.

In particular, one can compute the fraction of the vibration energy contained in a harmonic wave, arriving from the free field towards the suspension clamp with the damper, which is reflected towards the free field and the part which is absorbed by the damper (see Figure 3.2). It can be expressed by means of the coefficient of absorption calculated in [14] as a function of Z_{11} and the cable parameters T, m, as well as the distance l_1 and the frequency ω. It turns out that the coefficient of absorption is equal to one, i.e. all the incoming vibration energy is absorbed for a certain damper impedance $(Z_{11})_{\text{OPT}}$ which can easily be calculated. If the optimal complex damper impedance is written as

$$(Z_{11})_{\text{OPT}} = |(Z_{11})_{\text{OPT}}|\, e^{j\beta} \tag{3.19}$$

one has

$$|(Z_{11})_{\text{OPT}}| = \frac{\sqrt{Tm}}{\sin\left(\frac{\omega l_1}{\sqrt{\frac{T}{m}}}\right)} \tag{3.20}$$

and

$$\beta_{\text{OPT}} = \frac{\omega l_1}{\sqrt{\frac{T}{m}}} - \frac{\pi}{2}. \tag{3.21}$$

It should be noted that the optimal damper impedance depends on the frequency and on the damper location. It is not advantageous to have a damper impedance constant over a large frequency range, as some manufacturers seem to believe. This goal is often aimed for by building asymmetric dampers, i.e. dampers with unequal "masses" at the two ends of the messenger cable. If this type of construction is used, the translational motion of the clamp is strongly coupled to its rotational motion, and large bending strains in the cable may occur at the damper clamp; also the full Z matrix then does have to be used to describe the damper's dynamic behavior.

Relation (3.20) can be used together with (3.21) to choose a proper damper for a given cable. Of course, a damper with such ideal characteristics will not be generally available, but if the conditions (3.20) and (3.21) are approximately fulfilled, at least in part of the frequency range, the damper will work properly and not cause additional problems, as is sometimes the case if it is not well matched to the cable. From (3.20) it can also immediately be seen that the magnitude of the damper impedance should always be larger than the cable's impedance $\sqrt{Tm}$ for transverse waves.

3.1.5 Conclusions

Wind-excited vibrations in overhead transmission lines are highly worthy of study, in view of the large costs which are associated with cable fatigue damage. Even in the case of a single cable instead of a bundle of conductors, the problem is also interesting from an engineering point of view.

The least known quantities in the problem are the aerodynamic forces acting upon the cable, which depend on a large number of factors, such as vegetation under the line, etc. This implies that the mathematical model for the oscillating cable need not be too complex. On the other hand, the essentials of the cable's dynamic behavior as well as the proper matching between cable and damper should be included. It turns out that this can be done by using measured damper impedances and calculating the vibration levels for the cable modelled by a string with zero bending stiffness. Energy balance can be used to solve this problem with extremely small computational effort and sufficient precision. The bending strains, which are needed to limit the fatigue damage, can be calculated afterwards by means of simple formulae obtained by perturbation analysis.

Vibrations of a wind-excited transmission line provide an excellent example of mathematical modelling at different levels of one and the same engineering problem, showing how mathematical knowledge and good engineering insight can be combined in the difficult modelling process.

Chapter 4

Analytical Mechanics: Three Problems Revisited

4.1 Nonholonomic Systems: An Elementary but Common Mistake

A few weeks ago in a paper published by serious authors, in a renown scientific mechanical journal, a mistake occurred which was encountered several times in the literature. It is a known fact in the dynamics of non-holonomic systems that the action integral $\int_{t_0}^{t_1} L'\,dt$ for the augmented Lagrangian $L' = T - V - \mu f = L - \mu f$ is not stationary in the class of paths satisfying the equations of constraint (geometrically possible paths). Here T is the kinetic energy, V the potential energy, μ stands for the LAGRANGE multipliers and f for the constraints. These elementary facts are possibly well known by the CISM audience. On the other hand, we will not be very surprised if we find the same mistake again in the new literature (as several times in the past). This fact, which has been overseen by many authors during the last 100 years and as many times corrected afterwards, is explained for example in PARS [31], NEIMARK&FUFAEV [28] and in WHITTAKER [44]. These authors show that the equations of motion of a dynamical system with non-holonomic constraints cannot be obtained via calculus of variations by applying the multiplier rule in the usual way. This caused the search for more general variational principles as well as for new formalisms for obtaining the correct equations of motion for non-holonomic systems.

Since this fact falls into oblivion with regularity, it might be useful to illustrate it with an example borrowed from PARS [31] (see pages 529 and 530). Consider a particle P with position characterized by the cartesian coordinates (x, y, z) in inertial space and subjected to the single non-holonomic constraint

$$A_x\dot{x} + A_y\dot{y} + A_z\dot{z} + A = 0, \tag{4.1}$$

where A_x, A_y, A_z and A may be functions of x, y, z and t. The equations of motion are then given by

$$\begin{aligned} m\ddot{x} &= X + \lambda A_x, && (4.2a)\\ m\ddot{y} &= Y + \lambda A_y, && (4.2b)\\ m\ddot{z} &= Z + \lambda A_z, && (4.2c) \end{aligned}$$

both for (4.1) holonomic or non-holonomic, where X, Y and Z are the given forces and λ is a LAGRANGE parameter representing the constraint forces. Consider the particular constraint

$$z\dot{x} - \dot{y} = 0 \tag{4.3}$$

with $X = Y = Z = 0$, $V \equiv 0$. We assume that all the coordinates have been normalized, so that they are nondimensional and the particle has unitary mass. The correct equations of motion obtained from (4.2) are given by

$$\ddot{x} = \lambda z, \tag{4.4a}$$
$$\ddot{y} = -\lambda, \tag{4.4b}$$
$$\ddot{z} = 0, \tag{4.4c}$$

to which the equation of constraint (4.3) has to be added. The parameter λ stands for the LAGRANGE parameter characterizing the force of constraint corresponding to (4.3). The solution of this system of differential equations is simple and for the initial conditions

$$x(0) = 0, \quad y(0) = 0, \quad z(0) = 0, \tag{4.5a}$$
$$\dot{x}(0) = u, \quad \dot{y}(0) = 0, \quad \dot{z}(0) = w, \tag{4.5b}$$

with $w \neq 0$ and with the abbreviation

$$\sinh\theta(t) = wt \tag{4.6}$$

yields

$$x(t) = \tfrac{u}{w}\theta(t), \tag{4.7a}$$
$$y(t) = \tfrac{u}{w}(\cosh\theta(t) - 1), \tag{4.7b}$$
$$z(t) = \sinh\theta(t). \tag{4.7c}$$

Let us now compare the correct equations of motion (4.4) as well as the solution (4.7) with equations obtained from the calculus of variations via the stationarity of the action integral with the augmented Lagrangian

$$L' = \frac{1}{2}(\dot{x}^2 + \dot{y}^2 + \dot{z}^2) - \mu(z\dot{x} - \dot{y}). \tag{4.8}$$

The augmented Lagrangian L' leads to the differential equations,

$$\frac{d}{dt}(\dot{x} - \mu z) = 0, \tag{4.9a}$$
$$\frac{d}{dt}(\dot{y} - \mu) = 0, \tag{4.9b}$$
$$\frac{d}{dt}\dot{z} + \mu\dot{x} = 0, \tag{4.9c}$$

to which again the equation of constraint (4.3) must be adjoined. The motion given by (4.7) does not satisfy (4.9) for any choice of $\mu(t)$! In fact, from (4.6), (4.7c) follows $\ddot{z} \equiv 0$, so that (4.9c) leads to

$$\mu\dot{x} = 0. \tag{4.10}$$

Since $\dot{x}(t)$ does not vanish, one has

$$\mu(t) \equiv 0 \tag{4.11}$$

and (4.9a), (4.9b) assume the form

$$\ddot{x} = 0, \tag{4.12a}$$

$$\ddot{y} = 0. \tag{4.12b}$$

These differential equations are clearly not fulfilled by (4.7)! The motion (4.7) obtained by solving the correct differential equations (4.4) is therefore not the solution to EULER-LAGRANGE's equations corresponding to (4.8).

For a system described by the Lagrangian L in generalized coordinates $\mathbf{q} = (q_1, q_2, \dots, q_n)$ and by the additional constraints

$$f_r = \sum_{i=1}^{n} A_{ri}\dot{q}_i + A_r = 0, \quad r = 1, 2, \dots, s \tag{4.13}$$

the correct equations of motion are given by

$$\frac{d}{dt}\frac{\partial L}{\partial \dot{q}_i} - \frac{\partial L}{\partial q_i} = \sum_{r=1}^{s} \lambda_r A_{ri}, \quad i = 1, 2, \dots, n \tag{4.14}$$

together with (4.13). They are **not** given by

$$\frac{d}{dt}\frac{\partial L'}{\partial \dot{q}_i} - \frac{\partial L'}{\partial q_i} = 0, \quad L' = L - \sum_{r=1}^{s} \mu_r f_r, \tag{4.15}$$

unless the constraints are holonomic.

4.2 A simple Exercise in Liapunov Stability

In this section we will examine a question in the stability of mechanical systems, which has intrigued the author for many years. In fact, he was unable to give the answer himself; it was finally given in 1984 in [46]. The solution turned out to be rather simple but the author feels that the problem is of general interest due to the combination of engineering intuition with mathematics.

Consider a particle subjected to a central, attractive, time-independent force as shown in Figure 4.1.

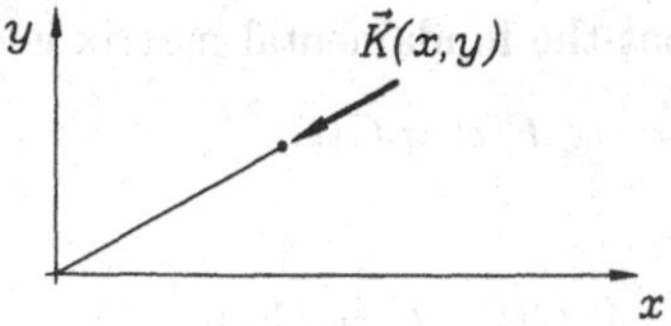

Figure 4.1: Particle subjected to a central attractive force

The force can then be written as

$$\vec{K}(x, y) = -f(x, y)\, x\, \vec{e}_x - f(x, y)\, y\, \vec{e}_y \tag{4.16}$$

and the differential equations in the planar case are

$$\ddot{x} = -f(x, y)\, x, \qquad \ddot{y} = -f(x, y)\, y, \tag{4.17}$$

if they are written in a non–dimensional form. The origin $x = 0$, $y = 0$ is an equilibrium position. The force field will be attractive if and only if $f(x, y) > 0$. Moreover let the function $f(x, y)$ be such that the existence and uniqueness of the solutions to (4.17) is guaranteed.

The following question can then be formulated: *Can the trivial equilibrium in a central attractive time–independent force field be unstable*? Here and in what follows LIAPUNOV's stability definition will be used. Probably most engineers would give a negative answer to the question formulated above. Particularly, if they are used to statical rather than dynamical problems, it may seem obvious to them that the central attractive force field, giving rise to a positive restoring force, always leads to a stable equilibrium. This question was first posed to me about twenty years ago by my former teacher, Professor Mauro Cesar from the university of Sao Paulo, Brazil. My feeling always was that it should be possible to find examples with instability and I tried to find them for several years, however without success.

Of course two remarks immediately can be made. First, in the case of a motion along a given straight line rather than in a plane, the answers are always *no*, i.e. the equilibrium will always be stable, as can easily be seen from the energy integral. Second, in the planar case, again instability will only be possible for a non–conservative force field, i.e. if

$$\frac{\partial}{\partial y} f(x, y)\, x \neq \frac{\partial}{\partial y} f(x, y)\, y \tag{4.18}$$

holds. Otherwise, a potential energy function exists with a minimum at the origin, so that this equilibrium position is stable due to the LAGRANGE–DIRICHLET theorem.

In order to examine the problem in all the other cases, we first consider a system of linear ordinary differential equations of the type $\dot{z} = A(t)z$, $z \in R^k$ or $\dot{Z} = A(t)Z$, $Z \in R^{k \times k}$ and remember the WRONSKI equation

$$\mid Z(t) \mid = \mid Z(t_0) \mid e^{\int_{t_0}^{t} \sum a_{ii} d\bar{t}} . \tag{4.19}$$

The stability behavior of this differential system follows from its fundamental matrix $F(t)$, which is defined by

$$\dot{F} = AF, \qquad F(0) = I . \tag{4.20}$$

In the particular case of periodic ordinary differential equations, i.e for

$$A(t + T) = A(t) , \tag{4.21}$$

the FLOQUET theorem tells us that the fundamental matrix is of the type

$$F(t) = C(t)e^{Bt} \tag{4.22}$$

with

$$C(t + T) = C(t) . \tag{4.23}$$

The characteristic numbers ρ are defined by

$$\mid e^{Bt} - \rho I \mid = 0 \tag{4.24}$$

or equivalently by

$$\mid F(t) - \rho I \mid = 0 \tag{4.25}$$

In general, these numbers will be complex and we have asymptotic stability if $\mid \rho_i \mid < 1$ for all characteristic numbers, and instability if at least for one characteristic number $\mid \rho_i \mid > 1$.

Consider the particular case of HILL's equation

$$\ddot{y} + p(t)y = 0\,, \tag{4.26}$$

which can be written as a first order system:

$$\dot{z}_1 = z_2\,, \qquad \dot{z}_2 = -p(t)z_1. \tag{4.27}$$

The fundamental solutions are

$$\begin{aligned} y(t) &= \phi(t), \qquad \phi(0) = 1, \qquad \dot{\phi}(0) = 0 \\ y(t) &= \psi(t), \qquad \psi(0) = 0, \qquad \dot{\psi}(0) = 1 \end{aligned} \tag{4.28}$$

and according to (4.25) the characteristic numbers satisfy

$$\begin{vmatrix} \phi(T) - \rho & \psi(T) \\ \dot{\phi}(T) & \dot{\psi}(T) - \rho \end{vmatrix} = 0\,. \tag{4.29}$$

This leads to

$$\rho^2 - \rho\left(\phi(T) + \dot{\psi}(T)\right) + \left(\phi(T)\dot{\psi}(T) - \psi(T)\dot{\phi}(T)\right) = 0\,. \tag{4.30}$$

Introducing the abbreviation

$$2a = \phi(T) + \dot{\psi}(T) \tag{4.31}$$

and observing that $\phi(T)\dot{\psi}(T) - \psi(T)\dot{\phi}(T) = 1$, as follows from (4.19), gives

$$\rho_{1,2} = a \pm \sqrt{a^2 - 1}\,. \tag{4.32}$$

Three different cases can now be distinguished. For $a^2 < 1$ one has $|\rho_1| = |\rho_2| = 1$, the characteristic numbers being both complex. If $a^2 > 1$ one obtains $\rho_1 > 1 > \rho_2$ and both characteristic numbers are real. Therefore, in both cases, $a^2 < 1$ and also $a^2 > 1$ no T–periodic solutions exist. The third case $a^2 = 1$ leads to $\rho_1 = \rho_2 = 1$ (real). In this case at least one T-periodic solution exists, for example $\phi(t)$. The other solution, $\psi(t)$, either also being T-periodic or of the type $\psi(t) = t\,\Theta(t)$ with $\Theta(t+T) = \Theta(t)$.

In what follows we will suppose $\phi(t)$ periodic and therefore $\dot{\psi}(T) = 1$. From the WRONSKI–determinant and the condition $a^2 = 1$ we obtain the following result.

The trivial solution of (4.26) is stable if and only if $\psi(t)$ is periodic, respectively $\psi(T) = 0$. This result will now be used to examine in more detail the particular case

$$\ddot{x} = -f(x)\,x, \qquad \ddot{y} = -f(x)\,y. \tag{4.33}$$

In what follows we assume $f(0) > 0$ (in the case $f(0) \leq 0$ the differential equation $\ddot{y} = -f(0)y$ leads to instability of the equilibrium).

Of course the first of the differential equations (4.33) represents a conservative system and the corresponding potential energy is

$$U(x) := \int_0^x f(\bar{x})\bar{x}\,d\bar{x}\,. \tag{4.34}$$

Due to the positiveness of $f(x)$ the trivial equilibrium position $x = 0$ of this differential equation is stable (LAGRANGE–DIRICHLET theorem), and in a sufficiently small neighborhood of $x = 0$, $\dot{x} = 0$ all the solutions $x(t)$ are periodic.

Let $\tilde{x}(t; x_0, 0)$ be such a periodic solution with period $T(x_0)$, then

$$dt = \frac{dx}{\sqrt{2\,(U(x_0) - U(x))}} \tag{4.35}$$

follows from the energy integral. We define

$$\phi(t) := \frac{1}{x_0}\tilde{x}(t; x_0, 0) \tag{4.36}$$

with

$$\phi\left(\frac{T}{2}\right) = \frac{x_0^-}{x_0}, \qquad \dot{\phi}\left(\frac{T}{2}\right) = 0, \tag{4.37}$$

where x_0^- is defined by Figure 4.2.

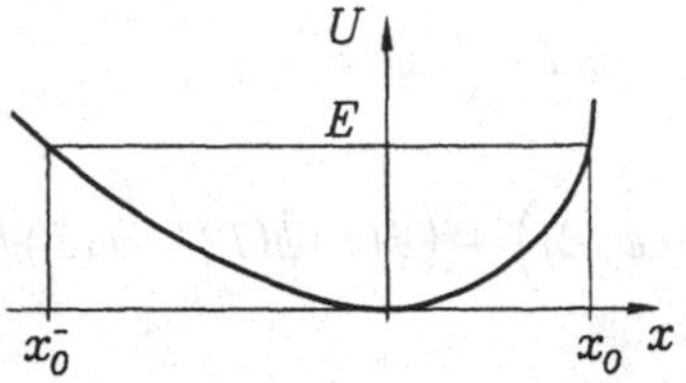

Figure 4.2: Potential Energy

If the time t_1 is defined by $\phi(\pm t_1) = 0$, then obviously

$$\dot{\phi}(\pm t_1) = \mp\sqrt{2U(x_0)}\,. \tag{4.38}$$

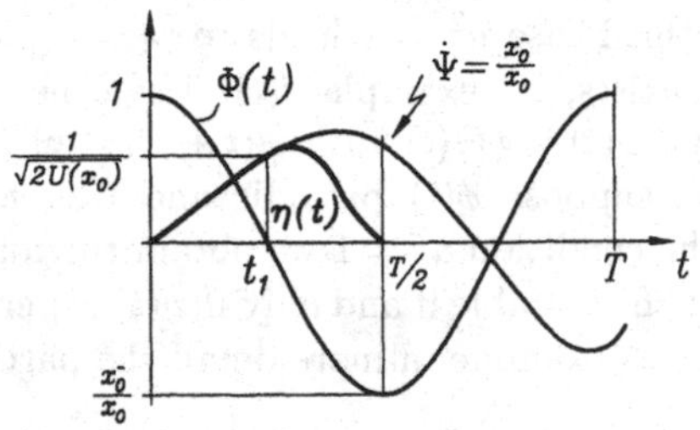

Figure 4.3: Solutions Φ and Ψ

If we substitute $x = \tilde{x}(t)$ into the second of the differential equations (4.33) we get

$$\ddot{y} + f(\tilde{x}(t))y = 0\,. \tag{4.39}$$

Obviously, this is a differential equation of the HILL type and one of its solution is known, namely the T–periodic solution

$$y(t) = \phi(t) = \frac{1}{x_0}\tilde{x}(t)\,. \tag{4.40}$$

From what was said previously about HILL's equation, we know that (4.39) is stable if and only if $\psi(T) = 0$. In what follows, we will examine if this is the case in general.

The general solution of (4.39) can be written as

$$y(t) = y(0)\phi(t) + \dot{y}(0)\psi(t)\,. \tag{4.41}$$

It is clear that $\psi(t+T)$ is also a solution and therefore can be written as

$$\psi(t+T) = \psi(T)\phi(t) + \psi(t)\,. \tag{4.42}$$

With $t = -T/2$ this gives

$$\psi\left(\frac{T}{2}\right) = \psi(T)\phi\left(-\frac{T}{2}\right) + \psi\left(-\frac{T}{2}\right)\,. \tag{4.43}$$

Since $\phi(t) = \phi(-t)$ we also have $\psi(t) = -\psi(-t)$, i.e. $\psi(t)$ is an odd function. Therefore, we get

$$2\psi\left(\frac{T}{2}\right) = \psi(T)\phi\left(-\frac{T}{2}\right) \tag{4.44}$$

and $\psi(T) = 0$ if and only if $\psi(T/2) = 0$.

Besides the energy integral for the first of the differential equations (4.33) we also have the integral of momentum of the differential system

$$\dot{y}x - \dot{x}y = c\,, \tag{4.45}$$

which is equivalent to

$$\dot{\psi}\phi - \dot{\phi}\psi = 1\,, \tag{4.46}$$

and can be found using the WRONSKIan. Therefore, we also have

$$\dot{\psi}\left(\frac{T}{2}\right) = \frac{x_0}{\dot{x}_0}\,. \tag{4.47}$$

Let us now define the solution $\eta(t)$ of the second differential equation in (4.33) by the inital conditions

$$\eta\left(\frac{T}{2}\right) = 0\,, \qquad \dot{\eta}\left(\frac{T}{2}\right) = \frac{x_0}{\dot{x}_0}\,. \tag{4.48}$$

From what was said above, we have stability if and only if $\eta(t) \equiv \psi(t)$ and this condition will be examined further. Using variation of constants we obtain

$$\psi(t) = \phi(t)\int_0^t \frac{d\bar{t}}{\phi^2(\bar{t})}\,, \qquad -t_1 < t < t_1\,, \tag{4.49}$$

and

$$\eta(t) = \phi(t)\int_{T/2}^t \frac{d\bar{t}}{\phi^2(\bar{t})}\,, \qquad t_1 < t < T - t_1\,, \tag{4.50}$$

and therefore also

$$\eta(t_1) = \psi(t_1)\,, \qquad \dot{\eta}(t_1) = \dot{\psi}(t_1)\,. \tag{4.51}$$

This leads to the algebraic condition

$$\int_0^x \frac{g(x)-g(\overline{x})}{\sqrt{\left(1-\frac{\overline{x}^2}{x^2}\right)\left(1-\frac{U(\overline{x})}{U(x)}\right)}\left[\sqrt{1-\frac{\overline{x}^2}{x^2}}+\sqrt{1-\frac{U(\overline{x})}{U(x)}}\right]} d\overline{x} =$$
$$\int_0^{h(x)} \frac{g(h(x))-g(\overline{x})}{\sqrt{\left(1-\frac{\overline{x}^2}{h(x)^2}\right)\left(1-\frac{U(\overline{x})}{U(h(x))}\right)}\left[\sqrt{1-\frac{\overline{x}^2}{h(x)^2}}+\sqrt{1-\frac{U(\overline{x})}{U(h(x))}}\right]} d\overline{x} \tag{4.52}$$

for x, which has to be satisfied identically for any x in some neighborhood of the origin, for the system to be stable. In (4.52) the abbreviations

$$h(x_0) = x_0^- , \qquad g(x) := \frac{U(x)}{x^2} \quad \text{for } x > 0 \tag{4.53}$$

were used and one has $g(0) = f(0)/2$. The general behavior of the function $h(x)$ is shown in Figure 4.4.

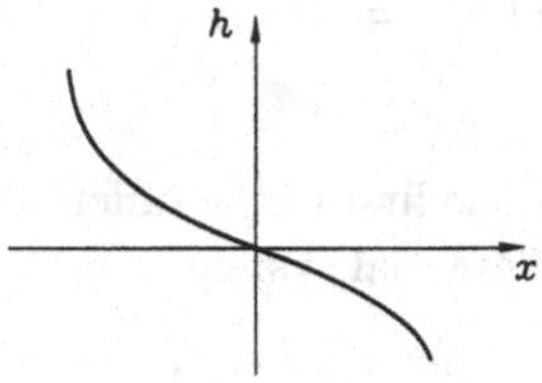

Figure 4.4: Behavior of the function $h(x)$

It is obvious that such an identity as (4.52) will only be fulfilled under very special conditions. It is fulfilled if the potential energy $U(x)$ is *even* since then one has

$$h(x) = -x , \qquad g(x) \text{ even}, \tag{4.54}$$

and

$$\psi\left(\frac{T}{2}\right) = 0 , \qquad g(\overline{x}) \equiv g(x) , \tag{4.55}$$

so that $g(x)$ is constant. In this case the potential energy is of the type

$$U(x) = g(0)x^2 \tag{4.56}$$

and the system is stable.

As shown in [46] in the general case, i.e. with $U(x)$ *not even*, (4.52) is not fulfilled so that the equilibrium is generically unstable. For example, the potential energy

$$U(x) = \frac{x^4}{4} + q\frac{x^3}{3} \tag{4.57}$$

leads to the differential system

$$\ddot{x} = -x^3 - qx^2 , \qquad \ddot{y} = -(x^2 + qx)y \tag{4.58}$$

with an unstable origin, as can be easily shown (q is an arbitrary non vanishing constant).

From [46], it also follows that substituting the function $f(x)$, for a generic function $f(x, y)$ of the two variables x and y, generically leads to instability in (4.17). Let us conclude with an example: In the differential system

$$\ddot{x} = -(e + x^2 + x^3)x\,, \qquad \ddot{y} = -(e + x^2 + x^3)y \tag{4.59}$$

the function $f(x)$ obviously is not even. Trajectories of the solution corresponding to

$$\begin{aligned} e &= 0.5 \\ x_0 = 0.5\,, &\quad \dot{x}_0 = 0, \\ y_0 = 0\,, &\quad \dot{y}_0 = 0.4 \end{aligned} \tag{4.60}$$

are given in Figure 4.5 in the x–y–plane.

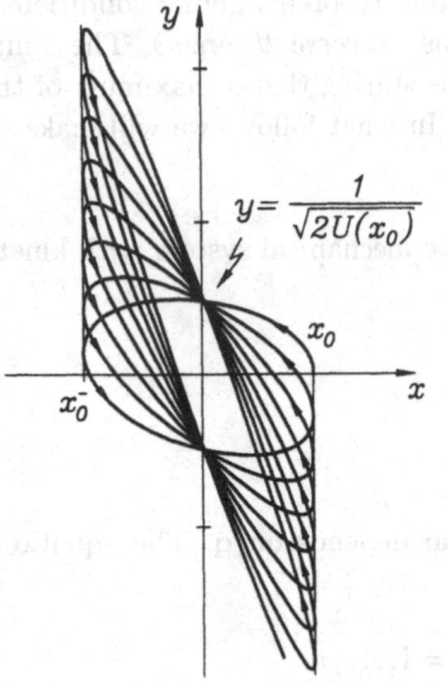

Figure 4.5: Trajectories of the solution for given parameter values

They were computed by numerical integration. It can clearly be seen that the periodic oscillation in the x–direction leads to unbounded parametric oscillations in the y–direction.

The author feels that this apparently very simple question, originally formulated by Professor Mauro Cesar and finally answered by Angelo Barone Neto, Gaetano Zampieri and Mauro Cesar is an excellent example on how intuition may mislead the engineers, if he is not skilled in dynamics.

4.3 Some Remarks on the Lagrange-Dirichlet Theorem and its Inverses

In the previous section we discussed a stability problem for which the engineering intuition did not lead to a correct answer, although the problem in itself is not difficult. In this section we wish to make some remarks on a classical problem in the stability of mechanical system which dates back to LAGRANGE and even today is not completely solved, although great progress has recently been made. This problem has in common with the previous one, the fact that the naive intuition suggests simple answers, which are mathematically wrong.

The LAGRANGE-DIRICHLET theorem states that in a discrete conservative mechanical system a minimum of the potential energy always implies stability of the equilibrium position at this minimum. It was first stated by LAGRANGE in his Mechanique Analytique in 1788 and later a correct prove was given by DIRICHLET (1805-1859). This proof is contained in the later editions of LAGRANGE'S Mechanique Analytique.

It seems intuitively obvious that the condition of minimum of the potential energy is necessary and sufficient in general for the type of system under consideration, although this is not the case, as follows from simple counter examples given by Wintner (1941), as well as earlier by Painlevé (1897). Often, theorems giving conditions of instability for conservative discrete systems are labeled as *inverse theorems.* The simplest form of such an inverse theorem would of course be one stating that a maximum of the potential energy leads to an unstable equilibrium position. In what follows we will make some remarks on the theorems and give simple examples.

Consider a discrete conservative mechanical system with kinetic energy

$$T(\mathbf{q},\dot{\mathbf{q}}) = \frac{1}{2}\dot{\mathbf{q}}^T\mathbf{A}\,\dot{\mathbf{q}} \qquad (4.61)$$

and potential energy

$$U(\mathbf{q}) \qquad (4.62)$$

where the matrix $\mathbf{A}$ in general depends on $\mathbf{q}$. The equations of motion follow from LAGRANGE'S equations

$$\frac{d}{dt}\frac{\partial L}{\partial \dot{q}_i} - \frac{\partial L}{\partial q_i} = 0 \qquad i = 1,\ldots,n \qquad (4.63)$$

with $L = T - U$ as

$$\ddot{\mathbf{q}} = \mathbf{f}(\mathbf{q},\dot{\mathbf{q}}). \qquad (4.64)$$

In what follows, we assume that the regularity conditions on T and U are such that existence and uniqueness is guaranteed for the solutions of (4.64).

First, let us recall LIAPUNOV's stability definition: *The motion* $\mathbf{q}(\mathbf{a},\mathbf{b},\mathbf{t})$ *is stable if for any* $\varepsilon > 0$ *there exists a* $\delta(\varepsilon) > 0$ *such that*

$$|\mathbf{q}_0 - \mathbf{a}| + |\dot{\mathbf{q}}_0 - \mathbf{b}| < \delta(\varepsilon) \quad \Longrightarrow$$
$$|\mathbf{q}(\mathbf{q}_0,\dot{\mathbf{q}}_0,t) - \mathbf{q}(\mathbf{a},\mathbf{b},t)| + |\dot{\mathbf{q}}(\mathbf{q}_0,\dot{\mathbf{q}}_0,t) - \dot{\mathbf{q}}(\mathbf{a},\mathbf{b},t)| < \delta(\varepsilon), \quad \forall\, t \geq 0. \qquad (4.65)$$

LIAPUNOV's stability therefore is defined as uniform (with respect to t) continous dependence on the initial conditions. Let us now assume that the system has an equilibrium position $\mathbf{q} \equiv \mathbf{c} \equiv \mathbf{0}$. A precise formulation of LAGRANGE'S theorem is that a strong, relative, not necessarily isolated minimum of the function $U(\mathbf{q})$ at $\mathbf{q} = \mathbf{0}$ implies stability of this equilibrium solution. The theorem is a trivial result of LIAPUNOV's first stability theorem with a positive definite function $T + U$ used as a LIAPUNOV function. As far as we know, in fact the LAGRANGE-DIRICHLET theorem inspired LIAPUNOV to formulate his direct method. Consider now the following example of a conservative one degree of freedom system

$$T = \frac{1}{2}\dot{q}^2, \qquad U = \frac{1}{4}q^4, \qquad \ddot{q} = -q^3. \tag{4.66}$$

From the surfaces corresponding to the energy integral, depicted in Figure 4.6 it is obvious that the stability of the trivial solution is guaranteed by the first integral.

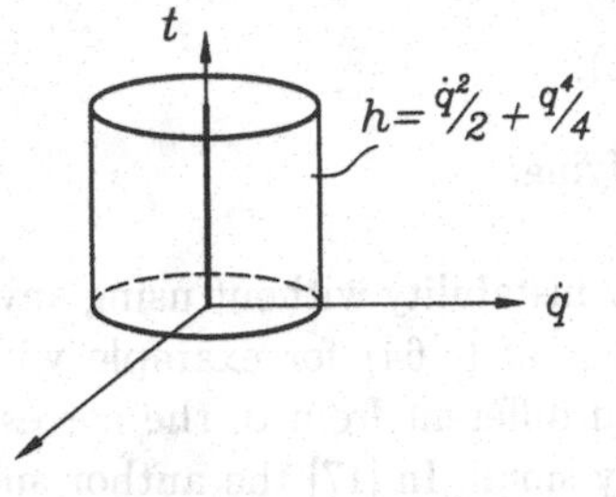

Figure 4.6: Surface of the energy integral

If we invert the sign of the function U one has

$$\ddot{q} = q^3 \tag{4.67}$$

instead of (4.66).

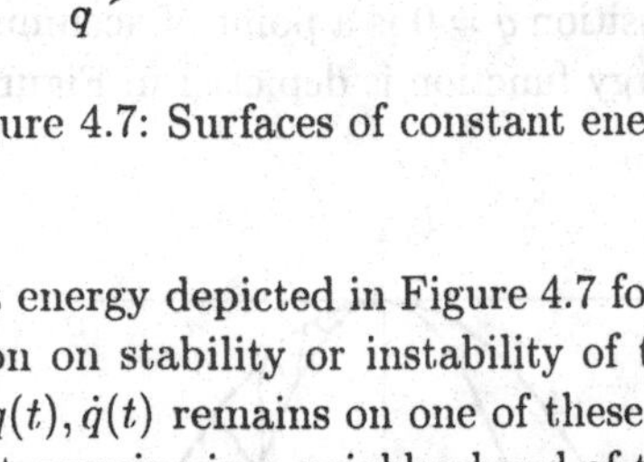

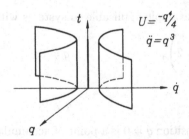

Figure 4.7: Surfaces of constant energy

The surfaces of constant energy depicted in Figure 4.7 for this case obviously no longer permit to draw any conclusion on stability or instability of the solution $q = 0$, since it is only known that the motion $q(t), \dot{q}(t)$ remains on one of these surfaces for all time. It is not immediately known whether it remains in a neighborhood of the t-axes or not.

This example shows that in general the instability theorems are much harder to prove than the stability theorems, since the first integral is no longer of any immediate use. For this reason during the last two hundred years most instability theorems applied only to rather particular situations. Consider e.g. the following theorem due to LIAPUNOV: If the potential energy $U(\mathbf{q})$ has a maximum at $\mathbf{q} = \mathbf{0}$ and is given by an expression of the type

$$U = U_m + U_{m+1} + \dots \qquad m \geq 2, \tag{4.68}$$

with U_m homogeneous polynomial of order m, and if the maximum follows from the terms of lowest order U_m, then $\mathbf{q} = \mathbf{0}$ is unstable. LIAPUNOV proved his theorem using his direct method and giving a suitable LIAPUNOV function. The theorem is however not applicable to such simple situations as the one corresponding to

$$U = -q_1^2 - q_2^4 \qquad (n = 2). \tag{4.69}$$

This of course is deeply unsatisfying.

A possibility to prove LIAPUNOV instability without using any LIAPUNOV function is to show directly the existence of solutions of (4.64) for example with initial conditions $\mathbf{q}(0) = \mathbf{0}$, $\mathbf{q}(t) = \mathbf{q}^*$, where $\mathbf{q}^*$ is fixed and different from $\mathbf{0}$, the corresponding t is arbitrary and the initial velocity $\dot{\mathbf{q}}(0)$ is arbitrarily small. In [17] the author succeeded, using a description of the dynamics of the system by JACOBI's principle

$$\delta \int_{\mathbf{q}_1}^{\mathbf{q}_2} 2\,(h - U(\mathbf{q}))\sqrt{\sum a_{ij} dq_i dq_j} = 0, \tag{4.70}$$

instead of the differential equations (4.64). HILBERT's idea of minimizing sequences was then used to prove that a motion, as defined above, exists, and the instability theorem demonstrated is the following: *If the potential energy* $U(\mathbf{q})$ *has a strong relative non–necessarily isolated maximum at* $\mathbf{q} = \mathbf{0}$*, then the equilibrium position* $\mathbf{q} \equiv \mathbf{0}$*, is unstable.*

This instability theorem is for example applicable to systems with potential energy such as

$$\begin{aligned} U(q) &= -q^6 \left(\cos\left(\frac{1}{q}\right)^2 - 2 \right), & q \neq 0, \\ U &= 0, & q \neq 0, \end{aligned} \tag{4.71}$$

where the trivial equilibrium position $q = 0$ is a point of accumulation of stable and unstable equilibria. Such a potential energy function is depicted in Figure 4.8.

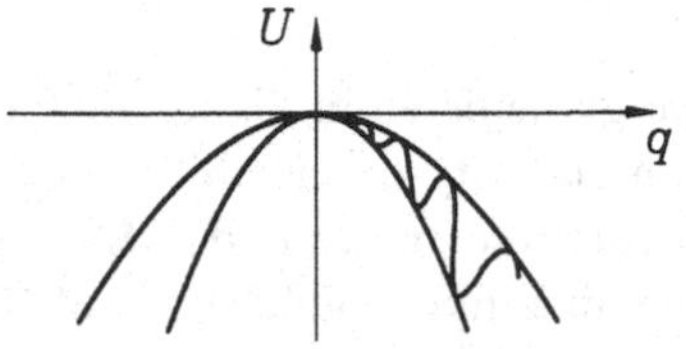

Figure 4.8: Potential energy function

Since then, similar approaches were used to generalize this idea in proving inverse theorems.
The papers of Rumjantsev, Salvadori, Mawhin, Tagliaferro, Kozlow and Palamadov have to be mentioned in this context. A review of the extensive russian literature is given in [30].
Nevertheless, the original question of whether a minimum of the potential energy is a necessary and sufficient condition for stability in a conservative system has not been completely answered up to now, as far as the author knows.
These few remarks, showing that an apparently trivial question may lead to interesting and still partially open problems of analytical mechanics, were made in order to stimulate the audience's further interest in the fascinating problem of stability in analytical mechanics.

Bibliography

[1] Anderson, K.; Hagedorn, P.: On the Energy Dissipation in Spacer Dampers in Bundled Conductors of Overhead Transmission Lines, *Journal of Sound and Vibration (1995)*, 180(4), 539–556.

[2] Chen, S.S.; Wambsganss, M.W.; Jendrzejczyk, J.A.: Added Mass and Damping of a Vibrating Rod in Confined Viscous Fluids, *Journal of Applied Mechanics*, Vol. 43, 1976, 325-329.

[3] Claren, R.; Diana, G.: Vibrazioni dei conduttori, *L'Energia Elettrica*, No. 10, 1966.

[4] Cole, J.D.: *Perturbation Methods in Applied Mathematics*, Blaisdell Publ. Comp., 1968.

[5] Cremer, L.; Heckl, M.: *Structure-Borne Sound*, Springer-Verlag, Berlin, Heidelberg, New York, London, 1988.

[6] Cremer, L.: *Vorlesungen über Technische Akustik*, Springer-Verlag, Berlin, Heidelberg, New York, 1971.

[7] Diana, G.; Falco, M.: On the forces transmitted to a vibrating cylinder by a blowing fluid, *Meccanica*, Vol. 6, 1971, 9–22.

[8] Doocy, E.S.; Hard, A.R.; Rawlins, C.B.; Ikegami, R.: *Transmission line reference book*, Wind induced Conductor Motion, Electric Power Research Institute, Palo Alto, Cal. 1979.

[9] Van Dyke, M.: *Perturbation Methods in Fluid Mechanics*, Parabolic Press, Palo Alto, California, 1975.

[10] Eckhaus, W.: *Asymptotic Analysis of Singular Perturbations*, North-Holland, Amsterdam, 1979.

[11] Eckhaus, W.: *Matched Asymptotic Expansions and Singular Perturbations*, North-Holland, Amsterdam, 1973.

[12] Farquharson, F.; Mehugh, R.E.: Wind tunnel investigation of conductor vibration with use of rigid models, *Trans. AIEE*, Vol. 75, 1956, Part III, 871–878.

[13] Hagedorn, P.: *Nonlinear Oscillations*, Clarendon Press, Oxford, 1988.

[14] Hagedorn, P.: Ein einfaches Rechenmodell zur Berechnung winderregter Schwingungen an Hochspannungsleitungen mit Dämpfern, *Ing.-Arch.*, Vol. 49, 1980, 161–177.

[15] Hagedorn, P.: On the computation of damped wind-excited vibrations of overhead transmission lines, *Journal of Sound and Vibration*, Vol. 83, 1982, 253–271.

[16] Hagedorn, P.; Wallaschek, J.: Traveling Wave Ultrasonic Motors, Part I: Working Principle and Mathematical Modelling of the Stator, *Journal of Sound and Vibration*, Vol. 155(1):31–46, 1993.

[17] Hagedorn, P.: Die Umkehrung der Stabilitätssätze von Lagrange-Dirichlet und Routh, Archive for Rational Mechanics and Analysis, Vol. 42, 1971, 4, 281-316.

[18] Hagedorn, P. & Mahwin, J.: A Simple Variational Approach to an Inversation of the Lagrange-Dirichlet Theorem, Archive for Rational Mechanics and Analysis, 120 (1992), 327-335.

[19] Horacec, J.; Zolotarev, I.: Acoustic-Structural coupling of Vibrating Cylindrical Shells with Flowing Fluid, *Journal of Fluids and Structures*, Vol. 5, 1991, 487-501.

[20] Johnson, K.L.: *Contact Mechanics*, Cambridge University Press, Cambridge, 1985.

[21] Kaplun, S.: *Fluid Mechanics and Singular Perturbations*, Ed. by Lagerstrom, P.A., Howard, L.N. and Liu, C.S., Academic Press, New York, 1967.

[22] Kevorkian, J.; Cole, J.D.: *Perturbation Methods in Applied Mechanics*, Applied Mathematical Sciences, Vol. 34, Springer-Verlag, New York, 1980.

[23] Nayfeh, A.; Mook, D.T.: *Nonlinear Oscillations*, John Wiley & Sons, New York, 1979.

[24] Morse, P.M.; Ingard, K.U.: *Theoretical Acoustics*, 1968, New York: MCGraw-Hill.

[25] Nascimento, N.; Hagedorn, P.: Stochastic field processes in the mathematical modelling of damped transmission line vibrations, Paper presented at the *Fifth International Conference on Mathematical Modelling*, Berkeley, 1985.

[26] Nascimento, N.: Stochastische Schwingungen eindimensionaler, kontinuierlicher mechanischer Systeme, Doctoral Thesis, TH Darmstadt, 1984.

[27] Nayfeh, A.H.: *Perturbation Methods*, John Wiley & Sons, New York, 1973.

[28] Neimark, J.F. & Fufaev, N.A.: *Dynamics of Nonholonomic Systems*, American National Society, Rhodes Island, 1972.

[29] O'Malley, R.E.: *Introduction to Singular Perturbations*, Academic Press, New York, 1974.

[30] Palamadov, V. P.: On stability of an Equlibrium in a Potential Field, Functional Analysis and its Application, Vol. 11 (1977), No. 4, 42-55 (in Russian).

[31] Pars, B.A.: *Analytical Dynamics*, Heinemann, London, 1968.

[32] Païdoussis, M.P.; Nguyen, V.B.; Misra, A.K.: A Theoretical Study of the Stability of Cantilevered Coaxial Cylindrical Shells Conveying Fluid, *Journal of Fluids and Structures*, Vol. 5, 1991, 127-164.

[33] Rawlins, C.B.: Power imparted by wind to a model of a vibrating conductor, Electrical Products Division, ALCOA Labs., Massena, NY, 1982.

[34] Schäfer, B.: Dynamical modelling of wind-induced vibrations of overhead lines, *International Journal of Non-Linear Mechanics*, Vol. 19, 1984, 455–467.

[35] Seemann, W.: Stresses in a Thin Piezoelectric Element Bonded to a Half-space, *Applied Mechanics Reviews*, to appear November 1997.

[36] Seemann, W.; Hagedorn, P.: On the Realizability of a Traveling Wave Linear Motor, *13th International Modal Analysis Conference*, Conference Proceedings, Nashville 1995, USA, pp. 1778–1784.

[37] Seemann, W.; Wolf, K.; Straub, A.; Hagedorn, P.; Chang, F.-K.: Bonding Stresses between Piezoelectric Actuators and elastic Beams, *Proceedings of the SPIE Conference on Smart Materials and Structures*, San Diego, 1997.

[38] Seemann, W.; Wauer, J.: Vibrating Cylinder in a Cylindrical Duct Filled with an Incompressible Fluid of Low Viscosity, *Acta Mechanica*, Vol. 113, pp. 93–107, 1995.

[39] Seemann, W.; Wolf, K.; Hagedorn, P.: Comparison of Refined Beam Theory and FEM for Piezo-Actuated Structures, *ASME Design Engineering Technical Conferences*, Paper-No. VIB-3838, Sacramento, 1997.

[40] Straubli, T.: Untersuchung der oszillierenden Kräfte am querangeströmten, schwingenden Kreiszylinder, Doctoral Thesis, ETH Zürich, 1983.

[41] Tiersten, H.F.: *Linear Piezoelectric Plate Vibrations*, Plenum Press, New York, 1969.

[42] Wallaschek, J.: Piezoelectric Ultrasonic Motors, *Journal of Intelligent Material Systems and Structures*, Vol. 6:71–83, 1995.

[43] Weidenhammer, F.: Eigenfrequenzen eines Stabes im zylindrischen Luftraum, *Zeitschrift für Angewandte Mathematik und Mechanik*, Vol. 55, 1975, T187-T190.

[44] Whittaker, E.T.: *A treatise on the Analytical Dynamics of Particles and Rigid Bodies*, Cambridge University Press, London, 1937

[45] Yang, C.-I.; Moran, T. J.: Finite Element Solution of Added Mass and Damping of Oscillation Rods in Viscous Fluids, *Journal of Applied Mechanics*, Vol 46, 1979, 519-523.

[46] Zampieri, G. & Neto, A. Barone: Attractive Central Forces may yield Liapunov Instability. Relatório Técnico, Dep. Mat. Apl. USP, Dec. 1984.

[47] Zharii, O: *Thin Piezoelectric Element on an Elastic Half-Space*, Private Communication, 1994.